Field Crops: Biotechnology

Field Crops: Biotechnology

Edited by **Wendel Mason**

Syrawood Publishing House
New York

Published by Syrawood Publishing House,
750 Third Avenue, 9th Floor,
New York, NY 10017, USA
www.syrawoodpublishinghouse.com

Field Crops: Biotechnology
Edited by Wendel Mason

International Standard Book Number: 978-1-68286-307-7 (Hardback)

Printed in China.

Contents

Preface

Every book is initially just a concept; it takes months of research and hard work to give it the final shape in which the readers receive it. In its early stages, this book also went through rigorous reviewing. The notable contributions made by experts from across the globe were first molded into patterned chapters and then arranged in a sensibly sequential manner to bring out the best results.

The text provides a detailed study on principles, methods and practices in plant breeding. It also discusses the applications of biotechnology for crop improvement and seed production of field crops. The book brings forth some innovative concepts and the latest tools and techniques being employed for crop production. It covers researches which shed light on the genetic, physiological and medicinal aspects of field crops. It aims to serve as a valuable reference to biotechnologists, agriculturists and interested readers. The book also aims to outline future researches and challenges in this field.

It has been my immense pleasure to be a part of this project and to contribute my years of learning in such a meaningful form. I would like to take this opportunity to thank all the people who have been associated with the completion of this book at any step.

Editor

THE DETERMINATION OF RELATIONSHIP BETWEEN VERTICILLIUM WILT (*Verticillium dahliae* Kleb.) AND EARLY MATURITY IN COTTON (*Gossypium hirsutum* L.)

Oktay ERDOGAN[1], Nedim OZBEK[1], Aydın UNAY[2], M. Erhan GORE[3]*

[1]Cotton Research Station, Aydin-TURKEY
[2]University of Adnan Menderes, Faculty of Agriculture, Aydin-TURKEY
[3]University of Abant Izzet Baysal, Vocational Higher School, Bolu- TURKEY
[*]Corresponding author: oktaye@gmail.com

ABSTRACT

The present study was conducted to investigate the relationship between earliness of crop maturity and Verticillium wilt. Ten cotton genotypes were evaluated in natural infection under field conditions of Nazilli Cotton Reseach Institute in 2006 and 2007. Foliar disease index (FDI), vascular disease index (VDI), days to flowering (DF), days to boll opening (DB) and seed cotton yield (SCY) were determined. In regard to FDI and VDI, the most tolerant genotypes were Carmen (standard cotton cultivar), NGC, GSN 12 and M25-G, while the most sensitive genotype was NP Ozbek-100. NMCH-11/4, NCCH-9/2 and NCCH-8/1 were other sensitive genotypes. The performances to verticillium wilt of other standard cotton cultivars, Nazilli 84 S and Sayar 314 were non-stabile. The maturity characteristics, DF and DB, were significant negative correlated with all Verticillium severity parameters, whereas these characteristics were significant positive correlated with SCY. Carmen cultivar should be recommended for infected areas and improving diseases tolerance in cotton breeding programs.

Key words: Cotton, verticillium wilt, early maturity, disease severity, yield

INTRODUCTION

Cotton is an agricultural product serving as raw material for textile and other various industries. Turkish cotton area and production are expected to reach 410.000 hectares and 600.000 MT (2.7 million bales) in 2011. Cotton is grown in Turkey in four major regions: Southeastern Anatolia, Aegean, Çukurova and Antalya. Aegean region has an area of 74.000 hectares with the total production of 94000 tons (Sirtioglu 2011).

In the Aegean Region, cotton disease, especially Verticillium wilt is one of the major constraints affecting yield and quality. Verticillium wilt, caused by the soil-borne fungus *Verticillium dahliae* Kleb., is found in many cotton fields of the Aegean Region. In Turkey, Verticillium wilt of cotton was first defined by Karaca et al. (1971). The losses of seed cotton yield due to Verticillium wilt were estimated 12 % and 10 % for Aegean and Çukurova, and Southeastern Anatolia regions, respectively (Sağır et al., 1995). The disease was responsible for significant yield losses (approximately 1.5 million bales) in the world cotton belt (Nemli 2003). Several factors, including variety selection, plant density, pathogen aggressiveness, inoculum density (microsclerotia per gram of soil), and environmental conditions influence Verticillium wilt development. Initial infections occur early in the growing season, following the germination of microsclerotia. The fungus infects through the roots, invades the vascular system resulting in a systemic infection. Prolific growth in the xylem vessels disrupts the plants ability to transport water and nutrients.

Cotton growers have generally created favorable conditions for Verticillium wilt, İncreased vegetative growth, late maturity and reduced boll load induced by excesiive nitrogen fertilization and frequent irrigation.

Moreover, crop rotation is not used in the Aegean Region. One of the main strategies for management of Verticillium wilt of cotton involves the use of Tolerant and/or resistant cultivars.

Bowman (1999) stated that the Texas A&M programs at College Station (MAR) and Lubbock have been breeding for Verticillium resistance since the 1960s. The Texas A&M program at Lubbock released 22 influential lines that have contributed to 59 cultivars of the 260 released between 1970 and 1990 (Bowman et al.,1996). That program does not produce cultivars, but impacts Texas High Plain's cultivars through release of storm proof material with improved earliness, fiber quality, and increased resistance to Verticillium wilt. Today, private and public breeders routinely screen for Verticillium wilt resistance.

Recently, evaluation and breeding for Verticillium wilt and root-knot nematode resistance and tolerance for drought and salt stresses have also become an integral part of the breeding effort (Bajaj et al., 2008; Zhang et al., 2006, 2008). In addition, extensive effort has been made to introduce desirable genes and traits from *G. barbadense* L. into Upland cotton (Zhang and Percy 2007).

Early maturity may also help in reducing losses from diseases such as Verticillium Wilt and boll rot (Verhalen et al., 1984; Frisbie et al., 1989). Earliness in cotton is defined as shorter time flowering from sowing, faster maturation and early yield potential (Poehlman and Sleper 1995). Galanopoulu (2006) revealed that early maturing varieties were generally more susceptible to the late infestation than the early and than late varieties to the late infestation, and similar reaction was also displayed between early and late sowings. It was reported that the highest yield losses were observed in plants showing symptoms before first flowering stage, and the effects of late stage symptoms on yield was negligible (Bejarano-Alcazar et al., 1997).

The objectives of this work were to determine the Verticillium wilt tolerance of standard cotton cultivars and improved genotypes by Nazilli Cotton Research Institute and the relationships between early maturity and Verticillium wilt severity parameters.

MATERIALS AND METHODS

In the study, three cotton standard cultivars as control (CARMEN, NAZILLI 84 S and SAYAR 314) and 7 genotypes that have different days to flowering and boll opening were used as material.

Field trials were conducted in field infested with the non-defoliating pathotype in 2006 and 2007 cotton growing seasons. The inoculum density of *Verticillium dahliae* Kleb. in naturally infested field was determined 69 ms/g (Erdoğan et al., 2011). El-Zik (1985) revealed that the fungal population density required for 50 % disease is approximately 22 p/g. Also, with 60 p/g symptoms appear early, about 50-60 days after planting, compared with 90 days with 5 p/g. The Experiments were arranged in a randomized complete block design with four replications. Each plot consisted of two rows of 12 m length, with 0.7 m between rows. Initial soil analysis showed that experimental area had clay soil, slight alkaline (pH: 7.85), low salinity and rich in lime, organic matter, potassium and phosphorous. The cultural managements such as irrigations, fertilizations and crop protection applications were programmed according to recommendations for Aegean Region cotton growing.

Observations of diseases were taken from the middle two rows of each plot with all of the plants on leaves at 5-10 %, 50-60 % cotton boll opening stage and 0-3 scale (0=healty plant; 1=50% foliar symptoms on leaves; 2= 51-75 % foliar symptoms on leaves; 3= 76-100 % foliar symptoms on leaves and plant died) was used for observations of diseases (Ünal and Aydın 1980).

Disease severity was evaluated for the middle two rows of each plot with all of the plants on stems at harvesting period and 0-3 scale (0= no discoloration xylem on trunk sectional area; 1=1-33 % discoloration xylem; 2= 34-67 % discoloration xylem; 3= 68-100 % discoloration xylem) was used for observations of diseases (Buchenauer and Erwin 1976). Disease Severity Index was calculated by using the following Formula with formula; $DSI=[(ax0)+(bx1)+(cx2)+(dx3)]/M$

Where a, b, c, and d refer the plant number with degree 0, 1, 2, and 3 respectively, and M refer total plant number.

Disease rates were calculated and obtained data were subjected to Arcsin for transformation (Karman 1971). Statistical analyses were performed using JMP 5.0.1 statistical software and the means were grouped by means of the LSD (0.05) test. Besides, discoloration of the interior of the stems was taken into account and plants were marked as healthy or diseased.

The number of days from sowing to flowering (DF) and the number of from sowing to boll opening (DB) were calculated by using first white flower and first boll opening observations, respectively. Seed Cotton Yield (SCY) data were recorded from two rows of each harvested two times by hand.

RESULTS AND DISCUSSION

Variance analysis showed that genotype and year were significant for all the investigated characteristics, and genotypes were significantly different for all observed characters, while year x genotype interactions were significant for FDI (5-10 %), FDI (50-60 %) and VDI (Table 1). The year x genotype interaction for Verticillium severity parameters are consistent with previous studies reported by Kheiri and Fatahi (2010) and Karademir et al. (2010). The results of interactions demonstrated environment altered Verticillium severity and virulence of pathogen was increased by favorable climatic conditions. Disease development and symptom expression are influenced by environment, including temperature, light and soil moisture etc. and by cultural management such as planting date, plant density, fertilization and irrigation (El-Zik and Frisbie 1985; El-Zik 1985). It can be said that all were not affected in the same way by Verticillium severity.

Standard cotton cultivar of Aegean Region, CARMEN had the lowest FDI (5-10 %), FDI (50-60 %) and VDI values (Table 2). The results of many studies demonstrated that Carmen cultivar was recommended for yield and fiber quality in infected area (Erdoğan et al., 2006; Göre et al., 2009; Karademir et al., 2010). NGC, GSN 12 and M25-G genotypes had characteristics similar to that of CARMEN for Verticillium tolerance. Remarkably, NP OZBEK-100 had the highest disease severity for each of the three characters. This cultivar was followed by by NCCH-9/2, NMCH-11/4, NCCH-8/1, NAZILLI 84 S (control) and SAYAR 314 (control), respectively.

Table 1. Results of variance analysis for the investigated characters.

Source of Variance	Df	FDI (5-10)	FDI (50-60)	VDI	DF	DB	SCY
Rep.	3	0.01	0.01	0.07	0.18	0.30	9416.22
Year (Y)	1	0.30*	0.67*	0.15*	10.51*	6.05*	11367.08*
Genotype (G)	9	0.23*	0.27*	0.18*	204.67*	901.87*	5970.68*
YxG	9	0.06*	0.07*	0.02*	0.57	0.38	475.93
Error	57	0.03	0.03	0.07	0.43	0.54	665.97
Total	79	0.04	0.05	0.033	23.83	103.26	1716.40

Df: degrees of freedom; * Significant at the 0.05 probability level.
FDI = foliar disease index, VDI = vascular disease index, DF; days to flowering, DB; days to boll opening, SCY: seed cotton yield,

Table 2. Data means of FDI (5-10 %), FDI (50-60%), VDI separated by LSD

	2006			2007		
Genotypes	FDI (5-10 %)	FDI (50-60 %)	VDI	FDI (5-10 %)	FDI (50-60 %)	VDI
NGC	0.38 f	0.58 e	0.59 c	0.51 c	0.59 c	0.62 c
GSN-12	0.44 ef	0.64 e	0.60 c	0.50 c	0.61 c	0.58 cd
NAZILLI 84 S (C)	0.96 d	1.13 d	1.02 b	0.73 ab	0.85 b	0.82 b
CARMEN (C)	0.26 g	0.54 e	0.45 c	0.36 d	0.47 d	0.44 e
NP ÖZBEK-100	1.64 a	1.90 a	1.49 a	0.83 a	0.90 ab	0.89 ab
NMCH-11/4	1.16 b	1.45 b	1.12 b	0.75 ab	0.86 b	0.85 ab
NCCH-9/2	1.15 b	1.57 b	1.09 b	0.77 ab	0.96 a	0.94 a
NCCH-8/1	1.07 bc	1.29 c	1.09 b	0.68 b	0.96 a	0.91 a
SAYAR 314 (C)	0.98 cd	1.14 d	1.07 b	0.78 ab	0.94 ab	0.94 a
M25-G	0.49 e	0.66 e	0.51 c	0.47 cd	0.56 cd	0.52 de
LSD	0.096	0.137	0.251	0.114	0.096	0.088

FDI = foliar disease index, VDI = vascular disease index

Between years and genotypes were significant for the number of days from sowing to flowering (DF) and boll opening (DB) (Table 1). Earliness of crop maturity in cotton genotype did not change due to years. DF ranged from 47.3 days (NP OZBEK-100) to 62.6 days (CARMEN), while the DB values were between 97.3 days and 121.1 days for same genotypes (Table 3). Comparing genotypes that differed in maturity times, the late maturity cotton genotypes were CARMEN, M25-G, NGC and GSN 12. The minimum values for DF and DB were obtained from NP OZBEK-100, NCCH-9/2 and NMCH-11/4, respectively.

Table 3. Data means of DF (day), DB (day), SCY (kg ha^{-1}) separated by LSD

Genotype	DF (day)	DB (day)	SCY (kg ha^{-1})
CARMEN (C)	61.6 a	121.1 a	3144 ab
M25-G	60.3 b	119.5 b	3001 bc
NGC	60.1 b	120.1 b	2991 bc
GSN-12	57.6 c	120.0 b	3278 a
SAYAR 314 (C)	55.8 d	116.1 d	2951 bc
NAZİLLİ 84 S (C)	55.3 d	118.0 c	3188 ab
NCCH-8/1	51.5 e	102.2 e	2845 cd
NMCH-11/4	50.3 f	98.2 f	2636 de
NCCH-9/2	49.3 g	97.7 fg	2591 de
NP ÖZBEK-100	47.3 h	97.3 g	2455 e
LSD (0.05)	0.659	0.733	365.40

DF; days to flowering, DB; days to boll opening, SCY: seed cotton yield,

The analysis of variance indicated that there were significant differences among genotypes for seed cotton yield (SCY). The maximum SCY values were obtained from GSN-12 (3278 kg ha^{-1}), NAZILLI 84 S (3188 kg ha^{-1}) and CARMEN (3144 kgha^{-1}), and followed by M25-G (3001 kg ha^{-1}) and NGC (2991 kg ha^{-1}), respectively. The lowest seed cotton yields were obtained from the early maturity cotton genotypes, NP OZBEK-100, NCCH-9/2 and NMCH-11/4.

Three Verticillium severity parameters, FDI (5-10 %), FDI (50-60 %), and VDI, were significantly positive correlated with each other (Table 4). In other words, These three Verticillium severity seem to confirm each other. However, the maturity characteristics, DF and DB, were negatively significant correlated with all Verticillium severity parameters, whereas these characteristics were positively significant correlated with SCY. It was showed that early maturity characteristics of cotton genotypes increased the Verticillium sensitivity, and late maturity genotypes had higher SCY values than early maturity genotypes. Davis and Cano-Rios (1981) emphasized that although earliness and Verticillium wilt tolerance combined in the same variety would be doubly profitable, earlier maturing cotton genotypes are generally more susceptible to Verticillium wilt. The results of this study confirmed those of Bejarano-Alcazar et al. (1997) who reported that yield increased with delay in the development of foliar symptoms during the crop season and effect of the wilt epidemics on yield was small or nil for plants that developed symptoms after opening of the first bolls.

Table 4. The correlation coefficients between observed characteristics.

Characteristics	DB	SCY	FDI (5-10 %)	FDI (50-60 %)	VDI
DF	0.8708**	0.4934**	-0.6952**	-0.6989**	-0.6385**
DB		0.5860**	-0.7141**	-0.7131**	-0.6653**
SCY			-0.2616**	-0.1957**	-0.1591**
FDI (5-10 %)				0.9652**	0.8615**
FDI (50-60 %)					0.8712**

*. ** significant at the 0.05and 0.01probability level.
FDI = foliar disease index, VDI = vascular disease index, DF; days to flowering, DB; days to boll opening, SCY: seed cotton yield

CONCLUSION

The present study reveals that the genotypes used as material in this research had different values for Verticillium severity parameters, earliness and seed cotton yield. The most tolerant genotypes to *Verticillium dahliae* Kleb. disease were CARMEN, NGC, GSN 12 and M25-G with late maturity and high yielding capacity. Especially, CARMEN and GSN-12 varieties can be recommended for infected areas. The significant correlations among maturity characters, seed cotton yield and Verticillium tolerance parameters showed that late maturity genotypes should be recommended for infected areas and improving diseases tolerance in cotton breeding programs.

LITERATURE CITED

Bajaj, S.,Wang, W., Hughs, S.E. Percy, R.G. Ulloa, M., Zhang, J. 2008. Evaluation of cotton germplasm and breeding populations for salt tolerance. National Cotton Council Beltwide Cotton Conference, January 8-11, 2008, Nashville, TN. Chapter 10.

Bejarano-Alcazar, J., M. A. Blanco-Lopez, J. M. Melero-Vara and R. M. Jimenez-Diaz, 1997. The influence of verticillium wilt epidemics on cotton yield in Southern Spain. Plant Pathology. 46(2):168-178.

Bowman, D.T., O.L. May, and D.S. Calhoun, 1996. Genetic base of upland cotton cultivars released between 1970 and 1990. Crop Sci. 36:577-581.

Bowman, D.T., 1999. Public cotton breeders-do we need them. J. Cotton Science. 3:139-152.

Buchenauer, H.D. and Erwin, C., 1976. Effect of the plant growth retardant Pydanon on Verticillium wilt of cotton and tomato. Phytopathology. 49:68-72.

Davis, D.D. and Cano-Rios, P. 1981. Breeding for early maturity and Verticillium wilt tolerance in upland cotton. Crop Sci., 21:319-321

El-Zik, K. M. and R. E. Frisbie, 1985. Integrated management systems for pest control and plant protection, in N. B. Mandava (ed.), CRC Handbook of Natural Pesticides: Methods. Vol. I. Theory, Practice, and Detection. CRC Press, Inc., Boca Raton, FL. 21-122 pp.

El-Zik, K.M., 1985.Integrated Control of Verticillium Wilt of Cotton. Plant Disease. 69: 1025-1032.

Erdoğan, O., Dündar, H., Göre, M.E., 2011. Determination of reactions of some cotton genotypes against verticillium wilt disease caused by Verticillium dahliae Kleb. Plant Protection Bulletin. 51 (2): 159-173.

Erdoğan, O., V. Sezener, N. Ozbek, T. Bozbek, I. Yavas and A. Unay, 2006. The effects of *Verticillium* wilt (*Verticillium dahliaee* Kleb.) on cotton yield and fiber quality. Asian Journal of Plant Sciences. 5: 867-870.

Frisbie, E., K. M. El-Zik and L. T. Wilson, 1989. Strategies and tactics for managing plant pathogens and nematodes. Integrated Pest Management Systems and Cotton Production. John Wiley and Sons. New York.

Galanopoulo, S., 2006. http://www.Ressources.ciheam.org./om/pdf/s14/CI01190.

Göre, E., O. K. Caner, N. Altın, H. Aydın, O. Erdoğan, F. Filizer and A. Büyükdoğerlioğlu, 2009. Evaluation of cotton cultivars for resistance to pathotypes of *Verticillium dahliaee*. Crop Protection. 28(3): 215-219.

Karaca, İ., Karcılıoğlu, A., ve Ceylan, S., 1971. Wilt disease of cotton in the Ege region of Turkey. The Journal Turkish Phytopathology. 1(1): 4-11.

Karademir, E., Karademir, Ç.,Ekinci, R., Baran, B., Sağır, A., 2010. Assessment of Tolerance Level of some Cotton (*Gossypium hirsutum* L.) Varieties against Verticillium wilt (*Verticillium dahliae* Kleb.). Not. Bot. Hort. Agrobot. Cluj. 38(1): 196-202.

Karman, M., 1971. General Information in Plant Protection Investigations, Evaluation and Performing of the Trials. Regional Agricultural Protection Institute, pp. 279, Bornova-İzmir.

Kheiri, A., Fatahi, M., 2010. Evaluation of Verticillium Wilt Tolerance in Different Cotton Cultivars. Journal of Research in Agricultural Science. 6: 57-61.

Nemli, T., 2003. Cotton Disease and Control Methods. Cotton Training Seminar, 14-17 October, pp. 103-111, İzmir-Turkey.

Poehlman, J.M.,Sleper, D.A., 1995. Breeding Field Crops. Iowa State Univ. Press, 378 p.

Sağır, A., Tatlı, F., and Gürkan, B., 1995. Studies on the Disease Shown Cotton Sowing Area in Southeastern Anatolia

Region. Southeast Anatolia Region Plant Protection Problems and Solution Suggestion Symposium, 27-29 April, pp. 5-9, Şanlıurfa-Turkey.

Sirtioglu, I., 2011. Turkey Cotton and Products Annual 2011.USDA Foreign Agricultural Services. Global Agricultural Information Network.

Ünal, M., and Aydın, G., 1980. Resistance to verticillium wilt in Cotton Breeding. Nazilli Cotton Research Institute, 1 (27): 23.

Verhalen, L. M., M. B. Bayles, and N. B. Thomas, 1984. "Cotton Varieties for Oklahoma: A Short Season Environment."In Summary Proceedings Western Cotton Production Conference, pp. 8-15, Oklahoma City OK, 13-14 August 1984.

Zhang, J.F. and R.G. Percy, 2007. Improving Upland cotton, *Gossypium hirsutum* by introducing desirable genes from *G. barbadense*. Proc. World Cotton Res. Conf.-4, 10–14Sept. 2007, Lubbock, TX, USA, DVD, Omin Press.

Zhang, J.F., C. Waddell, C. Segupta-Gopalan, C. Potenza, and R.G. Cantrell, 2006. Relationships between root-knot nematode resistance and plant growth in Upland cotton:galling index as a criterion. Crop Science. 46: 1581–1586.

Zhang, J.F., W. Wang, S. Bajaj, H. Gatica, S. Sanogo, R. Flynn, C. French, R. Percy, M. Ulloa, and E. Hughs, 2008. Verticillium wilt resistance in cotton: germplasm evaluation and inheritance. p 838–855. In Proc. Beltwide CottonConf. Nashville, TN. 8-11 Jan. 2008. Natl. Cotton Counc. Am., Memphis, TN.

DETERMINATION OF LAVENDER AND LAVANDIN CULTIVARS (*Lavandula sp.*) CONTAINING HIGH QUALITY ESSENTIAL OIL IN ISPARTA, TURKEY

Nimet KARA, Hasan BAYDAR*

Suleyman Demirel University, Agriculture Faculty, Department of Field Crops, Isparta, TURKEY
**Corresponding author: nimetkara@sdu.edu.tr*

ABSTRACT

The research was carried out during the 2009 and 2010 growing period with the aim of determining agricultural and technological characteristics of lavender cultivars. When the agricultural characteristics of the lavender and lavandin cultivars were examined, in both years the highest fresh stem flower yield was obtained from Dutch (5467 and 8204 kg ha^{-1}, respectively) and the highest dry stemless flower yield from Super A (1083 and 1463 kg ha^{-1}, respectively) cultivars. The highest essential oil content in both fresh stem flowers (the first year 2.00 %, the second year 1.90 %) and dry stemless flowers (the first year 9.62 %, the second year 8.87 %) was determined from Silver. Linalool, linalyl acetate and camphor were determined as the main components of essential oil in the lavender cultivar. The highest linalool content in fresh stem flowers was determined to be from Dutch (43.3 %) in the first year and from Vera (43.9 %) in the second year. The highest linalyl acetate content from Super A (42.5 and 19.8 %, respectively) and camphor content from Super A (19.8 %) in the first year and Dutch (10.0 %) in the second year were determined. The highest linalool content in dry stemless flowers from Dutch (46.5 and 47.0 %, respectively), linalyl acetate content from Super A (32.8 and 29.5 %, respectively) in both years and camphor content from Silver (12.6 %) in the first year and Dutch (10.9 %) in the second year were obtained.

Keywords: Lavender, lavandin, essential oil, essential oil components

INTRODUCTION

Lavandula, belonging to the *Lamiaceae* family, is a precious essential oil plant. *Lavandula,* which is cultivated around the world, is a significant perfume, cosmetic and pharmaceutical plant because of the high content and quality essential oil (Guenther, 1952). *Lavandula* oil is especially useful for use in nervous system stimulants, hypnotics, sedatives, tranquilizers and stress repellents. In addition, it has useful dermatological uses in the treatment sunburn and skin rashes (Cavanagh and Wilkinson, 2002), as well as strong antiseptic (disinfectant), and antibiotic (bacteria killing) effects (Lis-Balchin and Hart, 1999).

Lavandula is a plant which is fairly resistant to drought and temperature (Weiss, 1997). There are 39 *Lavandula* species (*Lavandula sp.*) which are mostly of Mediterranean origin. However, there are three important species within the genus producing lavender (*Lavandula angustifolia* Mill. = *L. officinalis* L. = *L. vera* DC), lavandin (*Lavandula intermedia* Emeric ex Loisel. = *L. hybrida* L.) and Spike lavender (*Lavandula spica* = *L. latifolia* Medik.) (Tucker, 1985). Lavender essential oil is higher quality than that produced by lavandins because lavender essential oil has lower camphor content than lavandin cultivars. Therefore, lavender essential oils are used in perfumes and aromatherapy, and lavandin essential oils are used in soap, detergents and cosmetics products (Lis-Balchin, 2002). However, lavandin essential oil yield is higher than that from lavenders (Beetham and Entwistle, 1982). *Lavandula sp.* is one of important medicinal and aromatic plants that are cultivated in Turkey. Isparta is the most significant lavandin (*L.x intermedia* Super A) production area of Turkey. In Isparta, lavandin is cultivated on an area of 250 hectares, with an annual production of 2.5 tons (Kara and Baydar, 2011).

In Isparta, average fresh stem flower and dry stemless flower yields per hectare of *L.x intermedia* Super A are 5-7.5 and 1-2.5 tons, respectively (Baydar and Erbas, 2007). Essential oil content of lavandin varies between 1.0-1.5 % in fresh stem flowers and between 5.0-6.0% in dry stemless flowers (Baydar, 2009). Quality and marketable value of lavender essential oil is measured according to essential oil composition. Generally, high quality lavender essential oil contains high levels of linalool and linalyl acetate, but the scent of essential oil deteriorates as the camphor content increases (Adam, 2006). Lavender essential oil has more than 100 components with linalool, linalyl acetate, 1,8-cineol and camphor as the major constituents. Linalyl acetate in lavender essential oil and linalool in lavandin essential oil are higher (Beetham and Entwistle, 1982). Generally, lavender essential oil is better quality because lavender essential oil has a lower camphor content than lavandin cultivars (Lis-Balchin, 2002; Baydar, 2009). The study was conducted with the aim to determine the lavender and lavandin cultivars that have high dry stemless flowers containing high quality essential oil (high linalyl acetate and low camphor) and high adaptation ability.

MATERIALS AND METHODS

Experimental Conditions

In the research, Raya, Munstead, Vera and Silver lavender cultivars belonging to *Lavandula angustifolia* species Mill. and Dutch, Giant Hidcote and Super A lavandin cultivars belonging to *Lavandula x intermedia* Emeric ex Lois. were used as plant materials. These lavender cultivars are among the important varieties that have high commercial value in world. The samples belonging to these cultivars were taken in the years 2009 and 2010 from the experiment founded in 2007 as four replications in a Randomized Complete Block Design at the Experimental Farm of the Agriculture Faculty of Suleyman Demirel University. The distance between rows was 1 m and intra row spaces were 50 cm (McGimbsey and Porter, 1999). Each plot area was 6 m in length and had 6 rows.

Lavender and lavandin cultivars were used in the experiment and their locations are shown in Table 1.

Table 1. Lavender and lavandin cultivars used in experiment

Species	Cultivar name	Commercial name	Origin (locations)
	Raya	Lavender	Bulgaria –Kazanlık
L. angustifolia	Munstead	"	France- Grasse
	Vera	"	Bulgaria-Kazanlık
	Silver	"	France- Grasse
	Dutch	Lavandin	Germany- Ahrensburg
L.x intermedia	Giant Hidcote	"	France-Grasse
	Super A	"	Turkey-Isparta

Some morphological characteristics and harvest dates belonging to the means of the experiment year of lavender and lavandin cultivars cultivated in Isparta ecological conditions are shown in Table 2 (Kara, 2011).

Table 2. Some morphological characteristics and harvest dates belong to lavender cultivars

Morphological characteristics		*L. angustifolia*				*L.x intermedia*		
		Raya	Munstead	Vera	Silver	Dutch	G. Hidcote	Super A
Flower colors	L	44.6	44.3	44.1	44.2	37.0	38.0	44.0
(h^0)	+a	22.8	19.7	18.4	19.6	19.9	24.4	20.3
	-b	20.3	17.8	16.0	16.7	16.7	21.9	17.7
Plant height (cm)		69.7	63.2	85.0	83.2	86.2	86.0	81.5
Growing form		Upright	Upright	S. Aslope	S. Aslope	S. Aslope	S. Aslope	S. Aslope
Ear number per plant		1217.8	955.7	860.3	632.1	941.7	821.7	600.5
Stem length with ear (cm)		30.3	24.1	33.6	36.8	35.7	33.2	35.8
Ear length (cm)		6.9	6.3	9.6	9.5	12.4	9.8	9.2
Flower number per ear (flower/ear)		34.9	29.9	63.0	64.1	65.7	61.2	73.5
Cluster number per ear (cluster/ear)		6.3	5.4	10.2	8.4	11.0	10.0	9.0
Harvest dates	2009	June 22	June 22	July 13	July 12	July 13	July 16	July 10
	2010	June 22	June 22	July 12	July 10	July 12	July 14	July 08

L: shine +a: red, -b: blue, S: Semi

Climatic Data and Soil Structure of the Experimental Area

Meteorological data for the growing seasons are shown in Table 3. Isparta Province is located at a 37° 45' N latitude, 30° 33' E longitude and 1050 m altitude. The long-term annual mean temperature, relative humidity, total annual precipitation, wind speed and sunshine duration per day in the area are 12.2 °C, 55 %, 493.4 mm, 2.4 m s^{-1} and 7.6 h, respectively (Anonymous, 2009-2010). With these climate characteristics, Isparta has features of the semi-arid climatic characteristics found in the Southwestern Anatolia region. The average temperatures and total precipitation in 2009 and 2010 were 12.9 and 13.7 °C and 666.7 and 710.4 mm respectively (Table 3). Meteorological data for 2009 and 2010 were higher compared to long-term meteorological data.

Table 3. Meteorological data of the experimental field*

Climatic factors	Years	Months												Total or Average
		Jan.	Feb.	March	April	May	June	July	August	Sept.	Oct.	Nov.	Dec.	
Average temperature (^{0}C)	2009	3.4	4.0	5.5	11.0	15.0	20.9	24.6	24.1	18.0	15.1	7.5	5.7	12.9
	2010	4.3	5.6	8.6	11.5	16.5	21.9	24.4	26.4	20.3	12.6	7.0	4.8	13.7
	Long years	1.7	2.6	5.9	10.5	15.5	20.1	23.4	25.8	18.3	12.8	6.9	3.0	12.2
Precipitation (mm)	2009	124.7	70.3	55.2	40.4	66.6	26.8	18.0	0.2	26.2	18.1	51.6	168.6	666.7
	2010	68.0	136.8	33.2	47.0	32.4	64.5	40.1	0.2	29.7	79.1	66.8	112.6	710.4
	Long years	64.2	54.9	52.8	58.8	46.0	27.8	12.8	0.3	15.4	38.0	51.5	70.9	493.4

* Meteorology Station, Isparta

Soil from a depth of 60 cm was sampled before setting the experiment and was subjected to physicochemical analysis. The soil was medium in nitrogen (19.8 kg NH_4^+ ha^{-1}), alkaline (pH 7.9), limy (1.3 $CaCO_3$ %), medium P (22 kg ha^{-1} P_2O_5) and high K_2O (850 kg ha^{-1}) contents.

Fresh Stem Flower and Dry Stemless Flower Yield

When plants reached blooming stage, 3 rows in the center of each plot were harvested manually in June and July (Table 2), and brought to the laboratory. The fresh stem samples were weighted and dried on the wire rack in shade at room temperature for a two-week period. Dry stem flowers were weighted and then dry stemless flowers were obtained by separating with hand from dried stem flower. Fresh stem, dry stem and dry stemless flower yield was calculated by converting from the parcel yield to hectare (Baydar and Erbas, 2007).

Essential Oils Distillation

200 g fresh stem flower in 1.5 L water and 50 g dry stemless flower samples in 1 L water from each plant cultivar were extracted by hydro-distilllation for 3 hours using Clevenger apparatus according to the standard procedure described in European Pharmacopoeia for determining the oil content (v/w %).

GC-MS Analysis

GC-MS (Gas Chromatography-Mass Spectrometry) analysis of the oil samples was performed on QP5050 GC-MS equipped with a Quadrapole detector. GC/MS analysis was employed under the following conditions: capillary column, CP-Wax 52 CB (50 m x 0.32 mm; film thickness = 0.25 μm); oven temperature program (60 ^{0}C increased to 220 ^{0}C at a rate of 2 °C/min and then kept at 220 ^{0}C for 10 min); total run time 60 min; injector temperature, 240 ^{0}C; detector temperatures, 250 ^{0}C; carrier gas, helium at a flow rate of 20 ml/min. Identification of constituents was carried out with the help of retention times of standard substances by composition of mass spectra with the data given in the NIST library (Stein, 1990). After lavender products were obtained according to the methods explained above, they were stored at 4 ^{0}C until GC-MS analyses. 1 μL of essential oil diluted with *n*-hexane was injected into the GC-MS system.

All the data were analyzed with analysis of variance (ANOVA) using the SAS Statistical Package Program. Means were compared using the DUNCAN test.

RESULTS AND DISCUSSION

Fresh Stem, Dry Stem and Dry Stemless Flower Yield

According to the results of variance analysis of data regarding fresh stem, dry stem flower and dry stemless flower yields, cultivar, year and year x cultivar interaction were statistically significant at the 0.01 level in both years (Table 4).

Fresh stem (6138 kg ha^{-1}), dry stem (3120 kg ha^{-1}) and dry stemless flower yields (1102 kg ha^{-1}) of the second year in the lavender were higher than those of first year (4626, 2197 and 807 kg ha^{-1}, respectively). Differences between the years might be due to the fact that 3-year-old plants continue to grow and that there was more rain in the second year as compared to the first. Stem number in plant can be higher than that subsequent year in perennial plants it is also reason higher yield in second years. Rainfall may affect crop yield and its components (Table 4).

Significant differences depending on years and cultivars in the fresh stem, dry stem flower and dry stemless flower yields of lavender and lavandin were found. The highest fresh stem and dry stem flower yields were determined in Dutch cultivar (5647 and 8204, 2885 and 4604 kg ha^{-1}, respectively), the lowest fresh stem and dry stem flower yields were obtained in Giant Hidcote cultivar (2905 and 3999 kg ha^{-1}, 1451 and 1955 kg ha^{-1}, respectively) in both years. Among the lavender and lavandin cultivars, the highest dried stemless flower yield was obtained from Super A (1083 and 1463 kg ha^{-1}, respectively), the lowest dry stemless flower yield was determined from Giant Hidcote (459 and 705 kg ha^{-1}, respectively) for both years (Table 4).

Ceylan et al. (1988) reported that the dry stemless flower yield of lavender varied between 1017-1948 kg ha^{-1}. Ceylan et al. (1996) determined that the dry stemless flower yield of lavender was 2340 kg ha^{-1}. Arabaci and Bayram (2005) and Atalay (2008) stated that the dry stemless flower yield of *Lavandula angustifolia* Mill. was between 1340-4430 kg ha^{-1} and 641-1134 kg ha^{-1}, respectively. Runham (1998) reported that the fresh stem flower yield of *Lavandula angustifolia* Mill. was 1651 kg ha^{-1}. Baydar (2009) found that 1 kg dry stemless flower was obtained by drying 5 kg fresh stem lavandin flowers (*L.x intermedia* Super A) in Isparta ecological conditions. In terms of dry stemless flower and fresh stem flower yield, our results are similar with the findings of the above-mentioned researchers.

Table 4. Yield and essential oil contents of lavender and lavandin cultivars

Cultivars	Fresh stem flower yield (kg ha^{-1})			Dry stem flower yield (kg ha^{-1})		
	2009	2010	Combined	2009	2010	Combined
***Lavander** (L. angustifolia)*						
Raya	5349 ab	6595 b	5972 b	2069 c	3355 c	2712 d
Munstead	4219 c	5764 c	4992 d	1664 d	2194 e	1929 f
Vera	4968 b	6422 b	5695 c	2384 b	3297 c	2841 c
Silver	4191 c	5333 d	4762 d	2079 c	2762 d	2421 e
***Lavandin** (L.x intermedia)*						
Dutch	5647 a	8204 a	6926 a	2885 a	4604 a	3745 a
G.Hidcote	2905 d	3999 e	3452 e	1451 e	1955 f	1703 g
Super A	5106 ab	6649 b	5878 bc	2847 a	3673 b	3260 b
Average	4626 B	6138 A		2197 B	3120 A	
LSD	55.427**	23.387**	21.193**	14.586**	13.146**	6.917**
CV value	5.88	1.23	3.00	3.26	1.75	2.31
Mean Square	34844.92	599872.2	429776.7	12009.67	79975.08	74281.1
	Dry stemless flower yield (kg ha^{-1})			**Fresh stem flower essential oil content (%)**		
***Lavander** (L. angustifolia)*						
Raya	850 c	1295 b	1073 c	1.25 cd	1.30 c	1.28d
Munstead	659 d	809 e	734 f	0.32 e	0.38 e	0.35 f
Vera	829 c	1187 c	1008 d	1.30 cd	1.37 c	1.34 d
Silver	804 c	927 d	866 e	2.00 a	1.90 a	1.95 a
***Lavandin** (L.x intermedia)*						
Dutch	962 b	1327 b	1145 b	1.67 b	1.55 b	1.61 b
G.Hidcote	459 e	705 f	582 g	1.07 d	1.12 d	1.10 e
Super A	1083 a	1463 a	1273 a	1.50 bc	1.48 bc	1.49 c
Average	807 B	1102 A		1.30	1.31	
LSD	6.053**	6.629**	3.162**	0.266**	0.134**	0.105**
CV value (%)	3.68	2.26	2.77	10.04	6.58	8.98
Mean Square	1637.92	18046.34	14793.58	1.12	0.87	1.90
	Dry stemless flower essential oil content (%)					
***Lavander** (L. angustifolia)*						
Raya	4.90 e	4.75 d	4.83 e			
Munstead	2.10 f	2.30 e	2.20 f			
Vera	5.07 de	5.10 d	5.09 e			
Silver	9.62 a	8.87 a	9.25 a			
***Lavandin** (L.x intermedia)*						
Dutch	6.10 c	6.97 c	6.54 c			
G. Hidcote	5.87 cd	5.12 d	5.50 d			
Super A	7.12 b	8.37 b	7.75 b			
Average	5.82	5.92				
LSD	0.844**	0.374**	0.325**			
CV value (%)	7.11	3.43	5.73			
Mean Square	21.00	23.33	42.46			

**: Significant at P<0.01 probability levels

Means in the same columns followed by the same letter (s) are not significantly different at the 1 % level

Fresh Stem and Dry Stemless Flower Essential Oil Content

According to the results of the variance analysis of data regarding fresh stem and dry stemless flower essential oil content of lavender, cultivar and year x cultivar interaction were statistically significant at the 0.01 level in the both years (Table 4).

The effect of the years on the essential oil content of lavender and lavandin was unsignificant. Significant differences depending on cultivars in the fresh stem and dry stemless flower essential oil content of lavender and lavandin were observed. The highest fresh stem and dry stemless flower essential oil contents were determined in Silver cultivar (2.00 and 1.90 %, 9.62 and 8.87 %, respectively), the lowest fresh stem and dry stemless flower essential oil contents were obtained from Munstead cultivar (0.32 and 0.38 %, 2.10 and 2.30 %, respectively) in both years. In the research, the dry stemless flower essential oil contents of the shorter Raya and Munstead cultivars were lower than those of the taller cultivars. The dry stemless flower essential oil content of Silver was significantly higher than the others cultivars.

Among the lavender and lavandin cultivars, while Silver, Dutch and Super A had high essential oil contents, Munstead and Giant Hidcote had lower essential oil contents (Table 4).

Some researchers stated that the essential oil of lavender must be at least 1.0-1.5 % (Wagner, 1980; Ceylan, 1996) and 0.5-1.0 % (Baytop, 1999). Ceylan et al. (1988) reported that the essential oil content of *L. officinalis* varied between 1.26-3.14 %. Renaud et al. (2001) found that dry stemless flower essential oil content varied between 7.1-9.9 % in lavandins and 2.8-5.0 % in lavenders. Baydar (2009) and Kara and Baydar (2011) reported that the essential oil content of *L.x intermedia* Super A cultivated in Isparta varied between 1.0-1.5 % and 2.24-2.35 %, respectively, in fresh stem flowers, 5.0-6.0 % and 7.50-8.60 % , respectively, in dry stemless flowers. Atalay (2008) and Arabaci and Bayram (2005) determined that the essential oil content of *Lavandula angustifolia* Mill. varied between 2.1-2.6 % and 1.54–2.34 %, respectively, in dry stemless flowers. In terms of dry stemless flower and fresh stem flower essential oil contents, our results agree with findings of the above-mentioned researchers.

Chemical Composition of the Essential Oil

Essential oil components and their rates obtained with GC-MS in lavender cultivars are shown in Table 5 and 6.

Linalool, linalyl acetate, camphor and terpineol-4-ol were determined as the main constituents in the fresh stem flower essential oil of lavender and lavandin cultivars. In the research, while the highest linalool content was determined in Dutch (43.3 %) during 2009 and in Vera (43.9 %) during 2010, the lowest linalool content was obtained from Raya in both years (28.5 and 37.3 %, respectively). The highest linalyl acetate content in the experimental years was determined from Super A (42.5 and 39.8 %, respectively), the lowest linalyl acetate content was obtained from Giant Hidcote in both years (4.35 and 3.76 %, respectively). Camphor wasn't determined in Raya and Munstead cultivars, the highest camphor content was obtained from Super A (19.8 %) during 2009 and Dutch (10.0%) during 2010. The highest terpinen-4-ol content was determined from Munstead (19.4 and 19.5 %, respectively), the lowest terpinen-4-ol content was obtained from Super A in both years (1.00 and 1.02 %, respectively) (Table 5).

Linalool, linalyl acetate, camphor and α-terpineol and borneol were determined as the main constituents in the dry stemless flower essential oil of lavender and lavandin cultivars. The highest linalool content was determined in Dutch (46.5 and 47.0 %, respectively), the lowest linalool content was obtained from Raya in both years (30.7 and 35.6 %, respectively). The highest linalyl acetate content was determined from Super A (32.8 and 29.5 %, respectively), the lowest linalyl acetate content was obtained from Giant Hidcote in both years (3.52 and 4.13 %, respectively). Camphor wasn't determined in Raya and Munstead cultivars, the highest camphor content was obtained from Silver (12.6 %) during 2009 and Dutch (10.9 %) during 2010. The highest α-terpineol content was determined from Munstead (19.2 and 20.6 %, respectively), the lowest α-terpineol content was obtained from Dutch (1.40 %) during 2009 and Silver (2.41 %) during 2010. The highest borneol content was determined from Giant Hidcote (7.83 and 9.54 %, respectively), the lowest borneol content was obtained from Munstead in both years (0.57 and 0.98 %, respectively) (Table 6).

Table 5. Essential oil composition (%) in fresh stem flower of lavender and lavandin cultivars

Essential Oil Components	RT	*Lavander (L. angustifolia)*						*Lavandin (L.x intermedia)*							
		Raya		Munstead		Vera		Silver		Dutch		Giant Hidcote		Super A	
		2009	2010	2009	2010	2009	2010	2009	2010	2009	2010	2009	2010	2009	2010
Myrcene	13.9	1.76	1.39	048	0.43	0.14	0.73	0.16	0.90	0.53	0.81	0.54	0.97	-	1.19
Limonene	16.1	0.47	0.46	0.73	0.62	1.29	1.31	0.77	1.27	0.95	1.43	1.03	1.80	1.01	1.06
Sabinene	16.7	--	1.92	-	1.32	-	13.3	-	12.9	-	16.7	-	16.9	-	6.52
Cymene	18.8	11.3	7.9	8.31	8.2	6.28	5.02	1.14	4.02	1.55	4.10	3.69	4.78	1.47	3.28
Butanoic acid hexyl ester	29.1	-	-	0.54	0.47	0.64	0.46	0.63	0.50	0.67	0.58	0.63	0.52	-	0.54
1-octen-3-ol	31.1	0.25	-	0.80	0.81	0.36	0.41	0.16	-	0.19	59.0	0.65	-	-	-
Linalool	37.4	28.5	37.3	42.5	37.8	38.1	43.9	41.8	43.6	43.3	42.5	34.9	42.8	34.8	41.8
Linalyl acetate	38.1	25.7	19.7	7.45	8.98	11.4	5.38	14.4	11.2	11.7	4.46	4.35	3.76	42.5	39.8
Neryl acetate	41.3	2.97	6.5	1.14	4.36	0.81	2.27	0.67	1.58	0.76	1.33	0.46	1.55	2.15	2.72
Terpineol-4-ol	41.6	3.96	4.94	19.4	19.5	3.27	5.86	1.64	2.56	2.06	3.44	2.41	2.90	1.00	1.02
Farnesene	48.8	1.46	1.72	3.12	2.69	1.56	2.53	0.42	0.78	0.59	1.58	1.83	1.61	0.44	1.30
Nerol	46.6	-	0.58	-	1.74	-	0.67	-	-	-	1.01	-	0.62	-	-
Crypton	46.7	0.31	0.21	1.72	1.64	-	-	-	-	-	0.74	0.75	0.82	0.32	0.55
α -terpineol	47.3	3.61	3.46	0.97	1.09	2.44	1.17	1.83	1.62	1.88	1.59	1.45	1.29	3.48	3.13
Borneol	47.5	0.90	2.6	0.49	0.66	3.52	5.06	3.38	6.57	3.28	5.96	7.61	10.1	1.72	5.13
Geranyl acetate	48.9	8.80	1.85	3.22	1.13	3.11	-	-	0.47	0.30	0.47	0.99	0.64	2.09	2.15
Benzaldehyde 4	53.0	-	0.85	-	0.71	-	-	-	-	-	0.29	-	-	-	0.58
Geraniol	55.7	1.65	2.21	0.63	0.89	0.78	-	0.61	0.84	0.61	0.40	0.23	-	1.54	1.25
Cadinol	73.6	-	0.77	-	1.19	-	0.83	-	-	-	0.44	-	0.46	-	0.30
Nonadecane	58.1	-	2.49	-	2.84	-	2.27	-	0.87	-	1.73	-	0.53	-	-
Heneicosane	68.9	-	1.29	-	1.60	-	1.96	-	0.48	-	1.08	-	-	-	-
α -pinene	8.2	-	-	-	-	-	0.46	-	0.50	-	0.47	-	0.53	-	-
β -pinene	11.6	0.19	-	0.46	-	0.13	0.57	0.11	0.69	0.33	0.59	0.61	0.60	-	-
Camphor	36.6	-	-	-	-	4.11	6.99	14.3	8.35	12.5	10.0	7.27	5.28	19.8	5.87
Eucalyptol	18.4	1.11	-	0.91	-	16.1	-	14.4	-	14.7	-	26.1	-	4.59	-
Sabinene hydrate	34.6	-	-	0.31	-	0.36	-	0.23	-	0.21	-	0.42	-	5.52	-
Caryophyllene	44.0	2.03	-	2.38	-	0.73	-	1.25	-	0.88	-	0.28	-	0.94	-
Lavandulol	49.0	0.54	-	2.28	-	-	-	-	-	-	-	-	-	-	-
Bisabolol	78.2	0.39	-	0.55	-	1.22	-	0.67	-	0.93	-	1.05	-	0.91	-

Table 6. Essential oil composition (%) in dry stemless flower of lavender and lavandin cultivars

Essential Oil Components	RT	*Lavander (L. angustifolia var)*						*Lavandin (L.x intermedia var)*							
		Raya		Munstead		Vera		Silver		Dutch		Giant Hidcote		Super A	
		2009	2010	2009	2010	2009	2010	2009	2010	2009	2010	2009	2010	2009	2010
Myrcene	13.9	0.15	0.79	-	0.26	0.12	0.41	-	0.82	-	0.60	0.20	0.59	-	0.99
Limonene	16.1	0.34	0.82	0.49	0.26	1.40	0.83	0.71	0.98	0.82	1.00	1.21	1.05	0.48	0.49
Sabinene	16.7	-	0.84	-	0.40	-	11.0		9.73	-	13.3	-	13.4	-	2.97
Cymene	18.8	10.8	4.82	7.45	4.76	4.52	2.35	1.44	2.98	1.71	3.03	3.93	2.18	1.54	2.12
Butanoic acid hexyl ester	29.1	-	-	0.54	0.67	0.69	0.66	0.53	0.72	0.59	0.59	0.80	0.61	0.57	0.40
1-octen-3-ol	31.1	0.35	0.79	0.93	0.89	0.46	0.96	0.19	0.43	0.32	0.89	0.33	0.65	-	-
Terpineol-4-ol	41.6	2.11	0.58	0.32	0.50	0.21	0.41	-	0.32	0.17	0.35	0.14	0.38	0.16	-
Linalool	37.4	30.7	35.6	46.1	39.5	44.5	46.8	44.9	46.8	46.5	47.0	43.1	45.7	36.5	39.1
Linalyl acetate	38.1	28.2	25.3	6.12	12.2	3.68	4.99	10.5	7.14	9.38	7.12	3.52	4.13	32.8	29.5
Neryl acetate	41.3	1.66	6.27	0.71	4.95	0.35	1.37	0.33	1.14	0.74	1.91	0.46	1.41	2.04	3.19
α -terpineol	47.3	4.72	8.45	19.2	20.6	2.81	3.99	1.87	2.41	1.40	3.13	3.02	3.47	-	-
Farnesene	44.8	1.19	2.48	1.90	3.42	0.86	0.97	0.52	0.53	0.65	0.83	0.92	1.17	0.30	0.09
Nerol	45.6	-	0.86	-	2.17	-	0.85	-	0.59	-	0.28	-	0.86	-	-
Crypton	46.7	0.29	-	1.95	1.69	0.55	0.81	0.33	0.50	0.63	0.53	0.87	0.97	0.28	-
Borneol	47.5	0.94	1.75	0.57	0.98	5.01	8.58	3.79	6.50	3.63	6.04	7.83	9.54	3.26	4.55
Geranyl acetate	48.9	8.56	3.50	3.11	1.20	0.92	-	1.12	1.32	1.24	0.36	1.08	0.75	2.00	3.8
Benzaldehyde 4	53.0	-	0.78	-	0.97	-	0.43	-	0.63	-	0.27	-	0.40	-	0.8
Geraniol	55.7	2.06	1.86	0.79	1.22	0.28	0.47	0.53	1.02	0.70	0.93	0.31	0.39	2.20	2.18
α -pinene	8.2	-	-	-	-	-	0.27	-	0.39	-	0.37	-	0.28	-	-
β -pinene	11.6	0.17	-	-	-	0.11	0.40	-	0.38	-	0.44	0.21	0.36	-	-
Camphor	36.6	-	-	-	-	8.14	9.22	12.6	8.99	11.7	10.9	5.34	8.35	4.30	-
Eucalyptol	18.4	0.48	-	0.58	-	19.8	-	17.4	-	15.6	-	21.7	-	3.92	-
Sabinene hydrate	34.6	-	-	0.48	-	0.51	-	0.48	-	0.41	-	0.39	-	-	-
Caryophyllene	44.0	1.42	-	-	-	-	-	0.47	-	0.27	-	-	-	0.58	-
Lavandulol	49.0	0.40	-	2.56	-	0.85	-	0.33	-	0.59	-	0.94	-	-	-
Germakren D	51.0	0.73	-	0.23	-	-	-	0.47	-	0.34	-	-	-	0.83	-
Bisabolol	78.2	-	-	-	-	1.48	-	0.63	-	0.91	-	1.08	-	0.87	-

Generally, in the study, linalool content was higher than linalyl acetate content in all of the lavender cultivars. According to quality standards of lavender essential oil composition determined by the International Organization for Standardization (ISO 3515:2002), linalool, linalyl acetate, cymene, terpinen-4-ol and camphor should be between 25.0-38.0 %, 25.0-45.0 %, 4.0-10.0%, 2.0-6.0 %, and 0-0.5 %, respectively (Anonymous, 2002). Borneol, eucalyptol, cymene and α-terpineol content in dry stemless flowers of the cultivars were higher than fresh stem flowers. In this study, linalool, eucalyptol, α-terpineol and cymene contents of the cultivars were accordance with ISO 3515:2002 lavender oil standards. Linalyl acetate content in the cultivars, except for Raya and Super A, was lower than ISO 3515:2002 lavender oil standards. Camphor content of the cultivars, except for Raya and Munstead, did not meet ISO 3515:2002 lavender oil standards.

Baydar (2009) stated that lavender essential oil because of the low camphor content was better quality than lavandin essential oil, and that the camphor content in a quality lavender oil must be between 0.5 and 1 % in lavenders and between 5.0 and 10.0 in lavandins. Most researchers stated that the main components of lavender essential oil were linalool and linalyl acetate, and their contents varied between 15.0-35.0 % and 30.0-50.0 % (Wichtl, 1984), 25.1-59.8 % and 25.82-54.76 % (Arabaci and Bayram, 2005), 34.3-54.6 % and 24.0-29.0 % (Kara and Baydar, 2011) respectively. The abundance of linalool correlated with the transcription of the linalool synthase gene, suggesting that linalool production is in part regulated transcriptionally. However, the degree of correlation between linalool abundance and linalool synthase transcription differed between *L. angustifolia* and *L. x intermedia*, suggesting additional, and differing mechanisms that control linalool abundance in these species (Boeckelmann, 2008). Barazandeh (2002) found that essential oil of *L. latifolia* Medik. contained 41 components and its main components were linalool (30.6-31.9 %), 1,8 cineol (18.8-20.9 %) and borneol (8.9-10.1 %). Nogueira and Romano (2002) found that the highest essential oil components of *L. viridis* were 1,8-cineole (18.2-25.1 %), camphor (9.1-15.7 %), α-pinene (8.8-14.1 %), borneol (4.1-4.8 %), β-pinene (1.2-5.6 %), karen (1.0-6.5 %) and α-terpineol (0.8-4.2 %). Fakhari et al. (2005) determined that essential oil of *Lavandula angustifolia* Mill. contained linalool (32.8 %), linalyl acetate (17.6 %), lavandulyl acetate (15.9 %), α-terpineol (6.7 %) and geranyl acetate (5.0 %) as major components. Alatrache et al. (2007) stated that essential oil of *L. latifolia* contained 40 components and linalool 32.3%, camphor 12.4% and 1,8-cineole 11.7 %. In another study it was stated that the highest essential oil components of lavender were linalyl acetate (47.56 %), linalool (28.06%), lavandulyl acetate (4.34%) and α-terpineol (3.75 %) (Verma et al., 2010).

Most researchers stated that the essential oil composition of lavender and lavandin varied depending on the genotype of the plant (Lawrence, 1994), growth stage on the date of collection, parts of the plant, harvest times at different times on the day and year (Baydar, 2009; Avcı, 2010), drying conditions and extraction technology (Pinto et al., 2007).

CONCLUSIONS

In term of fresh stem, dry stem and dry stemless flower yields, Super A, Dutch lavandin and Raya, Vera lavender cultivars were determined as high-yielding cultivars. In the research, the highest essential oil content in both fresh stem flowers and dry stemless flowers were determined in Silver lavender and Super A lavandin cultivars. However, these two cultivars were unsuitable for lavender oil standard (ISO 3515:2002) because the linalyl acetate content was lower than linalool content, and had low cimene, terpinen-4-ol and high-camphor content in both fresh stem flowers and dry stemless flowers. Among the lavender and lavandin cultivars, Raya and Munstead were identified as quality cultivars according to ISO 3515:2002 lavender oil standards owing to high contents of linalyl acetate, cimene, terpinen-4-ol and the fact that it is camphor-free in both fresh stem flowers and dry stemless flowers.

Based on the results of the research, Super A and Dutch cultivars had higher fresh stem flower, dry stemless flower yields as compared to the other cultivars. Raya and Munstead cultivars were determined to be higher quality cultivars because of being camphor-free.

As a result, we could recommend Super A and Dutch for dry stemless flower yields, Silver for essential oil content, and Raya and Munstead cultivars for essential oil quality.

ACKNOWLEDGEMENTS

The authors gratefully acknowledge the financial support of the Scientific Research Projects Coordinating Office of Suleyman Demirel University (Project No: 1759-D-08). This research is a part of Ph.D. Thesis of Nimet KARA

LITERATURE CITED

Adam, K.L., 2006. Lavender production, products, markets, and entertainment farms. Retrieved November 5, 2006, from http:// Composition of lavenders and lavandins cultivated in Turkey 679 attra.ncat.org/attra-pub/lavender.html

Alatrache, A., B. Jamoussi, R. Tarhouni, M. Abdrabba, 2007. Analysis of the essential oil of *Lavandula latifolia* from Tunisia. J. of Essen. Oil Bearing Plants, 10(6), 446-452.

Anonimous, 2002. ISO 3515:2002 Lavander Oil Quality Standarts.

Anonimous, 2009-2010. Data of Meteorology Station, Isparta.

Arabaci, O., E. Bayram, 2005. The effect of nitrogen fertilization and plant density on some agronomic and quality traits of lavender (*Lavandula angustifolia* Mill.) under ecological conditions of Aydin. Adnan Menderes Uni. J. of Agric. Faculty 2(2),13-19. (in Turkish).

Atalay, A. T., 2008. The effect on yield and quality characters of organic and inorganic fertilisers applied different doses on lavander (*Lavandula angustifolia* Mill.) grown in Konya ecological contitions. Selcuk University Faculty of Agriculture, Konya (Turkey). Department of Field Crops. Master Thesis, p:46. (in Turkish).

Avcı, A.B., 2010. A new record and distribution of the endemic *Thymus cilicicus* Boiss. & Bal. for the squares B3 and C3 in Turkey. Turkish Journal of Field Crops 15(1), 40-42.

Barazandeh, M.M., 2002. Essential oil composition of *Lavandula* latifolia Medik from Iran. J. Essent. Oil Res. Mar/Apr, 14(2), 103-104.

Baydar, H., S. Erbas, 2007. Effects of harvest time and drying on essential oil properties in lavandin (*Lavandula x intermedia* Emeric ex Loisel.). I. International Medicinal and Aromatic Plants Conference on Culinary Herbs, 29 April - 4 May 2007, Antalya-Turkey.

Baydar, H., 2009. Lavender. Medicinal and Aromatic Plant Science and Technology (3th press). Suleyman Demirel University Press No: 51, 274-278, Isparta. (in Turkish).

Baytop, T., 1999. Therapy with Plants in Turkey (Past and Present). Nobel Tip Press, Istanbul. (in Turkish).

Beetham, J., T. Entwistle, 1982. The Cultivated Lavenders. Royal Botanic Gardens, Melbourne.

Boeckelmann, A., 2008. Monoterpene production and regulation in Lavenders (*Lavandula angustifolia* and *Lavandula x intermedia*). University of British Columbia, Master thesis, p:86.

Cavanagh, H., J.M. Wilkinson, 2002. Biological activities of lavender essential oil. Phytotheraphy Res. 16, 301-308.

Ceylan, A., A. Vomel, N. Kaya, N. Celik, E. Nigdeli, 1988. A investigation on effects to yield and qulity of plant space in lavender. Ege University Journal of Agriculture Faculty 25(2), 135-145. (in Turkish).

Ceylan, A., E. Bayram, N. Ozay, 1996. The effects of different doses of nitrogen and plant densities on some agronomic and technologic characteristic of lavender (*Lavandula angustifolia* Mill.). Turk. J. of Agric. and Forestry 20, 567-572.

Fakhari, A.R., P. Salehi, R. Heydari, S.N. Ebrahimi, P.R. Haddad, 2005. Hydrodistillation-headspace solvent microextraction, a new method foranalysis of the essential oil components of *Lavandula angustifolia* Mill. Journal of Chromatography A, 1098 14–18.

Guenther, E., 1952. The Essential Oils, R.E. Krieger Publication Coorparation 5,3-38.

Kara, N., H. Baydar, 2011. Essential oil characteristics of lavandins (*Lavandula x intermedıa* emeric ex loisel.) of Isparta province, kuyucak district, where lavender production center of Turkey. Journal of Selcuk Agricultre and Food Sciences 25 (4), 41-45. (in Turkish).

Kara, N., 2011. Determination of lavender (*Lavandula sp.*) cultivars suitable for essential oil production and studies on micropropagation possibilities. Suleyman Demirel University, Graduate School of Applied and Natural Sciences, Department of Field Crops, Ph.D. Thesis, 2011, 178 p. (in Turkish).

Lawrence, B.M., 1994. Essential Oils, Allured Publishing Corp., Wheaton, IL , Perf. Flav. 19, 33.

Lis-Balchin M., S. Hart, 1999. Studies on the mode of the action of the essential oil of lavender (*Lavandula angustifolia* Miller). Phytother Res. 13, 540-542.

Lis-Balchin, M., 2002. Lavander. The Genus *Lavandula.* Medicinal and Aromatic Plants– Industrial Profiles 29, 268.

McGimbsey, J.A., N.G. Porter, 1999. Lavender: A Grower's Guide for Commercial Production. New Zealand Institute for Crop&Food Research, New Zealand.

Nogueira, J.M.F., A. Romano, 2002. Essential oils from micropropagated plants of *Lavandula viridis*. Phytochem Analysis 13, 4-7.

Pinto J.E.B.P., J.C.W. Cardoso, E.M. De Castro, S.K.V. Bertolucci, L.A. De Melo, S. Dousseau, 2007. Morphophysiological aspects and essential oil content in brazilian-lavender as affected by shadowing. Horticultural Brasileira 25(2), 210-214.

Renaud E.N.C., D.J. Charles, J.E. Simon, 2001. Essential oil quantity and composition from 10 cultivars of organically grown lavender and lavandin. J. of Essential Oil Res. 13(4), 269-273.

Runham, S., 1998. Small scale study of yield and quality of oils from six herb species. ADAS Arthur Rickwood.

Stein, S.E., 1990. National Institute of Standards and Technology (NIST) Mass Spectral Database and Software, Version 3.02, Juen USA.

Tucker, A.O., 1985. Lavender, spike and lavandin. The Herbarist, 51, 44-50.

Verma, R.S., L.U. Rahman, C.S. Chanotiya, R.K. Verma, A. Chauhan, A. Yadav, A. Singh, A.K. Yadav, 2010. Essential oil composition of *Lavandula angustifolia* Mill. cultivated in the mid hills of Uttarakhand, India. J. Serb. Chem. Soc. 75 (3), 343-348.

Wagner, H., 1980. Pharmazeutische Biologie 2. Drogen Undihra Inhaltshoffe, Gustav Fisher Verlag-Stuttgard, p:124, New York.

Weiss, E.A., 1997. Essential Oil Crops. CAB International, New York, USA.

Wichtl, M., 1971. Die Pharmakognostich Chemische Analys., Band 12, p: 254, Frankfurt/M.

INFLUENCE OF VARIED PLANT DENSITY ON GROWTH, YIELD AND ECONOMIC RETURN OF DRIP IRRIGATED FABA BEAN (*Vicia faba* L.)

Naser Al-Suhaibani[1] *,Salah El-Hendawy*[1,2*], *Urs Schmidhalter*[3]

[1]*Plant Production Department, College of Food and Agriculture Sciences, King Saud University, Riyadh, Saudi Arabia*

[2]*Agronomy Department, Faculty of Agriculture, Suez Canal University, Ismailia, Egypt*

[3]*Department of Plant Sciences, Technische Universität München, Freising-Weihenstephan, Germany*

**Corresponding author: shendawy@yahoo.com*

ABSTRACT

Faba bean frequently alter the structure and size of their canopy as a consequence of environmental factors. This peculiarity has been taken into account to handle the little information regarding how growth and yield of faba bean respond to significant changes in plant density and how these changes affect yield potential and economic return. The seeds of two faba bean varieties (Giza 716 and Giza 843) were cultivated around the emitters of a drip irrigation system at five densities (4, 6, 8, 10 and 12 plants dripper^{-1}). The results showed that the relative growth rate, net assimilation rate, total dry weight per plant and leaf area per plant gradually increased with increasing plant density up to 8 plants dripper^{-1} and significantly decreased with 10 and 12 plants dripper^{-1}; however, increasing plant density increased the plant height and decreased the number of branches per plant. The highest leaf area duration and leaf area index were found at 8 plants dripper^{-1}, while the lowest were found at 4 plants dripper^{-1}. While yield components of individual plants tend to significantly decrease with increased plant density, seed yield per hectare significantly increased with increasing plant density up to 8 plants dripper^{-1} and slightly decreased with 10 and 12 plants dripper^{-1}. Crop production functions with respect to the number of plants per dripper versus all the attributes measured exhibited strong quadratic relationships with the exception of the yield components of individual plants, which exhibited a strong linear relationship. Seed yield per hectare showed a strong relationship with crop growth and vegetative growth parameters but not with the yield components of individual plants. Finally, when the plant densities were converted to seeding rates, simulations indicated that 118.0 (6 plants dripper^{-1}) and 118.8 kg ha^{-1} (8 plants dripper^{-1}) for branched (Giza 716) and non-branched (Giza 843) varieties, respectively, were adequate to lower seed cost without reducing net profit.

Keywords: Competition, lateral spacing, seeding rate, seed yield, yield components

INTRODUCTION

Plant density defines the number of plants per square meter, which in turn determines the area available to each individual plant. For most crops, plant density has a major influence on biomass, crop yield and economic profitability (Rafiei, 2009; Albayrak et al., 2011; Ciampitti and Vyn, 2011). In faba bean, according to Loss et al. (1998), plant density can affect canopy architecture, light conversion efficiency, duration of vegetative growth, dry matter production, seed yield and ultimately, the economic productivity of a crop. Therefore, optimising plant density, which may be defined by both the number of plants per unit area and the arrangement of plants on the ground, is a pre-requisite for obtaining higher productivity of faba bean. This is because the number of plants per unit area is an important determinant of final seed yield, it is the first yield component to be established at the early crop cycle, it is largely dictated and controlled by the farmer himself, and finally, it is largely unaffected by environmental change (Dantuma and Thompson, 1983). However, other yield components such as number and weight of pods and seeds per plant and 100 seed weight which are established at a later stage in the course of the crop cycle are significantly affected by environmental conditions. Furthermore, the contribution efficiency of these components in the final seed yield is also associated with the number of plants per unit area (López-Bellido et al., 2005). Therefore, varying plant density may be a viable alternative of manipulating the productivity of faba bean under different environmental conditions through their changes in physiological processes.

The available literature pertaining to the response of faba bean to plant density indicates that a wide range of densities is commonly used (Bianchi, 1979; Caballero, 1987; Pilbeam et al., 1991; Almeida et al., 1995; Loss et al., 1998; López-Bellido et al., 2005; Mathews et al., 2008; Khalil et al, 2010; Ragab et al., 2010; Bakry et al., 2011). According to the literature, the optimum plant density to obtain high productivity for different faba bean crop varieties can range from 10 to 100 plants m^{-2}. These finding indicate that the faba bean crop has the ability to alter plant size and canopy structure in response to changes in plant density. In addition, the marginal response in yield is very small at high densities for this crop. Therefore, when the marginal cost of an increase in plant density approaches the marginal return from the increase in yield, further increases in seeding rate are not warranted (Graf and Rowland, 1987). Unfortunately, studies about the optimum plant densities have been conducted when faba bean is cultivated by conventional practices. In such practices, the faba bean seeds are sown in rows, raised beds or ridges with different intra-plant and inter-plant competition. To date, no extensive studies have investigated the optimum plant density when faba bean is grown using a drip irrigation system. Drip irrigation system has a unique capacity to allow different plant distributions around the emitters and has many advantages in the cultivation of field row crops in arid and semiarid regions. However, the main limitation of applying this system to growing field crops is that the initial installation costs have been considered prohibitive for field row crops. Among the various components of a drip system, laterals costs (constituting more than 35% of the total cost of the system) play an important role in determining the cost of the system for annual and closely spaced crops such as faba bean. Because many vegetable crops are grown at lateral spacing of 1.4 m or more with an emitter spacing of 0.3–0.5 m, using such a design could be one of the most significant factors in reducing the overall investment costs of drip irrigation when it is used for field crop production. Furthermore, such design may be considered for faba bean production through crop rotation with vegetable crops. However, information on planting arrangement around emitters with large spacing between laterals, which creates significant intra-plant competition without any inter-plant competition, has not been investigated. Therefore, the objective of this study was to investigate the influence of different plant distribution patterns around the emitters of drip irrigation systems which creates only one competition between plants on growth, yield and economic return of two varieties of faba bean.

MATERIALS AND METHODS

Experimental site description

This study was conducted at the Experimental Farm of the Faculty of Agriculture at Suez Canal University, Ismailia, Egypt (30°58′ N latitude, 32°23′ E longitude and 13 m above sea level) during the winter seasons of 2011 and 2012. The soil texture of the experimental site is predominantly sandy throughout its profile (65.2% coarse sand, 26.9% fine sand, 4.6% silt and 3.3% clay) with an EC of 0.52 dSm^{-1} and a pH of 7.85.

Experimental design and treatments

A randomised complete block split-plot design with three replicates was used in each season. Varieties and plant densities were randomly assigned to the main and split plots, respectively. A layout of the experimental plots is shown in Fig.1. The drip irrigation system, with a design to reduce the high overall investment costs of drip irrigation systems, was divided into two main sectors with the two varieties of faba bean (Giza 716 and Giza 843) assigned to the two sectors. Within each sector, there were three replicates of the same variety.

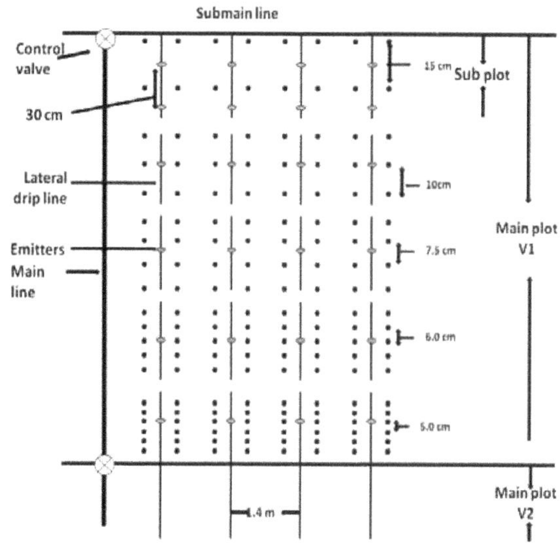

Fig. 1. Layout of one replicate of an experimental design that includes two varieties and five plant densities, showing locations of varieties and plant distributions around the emitters. 15, 10, 7.5, 6.0, and 5.0 cm indicated the distance between plants for 4, 6, 8, 10 and 12 plants $dripper^{-1}$ treatments, respectively.

Five plant densities (4, 6, 8, 10 and 12 plants $dripper^{-1}$) were randomly nested within each main plot of each variety as a split plot. Each split plot consisted of five polyethylene lateral drip lines (Twin-wall IV, 16 mm in diameter, and 0.3 m emitter spacing) with a length of 5 m. The lateral line was laid out along each faba bean row at 1.4 m. The split plot area was 35 m^2. To achieve the desired plant densities, 15, 10, 7.5, 6.0 and 5.0 cm were left between seeds for 4, 6, 8, 10 and 12 plants $dripper^{-1}$, respectively (Fig 1).

Agronomic practices

Seeds of faba bean varieties were sown on 15 October during both growing seasons. Before sowing, the seeds were inoculated with *Rhizobium leguminosarum* at the time of seeding. Therefore, a simulative dose of ammonium sulphate (20.5% N) (60 kg ha^{-1}) was added before the first irrigation. Phosphorus fertiliser was applied basally during seed-bed preparation at a level of 200 kg ha^{-1} as calcium super phosphate (15.5% P_2O_5). Potassium fertiliser was applied at a level of 100 kg ha^{-1} as potassium sulphate (48% K_2O) in one dose 40 days after sowing. Weed, pest, and disease control was performed in a timely manner. Hand harvesting was performed approximately 170 days after sowing.

Parameter assessments

At 70 and 100 days after sowing, 10 plants from each plot were harvested at random to determine the plant height, number of branches per plant, total dry weight per plant and leaf area per plant. Plant samples were separated into leaves and stems after the plant height was recorded. Leaf area was measured using a LI-3000 Area Meter (LI-COR, Walz Co., Lincoln, NE, USA). The leaf area index (LAI) was calculated by dividing the leaf area per plant by the area of ground occupied by the plant. After the leaf area was determined, the samples were dried in a forced-air oven at 75 °C for 72 h, and then their dry weights were determined. The total dry weight was obtained by the summation of the individual fractions.

Relative growth rate (RGR, g g^{-1} day^{-1}), net assimilation rate (NAR, g cm^{-2} day^{-1}) and leaf area duration (LAD, day) were derived using the following equations (Hunt, 1990):

$$RGR = 1/W \times \partial W/\partial T \quad (1)$$

$$NAR = 1/L_A \times \partial W/\partial T \quad (2)$$

$$LAD = ½ (LAI_1 + LAI_2) (T_2 - T_1) \quad (3)$$

where W, T, L_A and LAI are plant dry weight (g), time (day), leaf area (cm^2) and leaf area index, respectively.

After physiological maturity at approximately 170 days after sowing, an additional 10 plants from each plot were harvested at random to determine the number and weight of pods and seeds per main stem and branches. Seed yield was determined by hand harvesting an area of three rows, each 5.0 m in length (21.0 m^2 in total area), from each plot. Seed samples were collected from the yield samples to determine the moisture content, and seed yield was adjusted to a moisture content of 15.5%.

Economic return

Following Graf and Rowland (1987), the plant populations were converted to seeding rates (kg ha^{-1}) by multiplying plant population by the average seed weight for each variety. To simulate seed costs of one, two or three times that of the product price, the seeding rates were multiplied by a factor of 1, 2 or 3, respectively. The resulting values were subtracted from the yield to give comparable net yields. Then, the net yield was converted to a relative scale by dividing by a maximum yield of 2800 kg ha^{-1}. Finally, the relationship between seeding rates and relative net yield at different plant densities was calculated, while the seed cost to product price ratio ranged from 0:1 to 3:1. The 0:1 ratio represents when the seed cost was equal to the product price. However, the 1:1, 1:2 or 3:1 ratio represents the seed being one, two or three times that of the product price, respectively.

Statistical analysis

All measurements in this study were analysed using an analysis of variance (ANOVA) appropriate for a randomised complete block split-plot design with two varieties as the main plot, plant density as the subplots and replicates as blocks. Treatment means were compared using Duncan's multiple test. Probability levels lower than 0.05 was considered significant. Statistical analyses were performed using SAS statistical analysis package (SAS institute, Inc., Cary, NC). Polynomial and linear regression analyses were performed to investigate the relationship between plant density and different parameters of growth and yield. Regression analyses were performed using Microsoft Excel 2007.

RESULTS

Effect of plant density, variety and their interaction on crop growth parameters

The crop growth parameters, relative growth rate (RGR), net assimilation rate (NAR) and leaf area duration (LAD) were significantly affected by plant density. In both seasons, the highest RGR and NAR values were obtained at densities of 6 and 8 plants $dripper^{-1}$, followed by 4 plants $dripper^{-1}$, while 10 and 12 plants $dripper^{-1}$ produced the lowest values for both parameters (Table 1). Averaged over two seasons, planting with 10 and 12 plants $dripper^{-1}$ resulted in a significantly lower RGR by 34.8 and 47.3% and NAR by 38.8 and 47.9%, respectively, when compared to the medium plant density (8 plants $dripper^{-1}$). The differences in LAD between 6, 10 and 12 plants $dripper^{-1}$ were not significant, with the medium plant density (8 plants $dripper^{-1}$) and the low plant density (4 plants $dripper^{-1}$) producing the highest and lowest values for LAD, respectively(Table 1).

Table 1. Effects of varieties, plant population density and their interaction on the crop growth parameters relative growth rate (RGR), net assimilation rate (NAR) and leaf area duration (LAD) in 2011 and 2012.

	2011			2012		
	Giza 716 (V1)	**Giza 843 (V2)**	**Mean**	**Giza 716 (V1)**	**Giza 843 (V2)**	**Mean**
RGR (g g^{-1} day^{-1})						
D_1 (4 plants $driper^{-1}$)	0.040 b	0.041 b	0.040 b	0.040 b	0.040 a	0.040 b
D_2 (6 plants $driper^{-1}$)	0.045 a	0.044 a	0.045 a	0.043 b	0.043 a	0.043 a
D_3 (8 plants $driper^{-1}$)	0.045 a	0.046 a	0.045 a	0.047 a	0.040 a	0.043 a
D_4 (10 plants $driper^{-1}$)	0.027 c	0.030 c	0.028 c	0.026 c	0.034 b	0.030 c
D5 (12 plants $driper^{-1}$)	0.024 c	0.026 c	0.025 c	0.021 d	0.022 c	0.022 d
Mean	0.036 a	0.037 a		0.035 a	0.036 a	
NAR (g cm^{-2} day^{-1})						
D_1 (4 plants $driper^{-1}$)	4.53 ab	4.28 b	4.40 b	5.06 b	4.93 b	4.99 b
D_2 (6 plants $driper^{-1}$)	4.68 a	4.81 a	4.75 a	5.18 b	5.48 a	5.33 a
D_3 (8 plants $driper^{-1}$)	4.56 a	4.98 a	4.77 a	5.64 a	5.13 a	5.38 a
D_4 (10 plants $driper^{-1}$)	2.56 c	3.25 c	2.90 c	2.69 c	3.95 c	3.32 c
D5 (12 plants $driper^{-1}$)	2.69 c	3.08 c	2.88 c	2.35 d	2.38 d	2.36 c
Mean	3.80 a	4.08 a		4.18 a	4.37 a	
LAD (days)						
D_1 (4 plants $driper^{-1}$)	63.7 c	55.8 c	59.8 c	66.4 c	56.1 c	61.3 c
D_2 (6 plants $driper^{-1}$)	88.8 b	104.3 b	96.6 b	90.1 b	107.1 b	98.6 b
D_3 (8 plants $driper^{-1}$)	114.2 a	139.0 a	126.6 a	118.1 a	142.0 a	130.0 a
D_4 (10 plants $driper^{-1}$)	85.3 b	97.6 b	91.4 b	85.4 b	110.1 b	97.8 b
D5 (12 plants $driper^{-1}$)	79.4 b	97.9 b	88.7 b	86.1 b	108.9 b	97.5 b
Mean	86.3 b	98.9 a		89.2 b	104.8 a	

In a column, means with different letter denoted statistically difference between treatment groups according to Duncan's test ($P < 0.05$).

Although the differences between the two varieties in terms of RGR and NAR were not significant, the response of the two varieties to different plant densities was always significant with respect to these two parameters. However, the LAD for Giza 843 exceeded the values of Giza 716 by 12.7 and 14.9% in 2011 and 2012, respectively. Generally, for both varieties, the lowest RGR and NAR values were obtained at the high plant density (10 and 12 plants $dripper^{-1}$), but the influence of both densities on RGR and NAR for Giza 843 was less marked than that for Giza 716 (Table 1). The highest and lowest LAD values for both varieties was obtained at medium and low plant density, respectively, with no significant differences being detected between 6, 10 and 12 plants $dripper^{-1}$ (Table 1).

Effect of plant density, variety and their interaction on vegetative growth parameters

Results obtained over the 2-year study showed that the different vegetative growth parameters (i.e., plant height, number of branches per plant, total dry weight per plant, leaf area per plant and leaf area index) at 70 and 100 days after sowing were significantly affected by plant density (Tables 2 and3).The low plant density (4 plants $dripper^{-1}$) produced the lowest values for only two vegetative growth parameters (plant height and leaf area index) at two sampling times. Averaged over the two seasons, the low plant density resulted in decreases in plant height and in the leaf area index of 19.8 and 48.3% at 70 days after sowing and 35.8 and 25.4% at 100 days after sowing, respectively, when compared to the high plant density (12 plants $dripper^{-1}$), although the latter density decreased the branch number per plant by 46.0 and 54.3%,total dry weight per plant by 32.9 and 59.2% and leaf area per plant by 35.5 and 55.2%at 70 and 100 days after sowing, respectively, when compared to the low plant density. The highest total dry weight and leaf area per plant values at two sampling times were obtained at densities of 6 or 8 plants $dripper^{-1}$, which were not significantly different from one another (Tables 2 and 3).

Table 2. Effects of varieties, plant density and their combination on selected growth parameters at 70 days after sowing in 2011 and 2012

	2011			2012		
	Giza 716 (V1)	Giza 843 (V2)	Mean	Giza 716 (V1)	Giza 843 (V2)	Mean
Plant height (cm)						
D_1 (4 plants driper^{-1})	50.8 b	49.6 b	50.2 b	46.4 b	46.3 c	46.4 b
D_2 (6 plants driper^{-1})	52.7 b	51.8 ab	52.3 b	51.0 b	50.1 bc	50.6 b
D_3 (8 plants driper^{-1})	59.3 a	53.9 ab	56.6 a	62.0 a	56.5 ab	59.3 a
D_4 (10 plants driper^{-1})	59.4 a	53.4 ab	56.4 a	64.7 a	58.8 a	61.8 a
D5 (12 plants driper^{-1})	60.4 a	57.1 a	58.8 a	63.4 a	60.2 a	61.8 a
Mean	**56.5 a**	**53.2 a**		**57.5 a**	**54.4 a**	
Number of branches per plant						
D_1 (4 plants driper^{-1})	5.1 a	2.4 a	3.8 a	5.7 a	2.5 a	4.1 a
D_2 (6 plants driper^{-1})	4.7 a	2.5 a	3.6 a	5.1 a	2.3 a	3.7 a
D_3 (8 plants driper^{-1})	3.5 b	2.2 a	2.8 b	3.7 b	2.3 a	3.0 b
D_4 (10 plants driper^{-1})	2.9 bc	2.0 a	2.5 bc	3.6 bc	2.0 a	2.8 b
D5 (12 plants driper^{-1})	2.1 c	1.7 a	1.9 c	2.7 c	2.0 a	2.4 b
Mean	**3.7 a**	**2.2 b**		**4.2 a**	**2.2 b**	
Total dry weight (g plant^{-1})						
D_1 (4 plants driper^{-1})	12.7 a	9.7 b	11.2 a	14.8 a	12.1 b	13.4 a
D_2 (6 plants driper^{-1})	9.7 b	12.1 a	10.9 a	12.0 b	14.8 a	13.4 a
D_3 (8 plants driper^{-1})	9.2 b	11.7 a	10.5 a	11.0 b	15.8 a	13.4 a
D_4 (10 plants driper^{-1})	7.3 c	9.3 b	8.3 b	8.1 c	10.2 c	9.2 b
D5 (12 plants driper^{-1})	6.8 c	8.8 b	7.8 b	8.0 c	9.4 c	8.7 b
Mean	9.1 b	10.3 a		10.8 b	12.5 a	
Leaf area per plant (cm^2)						
D_1 (4 plants driper^{-1})	1481.3 a	1191.9 b	1336.6 b	1600.2 a	1276.4 b	1438.3 a
D_2 (6 plants driper^{-1})	1336.8 b	1558.1 a	1447.4 a	1461.8 b	1683.1 a	1572.4 a
D_3 (8 plants driper^{-1})	1318.1 b	1539.6 a	1428.9 ab	1424.8 b	1606.2 a	1515.5 a
D_4 (10 plants driper^{-1})	920.2 c	1063.4 bc	991.8 c	840.2 c	1130.0 bc	985.1 b
D5 (12 plants driper^{-1})	800.0 c	933.4 c	866.7 d	823.4 c	1023.4 c	923.4 b
Mean	1171.3 b	1257.3 a		1230.1 b	1343.8 a	
Leaf area index						
D_1 (4 plants driper^{-1})	1.41 d	1.14 e	1.27 d	1.52 d	1.22 d	1.37 d
D_2 (6 plants driper^{-1})	1.91 c	2.23 d	2.07 c	2.09 bc	2.40 c	2.25 c
D_3 (8 plants driper^{-1})	2.51 a	2.93 a	2.72 a	2.71 a	3.06 a	2.89 a
D_4 (10 plants driper^{-1})	2.19 b	2.53 c	2.36 b	2.00 c	2.69 bc	2.35 c
D5 (12 plants driper^{-1})	2.29 ab	2.67 bc	2.48 b	2.35 b	2.92 ab	2.64 b
Mean	2.06 b	2.30 a		2.14 b	2.46 a	

In a column, means with different letter denoted statistically difference between treatment groups according to Duncan's test ($P < 0.05$).

Table 3. Effects of varieties, plant population density and their combination on selected growth parameters at 100 days after sowing in 2011 and 2012

	2011			2012		
	Giza 716 (V1)	Giza 843 (V2)	Mean	Giza 716 (V1)	Giza 843 (V2)	Mean
			Plant height (cm)			
D_1 (4 plants driper^{-1})	84.2 b	83.6 b	83.9 b	80.8 c	79.6 c	80.2 d
D_2 (6 plants driper^{-1})	95.0 b	89.0 b	92.0 b	88.3 c	84.7 c	86.5 d
D_3 (8 plants driper^{-1})	118.5 a	111.7 a	115.1 a	117.8 b	114.3 b	116.1 c
D_4 (10 plants driper^{-1})	123.1 a	115.6 a	119.4 a	126.4 ab	119.0 b	122.7 b
D5 (12 plants driper^{-1})	127.0 a	119.1 a	123.1 a	135.3 a	130.8 a	133.1 a
Mean	109.5 a	103.8 a		109.7 a	105.7 a	
			Number of branches per plant			
D_1 (4 plants driper^{-1})	6.0 a	3.0 a	4.5 a	7.0 a	3.4 a	5.2 a
D_2 (6 plants driper^{-1})	5.1 ab	3.0 a	4.1 ab	5.9 b	3.4 a	4.7 a
D_3 (8 plants driper^{-1})	4.2 bc	2.3 a	3.3 bc	4.3 c	3.0 a	3.7 b
D_4 (10 plants driper^{-1})	3.3 cd	2.2 a	2.8 cd	3.7 cd	2.5 a	3.1 bc
D5 (12 plants driper^{-1})	2.3 d	1.7 a	2.0 d	2.9 d	1.9 a	2.4 c
Mean	4.2 a	2.5 b		4.7 a	2.9 b	
			Total dry weight (g plant^{-1})			
D_1 (4 plants driper^{-1})	41.8 a	33.5 b	37.6 b	48.8 a	39.8 b	44.3 b
D_2 (6 plants driper^{-1})	37.6 b	45.7 a	41.6 a	43.6 b	54.4 a	49.0 a
D_3 (8 plants driper^{-1})	35.4 b	46.3 a	40.9 a	44.8 b	52.3 a	48.6 a
D_4 (10 plants driper^{-1})	16.3 c	22.4 c	19.4 c	17.5 c	28.0 c	22.7 c
D5 (12 plants driper^{-1})	14.1 c	19.1 d	16.6 d	15.0 c	18.3 d	16.6 d
Mean	29.0 b	33.4 a		33.9 b	38.6 a	
			Leaf area per plant (cm^2)			
D_1 (4 plants driper^{-1})	2976.2 a	2716.9 b	2846.6 b	3046.2 a	2653.6 b	2849.9 b
D_2 (6 plants driper^{-1})	2809.6 a	3309.6 a	3059.6 a	2742.9 b	3312.9 a	3027.9 a
D_3 (8 plants driper^{-1})	2679.3 b	3326.1 a	3002.7 a	2709.4 b	3362.6 a	3036.0 a
D_4 (10 plants driper^{-1})	1468.6 c	1668.6 c	1568.6 c	1551.9 c	1951.9 c	1751.9 c
D5 (12 plants driper^{-1})	1051.9 d	1351.9 c	1201.9 d	1185.2 c	1518.6 c	1351.9 d
Mean	2197.1 b	2474.6 a		2247.1 b	2559.9 a	
			Leaf area index			
D_1 (4 plants driper^{-1})	2.83 c	2.59 d	2.71 d	2.90 c	2.53 c	2.71 c
D_2 (6 plants driper^{-1})	4.01 b	4.73 b	4.37 b	3.92 b	4.73 b	4.33 b
D_3 (8 plants driper^{-1})	5.10 a	6.34 a	5.72 a	5.16 a	6.41 a	5.78 a
D_4 (10 plants driper^{-1})	3.50 bc	3.97 c	3.73 c	3.69 b	4.65 b	4.17 b
D5 (12 plants driper^{-1})	3.01 c	3.86 c	3.43 c	3.39 b	4.34 b	3.86 b
Mean	3.69 b	4.30 a		3.81 b	4.53 a	

In a column, means with different letter denoted statistically difference between treatment groups according to Duncan's test ($P < 0.05$).

There were significant variations between the two varieties with respect to all vegetative growth parameters at two sampling times, with the exception of plant height (Tables 2 and 3). The Giza716 variety showed the highest number of branches per plant. The total dry weight per plant, leaf area per plant and leaf area index for Giza 843 were higher than those of Giza 716 by approximately 12.7, 7.6 and 11.7% at 70 days after sowing and 12.7, 11.7 and 15.0% at 100 days after sowing, respectively, as averaged over the two seasons (Tables 2 and 3). Although, plant density had no significant effect on the number of branches per plant for Giza 843, its average was reduced over two seasons for Giza 716 by 40.0 and 55.7% at 70 days after sowing and 45.1 and 60.2% at 100 days after sowing at 10 and 12 plants dripper^{-1}, respectively, when compared to the low plant density. For Giza 716, the total dry weight and leaf area per plant were significantly decreased with the increase in plant densities. For Giza 843, these parameters gradually increased with increasing plant density up to 8 plants dripper^{-1} and significantly decreased at densities of 10 and 12 plants dripper^{-1}; this result was observed in both growing seasons and at both sampling times. The highest and lowest LAI values for both varieties were obtained at 8 and 4 plants dripper^{-1}, respectively (Tables 2 and 3).

Effect of plant density, variety and their interaction on yield and yield components

While, all yield components (i.e., number of pods and seeds per main stem and branches, weight of pods per main stem and branches, total number of pods and seeds per plant and total weight of pods per plant) gradually decreased with increasing plant density, the seed yield per hectare was significantly increased with densities up to8 plants dripper^{-1} and thereafter slightly decreased at 10 plants dripper^{-1} with no significant differences observed between 4 and 12 plants dripper^{-1} (Tables 4 and 5). For instance, as averaged over the two seasons, as plant density was increased from 4 plants dripper^{-1} to 6, 8, 10 and 12 plants dripper^{-1}, all evaluated yield components were reduced by an average of 14.9, 34.6, 54.4 and 78.4%, respectively (Tables 4 and 5). However, compared with4 plants dripper^{-1}, the seed yield per hectare at 6, 8 and 10 and 12 plants dripper^{-1} was significantly increased by 39.4, 42.3, 33.7 and 14.4% in 2011 and by 29.4, 33.9, 28.6 and 14.5% in 2012, respectively (Table 5).

Table 4. Effects of varieties, plant population density and their combination on selected yield components in 2011 and 2012.

	2011			2012		
	Giza 716 (V1)	**Giza 843 (V2)**	**Mean**	**Giza 716 (V1)**	**Giza 843 (V2)**	**Mean**
		Number of pods per main stem				
D_1 (4 plants driper^{-1})	10.3 a	18.1 a	14.2 a	14.3 a	19.9 a	17.1 a
D_2 (6 plants driper^{-1})	9.0 ab	16.4 a	12.7 b	11.7 b	18.2 a	15.0 b
D_3 (8 plants driper^{-1})	7.1 bc	13.0 b	10.1 c	10.7 b	15.0 b	12.9 b
D_4 (10 plants driper^{-1})	6.2 c	10.2 c	8.2 d	8.0 c	11.2 c	9.6 c
D5 (12 plants driper^{-1})	3.1 d	4.5 d	3.8 e	3.3 d	4.2 d	3.8 d
Mean	7.2 b	12.5 a		9.6 b	13.7 a	
		Number of pods per branches				
D_1 (4 plants driper^{-1})	12.4 a	6.3 a	**9.3 a**	12.4 a	6.6 a	9.5 a
D_2 (6 plants driper^{-1})	8.6 b	6.5 a	**7.5 b**	9.3 b	6.1 a	7.7 b
D_3 (8 plants driper^{-1})	6.3 c	4.5 b	**5.4 c**	5.6 c	3.4 b	4.5 c
D_4 (10 plants driper^{-1})	4.4 d	2.7 c	**3.5 d**	4.7 d	2.2 b	3.5 d
D5 (12 plants driper^{-1})	3.1 e	2.3 c	**2.7 d**	2.3 e	2.6 b	2.4 e
Mean	7.0 a	4.5 b		6.9 a	4.2 b	
		Number of seeds per main stem				
D_1 (4 plants driper^{-1})	27.1 a	40.3 a	**33.7 a**	27.5 a	44.3 a	35.9 a
D_2 (6 plants driper^{-1})	27.8 a	38.0 a	**32.9 a**	29.2 a	38.7 a	34.0 a
D_3 (8 plants driper^{-1})	18.8 b	28.7 b	**23.7 b**	20.0 b	27.0 b	23.5 b
D_4 (10 plants driper^{-1})	13.6 c	18.7 c	**16.2 c**	14.6 c	18.7 c	16.7 c
D5 (12 plants driper^{-1})	5.2 d	8.4 d	**6.8 d**	6.6 d	8.9 d	7.7 d
Mean	18.5 b	26.8 a		19.6 b	27.5 a	
		Number of seeds per branches				
D_1 (4 plants driper^{-1})	26.8 a	23.8 a	**25.3 a**	32.5 a	26.7 a	29.6 a
D_2 (6 plants driper^{-1})	20.2 b	16.2 b	**18.2 b**	24.2 b	19.1 b	21.7 b
D_3 (8 plants driper^{-1})	17.2 b	14.4 b	**15.8 c**	21.4 b	16.7 b	19.0 c
D_4 (10 plants driper^{-1})	6.9 c	4.8 c	**5.8 d**	8.0 c	6.5 c	7.2 d
D5 (12 plants driper^{-1})	4.1 c	3.2 c	**3.7 e**	3.8 d	3.9 c	3.8 e
Mean	15.0 a	12.5 b		18.0 a	14.6 b	
		Weight of pods per main stem (g)				
D_1 (4 plants driper^{-1})	30.0 a	39.3 a	**34.7 a**	31.7 a	41.0 a	36.3 a
D_2 (6 plants driper^{-1})	26.3 b	33.3 b	**29.8 b**	29.7 a	35.0 b	32.3 b
D_3 (8 plants driper^{-1})	20.7 c	27.7 c	**24.2 c**	23.3 b	30.0 c	26.7 c
D_4 (10 plants driper^{-1})	18.7 c	25.7 c	**22.2 c**	20.7 b	26.9 c	23.8 d
D5 (12 plants driper^{-1})	8.7 d	8.7 d	**8.7 d**	10.0 c	9.3 d	9.7 e
Mean	20.9 a	26.9 a		23.1 b	28.4 a	

In a column, means with different letter denoted statistically difference between treatment groups according to Duncan's test ($P < 0.05$).

There were significant differences between the two varieties in yield and yield components. The Giza 843 variety produced higher yield components for the main stem than Giza 716, whereas the latter variety had higher yield components for branches than Giza 843 (Tables 4 and 5). For instance, as averaged over the two seasons, the number of pods and seeds per main stem as well as the pod weight per main stem for Giza 843 were 36.2, 29.9 and 20.5% higher, respectively, than those for Giza 716. The number of pods and seeds per branch as well as the weight of pods per branch for Giza 716was 37.4, 17.7 and 20.3% higher, respectively, than those for Giza 843 (Tables 4 and 5).

In addition, the response of the yield and yield components of the two varieties were influenced by plant density. For Giza 843, the highest seed yield was obtained at 8 or 10 plants dripper^{-1}; however, for Giza 716, the highest seed yield was obtained at 6 or 8 plants dripper^{-1}. The seed yield of Giza 716 at 10 plants dripper^{-1} was occasionally comparable to those at 4 plants dripper^{-1}, but for Giza 843, it was not (Table 5). At low plant density (4 plants dripper^{-1}), Giza 716 had seed yield values that were 30.6 and 24.1% higher than those of Giza 843; however, the latter had seed yield values that were 13.9 and 9.01% higher than those of Giza 716 at higher plant density (12 plants dripper^{-1}) in 2011 and 2012, respectively (Table 5).

Table 5 Effects of varieties, plant population density and their combination on selected yield components and seed yield in 2011 and 2012.

	2011			2012		
	Giza 716 (V1)	Giza 843 (V2)	Mean	Giza 716 (V1)	Giza 843 (V2)	Mean
Weight of pods per branches (g)						
D_1 (4 plants driper^{-1})	21.0 a	17.3 a	19.2 a	19.7 a	17.3 a	18.5 a
D_2 (6 plants driper^{-1})	18.3 a	15.0 a	16.7 b	17.3 a	16.3 a	16.8 b
D_3 (8 plants driper^{-1})	12.7 b	10.0 b	11.3 c	12.3 b	10.7 b	11.5 c
D_4 (10 plants driper^{-1})	8.7 c	5.0 c	6.8 d	9.7 c	5.7 c	7.7 d
D5 (12 plants driper^{-1})	4.0 d	1.7 d	2.8 e	4.0 d	2.7 d	3.3 e
Mean	12.9 a	9.8 b		12.6 a	10.5 b	
Total number of pods per plant						
D_1 (4 plants driper^{-1})	22.7 a	24.4 a	23.6 a	26.7 a	26.5 a	26.6 a
D_2 (6 plants driper^{-1})	17.6 b	22.9 a	20.3 b	20.9 b	24.3 a	22.7 b
D_3 (8 plants driper^{-1})	13.5 c	17.5 b	15.5 c	16.4 c	18.4 b	17.4 c
D_4 (10 plants driper^{-1})	10.6 d	12.9 c	11.8 d	12.7 d	13.4 c	13.1 d
D5 (12 plants driper^{-1})	6.3 e	6.9 d	6.6 e	5.6 e	6.8 d	6.2 e
Mean	14.1 b	16.9 a		16.5 a	17.9 a	
Total number of seeds per plant						
D_1 (4 plants driper^{-1})	53.9 a	64.1 a	59.0 a	59.9 a	71.0 a	65.5 a
D_2 (6 plants driper^{-1})	48.0 b	54.2 b	51.1 b	53.3 b	57.9 b	55.6 b
D_3 (8 plants driper^{-1})	36.0 c	43.0 c	39.5 c	41.4 c	43.7 c	42.5 c
D_4 (10 plants driper^{-1})	20.5 d	23.4 d	22.0 d	22.6 d	25.1 d	23.9 d
D5 (12 plants driper^{-1})	9.3 e	11.6 e	10.5 e	10.4 e	12.7 e	11.6 e
Mean	33.5 b	39.3 a		37.5 b	42.1 a	
Total weight of pods per plant (g)						
D_1 (4 plants driper^{-1})	51.0 a	56.7 a	**53.8 a**	51.3 a	58.3 a	54.8 a
D_2 (6 plants driper^{-1})	44.7 b	48.3 b	**46.5 b**	47.0 b	51.3 b	49.1 b
D_3 (8 plants driper^{-1})	33.3 c	37.7 c	**35.5 c**	35.7 c	40.7 c	38.2 c
D_4 (10 plants driper^{-1})	27.3 d	30.7 d	**29.0 d**	30.3 d	32.5 d	31.4 d
D5 (12 plants driper^{-1})	12.7 e	10.3 e	**11.5 e**	14.0 e	12.0 e	13.0 e
Mean	33.8 a	36.7 a		35.7 a	39.0 a	
Seed yield (kg ha-1)						
D_1 (4 plants driper^{-1})	1983.2 bc	1375.6 d	1679.4 c	2516.6 bc	1908.9 d	2212.7 c
D_2 (6 plants driper^{-1})	2829.8 a	2719.8 b	2774.8 b	3113.2 a	3153.2 b	3133.2 b
D_3 (8 plants driper^{-1})	2632.4 a	3193.2 a	2912.8 a	3085.7 a	3609.9 a	3347.8 a
D_4 (10 plants driper^{-1})	2167.4 b	2898.1ab	2532.7 b	2564.8 b	3634.0 a	3099.4 b
D5 (12 plants driper^{-1})	1815.7 c	2109.9 c	1962.8 c	2465.7 c	2709.9 c	2587.8 c
Mean	2285.7 b	2459.3 a		2749.2 b	3003.2 a	

In a column, means with different letter denoted statistically difference between treatment groups according to Duncan's test ($P < 0.05$).

Production functions

Production functions of number of plants per dripper versus crop growth parameters, vegetative growth parameters at 70 and 100 days after sowing, yield components and seed yield in both growing seasons are shown in Figs 2 and 3. When the production functions were done according to the crop growth parameters (RGR, NAR and LAD), vegetative growth parameters (total dry weight and leaf area per plant and leaf area index at 70 and 100 days after sowing) and seed yield ha^{-1}, the number of plants per dripper versus these parameters exhibited strong quadratic relationships with the exception of the total dry weight at 70 days after sowing, which instead showed a linear relationship (Figs. 2 and 3). However, the production functions of yield components (total number of pods and seeds per plant and total weight of pods per plant) versus the number of plants per dripper exhibited strong linear relationships (Fig. 3).

Correlations between seed yield and related traits

Averaged over two seasons, simple correlation coefficients between seed yield and its related traits are shown in Table 6. Seed yield correlated positively with crop growth parameters and vegetative growth parameters but showed no correlation with any of the yield components (Table 6).

DISCUSSION

In plant densities studies, intra- and inter-plant competition are two of the most important stresses affecting biomass production, crop yield and economic profitability. Increasing plant density per unit area increases both types of competition for production inputs such as solar radiation, water and nutrients. Low densities lead to inefficient use of these inputs. In this study, there was only one competition (intra-plant competition) whereas inter-plant competition was not a factor because there was sufficient distance between laterals in the drip design (1.4 m apart laterals). Intra-plant competition in this study arises from the distribution pattern of the plants around the drippers, which allowed different plant spacing of 15.0, 10.0, 7.5, 6.0 and 5.0 cm for treatments of densities 4, 6, 8,10 and 12 plants dripper^{-1}, respectively (Fig. 1). Notably, in this study, the widest inter-plant spacing combined with the narrowest intra-plant spacing (1.4 m × 0.05 m) gave the maximum plant density of only 28.6 plant m^{-2}. According to the literature, plant density under conventional cultural practices can range from 10 to 100 plants m^{-2} (López-Bellido et al., 2005; Mathews et al., 2008; Khalil et al, 2010; Ragab et al., 2010; Bakry et al., 2011). This finding raises the question of whether the plant densities in this study allowed for the potential maximum seed yield even with such great differences in densities compared to conventional cultural practices. Therefore, this study was conducted to understand the behaviour of growth and to better calculate the economic production function of two varieties of faba bean differing in branch capacity when grown in drip systems with only one competition parameter by increasing the space between laterals.

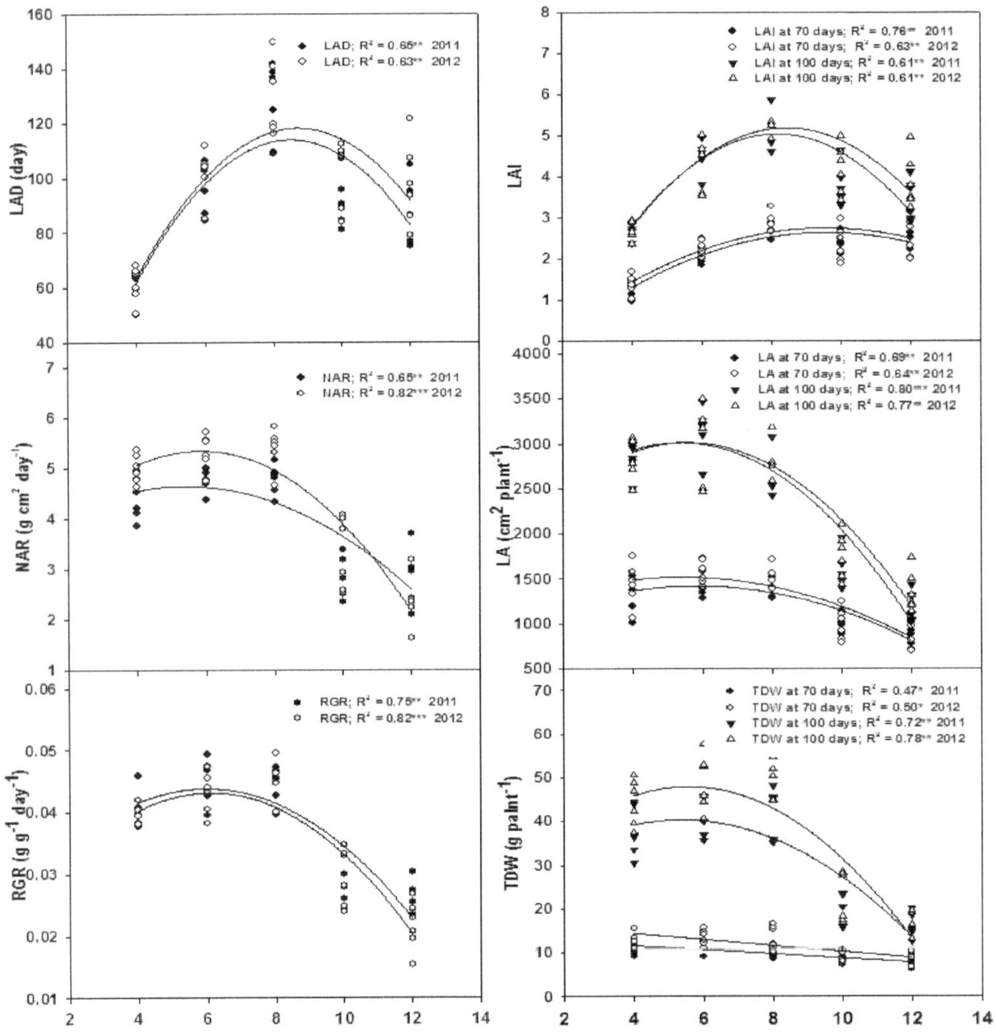

Fig. 2. Relationship between number of plants per dripper and crop growth parameters (relative growth rate, RGR; net assimilation rate, NAR and leaf area duration, LAD) and vegetative growth parameters at 70 and 100 days after sowing (total dry weight per plant, TDW; leaf area per plant, LA and leaf area index, LAI) in 2011 and 2012. Regression equations; *, ** and ***indicate significant at 0.05, 0.01 and 0.001P level, respectively.

Table 6. Simple correlation coefficients (lower right) and their significance levels (upper left) between seed yield and its related traits (data averaged over two seasons)

	Seed yield (kg ha-1)
Crop growth parameters	
RGR (g g-1 day-1)	0.53**
NAR (g cm-2 day-1)	0.51**
LAD (day)	0.70***
Vegetative growth parameters	
Total dry weight at 70 days after sowing (g $plant^{-1}$)	0.46**
Total dry weight at 100 days after sowing (g $plant^{-1}$)	0.52**
Leaf area per plant at 70 days after sowing (cm^2)	0.53**
Leaf area per plant at 100 days after sowing (cm^2)	0.50**
Leaf area index 70 days after sowing	0.49**
Leaf area index 100 days after sowing	0.76***
Yield components	
Number of pods per main stem	0.26 ns
Number of pods per branches	0.01 ns
Number of seeds per main stem	0.22 ns
Number of seeds per branches	0.07 ns
Weight of pods per main stem (g)	0.30 ns
Weight of pods per branches (g)	0.10 ns
Total number of pods per plant	0.19 ns
Total number of seeds per plant	0.17 ns
Total weight of pods per plant (g)	0.24 ns

*: significant at P < 0.01; **: Significant at P < 0.001; ns: non-significant.

Crop growth parameters such as RGR, NAR and LAD provide a framework for identifying potentially useful traits for yield improvement and to help us to understand the physiological response of growth and yield to plant density stress (Ball et al., 2001). These crop parameters reflect canopy development and efficiency of solar radiation interception, thereby reflecting the efficiency of photosynthesis and the final yield (Muchow et al. 1986). Furthermore, RGR is used as comprehensive trait for integrating morphological and physiological characteristics (McGraw and Garbutt, 1990). In this study, the highest RGR, NAR and LAD values were obtained at medium plant density (6 or 8 plants $dripper^{-1}$), while the highest plant density (10 or 12 plants $dripper^{-1}$) resulted in significantly lower values for RGR and NAR and higher values for LAD. However, the low plant density (4 plants $dripper^{-1}$) exerted moderate effects on RGR and NAR but produced the lowest values for LAD (Table 1). Increased crop growth parameters at medium densities (6 or 8 plants $dripper^{-1}$) may be because both densities exhibited the highest values for total dry weight and LAI, especially at 100 days after sowing, which corresponds to the start of flowering and the start of pod filling stages (Tables 2 and 3). This finding is consistent with the results of López-Bellido et al. (2005) for faba bean, who reported that the behaviour of crop growth indices in response to variation in plant density is generally linked to the dry matter and LAI values attained over various crop stages. However, the very narrowest intra-plant spacing (10 or 12 plant $dripper^{-1}$) accelerated leaf senescence and increased competition for light due to excessive shading between leaves. Whereas plants grown at very wide intra-plant spacing (4 plants $dripper^{-1}$) intercepted less light per plant during much of their life cycle than those grown at optimal plant density (Poulain et al., 1986; Polignand and Uggenti, 1989; Amato et al., 1992). The optimum plant density treatment that maintains adequate plant-to-plant spacing produces a complete canopy that is capable of maximising light interception. Therefore, a quadratic relationship has been found between the three crop growth parameters and plant density for the development of production functions (Fig. 2). The optimum plant density computed from the quadratic functions was approximately 6 plants $dripper^{-1}$ for RGR and NAR and approximately 9 plants $dripper^{-1}$ for LAD. These results are in partial agreement with those of El-Zahab et al. (1981), who observed an increase in RGR and NAR values at moderate densities, and Pilbeam et al. (1991) who reported that LAD is greater at higher plant densities during the vegetative stage.

Pilbeam et al. (1991) reported that until flowering in faba bean, the highest LAI values are achieved at the highest densities; however, the opposite holds true during pod development (i.e., the final LAI was greatest at lower densities). In this study, at 70 days after sowing, the highest LAI values were obtained at densities of 8 plants $dripper^{-1}$ followed by densities of 10 and 12 plants $dripper^{-1}$. However, at 100 days after sowing, a density of 8 plants $dripper^{-1}$ still produced the highest LAI values followed by a density of 6 plants $dripper^{-1}$, while a density of 4 plants $dripper^{-1}$ produced the lowest LAI values at both sampling times (Tables 2 and 3). Therefore, the optimum plant density that was computed from the quadratic functions of the relationship between LAI and plant density was approximately 10 plants $dripper^{-1}$ at 70 days after sowing and approximately 8 plants $dripper^{-1}$ at 100 days after sowing (Fig. 2). This finding indicates that during the early stages of vegetative growth, high plant densities will prompt higher LAI and apparently delay competition between plants. However, at later stages, high plant densities lead to rapid leaf senescence and competition between plants, which lead to decreases in LAI and dry matter accumulation. Therefore, the higher plant density (10 or 12 plants dripper^{-1}) resulted in a greater decrease in total dry weight and leaf area per plant measured at 100 days than at 70 days after sowing (Tables 2 and 3). In addition, the relationship between plant density and total dry weight or leaf area per plant at 100 days were higher than at 70 days after sowing (Fig. 2).

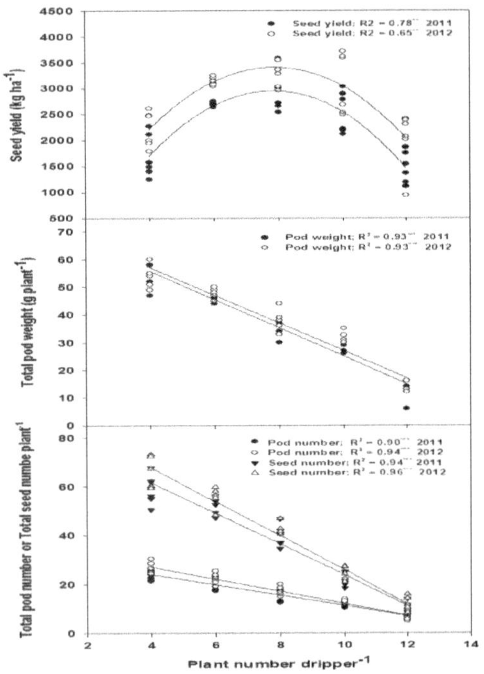

Fig. 3. Relationship between number of plants per dripper and total number of pods and seeds per plant, weight of pods per plant and seed yield per hectare in 2011 and 2012. Regression equations; ** and ***indicate significant at 0.01 and 0.001Plevel, respectively

In this study, increasing plant density resulted in a significant decrease in all yield components of individual plants, while seed yield increased with densities up to 8 plants dripper^{-1} and thereafter slightly decreased at 10 and 12 plants dripper^{-1} (Tables 4 and 5). Furthermore, final seed yield per hectare was significantly positively correlated with the crop growth and vegetative growth parameters, but not with the yield components of individual plants (Table 6). This finding shows that the final seed yield of faba bean did not reflect the response of yield components to plant density but did reflect the response of vegetative and crop growth parameters to plant density. The study of Achakzai and Taran (2011) agrees closely with the results from our study. Taken together, these results indicate that growth dynamic factors that occurred during the vegetative growth period could help explain the necessary management required to achieve optimal yields. Moreover, an increased understanding of how growth dynamic factors regulates faba bean yield in response to plant density could help us to explain the sensitive relationship between plant density and canopy development. This information may be especially relevant when applying a new agronomic practice of growing faba bean under less favourable environmental conditions. Mellendorf (2011) reported that the relationships between plant density and yield are explained by two concepts. First, maximum crop yield can only be achieved if the crop community is able to produce sufficient leaf area to provide maximum light interception during reproductive growth. Second, equidistant plant spacing maximises yield because it minimises plant-to-plant competitions. Therefore, the reduction in yield caused by high plant density (12 plant dripper^{-1}) could be explained by the competition between plants for this treatment begins during early vegetative growth. This early competition increases shading between leaves, leading to insufficient carbon fixation, increases respiration rate and increases intra-plant competition between vegetative and reproductive structures for assimilates. However, the reduction in yield caused by low plant density (4 plants dripper^{-1}) could be explained by the canopy development at the early stages for this treatment being insufficient to maximise light interception. Therefore, if the planting density is too high, plants may compete against each other, and the performance of individual plants may become a limiting factor for maximum crop yield. However, when the planting density is too low, each individual plant may perform at its maximum capacity, but there may be insufficient total plants to reach the optimum yield. In this case, total yield of the crop becomes a limiting factor.

This study also indicated that the effect of plant density on growth and yield of faba bean varied independently of faba bean genotypes. For Giza 716, which had more branching per plant than Giza 843, increasing plant density resulted in a significant decrease in total dry weight and leaf area per plant. For Giza 843, the both parameters increased with densities up to 8 plants dripper^{-1} and significantly decreased at densities of 10 and 12 plants dripper^{-1} (Tables 2 and 3). Seed yield of Giza 716 increased with densities up to 8 plants dripper^{-1} while for Giza 843, it increased with densities up to 10 plants dripper^{-1} (Table 5). These results indicate that the contribution of branches to seed yield decreased with increasing plant density due to excessive competition. This could be advantageous for Giza 716 under low density because plants compensate for open space by producing more branches. However, under high plant density, the branches contribute less than the main stem to seed yield.

Because the cost of faba bean seeds increases each year, there is a greater need to define optimum plant density to maximise yields and economic return. In this study, the optimal seeding rates for two varieties are presented in Fig 4. This figure simulates the relative net yields at the five plant densities when the seed cost to product price ratio ranges from 0:1 to 3:1. For Giza 716 and Giza 843, all the cost to product price ratios examined indicated that net yield increases up to a seeding rate of 118.0 (6 plants dripper^{-1}) and 118.8 kg ha^{-1} (8 plants dripper^{-1}), respectively. The relative net yields for Giza 716 at a 118.0 kg ha^{-1} seeding rate with a 1:1 ratio were occasionally comparable to those at a 156.8 kg ha^{-1} seeding rate with a 0:1 ratio. However, the relative net yields for Giza 843 at a 118.8 kg ha^{-1} seeding rate with a 1:1 ratio were occasionally comparable to those at a 148.8 kg ha^{-1} seeding rate with a 0:1 ratio. These results indicate that economically optimum-seeding rates could be 25% less than the optimum seeding rate due to increasing seed costs. Therefore, one might hypothesise that seeding rates below current practices could be used to lower seed costs without reducing net profit. This hypothesis is of increasing interest due to the increase in seed costs for faba bean producers.

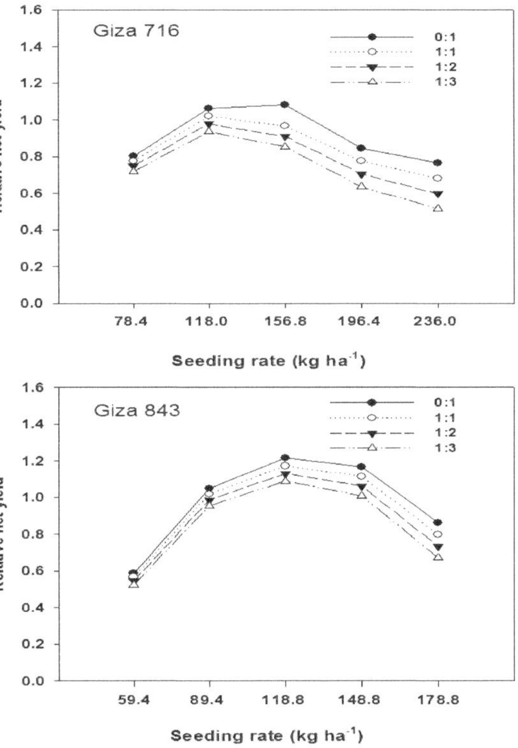

Fig. 4. Simulations of yield for Giza 716 and Giza 843 at five plant density when seed cost are nil, 1, 2 or 3 times of the product price which represent in the figure as 0:1, 1:1, 1:2 or 1:3, respectively.

ACKNOWLEDGMENTS

This project was supported by King Saud University, Deanship of Scientific Research, College of Food & Agriculture Sciences, Research Center.

LITERATURE CITED

Achakzai, A.K.K. and S.A. Taran. 2011. Effect of seed rate on growth, yield components and yield of mash bean grown under irrigated conditions of arid uplands of Balochistan, Pakistan. Pak. J. Bot. 43(2): 961-969.

Albayrak, S., M. Türk, and O. Yüksel. 2011. Effect of row spacing and seeding rate on hungarian vetch yield and quality. Turkish J. Field Crops.16 (1): 54-58.

Almeida, F.A.C., L. López-Bellido, M. Fuentes and J.E. Castillo. 1995. Effect of plant density on growth and yield of faba bean (*Vicia faba* L.) in Mediterranean conditions. In: AEP (Ed.), Proceeding of Second European Conference on Grain Legumes, 9–13 July, Copenhagen, Denmark, p. 167.

Amato, G., R. Cibella, D. Giambalvo and L. Gristina. 1992. Observations of the reproductive development in faba bean (*Vicia faba* L. var. equina) in relation to plant density. In: AEP (Ed.), Proceedings of the First European Conference on Grain Legumes. 1–3 June, Angers, France, pp. 245–246.

Bakry, B.A., T.A. Elewa, M.F. El karamany, M.S. Zeidan and M.M. Tawfik. 2011. Effect of row spacing on yield and its components of some faba bean varieties under newly reclaimed sandy soil condition. World J. Agric. Sci. 7 (1): 68-72.

Ball, A.R., W.R. McNew, D.E. Vories, C.T. Keisling and C.L. Purcell. 2001. Path analyses of population density effects on short-season soybean yield. Agron J. 93:187-195.

Bianchi, A.A. 1979. Results of three years of experimental trials on the cultural techniques of the horse bean for seeding (*Vicia faba* minor Beck). 2- Plant densities and distance between the rows. Riv. Agron., 13: 201-206.

Caballero, R. 1987. The effect of plant population and row width on seed yields and yields components of field beans. Res. Dev. Agric., 4: 147-150.

Ciampitti, I.A. and T.J. Vyn. 2011. A comprehensive study of plant density consequences on nitrogen uptake dynamics of maize plants from vegetative to reproductive stages. Field Crop Res. 121, 2–18.

Coelho, J.C. and P.A. Pinto. 1989. Plant density effects on growth and development of winter faba bean (*Vicia faba* L. var. minor). Fabis News let. 25, 26–30.

Dantuma, G. and R. Thompson.1983. Whole-crop physiology and yield components. In: Hebblethwaite, P.D. (Ed.), The Faba Bean (*Vicia faba* L.).Butterworths Publisher, London, pp. 143–158 (Chapter 6).

El-Zahab, A.A.A., A.A. Al-Babawy and A.S.Nidawy.1981. Density studies on faba beans (*Vicia faba* L.). Part II: growth parameters. Zeitschriftfür Acker- und Pflanzenbau 150 (4), 303–312.

Graf, R.J. and G.G. Rowland.1987. Effect of plant density on yield and components of yield of faba bean. Can. J. Plant Sci. 67, 1–10.

Hunt, R. 1990. Basic Growth Analysis: Plant Growth Analysis for Beginners. Academic Press, UK.

Khalil, S.K., A. Wahab, A. Rehman, F. Muhammad, S.Wahab, A.Z. Khan, M.Zubair, M.K. Shah, H. Khalil and R. Amin.2010. Density and planting date influence phonological development, assimilate partitioning and dry matter production of faba bean. Pak. J. Bot., 42(6): 3831-3838.

López-Bellido, F.J., L. López-Bellido and R.J. López-Bellido.2005. Competition, growth and yield of faba bean (*Viciafaba* L.). Eur. J. Agron. 23, 359–378.

Loss, S.P., K.H.M. Siddique, L.D. Martin and A. Crombie. 1998. Responses of faba bean (*Vicia faba* L.) to sowing rate in South-western Australia. Part II: canopy development, radiation absorption and dry matter partitioning. Aust. J. Agric. Res. 49,999–1008.

Malik, M.F.A., M. Ashraf, A.S. Qureshi and A.Ghafoor. 2007. Assessment of genetic variability, correlation and path analyses for yield and its components in soybean. Pak. J. Bot., 39(2): 405-413.

Mathews, P.W., E.L. Armstrong, C.J. Lisle, I.D. Menz, P.L. Shephard and B.C. Armstrong. 2008. The effect of faba bean plant population on yield, seed quality and plant architecture under irrigation in southern NSW. Proceeding of the 14th Australian Agronomy Conference. September, Adelaide South Australia. www.agronomy.org.au.

McGraw, J.B. and K.Garbutt.1990. The analysis of plant growth in ecological and evolutionary studies. Trends Ecol. Evol.5: 251–254.

Mellendorf, E.N. 2011. Soybean growth and yield response to interplant competition relief in various plant density environments. MSc Thesis, University of Illinois at Urbana-Champaign, USA.85 p.

Muchow, R.X., T.R. Sicclair, J.M. Benneth and L.C. Hammond. 1986. Response of leaf growth, leaf nitrogen and stomatal conductance to water deficits during vegetative growth of field grown soybean. Crop Sci. 26, 1190–1195.

Pilbeam, C.J., G. Duc and P.D. Hebblethwaite. 1990. Effects of plant population density on spring-sown field beans (*Vicia faba* L.) with different growth habits. J. Agric. Sci. 114, 19–33.

Pilbeam, C.J., P.D. Hebblethwaite, H.E. Ricketts and T.E.Nyongesa.1991. Effects of plant population density on determinate and indeterminate forms of winter field beans (*Vicia faba*).Part 1: Yield and yield components. J. Agric. Sci. 116, 375–383.

Polignand, G.B. and P. Uggenti. 1989. Variation analysis of physiological traits among different entries of faba bean (*Vicia faba* L.). Fabis Newslett. 25, 22–25.

Poulain, D., S. Keller and J. Le Guen. 1986. Canopy development and efficiency of foliar light interception in winter faba bean. Fabis Newslett. 16, 13–19.

Rafiei, M. 2009. Influence of tillage and plant density on mungbean. Am.-Eurasian J. Sustain. Agric., 3(4): 877-880.

Ragab, A.A., A.T. Emanand S.M. Abd El-Rasoul. 2010. A comparison between traditional and recent bioinocula on growth and productivity of faba bean (*Vicia faba* L.) grown in calcareous soil. Intern. J. Academic Res., 2(4): 245-253.

Siddique, M., M.F.A. Malik and S.I. Awan. 2006. Genetic divergence, association and performance evaluation of different genotypes of mungbean (*Vigna radiata*). Int. J. Agric. Biol., 8(6):793-795.

4

DETERMINATION OF FATTY ACID COMPOSITION ON DIFFERENT FALSE FLAX (*Camelina sativa* (L.) Crantz) GENOTYPES UNDER ANKARA ECOLOGICAL CONDITIONS

Duran KATAR[1]

[1]*Eskisehir Osmangazi University, Ali Numan Kırac Campus, Faculty of Agriculture, Eskisehir-TURKEY*
** Corresponding Author: durankatar6060@gmail.com*

ABSTRACT

This research was conducted to determine fatty acid composition of different false flax (*Camelina sativa* (L.) Crantz) genotypes (Vinimik 17, PI 304269, CR 476/65, CR 1674/90, Ames26665, Ames26667, Ames26673, Ames26676, Ames26680, Ames26686 and Ames28372) in terms of oleic, linoleic, linolenic, steraic, eicosenoic and erucic acids in 2010 and 2011. Results showed that having highest linolenic acid content, Ames28372 could be suggested to be used as oil crops for oil industry or as medicine plant / as biodiesel fuels. Though the content of erucic acid in camelina genotypes used in this study was too beyond the permitted value of 1 %, they could be safely used with erucic acid level decreased under codex limit by breeding programs.

Key words: False flax (*Camelina sativa* (L.) Crantz), oil content, fatty acid composition

INTRODUCTION

False flax (*Camelina sativa* (L.) Crantz), as an ancient-seed crop, has been cultivated since ancient times not only in Europe but also Asia (Zubr, 1997; Koncıus and Karcauskıene, 2010). This crop had been cultivated and used for human nutrition, medicine and illumination by 1940's, and then it has replaced with canola (Crowley and Fröhlich, 1998). The adaptability of false flax genotypes to the different environment causes considerable variation in the oil content and fatty acid profile (Koncıus and Karcauskıene, 2010). Even though, spring-habit false flax genotypes exist, there are also winter-habit genotypes (Crowley and Fröhlich, 1998). Being a green plant, false flax has great potential to be used as oil crop with less chemical fertiliziers, and this features makes false flax superior crop if compared to the other oil crops (Putnam et al., 1993; Kurt and Seyis 2008).

With almost high seed yield and oil content, false flax has been collected attention of oil enterprises on itself (Kurt and Seyis, 2008; Koncıus and Karcauskıene, 2010). In recent years, importance of false flax originated omega-3 concept has been put on agenda. Breeding programs have intensified on genotypes having high content of omega-3 in some countries, including mainly Germany. Moreover, in the breeding programs, special-purposed genotypes for high level of linolenic acid, erucic acid level in genotypes were limited to zero percent, edible purposed genotypes having less linolenic acid (Karvonen et al., 2002).

Vegetable oils have been used for different purposes, including human nutrition, pharmacology, industry, biodiesel etc. and composition of fatty acids plays important role determining usability/coherence of vegetable oils (Grombacher et al., 1993). Due to less oxidation stability, once some vegetable oils having higher linolenic content are not preferred to be used as edible oil, with easily combustibility, use of them are increasing more and more as biodiesel fuels (Frohlic and Rice, 2005; Abromovic et al., 2007; Sabzalian et al., 2008).

The suitability of a vegetable oil for a particular use such as nutritional, industrial or pharmaceutical use are determined by its fatty acid profile which is formed by both genotypic potential and environmental conditions (Sabzalian et al., 2008; Koncıus and Karcauskıene, 2010). It is therefore important that new genotypes having suitable oil acid composition are improved, which has been achieved in the other oil crops (Putnam et al., 1993; Vollman and Rajcan, 2009; Vollman et al., 1996; Zubr, 2003; Urbaniak et al., 2008; Koncıus and Karcauskıene, 2010). A number of studies related to fatty acid composition in camelina oil have been reported (Putnam et al., 1993; Budin et al., 1995; Agegnehu and Honermeier, 1997; Angelini and Moscheni, 1998; Zubr and Matthaus, 2002; Francis and Campbell, 2003; Abromvic and Abram, 2005; Ehrensing and Guy, 2008; Hrastar and Kosır, 2011; Sipalova et al., 2011). The purpose of this study was to determine and evaluate fatty acid profile of eleven false flax (*Camelina sativa* (L.) Crantz) genotypes under Ankara ecological condition.

MATERIALS AND METHODS

This study was carried out in research area of Central Research Institute for Field Crops in Turkey in 2010 and 2011 years. Genotypes, Vinimik 17, PI 304269, CR 476/65, CR 1674/90, Ames26665, Ames26667, Ames26673, Ames26676, Ames26680, Ames26686 and Ames28372 were gathered from Thrace Agricultural Research Institute.

Soil characteristics were given in Table 1. Research area had lime-loamy soil structure, 8.06 pH, 0,041 % salt, 1.57 % organic matter and 2.65 % lime. Besides, climatic conditions during two years were given Table 2.

Table 1. Soil Characteristics of Research Area

Structure	Lime (%)	Total Salt (%)	Plant-Available Phosphorus (P_2O_5) (kg/da)	Plant-Available Potassium (K_2O) (kg/da)	pH	Organic Matter (%)
Clay-Loam	2.65	0.041	11.41	215.233	8.06	1.57

Source: Soil Fertilizer and Water Resources Research Institute.

Table 2. Climatic Data of the Study Area in Ankara

Years	January	February	March	April	May	June	July
Total Rainfall (mm)							
2010	56.2	39.4	41.0	13.8	22.0	76.0	20.2
2011	28.0	5.0	42.0	35.0	86.0	37.0	13.0
1975-2010	39.2	33.6	36.1	50.0	49.7	35.1	16.0
Mean Temperature (°C)							
2010	1.2	4.0	7.0	9.4	15.0	19.0	21.0
2011	0.2	-0.6	3.0	8.0	12.0	17.0	23.0
1975-2010	0.3	2.1	6.2	11.3	16.0	20.2	23.5
Years	**August**	**September**	**October**	**November**	**December**	**Tot/Mean**	
Total Rainfall (mm)							
2010	0.0	3.0	16.5	26.4	65.6	379.9	
2011	0.2	0.0	81.6	24.0	50.0	401.8	
1975-2010	12.4	18.9	32.5	36.0	42.6	402.1	
Mean Temperature (°C)							
2010	25.5	16.7	14.5	5.2	3.4	11.8	
2011	21.0	17.0	12.3	8.7	4.6	10.5	
1975-2010	23.2	18.7	13.0	6.8	2.2	12.0	

[1/]Data were taken from Ankara Regional Meteorological Service.

Total rainfall in 2010 and 2011 (379.9 mm and 401.5 mm, respectively) were lower than long-term-year rainfall (402.1 mm). Total rainfalls in 2010 and 2011 crop growing periods (March-June), were 172 mm and 213, respectively. Besides, rainfalls, important for yield in early development stage (in May), were 22 mm in 2010 and 86 mm in 2011. Mean temperatures (Table 2) in 2010 and 2011 (11.8 °C and 10.5 °C, respectively) were lower than long-term-year temperature (12.0 °C). This study was conducted in randomized complete block design with three replications. Plot size and row spacing were 1.2 x 5.0 = 6.0 m^2 and 30 cm, respectively. Plants were sown in the middle of March (12 March in the first year and 15 March in the second year). No fertilizer was given and genotypes were harvested in the middle of June (15 June in the first year and 20 June in the second year). One row at both sides and 0.5 m in both edges were removed and 5 m^2 harvested area was evaluated.

The seeds obtained from the experimental field were properly grounded and the oil was extracted by n-hexane in a Soxhlet extractor for four hours. Recovered crude oils were taken to dry out in a rotator at 35 °C. The oil content was calculated on dry mass basis.

Fatty acids were etherified as methyl esters and analysed by Agilent 6890N Network with equipment with DB - 23 capillary columns (JW Scientific 122 - 2362 DB - 23; 60.0 m x 250 µm x 0.25 µm) GC and FID detector. Helium was used as carrier gas at a flow rate of 1 mL / min. Temperature of injector and detector were 260 °C and 240 °C, respectively. Column temperature was kept at 220 °C for 69 min. Samples of 0.5 µl was injected by hand and in the split mode (20:1). FAMEs were identified by comparison of their retention times with those of reference standards. The content of fatty acids was calculated from corresponding integration data. Results of the erucic acid (%), the eicosenoic acid (%), the linolenic acid (%), the linoleic acid (%), the oleic acid (%), and oil content (%) were analyzed in TARIST, Minitab 15 pocked statistical programs.

RESULTS AND DISCUSSION

In this study, oil content and levels of fatty acids (erucic acid, linolenic acid, linoleic acid, oleic acid and eicosenoic acid) in *Camelina sativa* genotypes were analyzed. Variance analyses and means were presented in Table 3 and 4. Table 3 showed that differences between years in linolenic acid and oil content were found as significant at 1%. Differences in years for oleic acid and eicosenoic acid were significant at 5 %, while it was insignificant ($p > 0.05$) in linoleic acid. Genotype x year interactions were found as significant at 1 %. Responses versus years among genotypes differently reflected. So, these phenomena made interactions significant. Bigger F values in genotypes (Table 3) assigned that differences in genotypes had more significant effect than yearly differences.

Table 3. Variance Analysis Table of Some Fatty Acid, evaluated.

Source of Variation	D.F.	Oleic acid (%) Sum of Means	Oleic acid (%) F Value	Linoleic acid (%) Sum of Means	Linoleic acid (%) F Value	Linolenic acid (%) Sum of Means	Linolenic acid (%) F Value
Replication	2	0.002	0.007	0.030	0.020	0.006	0.402
Years	1	22.563	76.237*	23.045	15.877ns	79.333	5075.089**
Error$_1$	2	0.296		1.452		0.016	
Genotypes	10	1.918	227.262**	8.568	303.992**	12.905	841.393**
Year x Gen.	10	1.088	128.899**	3.843	136.347**	11.762	766.843**
Error$_2$	40	0.008		0.028		0.015	
Mean	65	0.824		2.327		5.026	
C.V.(%)		0.296		1,452		0,016	
Source of Variation	**D.F.**	**Eicosenoic acid (%) Sum of Means**	**Eicosenoic acid (%) F Value**	**Erucic acid (%) Sum of Means**	**Erucic acid (%) F Value**	**Oil Content (%) Sum of Means**	**Oil Content (%) F Value**
Replication	2	0.001	0.008	0.001	0.60	1.53	2.53
Years	1	12.143	87.495*	1.979	108.708**	788.33	1304.65**
Error$_1$	2	0.139		0.002		0.60	
Genotypes	10	0.282	60.811**	0.265	28.294**	51.16	41.48**
Year x Gen.	10	0.809	174.571**	0.362	38.649**	41.72	33.83**
Error$_2$	40	0.005		0.009		1.23	
Mean	65	0.362		0.133		27.24	
C.V.(%)		0.139		0.002		18.36	

C.V. = coefficient of variation; ns = nonsignificant; *: significant at $P < 0.05$; **: significant at $P < 0.01$

Means of erucic acid (%), linolenic acid (%), linoleic acid (%), oleic acid (%), eicosenoic acid (%) and oil content (%) in false flax genotypes were given in Table 4. Erucic acid level in the first year (3.389%) was found as lower than the second year (3.043%). Ames26686 in the first year (3.900%), Ames26676 in the second year (3.487%) and Ames26673 in the mean of both years (3.833%) had the highest erucic acid content. Lowest erucic acid contents belonged to PI 304269 (2.290 % in the second year), CR 476/65 (2.920 % in the first year and in mean of both years). First year had the higher linolenic acid than the second year. The highest linolenic acids were taken from Ames26665 in the first year, Ames228372 in the second year. The lowest linolenic acid levels belonged to PI 304269 in the first year and Ames26676 genotype in the second year. As a means of years, in joint analysis, Ames28372 genotype had the highest linolenic acid, as the lowest one belonged to Ames26676 genotype (Table 4). The highest linoleic acid was determined in Ames26676 genotype and the lowest value belonged to CR 1674/90 genotype in the first, second years and joint analysis. In oleic acid, the first year had the higher oleic acid than the second year. In the first year CR 476/65 and Ames28372 genotypes had the highest and lowest oleic acid levels, respectively. In the second year and joint analysis, the highest and the lowest oleic acid levels belonged to CR 1674/90 and Ames26680, respectively (Table 4). Eichosenoic acid level in the first year was higher than the second year. While CR 1674/90 genotype had the highest eicosenoic acid level and lowest one was found as Ames26676 genotype in the second year. The first year and joint analysis draw same trend that CR 476/65 genotype had the highest eicosenoic acid level and lowest level was determined in Vinimik 17 genotype (Table 4).

Oil content in genotypes was higher in the second year than that of the first year. The highest oil content in the first year, the second year and joint analysis were found in Ames26667, CR 1674/90 and Vinimik 17 genotypes, respectively. Moreover, the lowest oil content in the first, second years, in joint analysis were in CR476/65, Ames28372 (in the second year and joint analysis), respectively (Table 4).

Table 4. Means of Some Fatty Acid in False Flax (*Camelina sativa*).

	Oleic acid (%)			Linoleic acid (%)			Linolenic acid (%)		
	1st Year	2nd Year	Mean	1st Year	2nd Year	Mean	1st Year	2nd Year	Mean
Vinimik 17	16.770C	14.733EF	15.752EF	20.333B	19.787D	20.060DE	31.710C	30.760D	31.235D
PI 304269	16.853BC	15.380D	16.117D	20.140B	20.050D	20.095D	29.920E	31.850C	30.885E
CR 476/65	17.587A	15.550CD	16.568C	19.650C	19.873D	19.762F	31.113D	29.833F	30.473F
CR 1674/90	17.027B	17.047A	17.037A	18.167G	18.740F	18.453H	31.667C	32.250B	31.958B
Ames26665	16.103D	14.920E	15.512G	18.413EFG	21.567C	19.990DEF	34.363A	29.623F	31.993B
Ames26667	16.107D	15.623C	15.865E	18.307FG	21.303C	19.805EF	34.173AB	29.117G	31.645C
Ames 26673	17.527A	15.917B	16.722B	19.013D	22.957B	20.985C	31.930C	26.163H	29.047G
Ames26676	17.030B	14.597FG	15.813E	21.247A	23.570A	22.408A	31.223D	24.853I	28.038H
Ames26680	15.750E	14.483G	15.117H	21.230A	21.490C	21.360B	31.060D	30.153E	30.607F
Ames26686	16.170D	15.917B	16.043D	18.777DE	18.443F	18.610H	31.807C	31.650C	31.728C
Ames28372	15.690E	15.583C	15.637FG	18.637EF	19.133E	18.885G	34.063B	32.657A	33.360A
Mean	16.601a	15.432b	16.017	20.628	19.447	20.037	32.094A	29.901B	30.998
L.S.D.(%)	Year: 0.576,Gen: 0.143, YearxGen=0.203			Year: 1.276,Gen: 0.262, YearxGen=0.371			Year: 0.305,Gen: 0.193, YearxGen=0.273		
	Eicosenoic acid(%)			**Erucic acid (%)**			**Oil Content (%)**		
	1st Year	**2nd Year**	**Mean**	**1st Year**	**2nd Year**	**Mean**	**1st Year**	**2nd Year**	**Mean**
Vinimik 17	12.980D	14.600BC	13.790F	3.397C	2.993CDE	3.397DE	28.13	37.30	32.72A
PI 304269	12.997D	14.590CD	13.793F	3.403C	2.290F	3.403DE	25.63	35.53	30.58BC
CR 476/65	13.353C	14.860A	14.107BC	2.920E	2.930CDE	2.920F	22.17	37.07	29.62CD
CR 1674/90	14.363A	14.443DE	14.403A	3.647B	2.830DE	3.647CDE	24.50	39.77	32.13AB
Ames26665	13.367C	14.743AB	14.055CD	3.153D	3.200BC	3.153DE	23.50	33.77	28.63DE
Ames26667	13.263C	14.733ABC	13.998DE	3.093DE	3.253B	3.093DE	28.60	30.70	29.65CD
Ames 26673	14.267A	14.073F	14.170B	3.833AB	3.303AB	3.833A	24.43	30.67	27.55EF
Ames26676	13.983B	14.007F	13.995DE	3.387C	3.487A	3.387AB	26.40	28.43	27.42EF
Ames26680	13.990B	14.343E	13.995B	3.417C	3.327AB	3.417BC	25.07	27.90	26.48F
Ames26686	14.327A	14.610BC	14.468A	3.900A	2.743E	3.900BCD	23.63	25.60	24.62G
Ames28372	13.287C	14.610BC	13.948E	3.133DE	3.117BCD	3.133E	22.70	24.07	23.38G
Mean	14.510A	13.652B	14.081	3.389A	3.043B	3.216	24.98 B	31.89 A	28.43
L.S.D.(%)	Year: 0.395,Gen: 0.106, YearxGen=0.150			Year: 0.104,Gen: 0.151, YearXGen=0.214			Year:1.90 ,Gen:1.74 , YearxGen=2.45		

A: denotes significant at 1 %, a: denotes significant at 5 %

Correlation analysis was given in Table 5. As seen in Table 5, while relationship between oil content and erucic acid, oleic acid and ecosenoic acid, linoleic acid and linolenic acid, oleic acid and linoleic acid, oil content and oleic acid were found as negative and significant at 1 %; relationship between oil content and linolenic acid, linolenic acid and eicosenoic acid were determined as negative and significant at 5 % (Table 5). Relationship between fatty acids was mostly negative and no significant relationship was observed between fatty acids and oil content but erucic acid that had significant and negative relationship with oil content. On the other hand, by evaluating regression coefficients of genotypes for oil content and fatty acids, adaptabilities of genotypes were given in Figure 1.

Table 5. Correlations among Parameters False Flax (*Camelina sativa*) Genotypes.

	Erucic Acid	Ecosenoic Acid	Linolenic Acid	Linoleic Acid	Oleic Acid
Ecosenoic acid	-0.176ns				
Linolenic acid	-0.175ns	-0.251*			
Linoleic acid	0.129ns	0.110ns	-0.856**		
Oleic acid	0.218ns	-0.405**	0.348**	-0.445**	
Oil content	-0.477**	0.492**	-0.258*	0.201ns	-0.340**

ns = nonsignificant; *: significant at $P < 0.05$; **: significant at $P < 0.01$

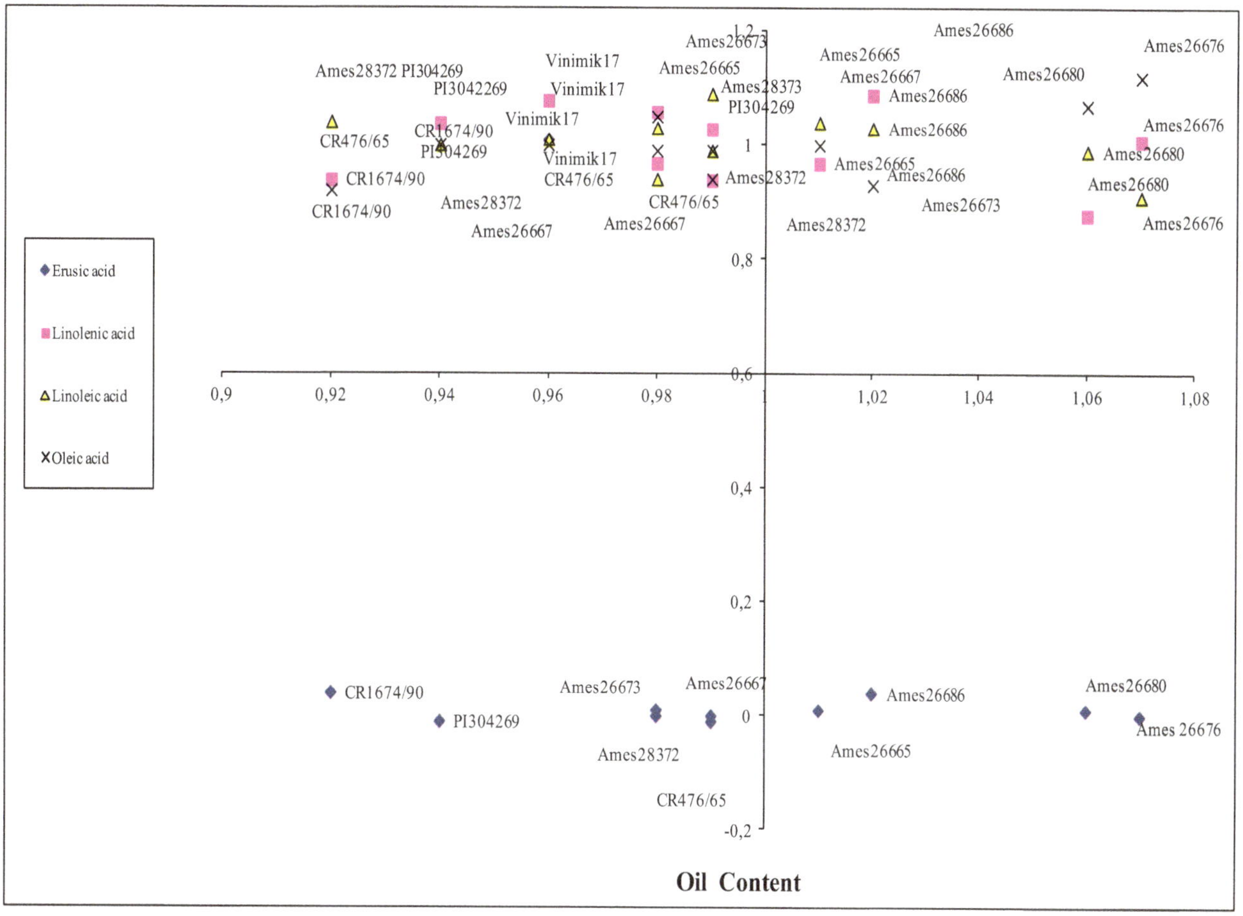

Figure 1. Regression coefficients of genotypes for oil content and fatty acids.

Regression coefficient is a measurement of reactions to different components/conditions in genotypes (Grabowski, 1998; Joshi and Press, 2007). Genotypes, having regression coefficient bigger than 1, can be well adapted to appropriate environmental conditions and high amount of fatty acids could be obtained. If $r < 1$, performance of genotype is low in poor conditions. Regression coefficient of genotype should therefore be as close as to 1. Vinimik 17, CR1674/90, PI 304269, Ames28373, Ames26680 and Ames26676 were observed more stabile than the other genotypes for components, evaluated.

CONCLUSION

Fatty acid profile of a oil crop determines its usability/suitability for nutritional, industrial or pharmaceutical value. This situation encourages looking for new sources of oil or new fatty acid profiles in different plant species (Sabzalian et al., 2008). Fatty acids are known as important ingredients, influenced by both environment and genotype and form quality of oil so they lead usage of oils (the purpose of industrial or edible consumption) (Crowley and Frohlich, 1998). Having fatty acids such as erucic, linolenic, linoleic, oleic acids in different levels, vegetable oils could be used as edible/ industrial purposes (Crowley and Frohlich, 1998; Geçgel et al., 2007).

In our study, years and genotypes significantly differed from each other in all characters; this is more likely to assign that responses of genotypes significantly varied in years and genotypes. Significant year x genotype interactions confirm this phenomena. So, similarly Seehuber (1987), Crowley and Frohlich (1998) and Francis and Campbel (2003) concluded that level of fatty acids profile and oil contents could significantly be sensitive to differences of years and genotypes. Seehuber (1987) and Francis and Campbel (2003) found that genotypic differences are more important than the effect of environmental conditions for fatty acids and oil contents. Valeri and Meirelles (1997), Tat and Vangerpen (1999) and Mc Crady (2007) revealed that plenty of genotypes with higher oil content and lower level of erucic acid were developed and served to use of industrial/edible purpose. Besides, legally admissible level of erucic acid in genotypes is accepted as 1-2 % of total fatty acid composition (Putnam et al. 1993; Taşan and Geçgel 2007).

Erucic acid, a monounsaturated fatty acid, is produced in wide range of oil crops especially in brassica family, and is known as brassidic acid (Vollman and Rajcan 2009). Permitted level of erucic acid in edible oils by codex is 1 % (Budin et al. 1995; Taşan and Geçgel 2007) and erucic acid level was found as higher than this level of 1 % in our study (Table 4). This stress that the genotypes have erucic acid levels over 1 %, and ranging from 2.920

to3.833 %, no usability of genotypes used are possible as edible oil crop, all they could be used in industrial purpose. Ames26673 and Ames26676 genotypes had the highest erucic acid levels with 3.833 % and 3.387 %, respectively. Because of higher level of erucic acid than acceptable level for edible purpose (max 2 % of total fatty acid cprofile), it could only be used as a surfactant/lubricant/precursor to biodiesel (Budin et al. 1995; Orlovius 2003; Taşan and Geçgel 2007). Erucic acid is monosaturated fatty acid and its content differs with not only years but also genotypes (Crowley and Frohlich 1998). Besides, erucic acid content in vegetable oil leads suitability of it for edibility (Sampath 2009) and in recent years, genotypes having almost zero erucic acid level were developed and presented to commercial use (Taşan and Geçgel 2007; Hrastar and Kosır 2011).

Oleic, linoleic and linolenic acids, essential components of fatty acid profile, should be taken regularly for health, besides vegetable oil, having rich amount of oleic, linoleic and linolenic acids, are more likely to be suitable for cooking (withstands up to more than 300°C) purposes (Hrastar and Kosır 2011). It was revealed that false flax is rich in oleic, linoleic and linolenic acids (Abromovic and Abram 2005, Hrastar and Kosır 2011). Common vegetable oils such as olive oil, corn and sunflower oils have less than 1 % of linolenic acid; whereas rapeseed or soybean oils have almost 8 %, and linseed oil with up to 60 % is the richest source of linolenic acid (Abromovic and Abram, 2005). Linolenic acid is the most abundant fatty acid in camelina oil and is almost 25-30 % (Zubr and Matthaus 2002; Abromovic and Abram 2005; Hrastar and Kosır 2011; Sipalova et al. 2011) and linoleic and oleic acids are, abundant in false flax, are classified as unsaturated omega-6 and monounsaturated omega-9 fatty acids, and they must be consumed properly for health (Crowley and Frohlich, 1998, Geçgel et al., 2007). Similar to in our study, levels of linolenic , linoleic and oleic acids were found as 30.998 %, 20.037 % and 16.017 %, respectively. It could be assumed that genotypes used are rich enough for these fatty acids. Meanwhile, Ames28372 genotype in linolenic acid, Ames26676 genotype in linoleic acid and CR1674/90 genotype in oleic acid were the richest genotypes. Many oil crops are of eichosenoic acid that is a monounsaturated fatty acid (Abromovic and Abram, 2005). With 14.468 %, Ames26686 genotype had the highest eichosenoic acid level in the study.

Oil content and its fatty acid composition are important characters for genotypes, since value of genotype is measured by the level of oil content and its fatty acid composition and this value determines the acceptability/usability of genotypes (Ehrensing and Guy 2008; Sampath 2009). Vinimik 17 possessed advantage for oil content, since the highest oil content (32.72 %) belonged to this genotype. The results showed that genotypes, used in this study were determined as industrial purpose. Having the highest levels of linolenic, linoleic, oleic acids and oil content, stability, Vinimik 17, CR 1674/90 and PI 304269 genotypes could be suggested to be used in oil industry. However more detailed researches are needed for further studies.

LITERATURE CITED

Abromovic, H., V. Abram, 2005. Physico-chemical properties, composition and oxidative stability of *Camelina sativa* oil. Food Technol.Biotechnol. 43 (1): 63–70.

Abromovic, H., B. Butinar, V. Nikolic, 2007. Changes occurring in phenolic content, tocopherol composition and oxidative stability of *Camelina sativa* (L.) Crantz oil during storage. Food Chemistry 104. 903-909.

Agegnehu, M., B. Honermeier, 1997. Effects of seeding rate and nitrogen fertilization on seed yield, seed quality and yield components of false flax (*Camelina sativa crtz*). Die Bodenkultur 48 (1).

Angelini, L., E. Moscheni, 1998. Camelina (*Camelina sativa [L.] Crantz*). In: Oleaginose Non Alimentari, pp. 82–85. Ed. G. Mosca. Bologna: Edagricole.

Budin, J.T., WM. Brene, D.H. Putnam, 1995. Some compositional properties of camelina (*Camelina sativa* l. crantz) seeds and oils. J. Am. Oil. Chem. Soc. 72(3): 309-315.

Crowley, J.G., A. Fröhlich, 1998. Factors affecting the composition and use of camelina. A Teagasc publication. Crops Research Centre, Oak Park, Carlow, Ireland.

Ehrensing, D.T., S.O. Guy, 2008. Cameliana. 'Oilseed Crops'Oregon State University. EM 8953-E.

Francis, C.M., M.C. Campbell, 2003. New high quality oil seed crops for temperate and tropical Australia. A report for the Rural Industries Research and Development Corporation. Crawley,Australia.

Frohlic, A., B. Rice, 2005. Evaluation of *Camelina sativa* oil as feedstock for biodiesel production. Ind. Crops Prod. 21(1). 25-31.

Geçgel, U., M. Demirci, E. Esendal, 2007. Fatty acid composition of oil from developing seeds of different genotypes of safflower (*Carthamus tinctorius* L.). J. Am. Oil Chem. Soc. 84. 47-54.

Grabowski, M.S., R.L. McCormick, 1998. Combustion of fat and vegetable oil derived fuels in diesel engines. Prog. Energy Combust. Science 24.125-164.

Grombacher, A., L. Nelson, D. Baltensperger, 1993. Crambe production. In Field Crops File G1126 University of Nebraska Extension Services.

Hrastar, R., I.J. Kosır, 2011. *Camelina sativa* (L.) Crantz as an alternative oilseed. J. Hmeljarski bilten, Vol. 18 pp. 85-92.

Joshi, R.M., M.J. Pregg, 2007. Flow properties of biodiesel fuel blends at low temperatures, Fuel 86. 143-151.

Karvonen, H.M., A. Are, N.S. Tadpole, I. Saline, M.I.J. Uusitupa, E.S. Sarkkinen, 2002. Effect of alpha-linoleic acid-rich *Camelina sativa* oil on serum fatty acid composition and serum lipids in hypercholesterolemic subjects. Metab. 51. 1253–1260.

Koncıus, D., D. Karcauskıene, 2010. The effect of nitrogen fertilizers, sowing time and seed rate on the productivity of *Camelina sativa.* Agrıculture,Vol. 97, No.4, p: 37-46.

Kurt, O., F. Seyis, 2008. An alternative oilseed crop: Camelina (*Camelina sativa* (L.) Crantz). Journal of OMU Agr. Fac. 23 (2). 116-120.

McCrady, J.P., 2007. Biodiesel Property Definition for Fuel Anomization and Combustion Modeling. Master's Thesis. University of Ilinois at Urbane Champaigh. Department of Agricultural and Biological Engineering. Urbana.

Orlovius, K., 2003. Oil seed rape (Fertilizing for High Yield and Quality), IPI Bulletin. Basel, 16.125.

Putnam, D.H., J. T. Budin, L.A. Field, and W.M. Breene. 1993 Camelina: A promising low-input oilseed. In J. Janick and J.E. Simon (Eds.), *New crops* (pp. 314-322). New York: Wiley.

Sabzalian, M.R., G. Saeidi, A. Mirlohi, 2008. Oil content and fatty acid composition in seeds of three safflower species. J. Am. Oil Chem. Soc. 85.717-721.

Sampath, A., 2009. Chemical Characterization of Camelina Seed Oil. The degree of Master of Science. New Brunswick, New Jersey. p. 180,

Seehuber, R., J. Vollman, M. Dambroth ,1987. Application of the single seed-descent method in false flax to increase the yield level. Landbauforsch. Voelkenrode (Germany) 37: 132-136.

Šipalova, M., T. Losak, J. Hkusek, J. Vollmann, J. Hude, R. Filipcik, M. Macek, S. Kracmar, 2011. Fatty acid composition of *Camelina sativa* as affected by combined nitrogen and sulphur fertilisation. Afr. J. Agric. Res. 6(16). 3919-3923.

Taşan, M., Ü. Geçgel, 2007. Examination of fatty acid compositions in mixtures of vegetable oils. J. of Tekirdağ Agricultural Faculty 4(1).

Tat, M.E., J.H. Van Gerpen, 1999. The kinematic viscosity of biodiesel and its blends with diesel fuel. J. Am. Oil Chem.Soc. 76(12).1511-1513.

Urbaniak, SD., C.D. Caldwell, V.D. Zheljazkov, R. Lada, L. Luan, 2008. The effect of cultivar and applied nitrogen on the performance of *Camelina sativa* L. in the Maritime Provinces of Canada. Can.J.Plant Sci. 88.111-119.

Valeri, D., A.J.A. Meirelles, 1997. Viscosities of fatty acids, Triglycerides, and their binary mixtures. J. Am. Oil Chem. Soc. 74:1221-1226.

Vollman, J., A. Damboeck, A. Ecki, H. Schrems, P. Runckenbauer, 1996. Improvement of *Camelina sativa*, an underexploited oilseed. In Progress in New Crops (Eds: J. Janick). ASHK Press, Alexandria, pp.357-362.

Vollman, J., I. Rajcan, 2009. Oil Crops. Volume 4 ISBN 978-0-387-77593-7 e-ISBN 978-0-387-77594-4 DOI 10.1007/978-0- 387-77594-4 Springer Dordrecht Heidelberg London New York.

Zubr, J., 1997. Oil-seed crop: *Camelina sativa*. Indust. Crops Prod. 6: 113-119.

Zubr, J., B. Matthaus, 2002. Effects of growth conditions on fatty acids and tocopherols in *Camelina sativa* oil. Ind. Crops Prod. 15: 155-162.

Zubr, J., 2003. Qualitative variation of *Camelina sativa* seed from different locations. Indust. Crops Prod. 17: 161-169.

DETERMINATION OF SUITABLE CORIANDER (*Coriandrum sativum* L.) CULTIVARS FOR EASTERN MEDITERRANEAN REGION

*Memet Inan[1], Saliha Kirici *[2], E. Sultan Giray[3], Murat Turk[4], Hassan Taghikhani[5]*

[1] *Kahta Vocational School of Higher Education, Adıyaman University, Adıyaman, Turkey*
[2] *Department of Field Crops, Agriculture Faculty, Cukurova University, Adana, Turkey*
[3] *Department of Chemistry, Letter and Science Faculty, Cukurova University, Adana, Turkey*
[4] *Ceyhan Vocational School of Higher Education, Cukurova University, Adana, Turkey*
[5] *Department of Field Crops, Institute of Science, Cukurova University, Adana, Turkey*
Corresponding author: kirici@cu.edu.tr

ABSTRACT

Coriander, *Coriandrum sativum* L., is a major medicinal and aromatic plant having aromatic carminative, stomachic, and antispasmodic properties. There are six improved cultivars in Turkey. In the research, it was aimed to adaptability trial of the new cultivars for Eastern Mediterranean. Therefore, field experiments were carried out during 2008- 2009 and 2009- 2010 growing seasons. Plant and agronomic properties were evaluated with oil content and composition. Oil compositions were analyzed with GCMS. As a result of the research, all investigated characters were varied significantly, except linalool which is major component in coriander fruit oil. With respect to fruit yield and essential oil content, the best results were obtained from Arslan and Erbaa cultivars, respectively. Linalool contents varied between 84.60 and 90.10 %; and maximum data of the component were obtained from Gurbuz with 90%.

Key Words: *Coriandrum sativum* L., essential oil content, linalool, yield

INTRODUCTION

Coriander (*Coriandrum sativum* L.), which is a spice plants belonging to the Apiaceae (Umbelliferae) family, is well known aromatic plant and name as "aşotu" and "kişniş" in Turkish it was grown at Göller Region, Ankara, Eskişehir, Erzurum, Gaziantep and Konya in Turkey (Ayanoglu et al., 2002). Maturated fruits and leaves (fresh and dried) were used as spice and oil distillated from maturated fruit is used mainly as a flavoring agent in pharmaceutical preparations. Fruits are used as aromatic and carminative and in laxative preparations to prevent griping (Kirici et al., 1997; Leung and Foster, 2003). Coriander has been used in medicines for thousands of years, various parts of this plant such as leaves, flower, seed, and fruit, possess diuretic, sedative, anti-diabetic, anti-mutagenic, antioxidant, anti-microbial, anti-convulsant, hypnotic and anthelmintic activities (Nadeem et al., 2013). Essential oil and especially its main component, linalool is an extremely important raw material in the perfume and cosmetic products, at the same time because of its bactericidal and fungicidal effects is also used in food and pharmaceutical products as a preservative. The essential oil content of the dried fruits varies from very low (0.03 %) to a maximum report of 2.7 %. The predominant constituent of essential oil of coriander is linalool, which forms approximately two-thirds of the oil (Shahwar et al., 2012).

In general, the coriander cultivation was on the basis of population of seed. Therefore, in the agricultural policies of the region winter product and the alternative coriander plant, six varieties which have been registered very important to determine yield potential in the Eastern Mediterranean region. In this study, the content and composition of the essential oil and yield components of the registered six varieties of coriander at Eastern Mediterranean Region conditions were investigated.

MATERIALS AND METHODS

Plant material: Six improved coriander cultivar, Arslan, Erbaa, Gamze, Gürbüz, Kudret-K and Pel-Mus were used in the study. The releasing dates and organizations of registered coriander varieties were shown in Table 1.

Field experiment: The experiments were arranged in a randomized complete block design with four replications for during 2 years (2008-2009 and 2009-2010) as a winter plant at Cukurova University, Faculty of Agriculture, and Department of Field Crops, on the Research and Application area. Fruits were sown at 12th November both years. Distance between rows was 25 cm and each plot consists of six row, plots were 4 x1.5 meters. Each plot received 40 kg ha^{-1} of nitrogen and phosphorus as basic fertilizer at planting, 20 kg ha^{-1} nitrogen fertilization at 2th April, 2009 and 30 March 2010 were done. Meteorological data for the growing seasons are shown in Table 2. Soil from a depth of 30 and 60 cm were sampled before setting the experiment, and were subjected to physicochemical analysis. Soils from depth 30 and 60 cm were total nitrogen 0.14% and 0.12 %, medium P_2O_5 (44.2 and 36.3 kg ha^{-1}), alkaline (pH of 7.68 and 7.66), salt 0.055 and 0.052 %, $CaCO_3$ 38.5 and 38.6%, respectively.

Table 1. Coriander *(Coriandrum sativum* L.*)* varieties

Variety Name	Registration Date	Owner*
Kudret-K	09.04.2004	Ondokuz Mayıs Univ. Agriculture Fac.,
Pel-Mus	09.04.2004	Ondokuz Mayıs Univ. Agriculture Fac.
Gamze	09.04.2004	Blacksea Agri.Research Institute
Erbaa	09.04.2004	Blacksea Agri.Research Institute
Arslan	07.04.2005	Ankara Univ. Agriculture Fac.
Gürbüz	07.04.2005	Ankara Univ. Agriculture Fac.

* Ondokuz Mayıs Univ. Agriculture Fac., 55139 Kurupelit / Samsun-TURKEY
*Black Sea Agricultural Research Intitute P.B. 39 55300 Tekkeköy /Samsun – TURKEY
*Ankara Univ. Agriculture Fac. 06110 Dışkapı /Ankara-TURKEY

Table 2. Means temperature and precipitation values at the study area

Months	Years	Temperature (⁰C)			Precipitation (mm)
		Min.	Max.	Mean	
November	2008	13.1	22.2	17.2	29.5
	2009	8.4	25.4	15.6	9.9
December	2008	6.4	15.7	10.6	46.7
	2009	6.0	20.6	13.3	79.5
January	2009	-2.0	18.8	9.6	144.3
	2010	-0.9	20.5	11.9	87.9
February	2009	3.8	20.1	11.0	78.0
	2010	1.3	22.3	12.6	66.3
March	2009	4.9	22.8	12.9	91.9
	2010	9.0	33.3	18.5	5.8
April	2009	9.3	29.3	17.7	29.2
	2010	15.2	35.5	23.8	58.7
May	2009	27.0	15.8	21.0	19.0
	2010	19.9	39.1	27.6	0.8
June	2009	18.9	38.4	27.2	0.0
	2010	24.8	43.1	31.1	1.5
*T.Prec. mm	2009/2010				438.6/310.4

Source:http://www.cukurova.edu.tr/Content/Asp/Turkish/cuMeteo Annual reports .asp
*T.Prec.: Total Precipitation

The soil textures were clay. The necessary observations from the germination to harvest were recorded. When the seeds had ripened, the process of harvesting started. The plots were harvested by plot harvest machine in 17 June 2009 and 16 June 2010. At harvest, plant height (cm), number of umbrella per plant and number of fruit per umbrella were recorded on ten plants randomly chosen in each plot, and thousand seed weight (g) and fruit yield (kg ha^{-1}) were obtained from the whole plot after separated side rows.

Isolation of essential oil

The fruits (40 g) of coriander that grounded in a blender separately were subjected to hydro-distillation using Clevenger-type glass apparatus for 3 hours for isolation of oils in each plot. The results are presented in ml 100 g^{-1}. The oils were stored in glass vials and kept -18°C at the deep frozen until GC-MS analyses.

GC–MS analysis

Qualification and quantification were carried out by using a Finnigan-Trace GC–MS equipped with an auto sampler. One microlitre of sample volume was injected using split method with 50 split ratio. Chromatographic separations were accomplished with a Zebron ZB-5 capillary column (5% phenyl–95% dimethylpolysiloxane, 0.25 mm i.d.×60 m, film thickness 0.25 m). Analysis was carried out using helium as the carrier gas, flow rate 1.0 mL/min. The column temperature was programmed from 40 to 260°C at 3°C /min. The injection port temperature was 200°C. The ionization voltage applied was 70eV, mass range *m*/*z* 41–400 a.m.u. The separated components were identified tentatively by matching with GC–MS results of National Institute of Standards and Technology (NIST) 05 and Wiley mass spectral library data. The quantitative determination was carried out based on peak area integration.

Statistical Analysis

The data were analyzed by the computer MSTAT-C package program. Results of experiments were also analyzed according to the Randomized Complete Block Design. Two years' were analyzed using split-plot design: main plot was the year, sub-plot was the cultivar. Significant differences among the mean values are given separately for each year of the trial.

RESULTS AND DISCUSSION

Data of agronomical and quality (essential oil content and composition) properties in the research were subject to variance analysis and the results were summarized in Table 3.

Table 3. Statistical analysis (mean of squar) for investigated characters

Source of Variance	Plant Height		Number of Umbrella per plant		Number of fruit per Umbrella		Thousand seed weight		Fruit yield		Essential oil content		Linalool Content	
	2009	2010	2009	2010	2009	2010	2009	2010	2009	2010	2009	2010	2009	2010
Cultivar	501**	538**	1134**	47.7*	31.49*	30.00	32.18**	23.46**	2932130**	148553	0.094**	0.020*	3.71	8.38
Error	53.1	1158	8.6	11.1	9.39	13.37	0.52	0.25	483674	128187	0.012	0.007	7.33	28.14
C.V. (%)	6.0	7.8	18.2	25.0	8.95	18.31	8.76	4.37	22.47	15.82	21.6	30.6	3.06	6.18
Combined Years														
Year	1149.4		90.58		2443.88**		131.63**		8305440.05**		0.758**		83.74	
Error1	212.3		24.0		8.673		0.95		139574.8		0.016		21.26	
Cultivar	189.9**		158.9**		44.71**		54.64**		2124054.2**		0.074**		9.66	
Year x Cultivar	418.9**		23.2		16.85		1.00**		956679.96*		0.040**		2.43	
Error 2	65.1		9.8		11.38		0.38		305925.26		0.009		17.73	
C.V.(%)	6.9		21.3		12.45		6.29		20.65		25.01		4.83	

DF: Degrees of freedom, C.V.: Coefficient of variation
* indicates the significance at 0.05 probability level
** indicates the significance at 0.01 probability level

Plant Height

The effect of cultivar and years on plant height are significant statistically ($p<0.001$) with year x cultivar interaction. The highest values of plant height were obtained from Kudret-K, Gamze and Erbaa cultivars in the first year, and also combined years values (Table 4). According to the mean of years, the highest plant height value was obtained from Kudret-K with 121.9 cm while the lowest value was in Arslan with 109.2 cm in the research. Plant height varies according to the grown region's ecology and plant genotypes. Plant height values were similar to the findings of some researchers (Mert and Kirici, 1998; Kirici, 1999; Kizil and Ipek, 2004) which indicate that plant height varies between 65.5 cm and 119.4 cm.

Table 4. The means of some morphological characters and thousand seed weight, fruit yield, essential oil and linalool contents obtained from Coriander cultivars in the trial years.

	2009	2010	Com.Years*	2009	2010	Com.years	2009	2010	Com.years
Cultivars	**Plant Height (cm)**			**Number of Umbrella per Plant**			**Number of Fruit per Umbrella**		
Arslan	102.7de	115.8 bc	109.2 b	7.90 c	9.67 b	8.78 c	29.68 c	15.30	22.49 b
Erbaa	127.5 a	112.6 cde	120.1 a	20.38 ab	17.60 a	18.99 a	34.52 ab	21.72	28.13 a
Gamze	129.7 a	111.2 cde	120.4 a	15.97 b	15.30 a	15.64 b	33.85 abc	19.60	26.73 a
Gürbüz	113.2 cde	115.5 bc	114.3 ab	11.07 c	8.75 b	9.91 c	33.42 bc	22.77	28.10 a
Kudret-K	129.7 a	114.1 cd	121.9 a	17.82 b	13.65 ab	15.74 b	38.15 a	18.70	28.42 a
Pel mus	127.1 ab	101.8 e	114.5 ab	23.45 a	15.13 a	19.29 a	35.80 ab	21.70	28.75 a
LSD (5%)	11.6 (int.), 8.24 (com.year)			4.42 (1.y. cul.)	5.03 (2.y.cul.)	3.21 (com.y.)	4.61(1.y. cul.)	ns	3.44(com.y.)
	Thousand Seed Weight (g)			**Fruit Yield (kg ha-^{1})**			**Essential Oil Content (%)**		
Arslan	13.66 b	16.13 a	14.89 a	4696.2 a A	2526.2 BCDE	3611 a	0.25 c D	0.21 b D	0.23 c
Erbaa	6.70 g	10.05 de	8.37 c	2924.4 b BCD	2313.2 CDE	2685 bc	0.59 ab AB	0.40 a C	0.49 a
Gamze	7.60 f	11.50 c	9.55 b	2465.7bBCDE	2170.0 DE	2756 b	0.59 ab AB	0.21 b D	0.40 ab
Gürbüz	7.89 f	11.75 c	9.82 b	3198.1 b B	2314.7 CDE	2756 b	0.44 b C	0.24 b D	0.34 b
Kudret-K	5.41 h	9.13 c	7.36 d	2285.0 b CDE	1945.7 E	2115 c	0.69 a A	0.26 b D	0.47 a
Pel mus	8.09 de	10.50 d	9.29 b	2998.7 b BC	2306.5 CDE	2652 bc	0.53 ab BC	0.26 b D	0.40 ab
LSD (5%)	0.89 (intr.)		0.62 (com.year)	1048(1.y. cul.) 798.7 (intr. Capital letter)	ns	564.8 (com.year)	0.16(1.y. cul.) 0.14 (intr. Capital letter)	0.13(2.y. cul.)	0.097 (com. year)

*Means followed by the same letter are not significantly different from each at 5% probability level ($p< 0.05$). ns: non significant

Number of Umbrella Per Plant

Number of umbrella per plant varied between 7.90 and 15.13 in 2009, the variation was significant statistically. The highest number of umbrella per plant (23.45) was obtained from the Pel-Mus cultivar in first year of experiment (2009). Similar to first year, there are significant statistically variation in 2010, but the highest value was obtained in Erbaa cultivar (Table 4). Rainfall of the first year was higher than that of second year (Table 2), for this reasons number of umbrellas per plant were different between the first and second years. The cultivars were statistically significant in the first and second year and combined years, but, it was not significantly different in the year x cultivar interaction (Table 3). The results for the number of umbrella per plant was an important character for seed and essential oil yield, and have been agreed with results of Kirici et al. (1997) (13.5- 16.8); Mert and Kirici (1998) (11.5- 17.7) and Kizil (2002) (11.3- 16.7).

Number of Fruit Per Umbrella

In the 2009 growing season, the highest number of fruit per umbrella (38.15) was obtained from the Kudret-K cultivar; however, results of other cultivars were similar except Arslan cultivar (Table 4). In the 2010 growing season, the highest one was recorded as 22.77 in Gürbüz cultivar. Amount of rainfall in the first year was higher than second year (Table 2), for this reasons number of fruit per umbrella was different between the first and second years. The cultivars were statistically significant in the first year and combined years, but, it was not significantly different in the second year and year x cultivar interaction (Table 3). The number of fruit per umbrella is an important yield component as much as the number of umbrella per plant. In terms of number of fruit per umbrella, Kirici et al. (1997), Mert and Kirici (1998) and Kizil (2002) reported similar results (23.5-37.1).

Thousand fruit weight

During the research, the highest thousand fruit weight was obtained from Arslan cv. as 13.66 g and 16.13 g, in 2009 and 2010 respectively (Table 4). The lowest values were detected at Kudret –K cv. as 5.41 g in the first year and 9.13 g in the second year. In general, the highest values were obtained from the second year, because of the great number of fruits per umbrella in the first year, the assimilate available for fruit filling was limited reducing fruit size (Arganosa et al., 1998). Effects of cultivars on thousand fruit weight in both years and combined year and year x cultivar interaction were found statistically significant (Table 3). The thousand fruit weight of coriander was within the range (8.52 – 11.05 g and 12.51 – 13.90 g) reported by Mert and Kirici (1998) and Kizil and İpek (2004) for Turkish coriander, the range for Canadian small-fruit cultivars (7.1- 7.3 g) and large-fruit cultivars (9.2- 9.9 g) reported by Arganosa et al. (1998). Kizil (2002) reported that thousand fruit weights of five coriander lines varied between 12.5 g and 13.6 g. According to the results, thousand fruit weight of coriander was affected by ecological conditions, fruit sizes and varieties.

Fruit Yield

The highest fruit yields both experimental years were obtained from Arslan cv. as 4696.2 kg ha^{-1} and 2526.2 kg ha^{-1}, respectively (Table 4). The lowest values were detected at Kudret-K as 2285.0 kg ha^{-1} in the first year and 1945.7 kg ha^{-1} in the second year. Effects of cultivar in both years, in the combined years, cultivar and year x cultivar interaction on dry leaf yield were found statistically significant (Table 3). The highest yield values were obtained from first year in the all coriander cultivars. Therefore, the climatic differences between years could also explain the responses of all coriander cultivars were obtained high fruit yield to the conditions that prevailed during the 2009 growing season. Kirici et al. (1997) and Mert and Kirici (1998) found that fruit yields of coriander varied 965 – 1780 kg ha^{-1} in the similar ecological conditions. Kizil and Ipek (2004) reported that fruits yield changed between 1568 – 2145 kg ha^{-1} and 1282 – 1486 kg ha^{-1}, respectively. Arganosa et al. (1998) reported that yield varied according to large (1915- 2061 kg ha^{-1}) and small sizes (1602- 1715 kg ha^{-1}) seed coriander. In terms of fruit yields our results are higher than the finding of the above-mentioned researches. The reason for this, plantations which have done during the winter, longer duration of vegetation and growing regions have played an important role in these differences.

Essential Oil Content

The content of essential oils of coriander fruits from different varieties were varied from 0.21- 0.69 % in both years (Table 4). In the experiment years, the highest values were obtained from Kudret-K (0.69 %) and Erbaa and Gamze (0.59 %) and the lowest values were obtained from Aslan (0.25 %) cv. According to the two-year averages the highest essential oil was obtained in Kudret-K variety (0.47 %) while the lowest essential oil content was obtained in Arslan cv. (0.23 %). Shahwar et al. (2012) reported that content of essential oil of coriander fruits varies from 0.03–2.6 %. According to pharmacopoeia; coriander fruits contained essential oil less than 0.5 % (Wagner et al., 1984). Some researchers reported that essential oil varied between 0.22 – 1.1 % (Mert and Kirici, 1998). The seed of Tunisian coriander had the highest essential oil yield with 0.68 % (Sriti et al., 2009). These essential oils variations can be attributed to some factors like climatic conditions, species and growing conditions (Shahwar et al., 2012).

Essential Oil Components

The essential oils from fruits of coriander cultivars analyzed by GC-MS, 22 compounds were identified. Both two years and all varieties as main component was linalool content ranged between 84.60 and 90.10 %. During two years, the highest mean values of linalool content (90.10 % and 87.80 %) were obtained from Gürbüz cultivar (Table 5). The lowest ones were obtained from Pel-Mus cv. (87.36 %) at the first year and Arslan cv. (84.60 %) at the second year. In regard to mean values of year content of linalool was over 86 %. In both years, all cultivars had similar contents of linalool, because there were no significant difference among mean values of cultivar and year and year x cultivar interaction. The other major components were γ-terpinene 1.37- 2.94 %, camphor 1.59- 2.97 %, nerol acetate 0.51- 2.87 % nerol 0.60- 1.98 % and α-pinene 0.85- 1.84 %. Composition of coriander seed essential oil was found to be varied at different researches. Shahwar et al. (2012) studied the chemical composition of coriander, and they found that the major volatile compounds in coriander seed were linalool (55.59 %), γ-terpinene (7.47 %), α-pinene (7.14 %), camphor (5.59 %), decanal (4.69 %), geranyl acetate (4.24 %), limonene (3.10 %), geraniol (2.23 %), camphene (1.78 %), and D-limonene (1.36 %). Leung and Foster (2003) reported that coriander fruits contain 0.2- 2.6 % (usually 0.4- 1.0 %) volatile oil, the major component of the oil is d-linalool, which is present in 55-74 %, depending on the ripeness of the fruits, geographical locations and other factors. Similarly,

Msaada et al. (2007) reported that the mature fruits essential oil was composed mainly of linalool (87.54 %). Bhuiyan et al. (2009), for fruits of coriander plants which collected from Bangladesh for a period of two years, have reported that essential oil was 0.42 % and the main components were linalool (37.65 %), geranyl acetate (17.57 %) and γ-terpinene (14.42 %). Gil et al. (2002) reported that oil composition was sensitive to the year variation in weather conditions as well as in soil environment. However, in our results; linalool content of coriander cultivars both years were stable.

Table 5. The means of essential oil components (%) obtained from fruits of coriander cultivars in the trial years*.

	Arslan		**Erbaa**		**Gamze**		**Gürbüz**		**Kudret-K**		**Pel-Mus**	
Components	**2009**	**2010**	**2009**	**2010**	**2009**	**2010**	**2009**	**2010**	**2009**	**2010**	**2009**	**2010**
α-pinene	1.24	0.85	1.84	1.38	1.10	1.73	0.99	1.12	1.01	1.57	1.17	1.17
camphene	-	0.12	0.18	0.10	0.16	0.19	0.09	0.07	0.04	0.16	1.16	0.13
sabinene	-	0.13	0.14	0.09	0.06	0.13	0.07	0.06	0.02	0.10	0.17	0.10
β-pinene	-	0.06	-	0.04	-	0.13	-	0.08	-	0.13	-	0.11
β-myrcene	-	0.34	0.49	0.41	0.36	0.41	0.30	0.27	0.32	0.32	0.38	0.56
cymol	1.04	0.59	0.67	0.57	0.50	0.75	0.58	0.68	0.32	0.66	0.78	0.77
limonene	0.23	0.56	0.41	0.55	0.35	0.76	0.29	0.62	0.35	0.66	0.42	0.70
Cis-ocimene	-	-	-	0.07	-	0.06	-	0.13	-	0.03	-	0.12
γ-terpinene	2.86	3.53	2.69	2.47	2.44	2.94	2.11	2.30	1.37	2.19	2.65	2.39
α-terpinolene	-	0.04	-	0.06	-	0.13	-	0.08	-	0.11	-	0.13
linalool	**88.18**	**84.60**	**88.05**	**85.17**	**89.13**	**85.46**	**90.10**	**87.80**	**88.25**	**87.59**	**87.36**	**84.67**
cis-sabinene hydrate	-	0.03	-	0.11	-	0.02	-	0.09	-	0.06	-	0.11
camphor	2.02	2.35	1.59	2.80	1.89	2.61	1.78	2.14	2.97	2.68	2.10	2.79
1-borneol	-	0.03	-	0.06	-	0.10	-	0.07	-	-	-	0.06
4-terpineol	-	0.05	0.3	0.21	-	0.10	0.10	0.13	0.09	0.10	0.14	0.17
p-menth-1-en-8-ol	-	0.07	0.09	0.18	-	0.17	0.14	0.18	0.32	0.21	0.33	0.27
n-decanal	-	0.02	-	0.15	-	0.08	-	0.03	-	0.11	-	0.15
nerol	1.52	1.56	1.11	1.85	0.60	1.29	1.13	1.35	1.52	1.06	1.40	1.98
thymol	0.47	1.80	-	0.15	1.97	0.03		-	-	-	-	-
nerol acetate	0.51	1.34	2.29	1.91	1.04	1.80	2.09	1.32	2.87	1.36	1.71	1.76
caryophyllene	-	0.20	-	-	0.29	-	-	-	-	-	-	-
2-dodecenal	-	0.03	0.14	0.34	0.04	0.23	0.09	0.10	0.28	0.15	0.41	0.32
total	**98.07**	**98.30**	**99.72**	**98.64**	**99.93**	**99.12**	**99.86**	**98.62**	**99.73**	**99.25**	**99.97**	**97.31**

*Values are mean of four replications

In our research, it was observed that the amount of linalool was 84-90%, which is higher than those of the previous studies as mentioned above. This might be due to the environmental conditions, locations and cultivars. Msaada et al. (2009) reported that linalool increase was concomitant with daily temperature. In the coriander essential oil must be contained linalool over 50-70% (Wagner et al., 1984). Indeed, linalool has a potential use as antisposmadic, immunostimulatory and antinociceptive, and is also used in the perfumery and liquor industries. For this reason, it suggests the exploitation of coriander seed as a low-cost renewable source of bioactive compounds for industrial processing in the fields of cosmetics, perfumes and nutraceuticals (Sriti et al., 2009).

CONCLUSION

In this study, Turkish coriander varieties were evaluated in climatic conditions of Eastern Mediterranean Region. As a results of studies, the maximum fruit yield was observed from Arslan cv. while oil content in Erbaa and Kudret-K varieties. All varieties in the studies have high linalool contents over 85% and the component had similar variation according to varieties and years. Because Aslan variety has high fruit yields, the variety is recommended for grower's commercial productions.

ACKNOWLEDGEMENTS

The study (ZF2009BAP17) was financially supported by the Çukurova University Scientific Research Projects Coordination Unit (ÇU-BAP). The authors wish to thank to the ÇU-BAP for financial support and Prof. Dr. A. GUMUSCU for cultivars.

LITERATURE CITED

Arganosa, G.C., F.W. Sosulski, A.E. Slikard. 1998. Seed yield and Essential oil of Northern-Grown Coriander (*Coriandrum sativum* L.), Journal of Herbs, spices&Medicinal Plants, 6 (2):23-32.

Ayanoglu, F., A. Mert, N. Arslan and B. Gurbuz. 2002. Seed yields, yield components and essential oil of selected coriander (*Coriandrum sativum* L.) lines. In Breeding Research on Aromatic and Medicinal Plants (Ed. C. Johnson and C.Franz), The Haworth Press, p:71-76.

Bhuiyan, N.I., J. Begum, M. Sultana. 2009. Chemical composition of leaf and seed essential oil of *Coriandrum sativum* L. from Bangladesh. Bangladesh J. Pharmacol, 4: 150-153.

Gil, A., E.B. De La Fuente, A.E. Lenardıs, M.L. Pereira, S.A. Suaärez, A. Bandoni, C. Van Baren, P.D.L. Lira, C.M. Ghersa. 2002. Coriander essential oil composition from two genotypes grown in different environmental conditions, J. Agric. Food Chem. 50: 2870-2877.

Kirici, S., A. Mert, F. Ayanoglu. 1997. The effect of nitrogen and phosphorus on essential oil content and yield values of coriander (*Coriandrum sativum* L.) at Hatay ecology, 2nd Field Crops Congrees of Turkey, September, 22-25, 1997, Samsun/Turkey Proceedings 347-351.

Kirici, S. 1999. Influence of seedlings rate region on morphological properties of coriander (*Coriandrum sativum* L.) collected from different locations . Cukurova Universtiy Journal of Agricultural Faculty, 14 (1) :33-40.

Kizil, S. 2002. The effects of different seed rates of selected coriander (*Coriandrum sativum* L.) lines on yield, yield components nd essential oil rate. Turkish Journal of Field Crops 7(2):99-105

Kizil, S. and A. Ipek. 2004. The Effects of Different Row Spacing on Yield, Yield Components and Essential Oil Content of Some Coriander *(Coriandrum sativum* L.) Lines. Journal of Agricultural Sciences, 10 (3): 237-244.

Leung, A.Y., S. Foster. 2003. Encyclopedia of Common Natural Ingredients used in food, drugs and Cosmetics (second Edition), Wiley-Interscience, pp: 193-195.

Msaada, K., K. Hosni, M.B. Taarit, T. Chahed, M.E. Kchouk, B. Marzouk. 2007. Changes on essential oil composition of coriander (*Coriandrum sativum* L.) fruits during three stages of maturity, Food Chemistry 102:1131-1134.

Msaada, K., M.B. Taarit, K. Hosni, M. Hammami, B. Marzouk. 2009. Regional and maturational effects on essential oils yields and composition of coriander (*Coriandrum sativum* L.) fruits. Scientia Horticulturae 122: 116-124.

Mert, A. and S. Kirici. 1998. To determine the yield and yield charecters of coriander (*Coriandrum sativum* L.) populations. Proceedings of XII.th International Symposium on Plant Originated Crude Drugs, New Trends and Methods in Natural Products Research. May, 20-22, 1998 Ankara, Abstract Book p. 112.

Nadeem, M., F. M. Anjum, M. I. Khan, S. Tehseen, A. El-Ghorab, J. I. Sultan. 2013. Nutritional and medicinal aspects of coriander (*Coriandrum sativum* L.) A review. British Food Journal, 115 (5): 743-755.

Shahwar, M.K., A. H. El-Ghorab, F. M. Anjum, M. S. Butt, S. Hussain, M. Nadeem. 2012. Characterization of coriander (*Coriandrum sativum* L.) seeds and leaves: volatile and non volatile extracts. International Journal of Food Properties, 15:736–747.

Sriti, J., T. Talou, W.A. Wannes, M. Cerny, B. Marzouk. 2009. Essential oil, fatty acid and sterol composition of Tunisian coriander fruit different parts, J. Sci. Food Agric.; 89: 1659–1664.

Wagner, H., S. Bladt, E.M. Zgainski. 1984. Plant Drug Analysis. Spring-Verlag Berlin, p: 11.

GENOTYPIC VARIATIONS ON YIELD, YIELD COMPONENTS AND OIL QUALITY IN SOME CAMELINA (*Camelina sativa* (L.) Crantz) GENOTYPES

Duran KATAR[1], Yusuf ARSLAN[2], Ilhan SUBAŞI[2]*

[1]Eskişehir Osmangazi University, Faculty of Agriculture, Ali Numan Kıraç Campus, Eskişehir, TURKEY
[2]Central Research Institute for Field Crops, Ankara, TURKEY
**Corresponding authorl: durankatar@gmail.com*

ABSTRACT

This research was conducted to determine performance of different false flax (*Camelina sativa* (L.) Crantz) genotypes (Vinimik 17, PI304269, CR 476/65, CR 1674/90, Ames 26665, Ames 26667, Ames 26673, Ames 26676, Ames 26680, Ames 26686 and Ames 28372) in terms of the plant height, the branch number, 1000 seed weight, the seed yield, oil content, oil yield and fatty acid components. Results showed that having highest seed yield, and oil yield, Vinimik 17, Ames 26665, Ames 26667 and Ames 26673 could be suggested to be used as oil crops for oil industry or as medicine plant. Having low saturated fat content this crop seems quality edible oil. Though genotypes are higher in erucic acid than 2% (maximum limit for canola quality edible oil), Genotypes having less erucic acid content should be developed by breeding programs/biotechnology.

Key words: False flax, seed yield, yield components, oil content, oil acids, genotypes

INTRODUCTION

False flax (*Camelina sativa* (L.) Crantz) is a member of the Cruciferae family, named as false flax or gold of pleasure (Abramovic et. al., 2007; Karaman et al., 2011). False flax (*Camelina sativa* (L.) Crantz), known as an oilseed crop, has been cultivated and used for human consumption for 3000 years (Zubr, 1997; Crowley and Fröhlich, 1998). Among six species of false flax (*C. sativa, C. laxa, C. rumelica, C. microcarpa, C. hispida, C. anomala*) *C. sativa* has solely economic importance (Kurt and Seyis, 2008). Studies revealed that linolenic acid in false flax is about 2-4 times higher than canola, soybean and olive oil (Korsrud et al., 1978; Crowley and Fröhlich, 1998; Karvonen et al., 2002; Kurt and Seyis, 2008). Containing high proportion of oil content and some important amino acids, false flax (*Camelina sativa* (L.) Crantz) could be taken into account as substantial input for food and industrial purposes (Karvonen et al., 2002; Eeva-Liisa et al., 2007). False flax could cause tremendous increase in omega-3 fatty acid that is important for human health when is used for poultry sector (Rokka et al. 2001; Eeva-Liisa et al., 2007). Moreover, false flax seems as essential source not only for biodiesel production and but industrial machinery. High level iodine in methyl ester keeps this oil undeteriorated and this phenomenon in this oil creates opportunity to be used in industrial use (Frohlic and Rice, 2005). So, high level adaptability, lower nutrient requirement, higher competition ability to other oil crops, higher oil quality for human health/consumption make false flax promising crop in the world (Korsrud et al., 1978; Crowley and Fröhlich, 1998; Kurt and Seyis, 2008). Though, false flax (*Camelina sativa* (L.) Crantz) is summer-habit plant, winter resistant types of false flax exist (Crowley and Fröhlich, 1998). Once, oil content is about 42 % in summer-habit types, 45 % in winter-habit types in false flax (*Camelina sativa* (L.) Crantz), This study was carried out to determine the performance of eleven false flax genotypes for yield, yield components and quality of oil acid components.

MATERIALS AND METHODS

This study was carried out in research area of Central Research Institute for Field Crops in Turkey in 2010 and 2011 years. Genotypes, Vinimik 17 (1), PI 304269 (2), CR 476/65 (3), CR 1674/90 (4), Ames 26665 (5), Ames 26667 (6), Ames 26673 (7), Ames 26676 (8), Ames 26680 (9), Ames 26686 (10) and Ames 28372 (11) were used in the study. All genotypes are alternative habit genotypes but were sown in fall as a winter-habit genotype. Soil characteristics were given in Table 1. Soils in the research area had lime-loamy structure with 8.06 in pH, 0,041 % in salt, 1.57 % in organic matter and 2.65 % in lime. Besides, climatic conditions consisting of 2010-2011 years and long term (1975-2010) were given Table 2. Total rainfall in 2010 was 379.9 mm so do 401.8 mm in 2011, it was also 402.1 mm in long term (1975-2010) years. Temperature in 2010, 2011 and long term years were 11.8°C, 10.5°C and 12.0°C, respectively, so did 60.5%, 60.0% and 60.1%, respectively in humidity.

Table 1. Physical and Chemical Characteristics of Soil in Research Area

Structure	Lime (%)	Total Salt (%)	Plant-Available Phosphorus (P_2O_5) (kg/da)	Plant-Available Potassium (K_2O) (kg/da)	pH	Organic Matter (%)
Clay-Loam	2.65	0.041	11.41	215.233	8.06	1.57

Source: Soil Fertilizer and Water Resources Research Institute.

Table 2. Rainfall, Temperature and Humidity in 2010, 2011 and Long Term Years (1975-2010) in Ankara Climatic Conditions

Years	January	February	March	April	May	June	July
				Total Rainfall (mm)			
2010	56.2	39.4	41.0	13.8	22.0	76.0	20.2
2011	28.0	5.0	42.0	35.0	86.0	37.0	13.0
1975-2010	39.2	33.6	36.1	50.0	49.7	35.1	16.0
				Mean Temperature (°C)			
2010	1.2	4.0	7.0	9.4	15.0	19.0	21.0
2011	0.2	-0.6	3.0	8.0	12.0	17.0	23.0
1975-2010	0.3	2.1	6.2	11.3	16.0	20.2	23.5
				Mean Humidity (%)			
2010	58.8	59.5	60.1	61.2	60.5	58.6	57.4
2011	59.7	62.1	62.4	60.8	60.7	58.9	58.4
1975-2010	58.2	59.4	61.2	60.8	60.3	59.1	60.0

Years	August	September	October	November	December	Tot/Mean
			Yearly Total Rainfall (mm)			
2010	0.0	3.0	16.5	26.4	65.6	379.9
2011	0.2	0.0	81.6	24.0	50.0	401.8
1975-2010	12.4	18.9	32.5	36.0	42.6	402.1
			Yearly Mean Temperature (°C)			
2010	25.5	16.7	14.5	5.2	3.4	11.8
2011	21.0	17.0	12.3	8.7	4.6	10.5
1975-2010	23.2	18.7	13.0	6.8	2.2	12.0
			Yearly Mean Humidity (%)			
2010	62.5	61.7	65.2	59.4	61.4	60.5
2011	60.4	61.2	57.8	57.3	60.8	60.0
1975-2010	61.3	63.1	60.7	57.9	59.2	60.1

[1/]Data were taken from Ankara Regional Meteorological Service.

This study was conducted in Randomized Complete Block Design with three replications. Plot size was 1.2 m x 5.0 m = 6.0 m^2 and distance between plants was also 30 cm (Kara, 1994). Genotypes were sown in 12-15 of March (Koncıus and Karcauskıene, 2010). No fertilizer was applied and genotypes were harvested in 15-20 of June (Vollman et al., 2007). One row at both sides and 0.5 m in both edges were discarded and harvest area was 5 m^2 (Vollman et al., 2007; Gugel and Falk, 2006). In the study, plant height, branch number per plant, 1000 seed weight, yield, oil content and oil yield were measured and analysed (Kara, 1994; Rokka et al. 2001; Gugel and Falk, 2006; Vollman et al., 2007; Eeva-Liisa et al., 2007;). For oil measurements, the seeds were properly grounded and the oil was extracted with n-hexane in a Soxhlet extractor for 4 h. Recovered crude oils were taken to dry out on a rotator at 35 ^{0}C. The oil content was calculated on dry mass basis. Fatty acids were etherified as methyl esters and analysed by Agilent 6890N Network with equipment with DB-23 capillary column (JW Scientific 122-2362 DB-23 ; 60.0 m x 250 µm x 0.25 µm) GC and FID detector. Helium was used as carrier gas at a flow rate of 1 mL/min. Injector and detector temperature were 260 ^{0}C and 240 ^{0}C, respectively. Column temperature was kept at 220 ^{0}C for 69 min. Samples of 0.5 uL was injected by hand and in the split mode (20:1). FAMEs were identified by comparison of their retention times with those of reference standards. The content of fatty acids was calculated from corresponding integration data (Rahamatalla, et al., 2001). Statistical analyses were made in TARİST, Minitab 15 pocked statistical programs (Düzgünes et al., 1987).

RESULTS AND DISCUSSION

Having rich oil content, false flax is getting an important crop for oil industry and high yielding genotypes having high quality fatty acids could help to meet deficit in oil industry (Budin et al., 1995). In our study, some plant characteristics, plant height (cm), the branch number per plant, 1000 seed weight (g), the seed yield (kg/da), oil content (%), oil yield (kg/da) were analyzed. False flax genotypes, Vinimik 17 (1), PI 304269 (2), CR 476/65 (3), CR 1674/90 (4), Ames 26665 (5), Ames 26667 (6), Ames 26673 (7), Ames 26676 (8), Ames 26680 (9), Ames 26686 (10) and Ames 28372 (11) were used and their variance analysis tables and means of characters were given in Table 3 and 4.

Table 3. Variance Analyses Table of Characters in False Flax (*Camelina sativa* (L.) Crantz) Genotypes

Source of Variation	D.F.	Plant Height (cm)		Branch Number per Plant		1000 Seed Weight (g)	
		Sum of Means	F Value	Sum of Means	F Value	Sum of Means	F Value
Replication	2	17.32	1.17ns	0.43	0.27ns	0.04	1.81ns
Years	1	3053.20	412.53**	493.64	314.21**	0.38	18.56*
Error$_1$	2	7.40		1.57		0.02	
Genotypes	10	49.51	6.97**	1.41	1.04ns	0.52	26.71*
Year x Gen.	10	13.85	1.95ns	2.18	1.61ns	0.01	0.73ns
Error$_2$	40	7.11		1.35		0.02	
Mean	65	61.59		9.0		0.10	
C.V.(%)		13.48		31.88		27.63	
Source of Variation	**D.F.**	**Yield (kg/da)**		**Oil Content (%)**		**Oil Yield (kg/da)**	
		Sum of Means	**F Value**	**Sum of Means**	**F Value**	**Sum of Means**	**F Value**
Replication	2	2.94	0.05ns	1.53	2.53ns	4.36	2.42ns
Years	1	23743.99	420.22**	788.33	1304.65**	4099.02	2273.61**
Error$_1$	2	56.50		0.60		1.80	
Genotypes	10	1461.44	12.87**	51.16	41.48**	179.73	13.53**
Year x Gen.	10	496.50	4.37**	41.72	33.83**	119.88	9.02**
Error$_2$	40	113.57		1.23		13.28	
Mean	65	738.24		27.24		117.52	
C.V.(%)		36.29		18.36		18.68	

ns: not significant, *: significant at 5%, **: significant at 1%.

Table 4. Means of Evaluated Characters in False Flax (*Camelina sativa* L. Crantz) Genotypes

	Plant Height (cm)			Branch Number per Plant			1000 Seed Weight (g)			
	1st Year	2nd Year	Mean	1st Year	2nd Year	Mean	1st Year	2nd Year	Mean	
Vinimik 17	59.17	73.43	66.30**A**	7.57	12.87	10.22	0.777	0.860	0.82**de**	1
PI 304269	53.43	63.36	58.40**B**	7.30	12.60	9.95	0.787	0.843	0.82**de**	2
CR 476/65	49.93	65.84	57.88**B**	6.27	11.97	9.12	0.770	0.827	0.70**e**	3
CR 1674/90	53.93	63.20	58.57**B**	6.77	12.10	9.43	1.263	1.480	1.37**ab**	4
Ames26665	50.53	64.57	57.55**B**	6.63	11.87	9.25	0.727	0.967	1.24**de**	5
Ames26667	52.23	63.20	57.72**B**	6.30	12.23	9.27	1.183	1.293	1.24**bc**	6
Ames 26673	51.70	63.67	57.68**B**	7.27	10.73	9.00	1.353	1.580	1.47a	7
Ames26676	49.87	65.27	57.57**B**	6.90	11.13	9.02	1.467	1.490	1.48a	8
Ames26680	45.73	63.97	54.85**B**	4.80	12.77	8.78	1.317	1.473	1.39**ab**	9
Ames26686	50.27	62.20	56.23**B**	6.13	13.00	9.57	1.377	1.530	1.45**ab**	10
Ames28372	48.86	66.60	57.73**B**	7.73	12.57	10.15	0.860	1.203	1.03**cd**	11
Mean	51.42**B**	65.02**A**	58.23	6.69**B**	12.17**A**	9.43	1.08**B**	1.23**A**	1.16	
L.S.D.(%)	Year: 6.647,Gen: 4.16			Year: 3.06			Year: 0.15,Gen: 0.22			
	Seed Yield (kg/da)			**Oil Content (%)**			**Oil Yield (kg/da)**			
	1st Year	**2nd Year**	**Mean**	**1st Year**	**2nd Year**	**Mean**	**1st Year**	**2nd Year**	**Mean**	
Vinimik 17	62.46	126.63	94.57**A**	28.13	37.30	32.72**A**	17.57	47.24	32.45**A**	1
PI 304269	47.50	75.50	61.50**CD**	25.63	35.53	30.58**BC**	12.13	26.81	19.44**D**	2
CR 476/65	46.82	78.50	62.66**CD**	22.17	37.07	29.62**CD**	10.39	29.15	19.71**D**	3
CR 1674/90	48.26	77.76	63.01**CD**	24.50	39.77	32.13**AB**	11.88	30.97	21.45**CD**	4
Ames26665	63.27	120.37	91.87**AB**	23.50	33.77	28.63**DE**	14.89	40.66	27.78**AB**	5
Ames26667	61.10	112.63	86.87**AB**	28.60	30.70	29.65**CD**	17.47	34.87	26.17**BC**	6
Ames 26673	68.81	130.67	99.74**A**	24.43	30.67	27.55**EF**	16.72	40.08	28.41**AB**	7
Ames26676	63.15	71.43	67.29**CD**	26.40	28.43	27.42**EF**	16.63	20.23	18.43**D**	8
Ames26680	44.07	80.40	62.23**CD**	25.07	27.90	26.48**F**	11.10	22.45	16.77**D**	9
Ames26686	47.31	67.20	57.25**D**	23.63	25.60	24.62**G**	15.35	17.22	16.28**D**	10
Ames28372	62.16	91.10	76.63**BC**	22.70	24.07	23.38**G**	14.12	21.92	18.02**D**	11
Mean	55.90**B**	93.84**A**	74.87	24.98**B**	31.89**A**	28.43	14.39**B**	30.1**A**	22.27	
L.S.D.(%)	Year: 18.37,Gen: 16.64			Year: 1.90,Gen: 1.74 Ye.x Gen:2.45			Year: 3.28,Gen: 5.69 Ye.x Gen:8.05			

A: denotes significant at 1%, a: denotes significant at 5%

In all parameters, differences between years were found as significant at 1% (only significant at 5% in 1000 seed weight). Besides, differences among genotypes and year x genotype interactions were also significant at 1 % in plant height, seed yield, oil content and oil yield. Differences among genotypes in branch number per plant and 1000 seed weight were insignificant and significant at 5%, respectively. Though year x genotype interactions in plant height, branch number per plant, 1000 seed weight were determined as insignificant, interaction between genotype and year were found as significant at 1% (Table 3).,

Means of plant height (cm), the branch number, 1000 seed weight (gr), the seed yield (kg/da), oil content (%), oil yield (kg/da) in False Flax (*Camelina sativa* L. Crantz) genotypes were shown in Table 4. Plant height is considered as one of the important plant characteristics (Vollman et al. 1996) and is highly relevant to yield and some yield components (Vollman and Rajcan, 2009). The highest plant height was taken from Vinimik 17 (66,30 cm) and with 54,85 cm lowest one belonged to Ames 26680 (Table 4). Plant height in false flax is accepted as stable for each genotype but it could be influenced by yearly climatic variations (Vollman et al. 1996; Urbaniak et al., 2008, Kaleem et al., 2010). This means that homogenous influence in genotypes occurred versus environmental and yearly differences. Similar to this, studies accepted that plant height is a genotypic stable character and changes with changes in environmental climatic conditions differences (El-Moueı et al., 2011, Kaleem et al., 2010) but mostly homogenous changes occur among genotypes versus ambient conditions (Vollman and Rajcan, 2009; El-Moueı et al., 2011). These results could explain why significantly differences among genotypes and insignificant genotypes x year interactions. Correlations between branch number, seed yield, oil content and oil yield were determined as significant at 1% (Table 5).

Table 5. Correlation Matrix of Characters in False Flax (*Camelina sativa* (L.) Crantz) Genotypes.

	Plant Height	Br. Nu.per PL	1000 Seed We.	Yield	Oil Content
Br.Nu.per Pl.	0.878**				
1000 Seed We.	0.032ns	0.145ns			
Yield	0.703**	0.645**	0.088ns		
Oil Content	0.667**	0.611**	-0.028ns	0.552**	
Oil Yield	0.752**	0.668**	0.035ns	0.942**	0.761**

**: significant at 0,01 level

Branch number per plant in most oil crops is major character refers valuable characteristic for oil content and yield (Agegnehu and Honermerier, 1997; Vollman and Rajcan, 2009). First year had (6,69) lower branches number per plant than second year (12,17, Table IV). This difference shows that yearly variations had significant effect on branch number per plant. Studies revealed that branch number per plant shows little differences in most plants and these differences among genotypes are more or less insignificant. More effect on branch number per plant could belong to environmental variations, having more effect than genotypic character (Koncıus and Karcauskıene, 2010; Wysocki and Sirovatka, 2007; Agegnehu and Honermerier, 1997; Akk and Ilumae, 2005). Similar to this, insignificant differences occurred among genotypes in the study. On the other hand, correlations between seed yield, oil content, and oil yield and plant height with branch number per plant were positive and significant at 1 % (Table 5).

Another major yield component in false flax is known as 1000 seed weight (Vollman and Rajcan, 2009, Vollman et al. 1996, Koncıus and Karcauskıene, 2010) and variations and fluctuations in 1000 seed weight commonly occur and these changes could be originated from environmental conditions and/or genotypic characters (Vollman and Rajcan, 2009; Vollman et al. 1996; Akk and Ilumae, 2005). Similar effects were recorded in this study and obviously differences between years and genotypes were determined as significant ($p<0.05$). 1000 seed weight in the second year (1.23 g) was significantly higher than the first year (1.08 g). The highest 1000 seed weights were taken from Ames 26673 (1,479 g) and Ames 26676 (1,489 g). The CR 476/65 (0,70 g) had the lowest value (Table 4).

As in every plant cultivated, yield is expressed as amount and quality (Vollman and Rajcan, 2009). Yearly fluctuations in rainfall and temperature must have differently affected genotypes and this reason could make interaction important. A number of studies pointed out that the seed yield in dry conditions mostly depends upon rainfall in most plants and yearly fluctuations rainfall and temperature more often affect seed yield and relatively oil content and oil yield (Seehuber, 1984; Seehuber et al., 1987; Putnam et al. 1993; Diepenbrock et al., 1995; Vollman et al. 1996; Akk and Ilumae, 2005; Vollman and Rajcan, 2009). Yield in the second year (93,84 kg da^{-1}) was so higher than previous year (55.90 kg da^{-1}). Since, rainfall in 2010- 2011 and long term (1975-2010) were 379.9 mm, 401.8 mm and 402.1 mm, respectively. More rainfall in 2011 seems reliable more seed and oil yield. In the same way, total rainfall in year including tillering stage (in May) plays vital role for yield and false flax needs exorbitant water in this stage. Rainfalls during March-June (crop growing period) in the second year were 213 mm and 172 mm in the previous year. Rainfalls in May were only 22 mm in 2010 and 86 mm in 2011. Vinimik 17 had the highest seed yield (94.57 kg/da), whereas Ames 266686 genotype (57.25 kg/da) was the lowest one (Table 4). In dry conditions seed yield performance in genotypes are naturally be different (Vollman and Rajcan, 2009, Vollman et al. 1996, Budin et al, 1995). Correlations between plant height, branch number per plant, oil content and oil yield with seed yield were significant at 1% (Table 5).

With about 40-45% oil content and high level of some oil acids such as linolenic, linoleic and oleic acids, false flax (*Camelina sativa* (L.) Crantz) seems promising crop for industrial and edible oil (Korsrud et al., 1978; Crowley and Fröhlich, 1998; Koncıus and Karcauskıene, 2010). Studies reported that oil yield changes with changing seed yield and oil content (Alessi et al., 1981; Putnam et al., 1993; Kara, 1994; Budin et al., 1995; Crowley and Frohlick, 1998; Karahoca, 2002; Vollman and Rajcan, 2009; Tomas et al., 2011; Berti et al., 2011, Gugel and Falk, 2006). Oil content and oil yield draw similar trend like seed yield. The effect of years, differences among genotypes and year x genotype interaction were found as significant at 1% (Table 3). The second year had higher oil content (31.89 %) and oil yield (30.15 kg/da) than the first year (24.98 % in oil content and 14.39 kg/da in oil yield, respectively). Vinimik 17 was found as the highest yielding genotype in oil content (32.72 %) and in oil yield (32.45 kg/da). Moreover, Ames 28372 gave the lowest oil content (23.38 %) and oil yield (18.02 kg/da). Year x genotype interactions were significant at 1% in both oil content and oil yield. Genotypes were significantly affected by changes in rainfall and temperature. Basically, climatic changes, especially changes in rainfall and temperature that have tremendous potential on seed yield and oil content could greatly affect oil yield (Zubr, 2003; Vollman et al., 1996; Koncıus and Karcauskıene, 2010; Kara, 1994; Gugel and Falk, 2006). Correlations between plant height, branch number per plant, seed yield and oil yield with oil content; correlations between plant height, branch number per plant, yield and oil content with oil yield were determined at 1% (Table 5).

In view of results, cluster analysis including similarities/dissimilarities, were given in Figure 1. Vinimik 17 (1, Group 1) and Ames 28372 (11, Group 5) created their own groups alone. Moreover, the other genotypes formed three groups; namely, PI304269 (2,), CR 476/65 (3,) and CR 1674/90 (4) one group (Group 3); Ames 26665 (5), Ames 26667 (6) and Ames 26673 (7) one group (Group 2); Ames 26676 (8), Ames 26680 (9), Ames 26686 (10) one group (Group 4).

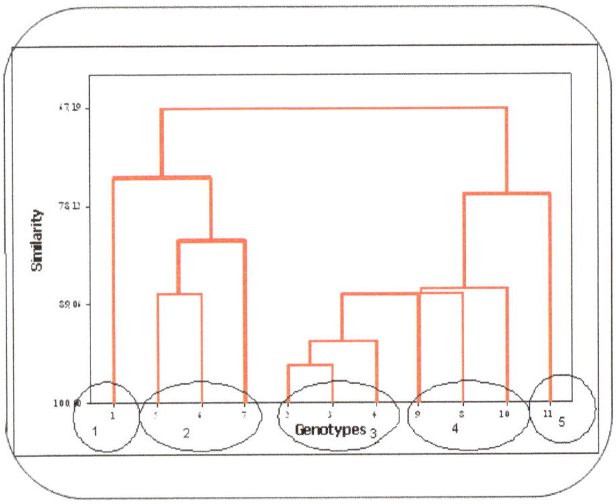

Figure 1. Cluster analysis of false flax (*Camelina sativa* (L.) Crantz) genotypes.

Oil content and its composition in oil crops is important feature (Abromovic et al., 2007; Urbaniak et al., 2008; Koncıus and Karcauskıene, 2010; Budin et al., 1995; Vollman and Rajcan, 2009; Velasco and Fernandez-Martinez, 2002,) and determine values of crops to be used in oil industry/medicine (Budin et al., 1995; Sabzalian et al. 2008; Vollman and Rajcan, 2009), False flax carries great potential for this. Compositions of oil fatty acids of genotypes in false flax were given Table 4. High contents of linolenic, linoleic and oleic acids, total saturated acid and polyunsaturated acid were determined in all genotypes. Similar to our findings, Budin et al., (1995) stated that false flax genotypes are higher in linolenic, linoleic and oleic acids, total saturated acid and polyunsaturated acid, lower in monounsaturated acid than that of canola and olive oils.

CONCLUSION

Results in Table 4, and 6, and Figure 1 showed that having highest seed yield, and oil yield, Vinimik 17, Ames 26665, Ames 26667 and Ames 26673 could be considered as an oil crops for oil industry or as medicine plant. Besides, false flax could be source of essential fatty acids, lowering cholesterol level in human diet. Having low saturated fat content this crop seems quality edible oil. Though genotypes are higher in erusic acid than 2% (maximum limit for canola quality edible oil), Genotypes having less erucic acid content should be developed by breeding programs/biotechnology. More studies to reveal characteristics and efficacy of oil acids in false flax genotypes and to develop genotypes having less erusic acid level under food codex level and rich in linolenic, linoleic and oleic acids, total saturated acid and polyunsaturated acid are hastily needed.

Table 6. Compositions of Oil Acids in False Flax (*Camelina sativa*) Genotypes.

	Genotypes													
Compositions	**Vniimk17**	**PI 304269**	**CR 476/65**	**CR 1674/90**	**Ames 26665**	**Ames 26667**	**Ames 26673**	**Ames 26676**	**Ames 26680**	**Ames 26686**	**Ames 28372**	**Mean**	**Canola Oil***	**Olive Oil****
Erusic Acid	3.20	2.85	2.95	3.26	3.19	3.16	3.57	3.44	3.39	3.35	3.13	3.23	0.21	0.00
Metil-cis 11.14.17 ech.	1.04	1.03	1.00	1.11	1.10	1.08	0.87	0.86	1.05	1.05	1.09	1.03	0.00	0.00
Echosadienoic Acid	1.78	1.78	1.77	1.69	1.77	1.74	1.58	1.76	1.88	1.78	1.75	1.75	0.00	0.00
Echosenoic Acid	13.80	13.80	14.11	14.40	13.96	14.00	14.17	13.99	14.18	14.45	13.93	14.07	0.22	0.00
Arachydic Acid	1.99	1.89	1.76	1.75	1.80	1.78	1.97	1.89	1.64	2.01	1.92	1.85	0.00	0.00
Linolenic Acid	**31.23**	**30.88**	**30.46**	**31.97**	**31.75**	**31.65**	**29.11**	**28.05**	**30.61**	**31.73**	**33.36**	**30.98**	**6.14**	**8.31**
Linoleic Acid	**20.16**	**20.73**	**20.34**	**18.46**	**19.96**	**19.95**	**20.99**	**22.39**	**21.33**	**18.59**	**18.87**	**20.16**	**20.25**	**10.01**
Oleic Acid	**15.79**	**16.15**	**16.67**	**17.05**	**15.53**	**15.92**	**16.72**	**15.77**	**15.07**	**16.06**	**15.55**	**16.03**	**55.65**	**29.05**
Stearic Acid	2.90	2.81	2.81	2.61	2.66	2.62	2.51	2.43	2.50	2.52	2.61	2.63	1.62	17.94
Palmitic Acid	6.21	6.06	6.35	5.92	6.47	6.29	6.25	6.82	6.77	6.50	6.34	6.36	3.64	25.38
Total Saturated Acid	**12.07**	**11.49**	**11.47**	**11.16**	**11.78**	**11.55**	**12.00**	**12.63**	**11.72**	**11.93**	**11.84**	**11.79**	**10.30**	**43.32**
Monounsaturated Acid	**33.55**	**33.56**	**34.51**	**35.46**	**33.45**	**33.86**	**35.28**	**34.14**	**33.52**	**34.76**	**33.48**	**34.14**	**62.64**	**33.37**
Polyunsaturated Acid	**51.39**	**51.61**	**50.80**	**50.43**	**51.71**	**51.60**	**50.10**	**50.44**	**51.94**	**50.32**	**52.23**	**51.14**	**28.02**	**23.20**

*(Akk and Ilumae,2005), **(Sirtori et al.,1992)

LITERATURE CITED

Abromovic, H., B.Butinar, and V. Nikolic, 2007. Changes occurring in phenolic content, tocopherol composition and oxidative stability of *Camelina sativa* (L.) Crantz oil during storage. *Food Chemistry* 104. 903-909.

Agegnehu, M.and B. Honermeier, 1997. Effects of Seeding Rate and Nitrogen Fertilization on Seed Yield, Seed Quality and Yield Components of False Flax (*Camelina sative Crtz*). Die Bodenkultur. 48 (1).

Akk, E. and E. Ilumae, 2005. Possibilities of growing *Camelina sativa* in ecological cultivation. Saku, Estonia, p:28-33.

Alessi, J., Power, J. F. and D.C.Zimmerman, 1981. Effect of Seeding Date and Populatıon on Water- Use Efficiency and Safflower Yield. *Agronomy Journal* Vol: 73. No: 5 Page: 783-787.

Berti M., R. Wilckens, S. Fischer, A. Solis, and B.Johnson, 2011. Seeding Date Influence on Camelina Seed Yield, Yield Components, and Oil Content in Chile Vol. 34, 1358-1365.

Budin, J.T., W. M. Brene, and D. H. Putnam, 1995. Some Compositional Properties of Camelina (*Camelina sativa* L. Crantz) Seeds and Oils. *Journal of the American Oil Chemists' Society*. Volume 72, Number 3, Page: 309-315.

Crowley, J. G. and A. Fröhlich,1998. Factors Affecting the Composition and Use of Camelina. Crops Research Centre, Oak Park, Carlow. ISBN 1 901138666.

Diepenbrock, W.A., J. Leon, and K. Clasen, 1995. Yielding ability and yield stability of lineseed in central Erope. *Agronomic J.* 87:84-88.

Düzgüneş, O., Kesici, T., Kavuncu, O. ve Gürbüz, F. 1987. Araştırma ve Deneme Metodları (İstatistik Metodları II). Ankara Üniversitesi Ziraat Fakültesi Yayınları:1021. Ders Kitabı, 295s.

Eeva-Liisa, P., T. Sini, V. Tuomo, E. Jarmo, and L. Karita, 2007. Effect of a *Camelina sativa* Expeller Cake on Performance and Meat Quality of Broilers. *Journal of the Science of Food and Agriculture*, Volume 87, Number 8, June 2007,pp. 1489-1494 (6).

El-Mouei, R.,W., Choumane and F. Dway, 2011. Characterization and Estimation of Genetic Diversity in Citrus Rootstocks**.** *Int. J. Agric. Biol., Vol. 13, No. 4*. 571-575.

Frohlic, A. and B. Rice, 2005. Evaluation of *Camelina sativa* Oil as Feedstock for Biodiesel Production. *Industrial Crops and Products* Volume 21, Issue 1, January 2005, Pages 25-31.

Gugel, R. K., K. K. Falk, 2006. Agronomic and seed quality evaluation of Camelina sativa in Western Canada. Can. J. Plant Sci. 86: 1047-1058.

Kaleem, S., F. U. Hassan, M. Farooq, M. Rasheed and A. Munir, 2010. Physio-morphic traits as influenced by seasonal variation in sunflower; A review. *Int. J. Agric. Biol.*; 12: 468-473.

Kara, K., 1994. Değişik Sıra Mesafelerinin Ketenciğin (*Camelina sativa* (L.) Crantz) Verim ve Verim Unsurları Üzerine Etkileri. *Tr. J. Of. Agricultural and Forestry* 18, 59-64. TÜBİTAK.

Karahoca, A., 2002. Çukurova Koşullarında Ketencik (*Camelina sativa*)'te Farklı Azot ve Fosfor Gübrelemesinin Tohum Verimi ve Yağ Oranına Etkileri. Çukurova Üniversitesi Fen Bilimleri Enstitüsü. Basılmamış Yüksek Lisans Tezi. Adana.

Karaman, Ş., M. Gülseven, N. Çömlekçioğlu and A. Ilçım. 2011. Fatty Acid Composition of *Matthiola longipetala ssp. Bicornis* from Turkey. Int.J. Agric. Biol. Vol. 13, No. 4, pp.581-585.

Karvonen H.M., Aro A., Tapola N.S., Salminen I.,Uusitupa M.I.J., Sarkkinen E.S. 2002. Effect of alpha-linoleic acid-rich Camelina sativa oil on serum fatty acid composition and serum lipids in hypercholesterolemic subjects. Metabolism, 51, 1253–1260.

Koncıus, D. and D. Karcauskıene, 2010. The effect of nitrogen fertilizers, sowing time and seed rate on the productivity of Camelina sativa . Agriculture. Vol. 97, No.4 p.37-47.

Korsrud, G.O., M.O.Keith.and J.M. Bell, 1978. A Comparison of the Nutritional Value of *Crambe* and *Camelina* Seed Meals with egg and Casein. Can. *J. Anim. Sci.* 58: 493-499.

Kurt, O. ve F. Seyis, 2008. Alternatif Yağ Bitkisi: Ketencik (*Camelina sativa* (L.) Crantz). OMU. *Zir. Fak. Dergisi*, 2008, 23 (2): 116-120.

Putnam, D.H., J. T. Budin, L.A. Field and W.M. Breene, 1993. Camelina : a promising low-input oilseed. P.314-322. İn. J. Janick and E. Simon (eds) , New Crops. Wiley, New York.

Rahamatalla, A. B., Babiker , E. E., Krishna, A. G., El Tinay, A. H., 2001. Changes in Fatty Acids Composition During Seed Growth and physicochemical characteristics of oil extracted from four safflower cultivars. Plant Foods for Human Nutrition, 56: 385-395

Sabzalian, M.R., G. Saeidi and A. Mirlohi, 2008. Oil Content and Fatty Acid Composition in Seeds of Three Safflower Species. *J. Am. Oil Chem. Soc.* 85:717-721.

Seehuber, R., 1984. Genotypic Variation for Yield-and Quality-Traits in Poppy and False Flax. *Seifen-Anstrichmittel* 86:177-180.

Seehuber, R., J.Vollman, and M. Dambroth, 1987. Application of the single seed-descent method in false flax to increase the yield level. *Landbauforsch. Voelkenrode* (Germany) 37: 132-136.

Tomas, L., H. Jaroslav, M. Jiri, V. Johann, P. Jiri, F. Radek, V. Ladislav, D. Ladisla and M. Anna, 2011. Effect of Combined Nitrogen and Sulphur Fertilization on Yield and Qualitative Parameters of *Camelina sativa* [L.] Crtz. (false flax). Vol:6, 313-321.

Urbaniak, S.D., C. D. Caldwell, V. D. Zheljazkov, R. Lada and L. Luan, 2008. The effect of cultivar and applied nitrogen on the performance of *Camelina sativa* L. in the Maritime Provinces of Canada. Can.J.Plant Sci. 88:111-119

Velasco, L. and J. M. Fernandez-martinez, 2002. Breeding Oilseed Crops for İmproved Oil Quality, Journal of Crop Production, 5:1-2,309-344.

Vollman, J., A. Damboeck, A. Ecki, H. Schrems and P. Runckenbauer, 1996. İmprovement of Camelina sativa, an underexploited oilseed. In Progress in New Crops , *J. Janick* (ed). ASHK Press, Alexandria, S.357-362.

Vollman, J., T. Moritz, C. Kargl, S. Baumgartner, H. Wagentristi, 2007. Agronomic Evaluation of Camelina Genotypes Selected for Seed Quality Characteristics. Industrial Crops and Products. Volume 26, Issue 3. p: 270-277.

Vollman, J. and I. Rajcan, 2009. *Oil Crops*, 1st edition, Volume 4 ISBN 978-0-387-77593-7 e-ISBN 978-0-387-77594-4 DOI 10.1007/978-0- 387-77594-4 Springer Dordrecht Heidelberg London New York

Wysocki, D. and N. Sirovatka, 2007. Camelina a potential oilseed crop for semiaridic Oregon.// Agronomy Abstracts.- http://extention. Oregonstate.edu/catalog/htm//sr/sr 108-e/sr 1083-09. pdf (accessed 17.01.2012)

Zubr, J., 1997. Oil-seed crop: *Camelina sativa*. İndustrial Crops and Products 6, p 113-119.

Zubr, J., 2003. Qualitative variation of *Camelina sativa* seed from different locations// Industrial Crops and Products. Vol.17, p. 161-169.

7

INBRED LINES OF DIFFERENT CYCLES OF SELECTION - DONORS OF FAVOURABLE ALLELES FOR THE IMPROVEMENT OF THE KERNEL ROW NUMBER OF F_1 MAIZE HYBRIDS

Goran TODOROVIĆ[1], Mile SEČANSKI[1], Tomislav ŽIVANOVIĆ[2], Rade PROTIĆ[3], Miroslav KOSTIĆ[4], Snežana JOVANOVIĆ[1] and Dragan BOŽOVIĆ[5]*

[1]*Maize Research Institute, Zemun Polje, Belgrade-Zemun, REPUBLIC SERBIA.*
[2]*Faculty of Agriculture, University Belgrade, REPUBLIC SERBIA.*
[3]*Institute for Science Aplication in Agriculture, Belgrade, REPUBLIC SERBIA.*
[4]*Institute for Medicinal Plant Research "Dr Josif Pančić", Belgrade.*
[5]*"Agros", Pančevo, REPUBLIC SERBIA.*
[*]*Corresponding author: gtodorovic@mrizp.rs*

ABSTRACT

Eight maize inbred lines of different cycles of selection, originating from synthetic populations BSSS and $BSCB_1$, were observed in the present study. The aim of the study was to evaluate inbreds and to determine which ones have the greatest relative values of favourable alleles for the improvement of the kernel row number in the elite single cross maize hybrid. Based on the estimated estimators: μG, UBND, PTC and NI, it was determined that the greatest number of favourable dominant alleles for the improvement of the kernel row number was detected in the inbred lines B73(C_5) and B84(C_7) originating from the synthetic population BSSS and in the inbreds B91(C_8) and B90(C_7) originating from the synthetic population $BSCB_1$. These inbreds are of the later cycles of recurrent selection and have proven to be better donors of favourable alleles than inbreds of earlier cycles of selection. The rank correlations among used estimators were positive and highly significant, while the highest values of correlations for the kernel row number were determined between μG and PTC, μG and UBND and PTC and NI. The greatest kernel row number was determined in the hybrid B73(C_5) x B84(C_7).

Key words: maize, inbred lines, donors, kernel row number, rank correlation

INTRODUCTION

The period of maize hybrids usage in the contemporary agricultural production is limited, meaning they have to be replaced with new hybrids having more favourable traits than the existing hybrids. One of the methods of the development of new hybrids is the improvement of existing hybrids, i.e. the improvement of one or both parental inbreds of the elite hybrid. The procedure involves the introduction of favourable alleles into the existing inbreds, which would be manifested through better agronomic properties of future hybrids.

The presentation and the development of the method for the identification of inbred lines that encompassing the favourable alleles that affected the expression of quantitative traits. These alleles were not present in the elite hybrid were given by Dudley (1984a). The selection of the inbred donors of favourable alleles is provided by the evaluation of the progenies, developed by crossing of each inbred, a potential donor with both parental components of the hybrid under the improvement (P_1 x D and P_2 x D, whereby P_1 and P_2 are parents of the single cross hybrid, while the D inbred is a donor of favourable alleles).

In many commercial programmes new inbred lines are developed from synthetic populations that were derived by crossing of parental inbreds of the existing elite hybrid (I_1 x I_2) to another inbred line, a donor, (Iw). By self-pollination of newly developed F_1 (I_1 x Iw) or after backcrossing, the process of the development of new inbreds begins. These inbreds will be tested with other parent of the elite hybrid (I_2), to isolate new inbreds that will, with an already mentioned inbred line, tester (I_2), result in a new hybrid that is more superior than the existing elite hybrid (I_1 x I_2), Bernardo (1990a). Mišević (1989b) studied 15 inbred lines as donors of favourable alleles for the improvement of seven elite hybrids by three estimators μG, UBND and PTC. Significant differences among inbred donors occurred when all three estimators were used. The greatest values of new favourable alleles were found in inbred lines belonging to heterotic groups different from parents of hybrids under improvement. The correlation among estimators were high, positive and significant, and the highest one exceeded 0.95 (between μG* and PTC).

Bernardo (1990a) estimated that selected Iw inbreds could be crossed to I_1 or I_2 in order to transfer new favourable alleles to a single cross hybrid. However, the favourable alleles already present in the single cross hybrid can be lost in the process of selection. It is important to consider the potential benefit of favourable alleles of Iw, as well as, the loss of favourable alleles already present in I_1 and I_2. The net improvement, i.e. NI statistics estimates the number of loci in which favourable alleles can be obtained minus the number of loci in which favourable alleles can be lost in the single cross hybrid. This estimator was compared with estimators μG, UBND and PTC. Results show a high correlation among all four estimators, whereby ranking of three inbred donors with the most common yield of the F_2 test crossing above the target hybrid was identical with ranking of NI estimator. Trifunović et al. (1999) also determined high rank correlations among four estimators, whereby they emphasised the advantage of the NI statistics within short-term programmes of selection, while the estimator Lplμ* was more advantageous within long-term programmes for the identification of sources of donors of favourable alleles for the improvement of the elite single cross hybrid.

Delić et al. (1996) has studied eight US SC hybrids as donors of favourable alleles for the improvement of grain yield of the elite hybrid Mo17 x L62 by the application of the following three methods: (1) method after Dudley (1987a), by which the value of the estimator Lplμ has been estimated, (2) XTC or the method of the average value of crossing of a donor to both parents of the elite hybrid (Sprague and Eberhart 1997), (3) UBND minimum method (Zanoni and Dudley 1989, Gerloff 1985). The calculated estimators of all three applied methods for the estimation of favourable alleles were compared on the basis of the Spearman's rank correlation coefficient.

According to Malvar et al. (1997), the estimation of favourable dominant alleles present in the donor but absent in the hybrid (μG), predicted three-way cross hybrid (PTC), minimum upper bound estimates (UBND), net improvement (NI) and general combining ability (GCA) are methods for the determination of maize inbred lines with favourable alleles. The correlations amont μG, UBND, PTC, NI and GCA were very high in all traits. Carena (2005) has estimated that improved maize populations and population hybrids could be profitable alternatives to commercial single cross hybrids, as well as, good elite sources of different inbred lines. The improved maize populations and population hybrids were estimated in experiments carried out in 29 environments. Data obtained under different environmental conditions show that at least 20% of estimated population hybrids did not differ from commercial single cross hybrids in relation to grain yield and the percentage of lodged plants. Attempts of selection aimed at elite populations and population hybrids led to the significant improvement of germplasm. These attempts should be supported in order to provide the development of elite sources of different inbred lines and the development of improved population hybrids for various markets.

The objective of the present study was to estimate which of observed inbred lines of different cycles of selections originating from synthetic populations BSSS and $BSCB_1$ had the highest relative values as donors of favourable alleles for the improvement of the commercial maize hybrid. The obtained results will be important in maize breeding programmes in selection inbred donors of favourable alleles for the kernel row number and improving parental components of the commercial single-cross hybrid.

MATERIALS AND METHODS

Maize inbred lines, obtained from different cycles of recurrent selection of two synthetic US populations (BSSS and $BSCB_1$) were selected. The inbred lines B14(C_0), B37(C_0), B73(C_5) and B84(C_7) were derived from the synthetic population Iowa Stiff Stalk Synthetic (BSSS), while the inbred lines B90(C_7), B91(C_8), B97(C_9) and B99(C_{10}) were developed from the synthetic population Corn Borer Synthetic No.1 ($BSCB_1$). Inbred lines derived from both synthetics were crossed to parental inbred lines of the hybrid ZP580 (ZPL1 and ZPL2). Two trials at two locations (Zemun Polje and Kraljevci) carried out during 2007 and 2008 were set up to estimate the relative values of loci. The first trial included the elite hybrid ZP 580 and hybrids developed by crossing parental inbreds of the hybrid ZP 580 to inbred donors from both synthetic populations. The second trial included inbred donors and parents of the hybrid ZP 580 that had been improved. The trial in four replications were set up according to the Randomised Complete-Block Design. The elementary plot size amounted to 3.36 m^2, while the sowing density was 59,500 plants ha^{-1}.

The estimation of a relative number of favourable alleles of studied inbreds in relation to the elite single cross hybrid (the hybrid selected to be improved) for observed traits was done according to the model coined by Dudley (1987 a.b). The model assumes that there is complete dominance $a = 1$ and $z = -\mu$, hence the genotypic value + +, + – and – – of genotypes is $z + 2\mu$; $z + \mu + a\mu$ and z, where z is the value of the least favourable genotype – –; μ is one half the difference in the genotypic value between + + and – – genotypes, while a is a degree of dominance. This model implies that μ is constant for all loci and that there is no epistasis.

Dominant alleles are favourable for the kernel row number (μ value is positive). Inbreds with higher value of μG can be considered a good source of favourable alleles. Positive and high values of μG for inbred donors mean the presence of new favourable alleles in inbreds, which are absent in inbred parents of the elite hybrid. Negative values of the relation μG - μD or μG - μF indicate that the number of loci for which inbred lines I_1 and I_2 carry favourable alleles, and Iw carries unfavourable alleles, is higher than the number of loci for which inbred lines I_1 and I_2 carry unfavourable alleles, and Iw carries favourable alleles, hence it is necessary to perform backcross of the F_1 generation (parent inbred x donor inbred) to parent inbred I_1 or I_2. If, on the other hand, the difference μG - μD or μG - μF is not significantly different from 0, the direct self-pollination of the F_2 generation is recommended in the process of selection. Positive values of the difference μG - μD or μG - μF lead to a conclusion that if the number of loci for which inbred lines I_1 and I_2 carry unfavourable alleles and Iw carries favourable alleles is higher than the number of loci for which inbreds I_1 and I_2 carry favourable alleles and the inbred Iw carries unfavourable alleles, backcross of the F_1 generation to the donor inbred (Iw) is recommended.

UBND (minimum upper bound estimates, Gerlof and Smith, 1988) is defined as a minimum of two expressions: (I_1 x Iw) - I_1 and (I_2 x Iw) - I_2, which have a prediction of 2μE + 2μG and 2μC + 2μG, respectively.

PTC (predicted three way cross, Hallauer and Miranda, 1988) is a predicted value of the three-way cross hybrid that is calculated as follows: [(I_1 x Iw) + (I_2 x Iw)] / 2 and have a prediction of N(z + μ) + μ(A + B + C + E + G - H), where N is a total number of loci that affect the observed trait.

NI (net improvement statistics, Bernardo, 1990 a, b) estimates the number of loci where favourable alleles can be added (minus the number of loci where they can be lost) in the elite hybrid. This estimator is calculated as a maximum value of the expression [(I_1 x Iw) - (I_1 x I_2)] / 2 or [(I_2 x Iw) - (I_1 x I_2)] / 2.

Heterosis (H) for kernel row number was calculated in relation to a superior parent.

RESULTS AND DISCUSSION

The highest (16.94), i.e. lowest (11.16) means for kernel row number was recorded in the inbred B73 (C_5), i.e. B14(C_0), respectively. The mean values of the kernel row number ranged from 13.53 in the hybrid ZPL2 x B99 to 15.57 in the hybrid ZPL2 x B73. The elite hybrid ZPL1 x ZPL2 had on average 14.04 kernel rows (Table 1).

Table 1. Mean values of the kernel row number of 10 maize inbred lines

Inbred lines	Kernel row number
ZPL1	13.19
ZPL2	12.67
B14(C_0)	11.16
B37(C_0)	13.03
B73(C_5)	16.94
B84(C_7)	15.83
B90(C_7)	13.34
B91(C_8)	12.54
B97(C_9)	13.30
B99(C_{10})	11.45

All observed inbred lines, except the inbred B99(C_{10}) of the synthetic population $BSCB_1$ had a significant effect (>2SE) of the estimator μG. The highest values of this estimator were recorded in the inbreds B73(C_5) and B84(C_7) of the synthetic population BSSS and B91(C_8) and B90(C_7) of the synthetic population $BSCB_1$ (Table 2).

The genetic descent of the donor inbreds to one of parents of the elite hybrid was evaluated according to the difference between estimators (μC+μF)-(μD-μE), DUDLEY (1984 a). Inbred lines B14(C_0) and B73(C_5) of the synthetic population BSSS had significant (>2SE) positive values, while the inbreds B37(C_0) and B14(C_0) had positive values, which point to their genetic descent to the inbred ZPL1 (Table 2 and Figure 3). The inbred lines B91(C_8), B97(C_9) and B99(C_{10}) of the synthetic population $BSCB_1$ had negative values of the difference between estimators (μC+μF)-(μD-μE) and were more related to the inbred line ZPL2, while the inbred line B90(C_7) was more related to the inbred line ZPL1 (Table 2 and Figure 3).

Table 1. Mean values of the kernel row number of 17 maize hybrids and heterosis (%)

Inbred lines	Kernel row number	Heterosis (%)
ZPL1 x ZPL2	14.04	6.4
ZPL1 x B14(C_0)	14.39	9.1
ZPL1 x B37(C_0)	14.11	6.9
ZPL1 x B73(C_5)	15.50	-8.5
ZPL1 x B84(C_7)	15.15	-4.3
ZPL1 x B90(C_7)	14.49	8.6
ZPL1 x B91(C_8)	14.96	13.4*
ZPL1 x B97(C_9)	14.56	9.5*
ZPL1 x B99(C_{10})	13.91	5.5
ZPL2 x B14(C_0)	14.95	17.9**
ZPL2 x B37(C_0)	14.09	8.1
ZPL2 x B73(C_5)	15.57	-8.1
ZPL2 x B84(C_7)	15.20	-3.9
ZPL2 x B90(C_7)	14.92	11.8*
ZPL2 x B91(C_8)	14.37	13.4**
ZPL2 x B97(C_9)	14.29	7.4*
ZPL2 x B99(C_{10})	13.53	6.8

*, ** significant at the probability levels of 0.05 and 0.01 respectively

The differences between the values of estimators μG and μD in the inbred B37(C_0) were not significant (>2SE), which indicates that self-pollination of the hybrid ZPL1 x B37 will be the better way of the development of the initial population for selection, because the inbred is genetically more related to the inbred ZPL1. The value of the estimator μF did not significantly differ from the value of the estimator μG, hence self-pollination of the hybrid B99 x ZPL2 is the best way to develop the initial population for selection, because the inbred B99(C_{10}) is more related to the inbred ZPL2 (Table 2).

Inbred donors B73(C_5), B84(C_7) and B14(C_0) of the synthetic population BSSS and the inbred line B90(C_7) of the synthetic population $BSCB_1$ had significant values of favourable alleles for the improvement of the kernel row number. These inbreds had significant (> 2SE) difference μG-(μD or μF) in favour of favourable dominant alleles. According to the estimation of relatedness that showed greater relatedness to the parent inbred ZPL1, it is necessary to back cross F_1 generations (ZPL1 x B73, ZPL1 x B84. ZPL1 x B14 and ZPL1 x B90) to inbred donors and then to initiate self-pollination with the aim to improve the inbred line ZPL1 (Table 2).

The inbred lines B91(C_8) and B97(C_9) of the synthetic population $BSCB_1$ show that they had significant values of favourable alleles for the improvement of the kernel row number in the elite hybrid ZP580. Both inbreds had a significant (>2SE) and a positive difference between the number of favourable dominant alleles and unfavourable recessive alleles and therefore they should be back-crossed to the F_1 generation (ZPL2 x inbred donor) with the aim to improve the inbred line ZPL2, parental components of the elite hybrid (Table 2).

Table 2. The estimation of the relative number of favourable alleles for the kernel row number

Donors (Iw)	μG^*	Relatedness $(\mu C^*+\mu F^*)-(\mu D^*+\mu E^*)$	$\mu G^*-(\mu D^*$ or $\mu F^*)$	Cross method
B14(C_0)	0.528*	0.822*	0.456*	(I_1 x Iw) x Iw
B37(C_0)	0.250*	0.234	0.037	Selfing
B73(C_5)	0.962*	0.334*	0.768*	(I_1 x Iw) x Iw
B84(C_7)	0.781*	0.309	0.581*	(I_1 x Iw) x Iw
B90(C_7)	0.545*	0.690*	0.440*	(I_1 x Iw) x Iw
B91(C_8)	0.657*	-0.328*	0.462*	(I_2 x Iw) x Iw
B97(C_9)	0.474*	-0.014	0.262*	(I_2 x Iw) x Iw
B99(C_{10})	0.149	-0.127	-0.063	Selfing

* significant at the level> 2SE (Standard Error)

Inbred donors were ranked according to the values of the estimator μG. Estimators UBND PTC and NI show that the inbred lines B73(C_5) and B84(C_7) were the best carriers of favourable alleles for the kernel row number (Table 3). High and significant values of rank correlations coefficients were determined among mentioned estimator (Table 4). Gained results are in accordance with results obtained by Vračarević (2000) and Trifunovića (2001).

Table 3. Estimation and the rank of μG, UBND, PTC and NI for the kernel row number

Inbred donor	μG	UBND	PTC	NI
B14(C_0)	0.528 (5)	1.198 (6)	14.667 (5)	0.454 (4)
B37(C_0)	0.250 (7)	0.925 (7)	14.100 (7)	0.037 (7)
B73(C_5)	0.962 (1)	2.312 (1)	15.537 (1)	0.768 (1)
B84(C_7)	0.781 (2)	1.962 (2)	15.175 (2)	0.581 (2)
B90(C_7)	0.545 (4)	1.300 (5)	14.703 (3)	0.440 (5)
B91(C_8)	0.657 (3)	1.706 (3)	14.670 (4)	0.462 (3)
B97(C_9)	0.474 (6)	1.374 (4)	14.425 (6)	0.262 (6)
B99(C_{10})	0.149 (8)	0.724 (8)	13.718 (8)	-0.063 (8)

Values of the rank are given within the ( )

Table 4. Rank correlation for the kernel row number

	μG	UBND	PTC	NI
μG	-	0.928**	0.976**	0.905**
UBND		-	0.880**	0.905**
PTC			-	0.928**
NI				-

** significant at the p<0.01 level

CONCLUSION

According to the estimators μG, UBND, PTC and NI from both synthetic populations the maize inbred donors of favourable alleles for the improvement of the kernel row number of the parental inbreds of the elite hybrid ZP 580 were identified.

The improvement of the kernel row number of inbred lines, parental components (I_1 and I_2) of the elite hybrid ZP 580 can be done with inbred lines B73(C_5) and B84(C_7) of the synthetic population BSSS and B91(C_8) and B90(C_7) of the synthetic population $BSCB_1$. These inbred lines also had the highest value of the estimator μG. The inbred line I_1 can be improved with inbred lines B73, B84 and B91, while the inbred line I_2 can be improved with the inbred line B91.

High and significant values of rank correlation were determined among observed estimators μG, UBND, PTC and NI.

ACKNOWLEDGEMENT

This paper presents results of the Project TR 31068 *„Improving maize and soya bean traits with molecular and conventional breeding"*, supported by the Ministry of Science and Technological Development of Republic Serbia.

LITERATURE CITED

Bernardo, R., 1990a. An alternative statistics for dentifying lines useful for improving parents of an elite single cross. Theor. Appl. Genet. 80:105-109.

Carena, M. J., 2005. Maize commercial hybrids compared to improved population hybrids for graih yield and agronomic performance. Euphytica. 141: 201-208.

Delić, N., Đorđević, J., Todorović, G., M. Babić, 1996. Ocena komercijalnih hibrida kukuruza *(Zea mays L.)* kao donora poželjnih alela za popravku prinosa zrna elitnog SC hibrida. Genetika. 28 (1): 9-16.

Dudley, J. W., 1984a. A method of indentifying lines for use in inproving parents of a single cross. Crop Sci. 24:355-357.

Dudley, J. W., 1987a. Modification of methods for identifying inbred lines useful for improving parents of an elite single crosses. Crop Sci. 27:944-947.

Gerlof, J. E., 1985. Choice of method for identifying germplasm with superior alleles. Ph. D. Diss. Iowa State Univ., Ames, USA (Diss. Abstr. 85-24656).

Gerlof, J. E. and O. S. Smith, 1988. Choice of method for identifuing germplasm with superior elleles. Theor. Appl. Genet. 76:209-216.

Hallauer, A. R. and J. B. Miranda, 1988. Quantitative genetics in maize breeding, 2nd ed. Iowa State University Press, Ames, Iowa, pp 361-362.

Mišević, D., 1989a. Evaluation of three test statistics used to identify maize inbred lines with new favorable alleles not present in an elite single cross. Teor. Apll. Genet. 77:402-408.

Mišević, D., 1989b. Identification of inbred lines as a source of new alleles for improvement of elite maize single crosses. Crop Sci. 29:1120-1125.

Malvar R. A., Pedro Revilla P., Cartea M. E., Butrón A. M. and A. Ordás, 2001. Cheching performance of field corn inbreds as donors of favorable alleles to improve early vigor and adaptation od sweet corn hybrids to European conditions. Maydica. 46:187-193.

Sprague, G.F. and S.A. Eberhart, 1997. Corn Breeding in Corn and Corn Improvement, Sprague G.F. ed. Am. Soc. Agron. Madison. Wis, pp. 305-363

Trifunović, S., 1999. Procena inbred linija i populacija kukuruza *(Zea mays L.)* kao donora poželjnih alela. Doktorska disertacija, Poljoprivredni fakultet, Novi Sad.

Vračarević, M., 2000. Identifikacija inbridovanih linija, potencijalnih donora poželjnih alela za popravku prinosa i komponenata prinosa hibrida kukuruza *(Zea mays L.)*. Magistarska teza. Poljoprivredni fakultet, Univerzitet u Beogradu.

Zanoni, U. and J. W. Dudley, 1989c. Use of F_2 Generations to Identify Inbreds With Favorable Alleles Not Present in an Elite Hybrid. Crop Sci. 29:583-589.

Zanoni, U. and J. W. Dudley, 1989a. Comparison of Different Methods of Identifying Inbreds Useful for Improving Elite Maize Hybrids. Crop Sci. 29:577-582.

THE EFFECT OF NATURAL GROWTH REGULATORS OBTAINED FROM *Ecklonia maxima* AND MINERAL NITROGEN ON TRUE PROTEIN AND SIMPLE SUGAR CONTENTS OF *Dactylis glomerata* L. AND *Festulolium braunii* (K. richt.) A. camus

*Agnieszka GODLEWSKA**, *Grażyna Anna CIEPIELA*

University of Natural Sciences and Humanities, Siedlce, POLAND,
**Corresponding author: agnieszka.godlewska@uph.edu.pl*

ABSTRACT

The objective of the study was to evaluate the effect of the biostimulant Kelpak on true protein and simple sugar contents of *Dactylis glomerata* and *Festulolium braunii* at different nitrogen rates. A field experiment was set up as a randomised sub-block design (split-split-plot) with three replicates. The following factors were examined: the biostimulant with the trade name Kelpak SL, nitrogen rates: 0, 50 and 150 kg N·ha^{-1}, pure-sown grass species and cultivars (monoculture). Protein compounds and simple sugars were determined in the dry matter of plants, and a carbohydrate/protein ratio was calculated. The *Ecklonia maxima* extract significantly increased true protein and simple sugar contents as well as the sugar/protein ratio in the tested plants. Nitrogen rates significantly influenced protein and sugar contents in the dry matter of *Dactylis glomerata* and *Festulolium braunii.* The carbohydrate/protein ratio averaged 1.07 and was within the range of values which are believed to be optimal for ruminants.

Key words: Biostymulator, *Dactylis glomerata, Festulolium braunii*, Kelpak

INTRODUCTION

Sustainable agricultural production has to compromise between the positive effect of crops on the soil environment and high yields of good quality (Kitczak, Czyż 2006). So far, plant production intensification has been mainly associated with increasing rates of mineral fertilisers, as a result of which the chemical composition of feed has deteriorated, the sward has thinned out and some animal species have migrated from these habitats (Jankowska et al. 2008). Recently, intensification of agriculture has slowed down due to an introduction of pro-ecological models (Szoszkiewicz et al. 2003) as a result of which rates of mineral fertilisers applied to grassland have been reduced. Plant biomass obtained from organically farmed grassland tends to be of poor quality so attempts have been made to find new solutions to produce high quality yields and, at the same time, reduce e.g. soil contamination. One of such solutions, which has been receiving more and more attention in recent years, is offered by regulators of plant growth and development (Matysiak, Adamczewski 2006). Many studies have revealed a wide range of beneficial effects of plant growth regulator applications on plants (Güllüoglu 2011). When applied in small quantities, such substances modify physiological functions in plants. Sea algae extracts contain not only phytohormones such as auxins and cytokinins (Durand et al. 2003) but also macroelements: Ca, K and P (Stirk et al. 2004), and microelements: Fe, Cu, Zn, B, Mn, Co and Mo (Khan et. al. 2009). Kelpak is a biostimulant produced from the seaweed *Ecklonia maxima* which is harvested off the shores of South Africa. Research has demonstrated that Kelpak has a beneficial influence on crop plants but its effect depends rather on the crop plant species and cultivar than the rate at which Kelpak is applied (Sultana et al. 2005). The biostimulant is used worldwide in the production of vegetables and fruit as well as cereals, rape and maize. However, there is a paucity of reports on an application of natural biostimulants to grassland.

The present work is an attempt to evaluate the effect of the biostimulant Kelpak on true protein and simple sugar contents in *Dactylis glomerata* and *Festulolium braunii* at different nitrogen rates.

MATERIALS AND METHODS

A field experiment was arranged as a randomised sub-block design (split-split-plot) with three replicates at the Siedlce Experimental Unit of the University of Natural Sciences and Humanities in late April, 2009. The plot area was 10 m^2. The soil of the experimental site represents average soils, Hortic Anthrosol (WRB). Prior to the experiment set-up the characteristics of the soil were as follows: neutral pH (pH in 1n KCl = 7.2), high humus content (3.78%), high available phosphorus and magnesium contents (P_2O_5 – 900 mg · kg^{-1}, Mg – 84 mg · kg^{-1}) and average total nitrogen and available potassium contents (N – 1.8 g · kg^{-1} d.m., K_2O – 19 mg · kg^{-1}). The following factors were examined:

- biostimulant with the trade name Kelpak SL applied at 2000 dm^3 · ha^{-1} and a control – no biostimulant,

- nitrogen application rate: 50 and 150 kg · ha^{-1}, and a control (0 kg · ha^{-1}),

- pure stands of grass species and cultivars grown in monoculture

- *Dactylis glomerata,* cv. Amila and Tukan,

- *Festulolium braunii*, cv. Felopa and Agula.

Kelpak is a growth stimulant unique on the Polish market which contains natural plant hormones: auxins (11 mg · l^{-1}) and cytokinins (0.03 mg · l^{-1}). It is an extract from the fastest growing seaweed (kelp) *Ecklonia maxima* harvested off the coast of South Africa.

The sowing rate of the plants was calculated based on the standards drawn up by IMUZ (the Institute for Land Reclamation and Grassland Farming) in Falenty (Jankowski et al. 2005) . In the growing season when the experiment was set up, nor the biostimulant were applied. The season was an introductory period when three weed-control cuttings were made. After the second cutting, mineral fertilisation was applied to all the plots at the rates of 30 kg · ha^{-1} N (ammonium nitrate) and 30 kg · ha^{-1} K_2O (potash salt). Phosphorus was not applied as the soil was rich in available forms of this element. Over the study period (2010-2012), the cutting regime was three harvests per year. Ammonium nitrate was applied three times per year. The total nitrogen amount was split into three equal rates which were applied before each cutting. Phosphorus and potassium needs of the mixtures were calculated taking into account the expected dry matter yields, the appropriate mineral (from the ruminant nutrition standpoint) contents of hay and soil P and K availability. Moreover, to determine phosphorus and potassium application rates, coefficients given by Fotyma and Mercik (1995) were used to convert the amounts of the nutrients taken up by grass and clover yields into the rates of phosphorus and potassium fertilisers. P and K fertilisation was applied to all the plots. Phosphorus was applied once as triple superphosphate at a rate of 40 kg · ha^{-1} P_2O_5 in the spring. The amount of potassium (160 kg · ha^{-1} K_2O) was split into three equal rates and applied prior to each cutting as 60% potash salt. The biostimulant was sprayed as an aqueous solution, the rate was 2000 cm^3 of biostimulant per hectare diluted in water to 400 dm^3. The spraying was performed before each cutting: the first application was three weeks before the first cutting, the second one two weeks after the first harvest, and the last one three weeks after the second harvest.

During each harvest, 0.5-kg green matter samples of grasses were taken from each plot to carry out chemical analyses. The samples were left to dry in a ventilated room. The airy dry matter was shredded and ground. The obtained material was subjected to chemical analysis to determine dry matter (by determining moisture content), protein compounds and simple sugars. The method of determination was near-infrared spectroscopy (NIRS) using a NIRFlex N-500 spectrometer and ready-to-use INGOT calibration applications.

Statistical analysis of the results included:

- variance analysis of a four-factor repeated (biostimulant, N doses, species, cultivar, cut) experiment. The experimental design was split-plot-split-block with the linear models as suggested by Trętowski and Wójcik (1991). Significance of differences between means was checked using the Tukey test at the significance level of $\alpha \leq 0.05$,

- correlation coefficient whose significance was checked using Student's t test.

RESULTS AND DISCUSSION

True protein content in the dry matter of the grass species tested averaged 94.6 kg·ha^{-1} (table 1). Statistical analysis of the results demonstrated that the biostimulant significantly affected true protein and simple sugar contents in the grasses examined depending on the nitrogen rate, cultivar, cut and study year. The average true protein content in plants increased by 11.3% following an application of the growth regulator regardless of the remaining factors examined in the experiment. According to Szabo et al. (2011), components in biostimulants such as auxins, gibberellic acid, cytokinins and amino acids increase the physiological activity of plants, for example protein synthesis. There are many reports in the world literature of the positive effect of sea algae extracts on protein content in various plant species (Sivasankari et al. 2006, Matysiak et al. 2012). However, no papers have been found concerning grasses. Both *Dactylis glomerata* and *Festulolium braunii* contained significantly more true protein when sprayed with Kelpak. Analysis of plant material showed that the biostimulant applied at different nitrogen rates significantly increased the concentration of true protein although the greatest increase, 12.4%, was recorded in plots which had not been N-fertilised. Statistically significant differences between true protein contents were found for each cut. The greatest accumulation of protein was in grass from the first harvest and it increased at consecutive cuts. Similar findings have been reported by Łyszczarz and Dembek (2003) as well as Jankowska et al. (2008). According to these authors, protein content in grass increases in successive re-growths due to poorer grass foliage at the first harvest compared with the following cuts. An application of Kelpak significantly increased the quantity of protein in grass at all the harvests. Statistical analysis revealed no significant effect of study years on protein amount in plants. In turn, the biostimulant significantly increased true protein content in the plant species tested in each study year.

Table 1. The effect of biostymulant on the true protein content and simple sugars in *Dactylis glomerata* and *Festulolium braunii* depending on nitrogen doses, cultivar, cut and years of study

Factor		True protein (g kg^{-1}s.m.)			Simple sugars (g kg^{-1}s.m.)		
		Treatment (A)		Mean	Treatment (A)		Mean
		I	II		I	II	
N dose (kg ha^{-1}) (B)	0	75.6	85.0	80.3	94.7	134.9	114.8
	50	88.3	97.1	92.7	89.0	120.2	104.6
	150	104.8	117.0	110.9	74.5	100.3	87.4
Dactylis glomerata (D)	Amila (C)	87.8	98.1	92.9	62.3	87.3	74.8
	Tukan (C)	81.7	91.6	86.6	62.7	82.4	72.6
Festulolium braunii (D)	Felopa (C)	97.8	107.4	102.6	117.7	155.4	136.5
	Agula (C)	90.9	101.7	96.3	101.6	148.8	125.2
Dactylis glomerata (D)	Mean	84.8	94.8	89.8	62.5	84.9	73.7
Festulolium braunii (D)	Mean	94.3	104.6	99.5	109.6	152.1	130.9
Cut (E)	1	76.3	84.1	80.2	100.7	136.4	118.5
	2	91.2	102.4	96.8	86.6	117.9	102.2
	3	101.2	112.6	106.9	71.0	101.2	86.1
Years (F)	2010	90.0	100.6	95.3	80.4	108.8	94.6
	2011	88.5	98.0	93.3	84.0	124.6	104.3
	2012	90.1	100.5	95.3	93.7	122.1	107.9
Mean		89.6	99.7	94.6	86.1	118.5	102.2
$LSD_{0.05}$		A – 2.8; B – 4.5; D – 4.1; E – 3.5; F – n.s.; A x B – 5.0; A x C x D – 7.9; A x D – 5.1 ; A x E – 5.5; A x F – 6.7 ; C x D – 4.2			A – 5.6; B – 8.6; D – 4.3; E – 8.5; F – 8.9; A x B – 12.8; A x C x D – 9.6; A x D – 6.2 ; A x E – 12.6; A x F – 13.3 ; C x D – 7.8		

I – control (without biostimulant); II – treatment with biostimulant
n.s. – not significant differences

The average simple sugar content in grasses amounted to 102.2 g·kg^{-1} d.m. Joubert and Lefranc (2008) have stated that active substances in sea algae extracts are similar to physiological activators in their mode of action because they may change the chemical composition of plants sprayed with such extracts. The biostimulant applied in the present study was found to influence the concentration of sugars in plants. Kelpak increased monosaccharide concentration in grasses by 37.6%, regardless of the remaining factors investigated in the study. Similar findings have been reported by many authors (Dobrzański et al. 2008, Szymczak-Nowak 2009, Pacholczak et al. 2012) who have stressed that sea algae extracts contribute to increased sugar contents in plants but, as with protein content, no literature data for grasses have been found. The effect of the biostimulant on sugar concentration was significant at each nitrogen level. At 0 N kg·ha^{-1}, sugars increased by as much as 42.4% following an application of Kelpak. Just like in the study by Kozłowski et al. (2001), seasonal changes in the carbohydrate content of grasses were observed. More simple sugars were determined in the grass of the first cut, which was most likely due to increased respiration of plants, induced by high temperatures, which makes use of sugars. Moreover, the three-year data of this study showed that significantly least sugars were determined in plants in the first study year, regardless of the remaining factors examined in the study. Moreover, statistical analysis revealed that the biostimulant contributed to significantly increased sugar contents determined for each cut and study year. The chemical composition of plants is species- and even cultivar-related (Falkowski et al. 2000), the finding confirmed in this study. Significantly more sugars were determined in *Festulolium braunii* regardless of the remaining factors. What is more, an application of the growth regulator significantly increased monosaccharide levels in both the grass species.

The true protein content in the dry matter of grasses significantly depended on the nitrogen rate (table 2). When the rate was increased from 0 to 50 kg·ha^{-1}, protein amount in *Dactylis glomerata* and *Festulolium braunii* increased by 18 and 12.9%, respectively, whereas at the rate of 150 kg N·ha^{-1} the respective increases were 41.7 and 34.8%. Also Ciepiela et al. (2008) have reported a significant increase in the protein content of *Dactylis glomerata* at the nitrogen rates of 60 and 90 kg·ha^{-1} compared with the rate of 0 kg N·ha^{-1}. Moreover, they found a significant association of nitrogen rates with the concentration of protein in the grass herbage at each harvest and for consecutive study years.

The nitrogen fertilisation regime significantly influenced the sugar content of each grass cultivar. Increasing nitrogen rates significantly reduced sugar content determined in the grass dry matter, which agrees with results reported by other authors (Lemaire and Salette 1982, Ciepiela 2004). The nitrogen fertilisation regime x cuts interaction and nitrogen fertilisation regime x study years interaction were found. During the three-year study period, increasing N rates significantly reduced the concentration of sugars in the grass dry matter from each cut and study year.

Table 2. The effect of nitrogen dose on the true protein content and simple sugars in *Dactylis glomerata* and *Festulolium braunii* in each cultivar, cut and study year

Factor		True protein ($g \cdot kg^{-1}$ s.m.) N dose ($kg \cdot ha^{-1}$) (B)			Simple sugars ($g \cdot kg^{-1}$ s.m.) N dose ($kg \cdot ha^{-1}$) (B)		
		0	50	150	0	50	150
Dactylis glomerata (D)	Amila (C)	79.4	90.3	109.1	83.6	76.8	64.0
	Tukan (C)	70.3	86.6	103.0	82.2	71.4	64.0
Festulolium braunii (D)	Felopa (C)	89.1	99.8	119.0	150.6	140.9	118.1
	Agula (C)	82.5	94.0	112.4	142.7	129.3	103.6
Dactylis glomerata (D)	Mean	74.9	88.4	106.1	82.9	74.1	64.0
Festulolium braunii (D)	Mean	85.8	96.9	115.7	146.7	135.1	110.8
Cut (E)	1	67.6	78.0	94.9	132.4	120.3	102.8
	2	82.4	95.3	112.8	114.5	105.1	87.0
	3	90.9	104.8	124.9	97.4	88.4	72.4
Years (F)	2010	82.4	92.5	111.0	107.1	96.5	80.1
	2011	77.8	91.5	110.5	118.4	108.3	86.2
	2012	80.7	94.0	111.1	118.8	109.0	95.9
$LSD_{0.05}$		B x C x D – 7.7; B x D – 4.9; B x E – 4.3; B x F – 6.9			B x C x D – 9.5; B x D – 9.3; B x E – 9.3; B x F – 10.0		

Festulolium braunii, a *Festuce pratensis* and *Lolium multiflorum* cross (Zwierzykowski et al. 1998), is of substantial nutritional value. It has got high protein and sugar levels (Borowiecki 2005), and in this respect is better than *Dactylis glomerata*. The results discussed here (table 3) showed a significantly higher true protein and simple sugar contents of *Festulolium braunii* compared with *Dactylis glomerata* harvested at each cut and in each study year. *Festulolium braunii* cv. Agula proved to be the best in terms of chemical composition.

Table 3. Content of true protein and simple sugars in two cultivar *Dactylis glomerata* and *Festulolium braunii* in each cut and study year

Factor		True protein content ($g \cdot kg^{-1}$ d.m.) Cut (E)			Years (F)			Simple sugars ($g \cdot kg^{-1}$ d.m.) Cut (E)			Years (F)		
		1	2	3	2010	2011	2012	1	2	3	2010	2011	2012
Dactylis glomerata (D)	Amila (C)	78 .2	95.6	105.0	94.1	91.4	93.4	97.0	75.2	52.3	72.4	76.0	75.9
	Tukan (C)	73 .7	88.6	97.5	86.2	85.8	87.9	89.1	75.1	53.5	71.2	71.6	74.9
Festulolium braunii (D)	Felopa (C)	87 .4	104.9	115.5	102.9	101.0	104.0	149.9	134.9	124.8	124.1	142.6	142.9
	Agula (C)	81 .3	98.2	109.3	98.1	94.9	95.9	138.2	123.7	113.6	110.7	127.1	137.9
Dactylis glomerata (D)	Mean	76 .0	92.1	101.3	90.2	88.6	90.6	93.0	75.1	52.9	71.9	73.8	75.4
Festulolium braunii (D)	Mean	84 .4	101.5	112.4	100.5	97.9	99.9	144.0	129.3	119.2	117.3	134.8	140.4
$LSD_{0.05}$		C x D x E – 6.0; C x D x F – 5.3; D x E – 5.5; D x F – 4.9						C x D x E – 10.3; C x D x F – 13.0; D x E – 8.9; D x F – 10.0					

The forage value of grassland herbage is often determined based on its carbohydrate/protein ratio which should not drop below 0.4, the optimal range for ruminants being 0.8 to 1.5. The average value of this ratio in the grass cultivars tested was 1.07 (table 4) and thus was within the above-mentioned limit. The most favourable value (1.59) was recorded for the plots which had not received nitrogen fertilisation but had been sprayed with Kelpak. The *Ecklonia maxima* extract clearly influenced the value of this indicator regardless of the nitrogen rate, cultivar, cut or study year. The highest values were determined for Kelpak-sprayed plants. Moreover, there was found an interaction between the biostimulant and the remaining factors. The ratios which were significantly highest were calculated for plants harvested from the Kelpak-sprayed plots which received the nitrogen rates of 0 and 50 kg·ha^{-1}. However, the increase was insignificant when the rate was increased to 150 kg N·ha^{-1}. When both grass species were compared, a higher sugar/protein ratio was calculated for *Festulolium braunii* although Kelpak significantly increased the value of this indicator for both the species. The sugar/protein ratio was also different between harvests. Ratios calculated for consecutive re-growths were significantly lower whereas increasing values were obtained for consecutive cuts when Kelpak had been applied. Analysis of the results showed that carbohydrate/protein ratios in the tested plants decreased in consecutive study years, the differences being statistically significant. The nitrogen rates/study years interaction was found in the first and second study year. The carbohydrate/protein ratio was not influenced by Kelpak application in the last study year. Nitrogen rates significantly affected the values of this indicator (table 5). The response of both the grass species was the same: the higher N ratio was calculated, the lower carbohydrate/protein ratio was obtained. The same tendency was observed for a nitrogen rates x cuts interaction and a nitrogen rates x study years interaction. Increased nitrogen rates contributed to significantly lower values of carbohydrate/protein ratio for each cut and study year.

Table 4. The effect of biostymulant on the value of carbohydrate/protein rations in *Dactylis glomerata* and *Festulolium braunii* in each nitrogen doses, cultivar, cut and study year

Factor		**Treatment (A)**		**Mean**
		I	**II**	
N dose (kg · ha^{-1}) (B)	0	1.25	1.59	1.42
	50	1.01	1.24	1.12
	150	0.71	0.86	0.78
Dactylis glomerata (D)	Amila (C)	0.71	0.89	0.80
	Tukan (C)	0.77	0.90	0.83
Festulolium braunii (D)	Felopa (C)	1.20	1.45	1.33
	Agula (C)	1.12	1.46	1.29
Dactylis glomerata (D)	Mean	0.74	0.90	0.82
Festulolium braunii (D)	Mean	1.16	1.45	1.31
Cut (E)	1	1.32	1.62	1.47
	2	0.95	1.15	1.05
	3	0.70	0.90	0.80
Years (F)	2010	0.89	1.08	0.99
	2011	0.95	1.27	1.11
	2012	1.04	1.21	1.13
Mean		0.96	1.19	1.07

$LSD_{0.05}$: A – 0.09; B – 0.10; D – 0.07; E – 0.10; F – 0.11.; A x B – 0.16; A x C x D – 0.17; A x D – 0.13 ; A x E – 0.15; A x F – 0.18 ; C x D – n.s.

I – control (without biostimulant); II – treatment with biostimulant
n.s. – not significant differences

Values of sugar/protein ratio calculated for both the cultivars of *Dactylis glomerata* and *Festulolium braunii* remained mostly unaffected by cuts and study years (table 6). Significant differences were determined for the grass species only. Compared with *Dactylis glomerata*, *Festulolium braunii* had higher values of the sugar/protein ratio for each cut over the whole study period. The values for *Festulolium braunii* were higher in each study year but the differences were not statistically significant.

Coefficients of linear correlation were calculated between true protein content, simple sugar content as well as carbohydrate/protein ratio for *Dactylis glomerata* and *Festulolium braunii* and the growth stimulant (table 7). The results indicated that the contents were significantly and positively correlated with an application of the stimulant for both the individual nitrogen rates and grass species, and regardless of these factors.

Table 5. The effect of N dose on the carbohydrate/protein ratios in *Dactylis glomerata* and *Festulolium brauni* in each cultivar, cut and study year

Factor		**N dose (kg · ha^{-1}) (B)**		
		0	**50**	**150**
Dactylis glomerata (D)	Amila (C)	1.05	0.85	0.59
	Tukan (C)	1.17	0.82	0.62
Festulolium braunii (D)	Felopa (C)	1.69	1.41	0.99
	Agula (C)	1.73	1.38	0.92
Dactylis glomerata (D)	Mean	1.11	0.84	0.60
Festulolium braunii (D)	Mean	1.71	1.39	0.96
Cut (E)	1	1.96	1.54	1.08
	2	1.39	1.10	0.77
	3	1.07	0.84	0.58
Lata (F)	2010	1.30	1.04	0.72
	2011	1.52	1.18	0.78
	2012	1.47	1.16	0.86

$LSD_{0.05}$:B x C x D – 0.20; B x D – 0.14; B x E – 0.16;B x F – 0.21

Table 6. Carbohydrate/protein ratios for two cultivars of *Dactylis glomerata* and *Festulolium braunii* in each cut and study year

Factor		Cut (E)			Years (F)		
		1	2	3	2010	2011	2012
Dactylis glomerata (D)	Amila (C)	1.24	0.79	0.50	0.77	0.83	0.81
	Tukan (C)	1.21	0.85	0.55	0.83	0.83	0.85
Festulolium braunii (D)	Felopa (C)	1.72	1.29	1.08	1.21	1.41	1.37
	Agula (C)	1.70	1.26	1.04	1.13	1.34	1.44
Dactylis glomerata (D)	Mean	1.22	0.82	0.52	0.80	0.83	0.83
Festulolium braunii (D)	Mean	1.71	1.27	1.06	1.17	1.38	1.41

$LSD_{0.05}$: C x D x E – n.s.; C x D x F – n.s.; D x E – 0 .13; D x F – 0.17

n.s. – not significant differences

Table 7. Association of the true protein content, simple sugars content and carbohydrate/protein ratio in *Dactylis glomerata* and *Festulolium braunii* with the biostimulant (correlation coefficient)

N dose (kg · ha^{-1})	Species: *Dactylis glomerata*	Species: *Festulolium braunii*	Independently of the species
	Biostimulant	Biostimulant	Biostimulant
	True protein		
0	0.867*	0.974*	0.933*
50	0.941*	0.974*	0.949*
150	0.937*	0.844*	0.901*
Independently of N dose	0.960*	0.888*	0.956*
	Simple sugars		
0	0.821*	0.489*	0.909*
50	0.777*	0.580*	0.915*
150	0.819*	0.937*	0.895*
Independently of N dose	0.811*	0.789*	0.925*
	Carbohydrate/protein ratio		
0	0.914*	0.865*	0.926*
50	0.901*	0.829*	0.918*
150	0.906*	0.939*	0.941*
Independently of N dose	0.931*	0.925*	0.945*

* significant coefficient at α = 0.05

CONCLUSIONS

1. The *Ecklonia maxima* extract significantly increased true protein content, simple sugar content and the sugar/protein ratio in the tested plants, regardless of their cultivar, nitrogen rates, cuts and study years.

2. Nitrogen rates significantly affected the protein and sugar contents in the dry matter of both *Dactylis glomerata* and *Festulolium braunii*. Higher nitrogen rates increased protein levels but also reduced sugar levels in the grass species tested.

3. The value of carbohydrate/protein ratio averaged 1.07 and fell within the range of values which are believed to be optimal for ruminants. The most favourable ratio (1.59) was calculated for plants harvested from non-fertilised Kelpak-sprayed plots.

LITERATURE CITED

Borowiecki J. 2005. Review on Festulolium braunii (K.Richter) A.Camus. Pamiętnik Puławski, 140:15-23. (in Polish)

Ciepiela G.A. 2004. Reaction of some grass species on the nitro gen fertilization applied in urea solution and in ammonium nitrate. Dissertation no 76, publication of the University of Podlasie, Siedlce (in Polish)

Ciepiela G.A., J. Jankowska, K. Jankowski, J. Jodełka 2008. The field quality from orchard Grass cultivated as a monoculture and in mixtures with different legumes. Pamiętnik Puławski, 147:5-13. (in Polish)

Dobrzański A., Z. Anyszka, K. Elkner 2008. Response of carrots to application of natural extracts from seaweed (Sargassum sp.) - Algaminoplant and from leonardite - Humiplant. Journal of Research and Applications in Agricultural Engineering, 53(3): 53-58. (in Polish)

Durand N., X. Briand, C. Meyer 2003. The effect of Marine bioactive substances (N PRO) and exogenous cytokinins on nitrate reductase activity in *Arabidospsis thaliana.* Physiologia Plantarum, 119:489-493.

Falkowski M., I. Kukułka, S. Kozłowski 2000. Chemical properties of meadow plants, publication of the Poznań University of Agriculture: 43-48. (in Polish)

Fotyma M., S. Mercik 1995. Fertilizationn and crop Technologies. [in]: Agricultural Chemistry. State Scientific Publishers, Warsaw: 233-295 (in Polish)

Güllüoglu, L., 2011. Effects of growth regulator applications on pod yield and some agronomic characters of peanut in Mediterranean region. Turkish Journal of Field Crops 16(2):210-214

Jankowski K., G.A. Ciepiela, J. Jodełka, R. Kolczarek 2005. Grassland. University of Podlasie Press: 119-122. (in Polish)

Jankowska J., G.A. Ciepiela, R. Kolczarek, K. Jankowski 2008. The influence of mineral fertilizer type and nitro gen dose on the Fielding and nutritive value of sward from a permanent meadow. Pamiętnik Puławski, 147:1-14. (in Polish)

Joubert J.M., G. Lefranc 2008. Seaweed phytostimulants in agriculture: recent studiem on mode of action two types of products from alga:growth and nutrition stimulants and stimulants of plant Demence reactions. Book of abstracts: Biostimulators in modern agriculture. Warsaw, p.16.

Khan W., U.P. Rayirath, S. Subramanian, M.N. Jithesh, P. Rayorath, D.M. Hodges, A.T. Critchley, J.S. Craigie , J. Norrie, B. Prithiviraj 2009. Seaweed extracts as biostimulants of plant growth and development. Journal of Plant Growth Regulation, 28: 386-399.

Kitczak T., H. Czyż 2006. Yielding of mixtures *Festulolium braunii (K.Richt) A.Camus with Trifolium repens L.* depending on the share of components and doses of nitro gen fertilization. Annales Universitatis Marie Curie-Sklodowska Lublin-Polonia, sectio E, LXI: 333-339. (in Polish)

Kozłowski S., B. Golińska, P. Goliński 2001. Sugars vs. Use value of meadow plants. Pamiętnik Puławski 125, 131-138, (in Polish)

Lemaire G., J. Salette 1982. The effect of temperature and fertilizer on the growth of two forage Grassem in spring. Grass and Forage Science, 37:191-198.

Łyszczarz R., R. Dembek 2003. Long-term studiem on earliness, Fields and nutritional value of Polish commaon orchard grass varietes. Biuletyn Hodowli i Aklimatyzacji Roślin, 225:29-42. (in Polish)

Matysiak K. K. Adamczewski 2006. Influence of bioregulator Kelpak on yield of cereals and other crops. Progress in Plant Protection 46(2): 102-108. (in Polish)

Matysiak K., S. Kaczmarek, D. Leszczyńska 2012. Influence of liquid seaweed extract of *ecklonia maxima* on Winter wheat cv Tonacja. Journal of Research and Applications in Agricultural Engineering, 57(4):44-47.

Pacholczak A., W. Szydło, E. Jacygard, M. Federowicz 2012. Effect of auxins and the biostimulator algaminoplant on rhizogenesis in stem cuttings of two dogwood cultivars (*Cornus alba* 'Aurea' and 'Elegantissima'). Acta sci. Pol., Hortorum Cultus 11(2):93-103.

Sivasankari S., V. Venkatesalu, M. Anantharaj, M. Chandrasekaran 2006. Effect of seaweed extracts on the growth and biochemical constituents of *Vigna sinensis*. Bioresource Technology, 97:1745-1751.

Szabo V., A. Sarvari, K. Hrotko 2011. Treatment of Stockplants with biostimulators and their effects on cutting propagation of prunus Marianna GF8-1. Acta Horticulturae 923:277-282.

Stirk W.A., G.D. Aurthur, A.F. Lourens, O. Novak, M. Strnad, J. van Staden 2004. Changes in cytokinins and auxin concentrations in seaweed concentrates when stored at an elevated temperature. Journal of Applied Phycology, 16:31-39.

Sultana V., S. Ehteshamul-Haque, J. Ara, M. Athar 2005. Comparative efficacy of Brown, Green and red Seaweeds in the control of Root infesting fungi and okra. International Journal of Environment Science and Technology, 21(1):75-81.

Szoszkiewicz J., J. Zbierska, R. Dembek, K. Szoszkiewicz, R. Staniszewski 2003. Phytosociological differentiation and agronomic value of meadow plant associations with legumes in the Wielkopolska and Kujawy regions. Bulletin of the Institute for Plant Breeding and Acclimatisation, 225:107-119. (in Polish)

Szymczak-Nowak J. 2009. Effect of biostimulators on health and yielding of sugar beet. Prog. Plant Prot./Post. Ochr. Roślin, 49(4):2031-2037. (in Polish)

Trętowski J., A.R. Wójcik 1991. Metodotogy for agricultural experiments. Publisher: Agricultural and Pedagogical University in Siedlce: 331-334 (in Polish)

Zwierzykowski Z., R. Tayyar, M. Brunell, A.J. Łukaszewski 1998. Genome recombination in intergeneric hybrids between tetraploid *Festuca pratensis* and *Lolium multiflorum*. Journal of Heredity, 89(4):324-328.

9

HYDROCYANIC ACID AND SUGAR CONTENT DYNAMICS UNDER NITROGEN AND SULPHUR APPLICATION TO FORAGE SORGHUM CULTIVARS

Ahmad SHER[1], Muhammad ANSAR[2], Abdul MANAF[2], Abdul QAYYUM[3], Muhammad Farhan SAEED[1] and Muhammad IRFAN[2]*

[1]*College of Agriculture, Bahauddin Zakariya University, Bahadur Campus Layyah, Pakistan*
[2]*Department of Agronomy, PMAS-Arid Agriculture University Rawalpindi, Pakistan*
[3]*Department of Agriculture, University of Haripur, 22620, Pakistan*
[*]*Corresponding author: ahmad_sher786@yahoo.com*

ABSTRACT

Sorghum crop provides excellent forage in dry land areas. It has ability to tolerate high temperature and grow under minimal soil moisture. Sugar content in Sorghum is of prime importance regarding, animal feed value as well as biofuel production. However, high level of HCN toxin production is a serious problem associated with this crop which could be lethal to animals if ingested in greater quantities. In this study hydrocyanic acid and sugar content dynamics were determined on two forage sorghum cultivars i.e. JS-2002 and Chakwal sorghum at different developmental stages (booting[a] and 50% heading[b]). Fertilizer treatments of nitrogen (0, 60 and 120 kg ha^{-1}) and sulphur levels (0, 20 and 40 kg ha^{-1}) were applied in replicates. The experiment was arranged in randomized complete block design. The results clearly depicted that increase in nitrogen levels gradually increased the HCN and sugar contents irrespective to sulphur and delayed harvesting times. Increase in HCN (21[a] & 22[b]%) and brix value (56[a] & 59[b]%) was found more at 120 kg N ha^{-1} when compared with 0 kg N ha^{-1} (control) treatment. Similarly increase in sulphur application caused increase in brix value however; inverse relationship with HCN content was recorded. Results also indicated that sorghum cultivar JS-2002 produced 27[a] & 40[b]% decline in HCN content while 8[a] & 18[b]% more sugar content when compared with Chakwal sorghum. Seasonal variations resulted maximum sugar production of 14.9, 25.6% and HCN content of 21.9 & 13.1 mg/100g during different year time (2010 and 2009 when compared with 2008, respectively) probably due to variation in seasonal rainfall. Three years field study indicate that JS-2002 produced less HCN with low dose of N application (0 kg ha^{-1}) and higher application of sulphur (40 kg ha^{-1}) while more brix value at 40 kg S and 120 kg N ha^{-1} under sub-tropical rainfed conditions.

Keywords: Brix value, HCN, nitrogen, sorghum cultivars, sub-tropical condition, sulphur

INTRODUCTION

Sorghum is one of the important summer fodder crops all over the country particularly in rainfed regions. It is nutritious, juicy, palatable and well-liked by the cattle. Because of its high tolerance to various stresses, it is extensively grown as a major source of fodder preferred over maize (Reddy *et al.,* 2004). Sorghum plants contain a cyanogenic glycoside called dhurrin. According to Lang (2001) when the plants cells damaged (insect attack, crust, hailing, frost injury), the available cyanogenic glycoside (dhurrin) is hydrolyzed by making two fermentation enzymes (*Ac/ac and Li/li*) which converted into hydrocyanic acid (HCN). Cyanogenesis is a character which genetically controlled, but it is often affected by different biotic and abiotic factors (Pederson *et al.,* 1996). The main factors responsible for HCN accumulation are plant species (Sudan grass have less HCN contents than sorghum), plant organs (leaves have higher HCN contents than other parts), suckers (tillers have more HCN potential), fertilizer (unevenness nitrogen and phosphorus in the soil causes the HCN contents to increase) and drought (dryland sorghum has more HCN contents than irrigated one due to water stress).

The available amount of cyanide poisoning in forage is important to consider for animal's physiological conditions. On the wet weight basis if HCN level in forage sorghum exceed 200 mg kg^{-1} is harmful for livestock (Fjell *et al.,* 1991).

No doubt nitrogen application is considered essential for efficient plant growth and cell division (Duli *et al.,* 2005; Saraswathy *et al.,* 2007). However, higher level of nitrogen application may increase HCN contents of forage sorghum and thus poisoning animals (Aziz and Abdel-Gwad, 2008). Application of nitrogen fertilizer also increases brix value (Pholsen and Sornsungnoen,

2004). Different growth stages of sorghum produce varying levels of sugar in stem quantitatively and qualitatively specifically maximum sugar content are accumulated at maturity stage (Parvatikar and Manjunath, 1991). Nitrogen fertilizer promotes sucrose content, protein percent and growth rate in sweet sorghum (Tsialtas and Maslaris, 2005).

Sulphur is an essential element of nutrition for plant and animal whose deficiencies have been increasing throughout the world (Platou and Irish, 1982). The use of S-free mineral fertilizers has created widespread deficiency of Sin soils (Chaubey *et al.,* 1992). Sulphur fertilization can increase the yield and the quality of forages in the areas of S deficient soils (Puoli *et al.,* 1991; Hallmark and Brown, 1994). Moreover, S fertilization is known to improve nitrogen (N) utilization efficiency by the crops (Schnug and Haneklaus, 1993).

Most of the scientific work had focused on nitrogen but little attention had been paid to sulphur. Keeping in view this scenario, the present study was undertaken to explore the hydrocyanic acid and sugar content dynamics under nitrogen and sulphur application to forage sorghum cultivars sown under sub-tropical dryland conditions of Pakistan.

MATERIALS AND METHODS

Experiment was carried out at University Research Farm (URF) Chakwal Road (33° 56' N, 72° 52' E, 498 m a.s.l.) , PMAS- Arid Agriculture University Rawalpindi, during 2008, 2009 and 2010 to determine the hydrocyanic acid and sugar content dynamics with varying nitrogen and sulphur application to forage sorghum cultivars. The treatments consisted of two forage sorghum cultivars i.e. JS-2002 and Chakwal sorghum were sown with each three nitrogen levels (0, 60 and 120 kg ha^{-1}) in the form of urea and sulphur levels (0, 20 and 40 kg ha^{-1}) as gypsum arranged in randomized complete block design replicated four times.

Before sowing, soil sampling was carried out at two depths i.e. 0-15 and 15-30 cm from experimental site for physico-chemical properties. The soil was sandy-loam with pH about neutral. The total N content and organic matter were low, particularly in the lower layers of soil i.e. 15-30 cm. While phosphorus status (less than 5 mg P kg^{-1}) was observed (Olsen method): poor in both soil layers. Mean temperature (°C), rainfall (mm) and relative humidity (%) data recorded during the study period is presented in Figure 1.

Determination of total cyanide through spectrophotometer

Leaf samples were chopped and ground with pestle and mortar. For calibration, a flat bottom plastic bottle with buffer of pH 6 and round filter paper disc were loaded. Grinded leaf sample (100 mg) was taken in the transparent bottle and poured 1 mL phosphate buffer solution in it. A yellow picrate paper was attached with the plastic stripe to absorb the HCN vapors and the upper end of stripe is further attached in the center of bottle lid with the help of tape and hanged in the

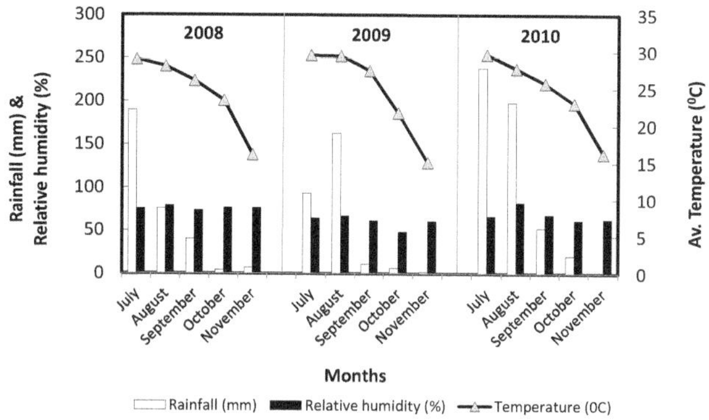

Figure 1. Rainfall, av. temperature and relative humidity recorded during crop growth period

bottle in such a way that picrate paper may not come in contact with liquid (leaf sample + buffer solution) lying in the bottom of bottle. A blank was also run having no leaves. Both the bottles were left for overnight at the room temperature. Next day the picrate paper was removed from plastic stripe cautiously. For about 30 minutes, the picrate paper was submerged into 5 mL distilled water with mild shaking. At 510 nm wavelength absorbance picrate solution was recorded by spectrophotometer as described by Bradbury *et al.* (1999). The hydrocyanic acid contents were calculated as under:

Total cyanide contents (mg kg^{-1}) =396 × absorbance reading.

Sweetness (%fresh stalk)

Sugar content (brix value) of forage sorghum was determined with the help of a refractometer (hand held brix refractometer, Sino Technology, Fujian, China), by the method of Yun-long *et al.* (2006).

Five plants were selected at random from each plot. The juice from the main stalk of each plant was extracted by simply twisting the stem. The extract of all five stalks was mixed and 0.5 ml of the mixed juice sample was applied to the automatic temperature compensating refractometer and the total sugar contents (as brix grades) were determined.

The HCN leaf content and sugar content were recorded at two growth stages viz. booting and 50 % heading.

The data collected were subjected to analysis of variance using the CoStat 6.3 software (CoHort Software, Monterey, CA, USA) to determine significance of the tested factors (cultivar, sulphur and nitrogen levels) and their interactions with sorghum traits. Means of significant effects were separated using the test at 5% level of significance.

RESULTS AND DISCUSSION

Hydrocyanic acid content

Hydrocyanic acid (HCN) content was observed in sorghum cultivars and results showed a significant difference (Table 1). The Chakwal sorghum produced the highest HCN content (22.6 and 15 mg/100 g) while the lowest was 11.5 and 9.0 mg/100 g by JS-2002 cultivar at booting and 50% heading stage, respectively. From the data, it appeared that 27-40% less HCN content was found in Js-2002 when compared with Chakwal sorghum at respective growth stages. The difference of HCN content may be due to the genetic makeup of different sorghum cultivars and climatic condition during the crop growth period at different years. The results are in agreement with findings of Hanuman *et al.* (2008) who reported that HCN content variation may depend on their genetic makeup of cultivars. Abusuwar and Hala (2010), Pandey *et al.* (2011), Sarfraz *et al.* (2012) and Sher *et al.* (2012) found higher HCN content in Abu Sabein, hybrid variety, Hegari and local sorghum, respectively than the other varieties. In our findings, Chakwal sorghum produced higher HCN content as compared with JS-2002. This decrease in HCN contents might be due to more green forage yield (data not shown) and sugar content.

Table 1. Interactive effect of Cultivar, Sulphur and Nitrogen application on hydrocyanic acid content (HCN) of sorghum forage in URF during summer 2008, 2009 and 2010

Treatments	**Growth stages**	
	Booting	**50% heading**
Cultivars (C)		
JS-2002	16.5 b	9.0 b
Chakwal sorghum	22.6 a	15.0 a
P	<0.001**	<0.001**
Sulphur Levels (S; Kg ha^{-1})		
0	28.4 a	17.1 a
20	17.9 b	11.0 b
40	12.3 c	8.0 c
P	<0.001**	<0.001*
Nitrogen Levels (N; Kg ha^{-1})		
0	17.8 c	11.0 c
60	19.3 b	11.7 b
120	21.5 a	13.4 a
P	<0.001**	<0.001**
Year (Y)		
2008	19.6 b	11.9 b
2009	21.9 a	13.1 a
2010	17.2 c	11.0 c
P	<0.001**	<0.001**
C × S		
P	1.00 ns	1.00 ns
C × N		
P	1.00 ns	1.00 ns
S × N		
P	<0.001**	<0.001**
C × S × N		
P	1.00 ns	1.00 ns
C × Y		
P	0.164 ns	0.748 ns
S × Y		
P	0.448 ns	0.605 ns
C × S × Y		
P	1.00 ns	1.00 ns
N × Y		
P	0.981 ns	0.995 ns
C × N × Y		
P	1.00 ns	1.00 ns
S × N × Y		
P	0.999 ns	0.992 ns
C × S × N × Y		
P	1.00 ns	1.00 ns
C.V. (%)	9.9	14.4

ns, *, ** mean non-significant, significant at $P \leq 0.05$ and $P \leq 0.01$, respectively
Different letters indicate significantly-different means (SNK test; $P \leq 0.05$)

Regardless to cultivars, increase in sulphur levels reduced the HCN content. Minimum HCN content of 12.3 and 8.0 mg/100g was recorded with the application of 40 kg S ha^{-1} whereas maximum was 28.4 and 17.1 mg/100g in control plot at booting and 50% heading stage, respectively. The increase rate of sulphur resulted is 53-57% less HCN content in comparison with control plots. The reason might be the sulphur as an important component of chlorophyll, amino acids, enzymes and seed formation (Sharma, 1991).

With the increase in nitrogen fertilizer HCN content also gradually increased in sorghum plants (Table 1). Fertilizing sorghum forage with 120 kg N ha^{-1} produced the highest HCN content (21.5 and 13.4 mg/100g) at booting and 50% heading stage, respectively while the lowest HCN content of 17.8 and 11.0 mg/100g was recorded in control treatment where no fertilizer was applied. The increase of HCN content (21-22%) was more with the application of 120 kg N ha^{-1} when compared with control treatment (0 kg N ha^{-1}). No doubt N is an important component for plant growth and development (Shehu *et al.*, 2010). In sorghum plant increase in N level significantly increased HCN content which can be harmful, therefore it is important to explore the genetic source of sorghum which has less HCN trigger towards high N application. The similar results have been reported by Sher *et al.* (2012) that with the application of @ 120 kg N ha^{-1} produced 64% more HCN content than control plot. Bahrani and Ghenateghestani (2004) also reported an increase of 55% higher HCN content with nitrogen application in forage sorghum.

Seasonal variations produced minimum HCN content of 17.2 & 11.0 mg/100g during 2010 as compared to 2009 probably as result of higher seasonal rainfall. A reduction of 22 and 16% less HCN content was recorded during 2010 than 2009 at booting and 50% heading stage, respectively. In our experiments the HCN content does not exceed the safe limit during entire crop growth duration. The toxin (HCN) can exceed the safe limit (>200 mg kg^{-1}) if animals are fed for long time on the pre- mature stages and forage produced under low rainfall conditions (Sher *et al.*, 2012).

Significant interaction of sulphur and nitrogen application was observed at both growth stages (booting & 50% heading) during three years of study (Figure 2). Maximum HCN content of 31.9 and 19 mg/100g was recorded with the application of 120 kg N ha^{-1} and in control plots where no sulphur was applied at respective growth stages, respectively.

Brix value (%fresh stalk)

Brix value is an important characteristic to determine the forage quality. Significantly differences were recorded in the brix value of forage sorghum cultivars exposed to varying nitrogen and sulphur applications grown under subtropical rainfed regions during 2008 to 2010 (Table 2). Sorghum cultivar JS-2002 produced highest brix value of 13.3 and 23% whereas Chakwal sorghum resulted lowest brix value of 11.3 and 21.3% at booting and 50% heading stage, respectively. The results indicated 8 and 18% higher brix value in JS-2002 than Chakwal sorghum at booting and 50% heading stage, respectively. Mahammed and Moataz (2009) found a significant difference of brix value among different genotypes. Similarly, Almodares *et al.*, (2007) reported that the difference of brix value might be due to genetic variability and stem thickness. Present work indicated that sorghum cultivar JS-2002 produced highest brix value than Chakwal sorghum.

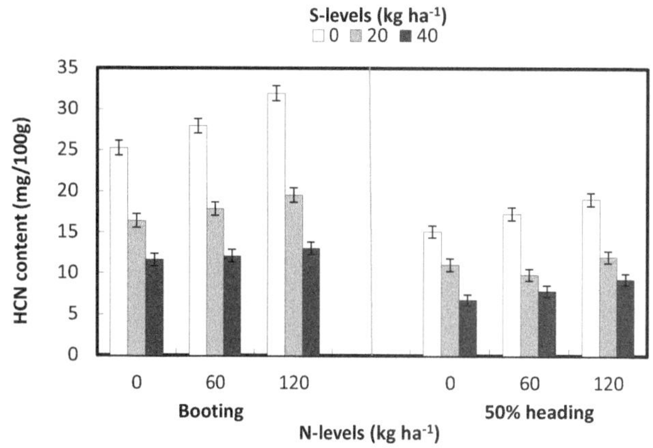

Figure 2. Effect of nitrogen and sulphur levels interaction on hydrocyanic acid content (HCN) at booting and 50% heading stage. Error bars represent the standard error (n = 5).

Increasing sulphur application progressively increased brix value of forage sorghum cultivars throughout the experimentation period under field conditions. Higher brix value of 13.3 and 23.8% was recorded with the application of 40 kg S ha^{-1} while lower (11.5 and 20.5%) in control treatment at respective growth stages, clearly indicating the positive affect of S application on forage sorghum. Regardless to sulphur, the highest brix value of 15 and 27.3% was observed with the application of 120 kg N ha^{-1}, which produced 56-59% higher brix value at booting and 50% heading stage, respectively than control treatment (9.6 and 17.2%) where no N was applied. Pholsen and Sukrsi (2007) found a range of brix value between 11 and 12.5% at N-K_2O rates. In our experiment, an average range of brix value (9.6 to 27.3%) was recorded at different nitrogen levels during study period. In contrast Almodares *et al.*, (2007, 2008) found no significant difference with the application of different nitrogen levels on brix value.

Seasonal variations produced maximum sugar content of 14.9, 25.6% at booting and 50% heading stage during 2010 as compared to 2009, respectively probably as result of seasonal rainfall variation as depicted in Figure 1.

Table 2. Interactive effect of Cultivar, Sulphur and Nitrogen application on brix value of sorghum forage in URF during summer 2008, 2009 and 2010

Treatments	Growth stages	
	Booting	50% heading
Cultivars (C)		
JS-2002	13.3 a	23.0 a
Chakwal sorghum	11.3 b	21.3 b
P	<0.001**	<0.001**
Sulphur Levels (S; Kg ha^{-1})		
0	11.5 c	20.5 c
20	12.1 b	22.2 b
40	13.2 a	23.8 a
P	<0.001**	<0.001*
Nitrogen Levels (N; Kg ha^{-1})		
0	9.6 c	17.2 c
60	12.1 b	22.0 b
120	15.0 a	27.3 a
P	<0.001**	<0.001**
Year (Y)		
2008	13.1 b	23.1 b
2009	8.8 c	17.7 c
2010	14.9 a	25.6 a
P	<0.001**	<0.001**
C × S		
P	0.800 ns	0.264 ns
C × N		
P	<0.046*	0.544 ns
S × N		
P	0.167 ns	0.106 ns
C × S × N		
P	0.616 ns	0.288 ns
C × Y		
P	<0.017*	<0.002*
S × Y		
P	<0.021*	<0.022*
C × S × Y		
P	0.717 ns	0.827 ns
N × Y		
P	<0.001**	<0.001**
C × N × Y		
P	0.672 ns	0.184 ns
S × N × Y		
P	0.370 ns	<0.047*
C × S × N × Y		
P	0.685 ns	0.968 ns
C.V. (%)	8.7	5.1

ns, *, ** mean non-significant, significant at $P \leq 0.05$ and $P \leq 0.01$, respectively
Different letters indicate significantly-different means (SNK test; $P \leq 0.05$)

Interaction between cultivar and nitrogen levels showed statistically significance difference in brix values at booting stage (Figure 3) while 50% heading stage remained non-significant. Maximum brix value 16.2% was recorded by JS-2002 with 120 kg ha^{-1} nitrogen fertilizer application while minimum was 8.8% in Chakwal Sorghum where no nitrogen was made. Significant interaction of C × Y, S × Y and N × Y were also found to be significantly different at booting and 50% heading stages (Figure 4.a, b, c). During 2010, the highest brix value of 16.1 and 26.7% was observed in JS-2002 while the lowest 8.1 and 17.4% was in Chakwal sorghum during 2009. Similarly, the highest brix value of 16.2 and 27.6% and 18.2 and 31.8% was recorded where sulphur and nitrogen were applied @ 40 and 60 kg ha^{-1} in 2010. Interactive effect of S × N × Y was found maximum brix value of 34.3% was recorded when sulphur and nitrogen was applied @ 40 and 60 kg ha^{-1}, respectively during the year 2010 while minimum 13.5% was in control plots of both nutrients in 2009 (Figure 5).

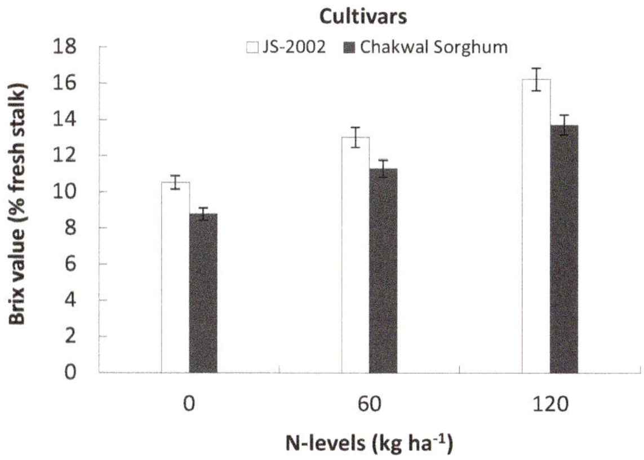

Figure 3. Effect of cultivar and nitrogen levels interaction on brix value (% fresh stalk) at booting stage. Error bars represent the standard error (n = 5).

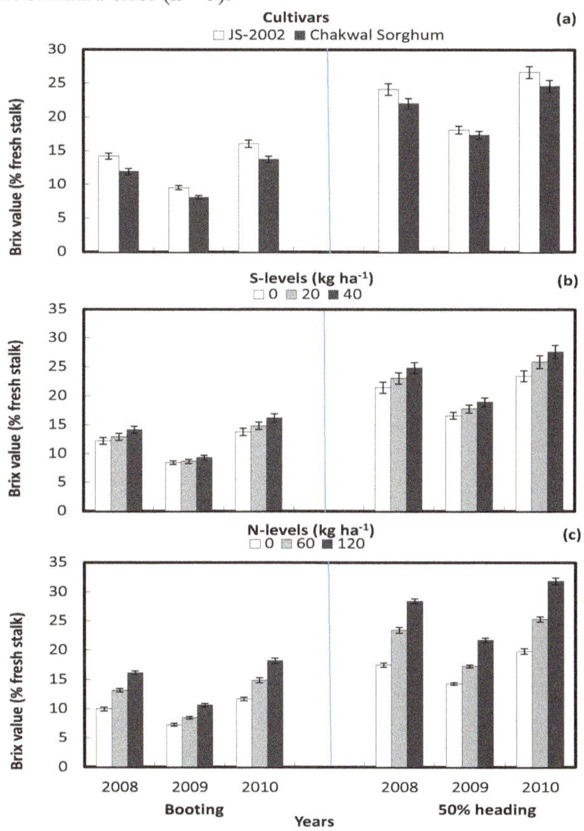

Figure 4. Effect of cultivar, sulphur and nitrogen levels interaction on brix value (% fresh stalk) at booting and 50% heading stage during 2008, 2009 and 2010. Error bars represent the standard error (n = 5).

CONCLUSIONS

The finding of three year field investigations clearly demonstrated that increased application of N resulted in increased content of HCN and brix in the forage sorghum. Therefore there is a need to explore sorghum genetic source which can result higher green forage yield with low HCN content. The application of S increased contents of brix while reduced the HCN content irrespective of sorghum cultivars. Therefore, our research work is a step towards investigation to explore balanced N and S fertilization ratio for reduced HCN content and increased sugar content in sorghum either for livestock or milling for industrial purposes.

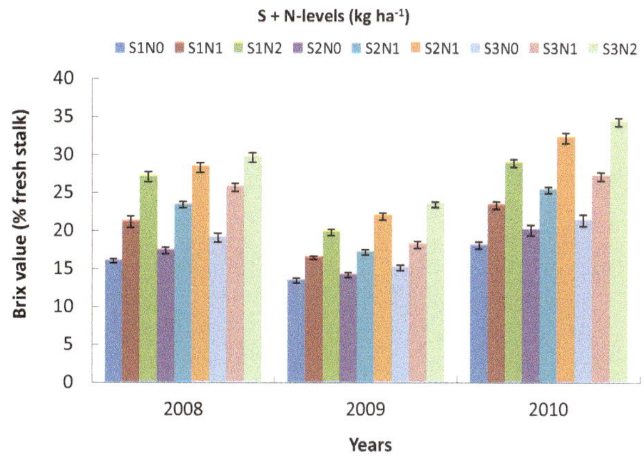

Figure 5. Effect of nitrogen and sulphur levels interaction on brix value (% fresh stalk) at 50% heading stage during three consecutive years (2008, 2009 & 2010). Error bars represent the standard error (n = 5).

LITERATURE CITED

Abusuwar, A.O. and Hala AEl-Zilal. 2010. Effect of chicken manure on yield, quality and HCN concentration of two forage Sorghum (*Sorghum bicolor* (L) Moench) cultivars. Agric and Biol. J. North America. Pp. 27-31.

Almodares, A., M.R. Hadi, M. Ranjbar and R. Taheri. 2007. The effects of nitrogen treatments, cultivars and harvest stages on stalk yield and sugar content in sweet sorghum. Asian J. Plant Sci. 6(2): 423-426.

Almodares, A., R. Taheri, I.M. Chung and M. Fathi. 2008. The effect of nitrogen and potassium fertilizers on growth parameters and carbohydrate contents of sweet sorghum cultivars. J. Environ. Biol. 29(6): 849-852.

Aziz-Abdel, T.K. and M.A.S. Abdel-Gwad. 2008. Yield and quality of tunis grass as affected by plant height at cutting and N fertilizer. Animal Agric. Sci. (Cairo). Pp. 157-169.

Bahrani, M.J. and A.D. Ghenateghestani. 2004. Summer forage sorghum yield, protein and prussic acid contents as affected by plant density and nitrogen top dressing. J. Agric. Sci. Tech. Pp. 73-83.

Bradbury, G.M., S.V. Egan and J.H. Bradbury. 1999. Picrate paper kits for determination of total cyanogens in cassava roots and all forms of cyanogens in cassava products. J. Sci. Food Agric. Pp. 593-601.

Chaubey, A.K., K.N. Dwivedi and R.S. Yadav. 1992. Effect of N, P and S on linseed. J. Ind. Soc. Soil Sci. 40: 758-761.

Duli, Z.K., V.R. Reddy, G. Kakani and V.R. Reddy. 2005. Nitrogen deficiency effects on plant growth, leaf photosynthesis and hyper spectral reflectance properties of sorghum. Europ J. Agron. 22: 391-403.

Fjell, D., B. Dale and T. Gene. 1991. Nitrate and prussic acid toxicity in forage causes, prevention and feeding management, pp: 1–4. Cooperative extension service, Kansas state university, Manhattan, Kansas, USA.

Hallmark, W.B. and L.P. Brown. 1994. Interactive effects of sulphur and potassium fertilization on Bermuda grass hay yields. Sulphur Agric. 18: 41-44.

Hanuman, S., S. Pushpendra and H.K. Sumerhya. 2008. Effect of fertility levels on fodder yield and HCN content of fodder sorghum (*Sorghum bicolor* (L) Moench) genotypes. Inter J. Tropical Agri. Pp. 417-420.

Lang, B. 2001. Sudan/sorghum forage management. Iowa state University.

Mohammed, M.I. and M.A. Moataz. 2009. Evaluation of newly developed sweet sorghum (*Sorghum bicolor*) genotypes for some forage attributes. American-Eurasian J. Agric. Environ. Sci. 6(4): 434-440.

Pandey, R.K., D. Kumar and K.M. Jadhav. 2011. Assessment of determinants for reducing HCN content in sorghum used for ruminant in Gujarat, India. Livestock Res. Rural Development. 23 (3), *Article #66*

Parvatikar, S.R. and T.V. Manjunath. 1991. Alternate uses of sweet sorghum, a new prospectus for juicy stalks and grain yields. J. Maharashtra Agric. 16: 352-354.

Pederson, G.A., T.E. Fairbrother and S.L. Greene. 1996. Cyanogenesis and climatic relationships in U.S. white clover germplasm collection and core subset. Crop Sci. Pp. 427–433.

Pholsen, S. and N. Somsungnoen. 2004. Effect of nitrogen and potassium rates and planting distances on growth, yield and fodder quality of a forage sorghum (*Sorghum bicolor* L. Moench). Pak. J. Bio. Sci. 7(10): 1793-1800.

Pholsen, S. and A. Suksri. 2007. Effect of phosphorus and potassium on growth, yield and fodder quality of IS 23585 forage sorghum cultivar (*Sorghum bicolor* L. Moench). Pak. J. Bio. Sci. 10(10): 1604-1610.

Platou, J.S. and R. Irish. 1982. The fourth major nutrient. The Sulfur Institute, Washington, DC.

Puoli, J.R., G.A. Jung and R.L. Reid. 1991. Effects of nitrogen and sulphur on digestion and nutritive quality of warm-season grass hays for cattle and sheep. J. Anim. Sci. 69: 843-852.

Reddy, B.V.S., S. Ramesh and P.S. Reddy. 2004. Sorghum breeding research at ICRISAT- goals, strategies, methods and accomplishments. International Sorghum and Millets Newsletter 45, 5-12.

Saraswathy, R., S. Suganya and P. Singaram. 2007. Environmental impact of nitrogen fertilization in tea eco-system. J. Environ. Biol. 28: 779-788.

Sarfraz, M., N. Ahmad, U. Farooq, A. Ali and K. Hussain. 2012. Evaluation of sorghum varieties/lines for hydrocyanic acid and crude protein contents. J. Agric. Res. Pp. 39-47.

Schnug, E. and S. Haneklaus. 1993. Impact of sulphur fertilization on fertilizer nitrogen efficiency. Sulphur Agric. 17: 8-12.

Sharma, B.K. 1991. Industrial chemistry. Krishna Prakashan. Goel publishing house Meerut.

Shehu, H.E., C.S. Ezekiel, J.D. Kwari and M.K. Sandabe. 2010. Agronomic Efficiency of N, P and K Fertilization in sesame (*Sesamum indicum*) in Mubi Region, Adamawa State, Nigeria. Nature and science. 8(8): 257-260.

Sher, A., M. Ansar, F.U. Hassan, G. Shabbir and M.A. Malik. 2012. Hydrocyanic acid contents variation amongst sorghum cultivars grown with varying seed rates and nitrogen levels. Int. J. Agric. Biol. 14: 720-726.

Tsialtas, J.T. and N. Maslaris. 2005. Effect of N fertilization rate on sugar yield and non-sugar impurities of sugar beets (*Beta vulgaris*) grown under Mediterranean conditions. J. Agron. Crop Sci. 191: 330-339.

Yun-long, B., Y. Seiji, I. Maiko and C. Hong-Wei. 2006. QTLs for sugar contents of stalks in sweet sorghum (*Sorghum bicolor* L. Moench). Agric. Sci. China. 5(10): 736-744.

MORPHOLOGICAL AND CHEMICAL CHARACTERS OF *Bituminaria bituminosa* (L) C.H. (Stirtion) GROWN NATURALLY IN THE MIDDLE BLACK SEA REGION

Erdem GULUMSER[*1], Zeki ACAR*[2]

[1] *Bozok University Agriculture and Natural Science Faculty, Yozgat, TURKEY*
[2]*Ondokuz Mayıs University Agricultural Faculty, Samsun, TURKEY*
**Corresponding author:* erdem.gulumser@bozok.edu.tr

ABSTRACT

In this study, morphological and chemical characters were investigated in *Bituminaria bituminosa* (L.) C.H. Stirtion genotypes collected from 24 different locations Middle Black Sea Region Turkey. The experiment was established in Samsun ecological conditions and carried out for two years (2009-2010). The investigated characters were; plant height, shoot number, main stem diameter, leaflet width and length, leaf number, inflorescence number, flower number per inflorescence, seed width and length, a thousand seed weight, dry matter yield, crude protein content, ADF, NDF, ADP values, ash, Ca, Mg, P, K, Fe, Zn and Mn contents. Only one cut was done in the first year as plant growth was relatively slow. Two cuts were performed in the second year due to greater regrowth ability. Plant height was 88.4 cm in 2009 and, was 80.5 and 74.2 cm for the first and second cutting in 2010 year. Main stem diameter in three cut in both years were recorded as 5.2 cm, 6.3 cm and 3.6 cm, respectively Chemical characters of the genotypes were examined in both leaves and stems, separately. Protein ratios in the leaves were found as 24.30%, 24.60% and 19.61%, respectively, These protein ratios remarkably decreased in the stems; 13.28%, 13.58% and 8.99%.

Key words: *Bituminaria bituminosa*, morphological and chemical characters.

INTRODUCTION

In Turkey, despite the significant decrease in animal number, especially goats and sheep in recent years, there is still a gap in good quality forage (Sabancı et al. 2010). This gap is very high especially during summer period due to the low pasture yield.

In Turkey, cultivated forage crops and plants existing naturally in pastures are cool season plants. Pasture plants are dormant or dry, because of high temperature and drought in June-September period which is resulting serious forage gap for the animals grazing on pasture.

Psoralea genus is a legume that this genus has 150 species (Hooker and Jackson, 1960). *Bituminaria bituminosa* (*Bb*) is a perennial species within Psoralea genus. It's origin is natural distribution of this plant are in coastal cities, located north west and south Anatolia. (Davis 1988; Akcin et al. 2010, Peters and Korkmaz. 1990; Kilinc et al. 1998).

Bb is cultivated in the Canary Islands and Morocco. In general, it grows on road sides, in shallow soils, in woodland and forests (Davis 1988). It can resist to -10°C (FAO 2010). Up to 1500 m altitude (FAO 2010). *Bb* is be able to green throughout the hot and dry summer period (Acar et al. 2001). *Bb* is a long day plant, flowering and set seeds mainly in the spring but flowering is going on the whole year except winter period (FAO 2010). *Bb* is improves the quality of pasture in the hot and dry periods that other species and grasses disappear or decrease their hay quality (Mendez and Fernandez, 1990).

According to global climate change scenario, Drought and high summer temperatures are expected in Turkey (Anon 2011). For this reason, we need new genotypes and plant cultivars which are resistant to summer drought and effective to increase pasture yield and quality. It is expected that *Bb* will be important for forage production and pasture improving in the future. However there are very few studies about its potential as forage and pasture plant in Turkey.

MATERIALS AND METHODS

This study was conducted for two years (2009-2010) in Experimental Field of Agriculture Faculty of Ondokuz Mayis University.

Plant Materials were collected from 4 provinces and 24 different locations; Samsun (Örencik, Kurupelit Kampüs 1 and 2, Kavak, Sarıyusuf Sapagı, Cataltepe, Cakallı, Nebyan 1 and 2, Atakent, Havza) , Sinop (Tangal, Gerze, Tıngıroglu, Ayancık and Yarımada), Ordu (Bacanak, Bolaman, Kumru 1 and 2, Kulak, Mesudiye, Caybası) and Zonguldak (Filyos) in 2008 and 2009. After the collection, seeds were cleaned and dried at 30 ^{0}C (Walker et al. 2007).

Each seed material was firstly sown into seed trays then, seedlings transplanted to the field with in 50x50 cm distances (Pecetti et al. 2007). First year, 14 plant transplanted to the field, but only 7 for them investigated (others did not reached at bolting). These seen genotypes, Orencik, Kurupelit Kampus 1, Bacanak, Kumru 1, Kulak, Mesudiye and Caybası). Second year, new 10 genotypes were transplanted to the field, and measurements and observations were done on totally 24 genotypes. Only one cutting was done in the first year as plant growth was relatively slow. Two cuts were performed in the second year due to greater regeneration. Both for one year and two years old plants.

Soil was clay with 1.36 % organic matter and pH: 6.45. Long term (1974-2009) annual rainfall and mean temperature of experimental area were 680.9 mm and 14.3 °C, respectively. Average temperature and total rainfall in growing periods of 2009 and 2010 ranged between 12.2 and 12.9 °C, and 793.6 and 732.9 mm, respectively (Anonymous 2010).

Five plants were randomly harvested in each plot at the 50% flowering stage. Morphological characters; plant height, branch number, main stem , leaflet width and length, leaf number, seed width and length, a thousand seed weight were determined. Then plant samples (leaf and stem) were dried at 60 °C in oven till constant weight, and ground to pass through 1 mm screen and mixed, leaf samples also include flowers. Chemical characters; dry matter, crude protein, ADF, NDF, ADP, Ca, Mg, K, P content of samples were determined by using Near Infrared Reflectance Spectroscopy (NIRS, "Foss 6500") with software package program 'IC-0904F' and Fe, Zn and Mn content of samples were determined by using Atomic Absorption Spectrophotometer (Kacar 1972). All the data were given as a mean, standard error and CV. Principal component analysis (PCA) was carried out using the statistical software package PASW (18) Statistics Data Editor program.

RESULTS AND DISCUSSION

The values about morphological and chemical characters of *Bb* genotypes are given in Tables 1, 2 and 3. Plant height was 88.4, 80.5 and 74.2 cm in the first year and the second year (first and second cuttings), respectively. Our results were generally comparable to findings of Davis (1965) who found that plant height ranged between 20 –140 cm.

Average main stem diameter in the first and second (for the first and second cuttings) years were determined as 5.2, 6.3, 3.6 cm, respectively. Average branch numbers of *Bb* genotypes changed from 12.1 to 14.7 cm (Table 1).

Table 1. According to morphological characteristics of *Bituminaria bituminosa* genotypes mean, standard error (Se) and coefficient of variation (CV) in 2009 and 2010 years.

Morphological Features	First Year		Second year first cutting		Second year second cutting	
	Mean±Se	CV	Mean±Se	CV	Mean±Se	CV
Plant height (cm)	88.4±7.3	17.0	80.5±4.3	12.6	74.2±5	15.3
Mean branch diameter (mm)	5.2±0.4	20.0	6.3±0.4	16.0	3.6±0.3	18.5
Branch number	13.7±1.8	31.2	12.1±1.3	25.5	14.7±2.0	29.6
Leaflet width (cm)	2.1±0.2	16.5	2.4±0.2	15.1	1.9±0.1	15.9
Leaflet length (cm)	4.5±0.2	9.9	4.5±0.2	11.5	3.5±0.2	15.2
Leaf number	254.7±46.1	39.8	261.4±22.7	20.9	158.7±2.0	29.6
Inflorances number	64.7±15.4	58.9	69.6±10.6	35.6	119.0±29.6	51.3
Flower number/Inflorances	11.2±1.1	20.5	16.5±1.1	16.0	5.6±0.4	16.3
Seed length (cm)	14.5±0.8	12.5	14.6±0.7	11.3	14.1±0.4	2.7
Seed width (cm)	3.3±0.2	13.6	3.8±0.1	7.50	3.3±0.1	7.1
Thousand seed weight (g)	29.50±1.02	8.47	30.89±0.50	7.77	31.42±0.43	6.52

Average leaf numbers were 254.7 in 2009, and 261.4 and 158.7 cm for the first and the second cuts in 2010. Inflorescence numbers of *Bb* genotypes were 64.7 in 2009, 69.6 and 119 for the first and the second cuts in 2010. Average flower numbers per inflorescence were 11.2, 16.5 and 5.6 for the consecutive years and cuts. Davis, (1988) found that flower numbers per inflorescence of *Bb* genotypes changed between 7 and 30. Average seed length were 14.5 cm in 2009, and 14.6 first cut and 14.1 second cutting cm in 2010 respectively. Thousand seed weights were 29.5, 30.89, 31.42 g in 2009 and the first and the second cuts of 2010, respectively.

Bb plants are mostly evaluated for grazing. Livestocks preferably graze its leaves and fresh sprouts, so chemical properties of leaves and stems were analysed separately.

Average crude protein ratios of genotypes were 24.30 %, 24.60 %, 19.61 % for leaf and 13.28 %, 13.58 %, 8.99 % for stem according to years and cuts (Table 2).

In previous studies, it was reported that crude protein of *Bb* genotypes changed between 15.9 % and 24.30% (Ventura et al. 1999, 2004; Acar et al. 2001; FAO, 2010; Gulumser et al. 2010; Gulumser 2011). protein ratio was higher in leaf than stems in all genotypes.

ADF contents of *Bb* genotypes were 21.20, 20.00, 15.40 % for leaf, and 45.27, 41.50, 43.90 % for stem in 2009 and 2010 years (first and second cut), respectively. It was reported that different researchers determined ADF content of different *Bb* genotypes between 21.9 % and 38.8 % and NDF contents between 25.0 and 50.1 % (FAO 2010). Average NDF contents were 25.85, 27.72 and 21.02 % for leaf and 52.00, 48.53 and 50.68 % for stem during the consecutive years and cuttings (Table 2).

Generally, leaf crude protein ratio was as higher as twice of stem (Table 2, 3). In terms of chemical content, differences between leaf and stem are expected consequences of anatomic and physiological structure of plants, especially due to higher cellulose content of stem cells.

Average Ca content of *Bb* genotypes were determined 1.94, 2.11, 2.61 % for leaf and 0.90, 1.13, 1.35 % for stem, and also Mg contents were 0.57, 0.56, 0.70 % for leaf and 0.19, 0.29, 0.32 % for stem during the consecutive years and cuttings.

Table 2. According to chemical properties of *Bituminaria bituminosa* genotypes (leaf)mean, standard error (Se) and coefficient of variation (CV) in 2009 and 2010.

Quality Traits	LEAF					
	FIRST YEAR		SECOND YEAR			
			FİRST CUTTİNG		SECOND CUTTING	
	Mean±Se	CV	Mean±Se	CV	Mean±Se	CV
DM (%)	87.19±0.40	0.13	87.00±0.10	0.71	86.97±0.05	0.31
CP (%)	24.30±0.44	4.85	24.60±0.02	5.21	19.61±0.31	7.44
ADF (%)	21.20±0.63	8.21	20.00±0.33	5.40	15.40±0.32	9.81
NDF (%)	25.85±0.87	8.93	27.72±0.31	5.59	21.02±0.45	1.53
ADP (%)	0.52±0.04	14.2	0.77±0.02	12.9	0.73±0.02	17.8
ASH (%)	-	-	10.02±0.35	17.2	10.30±0.51	23.1
Ca(%)	1.94±0.04	5.15	2.11±0.18	4.21	2.61±0.06	12.2
Mg (%)	0.57±0.02	10.5	0.56±0.01	11.2	0.70±0.01	8.28
K (%)	1.80±0.04	6.66	1.52±0.06	5.85	0.27±0.04	74.0
P (%)	0.39±0.008	5.89	0.36±0.004	6.38	0.26±0.006	11.1
Fe(ppm)	-	-	322.00±52.00	78.8	122.10±14.55	55.6
Zn (ppm)	-	-	104.00±10.06	47.5	96.00±11.62	56.2
Mn (ppm)	-	-	85.00±4.77	23.1	99.90±5.59	26.0

Table 3. According to chemical properties of *Bituminaria bituminosa* genotypes (Stem) mean, standard error (Se) and coefficient of variation (CV) in 2008 and 2009.

Quality Traits	STEM					
	FIRST YEAR		SECOND YEAR			
			FIRST CUTTING		SECOND CUTTING	
	Mean±Se	CV	Mean±Se	CV	Mean±Se	CV
DM (%)	87.91±0.09	0.28	87.50±0.15	0.84	88.62±0.1	0.56
CP (%)	13.28±0.45	9.03	13.58±0.25	5.21	8.99±0.27	14.34
ADF (%)	45.27±0.80	5.09	41.50±0.46	5.42	43.90±1.62	17.35
NDF (%)	52.00±0.97	4.96	48.53±0.51	5.21	50.68±1.04	9.64
ADP (%)	0.88±0.02	14.2	0.96±0.01	8.33	0.99±0.02	9.89
ASH (%)	-	-	7.30±0.19	13.3	6.60±0.30	22.00
Ca(%)	0.90±0.03	9.22	1.13±0.02	8.31	1.35±0.07	23.70
Mg (%)	0.19±0.01	14.2	0.29±0.007	12.7	0.32±0.02	22.81
K (%)	1.81±0.05	7.18	1.92±0.04	11.2	1.28±0.05	18.75
P (%)	0.33±0.003	2.72	0.32±0.004	7.50	0.25±0.008	15.60
Fe(ppm)	-	-	260.32±60.9	114	156.80±38.82	116.0
Zn (ppm)	-	-	55.60±5.85	51.6	37.30±5.59	48.23
Mn (ppm)	-	-	60.90±4.37	31.5	64.30±3.8	26.45

According to Principal Component (PC) analyses based on the investigated traits, PC1 corresponds to 51.88 of the total variation and PC2 to 21.80 % (totally 73.68 %) (Table 4). The obtained scatter plot using these first and second principal components is shown in Figure 1. Considering that all collected samples belong to the same species, namely *B. bituminosa*, it can be concluded that the genotypes collected from different locations exhibit remarkable differences regarding their morphological and chemical traits.

Table 4. Correlations between two components of *Bituminaria bituminosa* genotypes.

	Traits					
Components	**PH**	**BN**	**LN**	**FN/I**	**TSW**	**% Variation**
Component 1	0.896	0.917	0.970	- 0.091	- 0.018	51.88
Component 2	0.063	0.034	- 0.009	0.720	0.753	21.80

PH: plant height (cm); BN: branch number; LN: leaf number; FN/I: flower number/inflorescence; TSW: thousand seed weight (g).

Figure 1; plant height, branch number and leaf number features present the first component, and flower number per inflorescence and a thousand seed weight present the second component. According to principal component analyses, *Bb* genotypes influenced geographical regions and the genotypes collected from similar regions were in the same group (Figure 1).

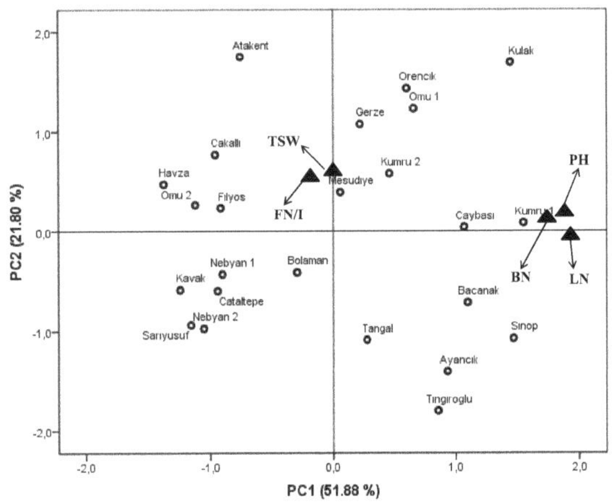

Figure 1. Dispersion of the *Bituminaria bituminosa* genotypes based on the first two principle components

CONCLUSION

Cool season forage crops are cultivated in Turkey. These cool season plants are either thoroughly dry or dormant in summer period. Therefore, the deficiency in substantial forage in high quality occurs from middle of June till end of September. So, the plants can be grown without irrigation or can kept freshness throughout the summer period are important for animal feeding when pasture quality decreased. Present study showed that *Bituminaria bituminosa* (L.) C.H. Stirtion can provide a fresh and high quality feed to animals during summer period.

In addition, *Bb* genotypes can be grown as fodder crops in stony, in clined, poor, eroded and marginal lands, where it is not suitable for other crops. However, further studies are needed test different genotypes under field and range condition. This research is one of the first studies about this area carried out in Turkey, and it will lighten the next studies.

LITERATURE CITED

Acar, Z., Ayan, İ., Gülser, C. Some morphological and nutritional properties of legumes under natural conditions. Pakistan Journal of Biological Sciences, 2001, 4(11): 1312-1315.

Akçin, A.A., Akçin, A., Kutbay, G. A study on flora of Cakmak Dam and its surroundings (Carşamba, Samsun/Turkey). Biological Diversity and Conservation, 2010, 3(1), 28-44.

Anonymous, http//www.antoloji.com/climate-change-2030. (07.13.2011).

Davis, P.H. Flora of Turkey and East Aegean Islands, Vol: 1-10, 1965-1988, Edinburg.

FAO. *Bituminaria bituminosa* C.H. Stirtion. http://www.fao.or/ag/AGP/agpc/doc/Gbase/new_species/tedera/bitbit.htm (01.01.2010).

Gülümser, E., Başaran, U., Acar, Z., Ayan, İ., Mut, H. Determination of some agronomic traits of *Bituminaria bituminosa* accessions collected from Middle Black Sea Region. Options Mediterraennes, 2010, no. 92, p. 105-108.

Gülümser, E. Characterization and determination of agricultural characterictics of *Bituminaria bituminosa* L. genotypes naturally grown at the Central Black Sea Region. 2011, OMU Institute of Science, Master's Thesis (unpublished).

Hooker, J.D., Jackson, D., 1960. Index Kewensis An Enumeration of The Genera and Species of Flowering Plants. Vol. II. Oxford: Oxford University Press. p. 643-645.

Kacar, B., 1972. Plant and Soil Chemical Analysis. II. Plant Analysis. Ank. Univ. Zir. Faculty. Yay. No.: 453 Implementation Guide No.: 155 Ankara.

Kılınç, M., Kutbay, H.G., Akçin, A. Flora of Samsun and Surrounding Kocadağ. XIV. National Congress of Biology, September 7 to 10, 1998, Vol I, pp: 95-111, Samsun.

Mendez, P., Fernandez, M., 1990. Interest forage varieties of *Bituminaria bituminosa* (L.) Stirton (tedera) of the Canary Islands. In 'XXX Scientific Meeting of the Spanish Society for the study of pastures'. Donostia-San Sabastian p. 264-272

Pecetti, L., Tava, A., Pagnotta, MA., Russi, L. Variation in forage quality and chemical composition among Italian accessions of *Bituminaria bituminosa* (L.) Stirt. Journal of the Science of Food and Agriculture, . 2007, 87. 985-991.

Peters, A. and Korkmaz, H. Gorge, Reservoir area Altınkaya, Sahinkaya, Bafra the region between the Dam Body (Down Reservoir Area), and Flora I. Surroundings of the road X. National Congress of Biology. July 18 to 20, 1990, Erzurum.

Sabancı C.O., Baytekin H., Balabanlı C., Acar Z. Opportunities for increasing forage crops production. Turkey Agricultural Engineering VII. Technical Conference, 11-15 January, Ankara, 2010, p. 343-360.

Ventura, M.R., Flores, M.P., Castanon, J.I.R. Nutritive value of forage shrubs: *Bituminaria bituminosa.* Acacia salicina and Medicago arborea. In 'Cahiers Options Mediterraneennes. 1999, 39: 171-173.

Ventura, M.R., Castanon, J.I.R., Pieltain, M.C., Flores, M.P. Nutritive value of forage shrubs: *Bituminaria bituminosa, Rumexlunaria, Acacia salicina, Cassia sturtii* and *Adenocarpus foliosus*. Small Rumin. 2004, Res. 52, 13–18.

Walker, D.J., Bernal, M.P., Correal. E., 2007. The influence of heavy metals and mineral nutrient supply on *Bituminaria bituminosa.* Water Air Soil Pollut. 184. 335–345.

EFFECT OF NITROGEN FERTILIZERS AND OIL ADJUVANTS ON NICOSULFURON EFFICACY

Robert Idziak, Zenon Woznica*

Poznan University of Life Sciences, Faculty of Agronomy and Bioengineering, Department of Agronomy Dojazd, Poznan
**Corresponding author: robertid@up.poznan.pl*

ABSTRACT

Effect of oil adjuvants and nitrogen fertilizers on weed control with nicosulfuron was investigated at two locations in the years 2006 – 2008. Treatments included nicosulfuron at 60 g ha^{-1} without adjuvant and at 30 g ha^{-1} with or without methylated seed oil (MSO) and petroleum (PO) and additionally with and without ammonium nitrate (AMN), urea (U) or urea – ammonium nitrate liquid fertilizer (UAN). Reduction of nicosulfuron rate from 60 to 30 g ha^{-1}, except *Echinochloa crus-galli;* significantly decreased *Chenopodium album, Viola arvensis, Polygonum convolvulus*, and *Polygonum aviculare* control. *E. crusgalli* control ranged from 93 to 99% regardless of herbicide and adjuvant treatment. However, addition of oil adjuvants, especially MSO strongly increased weed control, especially when used with AMN and UAN nitrogen fertilizers. Reduction of nicosulfuron rate from 60 to 30 g ha^{-1} strongly reduced grain yield of maize. However, addition of PO and MSO oil adjuvants, especially MSO applied in a mixture with AMN or UAN nitrogen fertilizers completely overcome this effect and allowed to obtain comparable grain yield of maize as obtained after application of recommended nicosulfuron rate (60 g ha^{-1}).

Key words: Methylated seed oil, mineral adjuvant, sulfonylurea herbicide, weed control.

INTRODUCTION

During the first weeks after sowing, maize (*Zea mays* L.) plants are particularly exposed to weed competition (Gosheh et al., 1996; Evans et al., 2003), including such persistent species as *Chenopodium album* L. and *Echinochloa crus-galli* (L.) Pal. Beauv. The presence of weeds during this period may result in considerable maize yield loss. According to Wilson and Westra (1991), the occurrence of *Panicum dihotomiflorum* Michx. in a maize field during the first two weeks after sowing, caused a 10% decrease in the crop yield.

Poor maize competitiveness with weeds makes human intervention necessary. The use of herbicides is the most effective method of weed control in maize fields. In the 1970s, sulphonylurea herbicides were discovered and introduced into the market. They soon became very popular as they control numerous persistent weed species, were safe for cultivated plants and the environment, and were characterised by low toxicity to mammals (Russell et. al., 2002).

Nicosulfuron belonging to sulfonylurea derivatives is an acetolactate synthase ALS inhibitor (EFSA, 2012; HRAC, 2012), which blocks the production of amino acids, such as valine, leucine and isoleucine, and, as a result, it interferes with the formation of proteins and other functional plant components. This herbicide was registered to post-emergent applications to control grass and some dicot weeds in maize. When used in recommended rates, nicosulfuron is safe for both maize plants and the environment. Nicosulfuron-containing herbicides are available in solid (Greek and Hale, 2005) and liquid formulations (Waligora and Szpurka, 2009). Nicosulfuron is a weak acid with the dissociation constant pK_a at a level of 4.3 (Greek and Hale, 2005; Regitano and Koskinen, 2008). Therefore, sulfonylurea herbicides are weakly absorbed by soil, especially with $pH > pK_a$ (Martins and Mermoud, 1999). Nicosulfuron may be used together with non-ionic surfactants, petroleum oils, methylated ester of plant oils, often (with addition of urea ammonium nitrate (UAN) or ammonium nitrate fertilizer (AMS)) or basic (high pH) blend adjuvants for postemergence weed control (EXTOXNET, 2010).

Adjuvants added to spraying liquid are used to improve the effectiveness of foliage-applied herbicidal treatment (Hazen, 2000; Penner, 2000). This improvement mostly occurs when these preparations are used at reduced rates and in disadvantageous environment conditions (Praczyk and Adamczewski, 1996). The action of adjuvants consists mainly in increasing the retention of spray droplets, plant surface wetability and absorption of herbicide from spray deposit on plant surface into their cells (Sanyal et al., 2006).

The objective of this study was to determine the effectiveness of different oil and nitrogen fertilizer adjuvants on efficacy of nicosulfuron applied at reduced rate and to determine this effect on maize grain yield.

MATERIALS AND METHODS

Field trials were conducted during the 2006, 2007 and 2008 growing seasons at Brody and Zlotniki, located near Poznan, Poland. Experimental fields were prepared according to good experimental practice – moldboard plowed in the fall, and shallow cultivated in the spring prior to planting. Mineral fertilization was adjusted to the plants' needs, taking into account the nutrient content in the soil. Phosphorus and potassium were applied in both locations prior to plowing in the fall at 60 and 117 kg ha^{-1}, respectively. A broadcast application of nitrogen fertilizers in Brody and Zlotniki, was at 140 kg ha and 160 kg ha^{-1}, respectively. Each year maize was planted in the first 10 days period of May. Soil texture, soil organic matter content, soil pH, corn hybrid, seeding date, seeding rate and herbicide application dates are presented in Table 1.

Table 1. Soil characteristics, corn hybryd, seeding date, seeding rate and spray dates for experiments conducted in Brody and Zlotniki, PL, in 2006 – 2008

Year	Location	Soil texture	Soil organic matter (%)	Soil pH	Corn hybrid	Seeding date	Seeding rate (seeds ha)	Spraying date
2006	Brody	sandy loam	1.5	6.1	Fido	April 20	75,000	May 17
	Zlotniki	sandy loam	1.1	5.3	Claryca	April 28	84,000	May 22
2007	Brody	sandy loam	1.5	6.0	PR39T45	April 17	80,000	May 21
	Zlotniki	sandy loam	1.1	5.3	Claryca	April 13	84,000	May 11
2008	Brody	sandy loam	1.2	6.1	PR39B56	April 18	80,000	May 16
	Zlotniki	sandy loam	1.0	5.8	PR39K	April 15	84,000	May 15

Field experiments at each localization were established as a randomized complete block design with 4 replication, on 2.5 x 10 m plots, consisting of 4 rows of plants with 70 cm spacing. Seeds were planted at a depth of 4 cm, with 19.5 cm spacing in a row.

Treatments included an untreated check and nicosulfuron (Accent 75 WG, DuPont Poland) at 60 g ha^{-1} (recommended rate) and 30 g ha^{-1} (reduced rate). Reduced rate of nicosulfuron was applied alone and with two oil origin adjuvants: 1) MSO – methylated seed oil of rapeseed oil fatty acids with a build-in system buffering pH of the liquid spray liquid at the level of 7.3 – 7.8 plus emulsifier (Atpolan BIO 80 EC, ZPH AGROMIX, Niepolomice, Poland) and 2) PO – emulsified petroleum oil (Atpolan 80 EC by ZPH AGROMIX, Niepolomice, Poland), both at 1.5 L ha^{-1}. Oil origin adjuvants were applied alone or in a mixture with ammonium nitrate (AMN), urea (U), both at 30 g ha^{-1} and urea – ammonium nitrate liquid fertilizer (UAN) at 4 L ha^{-1}, at the 4-5-leaf stage of maize. Herbicide treatments in Brody were performed using a CO_2-pressurized plot sprayer equipped with flat fan nozzles Tee Jet DG 11002, delivering 230 L ha^{-1} of water at 220 kPa, and in Zlotniki with a CO_2-pressurized plot sprayer equipped with flat fan nozzles Tee Jet 11003 VP, calibrated to deliver 200 L ha^{-1} of water at 200 kPa. Herbicide application were made with a 2 m boom equipped with five nozzles spaced 50 cm apart. Application dates and environmental conditions at application (air temperature and humidity) for both locations are listed in Table 2.

Table 2. Herbicide application dates and environmental conditions at application

Year	Location	Application date	Air temperature at treatment °C	Air humidity at treatment %	Mean air temperature 1 WAT* °C	Rainfall 1 WAT* mm
2006	Brody	May 25	23.6	70	19.2	0.1
	Zlotniki	May 22	21.0	60	13.4	14.6
2007	Brody	May 17	21.7	58	14.7	7.6
	Zlotniki	May 11	21.0	75	13.7	42.8
2008	Brody	May 21	24.9	68	20.8	31.9
	Zlotniki	May 15	19.0	51	14.0	7.5

*first 7 days after treatment

Six weeks after treatment weeds were cut off at the soil surface from two 0.7 m^{-2} rectangle in each plot and biomass of each species was determined. Assessment of herbicidal efficacy based on the Henderson-Tilton formula. Estimation of crop injury was rated 2 and 4 weeks after treatment using 0 to 100% scale (0% = no injury and 100% = plant death). Yield was measured at crop maturity by harvesting cobs from two middle rows of each plot and the weight and seed moisture content were recorded. Maize grain yield was expressed at 15% moisture level.

An analysis of variance was conducted on all data. Data were analyzed using procedures in ARM 8 (Agricultural Research Manager, Gylling Data Management, Inc.) manage and statistical summarize software. Variances were partitioned into the fixed effects of year and location, and the interaction of year and location by the fixed effect. Percent weed control data were transformed prior to analysis to correct for unequal variance and then back-trans-formed for presentation in the tables. Mean separation using Protected Tuckey`s Least Significance Difference was used to separate treatment means ($P = 0.05$).

RESULTS AND DISCUSSION

Air temperature at herbicide treatment, depending on years and locations, ranged from 19.0 to 24.9 °C, and relative humidity from 51 to 75% in Zlotniki and from 58 – 70% in Brody. The average temperature during the the first 7 days after application varied from 13.4 – 14.0 °C in Zlotniki and from 14.7 – 20.8 °C in Brody. The amount of rainfall during first week after treatment did not exceed 43 mm in Zlotniki and 32 mm in Brody (Table 2). Good absorption of the active ingredient of herbicides ensuring their efficacy takes place when physiological processes in weeds are sustained at a relatively high level. For most herbicides the optimal air temperature ranges from 10 to 25 ºC at appropriate humidity (Caseley, 1989; Joseph et al., 1996). Generally, the higher temperature the higher activity of herbicides is observed (Kells et al., 1984; Anderson et al., 1993). However, application of herbicide at too high temperatures may reduce their effectiveness due to wilting, closing of the stomata and quick drying of the spray liquid droplets, caused by limited herbicidal absorption (Ritter and Coble, 1981). However not only temperature and humidity influence herbicidal action, but also the weed. For example, development *Amaranthus rudis* Sauer and *Digitaria sanguinalis* L. Scop. was inhibited more effectively by mezotrion at 18 ºC than at 32 ºC, as opposed to, e.g. *Abutilon theophrasti* L. and *Xanthium strumarium* L. (Johnson and Young, 2002). Similar interactions are also pointed out by other authors (Kells et al., 1984). The weather conditions in all the years and in both location during application and several days after were favourable for the herbicidal action.

In total 20 weed species were identified at experimental plots. Weed community varied across locations and years and only weed species occurring during all years in both locations were analysed. There was no injury to corn with the herbicides and adjuvant evaluated (data not shown). The random effects of location, year and their interactions were not significant, therefore weed control and grain yield data were pooled only by treatment.

There were significant differences amongst treatments in herbicidal efficacy on prominent individual weed species based on their percent control by fresh mass reduction. Reduction of nicosulfuron rate from 60 to 30 g ha^{-1}, except *Echinochloa crus-galli*; significantly decreased or tended to decrease *Chenopodium album, Viola arvensis* Murr., *Polygonum convolvulus* L., *Polygonum aviculare* L. and *Geranium pusillum* L. control. For example, nicosufuron at recommended rate of 60 g ha controlled the most occurring weed *Chenopodium album* merely in 80% and at reduced rate of 30 g ha^{-1} a strong decrease in efficacy (to 57%) was observed (Table 3). Nicosulfuron effectively limits the occurrence of monocot weeds, for example *Setaria* species, as well as *Agropyron repens* (L.) Pal. Beauv. or *E. crus-galli*, and some dicot weeds such as *Amaranthus retroflexus* L. and *Fagopyrum esculentum* Moench but it does not control *Chenopodium album* and *Solanum nigrum* L. at satisfactory level (NDWCG, 2010). Our results indicate that nicosulfuron applied alone at recommended or reduced rate even without adjuvants greatly controled only *Echinochloa crus-galli*, control ranged from 93 to 99% and was not modified by herbicide and herbicide-adjuvant treatment. Auskalniene and Auskalnis (2006) and GWC (2008) also confirmed high sensitivity of this species to nicosulfuron. In most cases, even a reduction of nicosulfuron rate, especially with adjuvants, does not limit its effectiveness in relation to other annual grass weed species (Nalewaja et al., 1998; Green and Hale, 2005).

All adjuvant treatments provided greater control of weeds compared to the use of nicosulfuron at 30 g ha^{-1} without adjuvant. However, there were no statistical differences observed among MSO, PO, and PO + U (efficacy 70, 64 and 66%, respectively). The addition of nitrogen fertilizers increased or tended to increase control of most weed species however the herbicide effectiveness depended on added nitrogen adjuvant to nicosulfuron with MSO or PO. The mixture of MSO or PO with AMN or UAN was more effective than with U. Generally nicosulfuron applied with MSO and AMN or UAN greater controlled most weed species than with PO.

Table 3. Weed control as influenced by oil and nitrogen fertilizer adjuvants with herbicide treatment (average from Zlotniki and Brody 2006 – 2008)

Nicosulfuron rate	Adjuvant*	CHEAL	VIOAR	POLAV	ECHCG	POLCO	GERPU
g a.i. ha		% control					
Untreated check	-	847 (g m^{-2})	183 (g m^{-2})	180 (g m^{-2})	140 (g m^{-2})	139 (g m^{-2})	77 (g m^{-2})
60	None	80	97	88	98	93	96
30	None	57	83	59	98	74	82
30	MSO	70	95	73	99	91	91
30	MSO + AMN	87	92	86	93	93	95
30	MSO + U	76	93	74	96	87	96
30	MSO + UAN	83	97	81	98	88	94
30	PO	64	92	73	99	86	88
30	PO + AMN	79	96	78	98	95	99
30	PO + U	66	90	69	99	90	93
30	PO + UAN	80	96	84	97	85	94
LSD 0.05		15.8	9.6	22.2	ns	17.0	13.1

Note. ns, not significant at P = 0.05 level; * MSO, methylated seed oil at 1.5 L ha^{-1}; PO, petroleum oil at 1.5 L ha^{-1}; AMN, ammonium nitrate at 3.3 kg ha^{-1}; U, urea at 2.5 kg ha^{-1}; UAN, 28% N urea-ammonium nitrate liquid fertilizer at 4.0 L ha^{-1}
GERPU – *Geranium pusillum*; ECHCG – *Echinochloa crus-galli*; VIOAR – *Viola arvensis*; CHEAL – *Chenopodium album*; POLCO – *Polygonum convolvulus*; POLAV – *Pologonum aviculare*

The data relating to the effect of the herbicide mixture with adjuvants on grain yield indicate significant differences among treatments. The lowest grain yield of 3.5 t ha was recorded with the untreated check that was at par (5.0 t ha^{-1}) with the application of nicosulfuron at reduced rate (Figure 1). Significantly higher grain yield was harvested from plots when nicosulfuron at 60 g ha^{-1} and at reduced rates with all adjuvants were applied. The grain yield from the plots where nicosulfuron was applied with PO and MSO was comparable to the yield obtained after the application of 60 g ha^{-1}. The addition of AMN or UAN to a spray liquid, led to a further yield increase, however this trend was not statistically confirmed. The results of the research make it possible to conclude that the effectiveness of treatment does not depend only on the rate of the herbicide but also on the appropriately selected adjuvant. The research also showed that only a mineral adjuvant appropriately selected in respect of the active ingredient increased the efficacy of the treatment. Greek and Cahill (2003) found that the addition of adjuvants with buffering properties (pH adjuster) were able to enhance nicosulfuron action by maintaining the alkaline reaction of the spray liquid, as sulfonylurea herbicides are highly water soluble in an alkaline environment and their activity increases, especially if they are used with crop oil concentrate, modified seed oil, and hydrophilic nonionic surfactants. The results of various experiments show that oil adjuvants, just as in the present author's study, differ in respect of their influence on the herbicidal effectiveness of nicosulfuron. Our results are confirmed by Nalewaja et al. (1995) who pointed out that nicosulfuron was the most effective in a mixture with MSO adjuvant, next, when combined with seed oils, and its effectiveness was the lowest with paraffin oil.

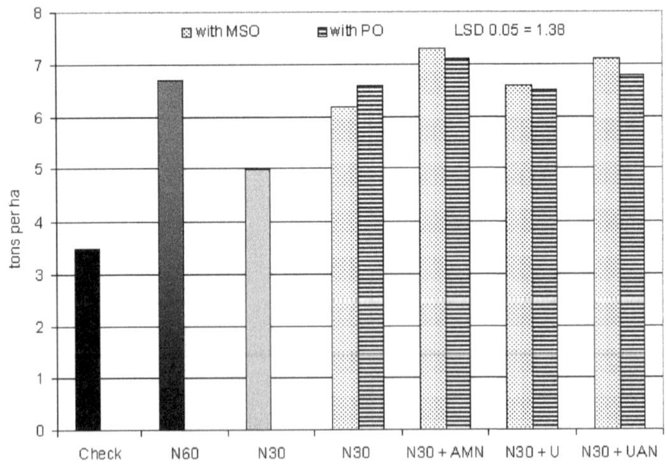

Figure 1. Grain yield of maize in correlation to herbicide treatments (average from Zlotniki and Brody 2006 – 2008) N60, nicosulfuron at 60 g ha^{-1}; N30, nicosulfuron at 30 g ha^{-1}; MSO, methylated seed oil at 1.5 L ha^{-1}; PO, petroleum oil at 1.5 L ha^{-1}; AMN, ammonium nitrate at 3.3 kg ha^{-1}; U, urea at 2.5 kg ha^{-1}; UAN, 28% N urea- ammonium nitrate liquid fertilizer at 4.0 L ha^{-1}

It is widely known that addition of nitrogen fertilizers to spray mixtures of certain foliar-applied herbicides enhances herbicide efficacy (Joost, 1998). The most commonly used are ammonium sulfate, ammonium nitrate and also urea – ammonium nitrate liquid fertilizer (Woznica, 2003). According to Gronwald at al. (1993) direct impact of mineral adjuvants may be cause by the effect of ammonium ions on the foliar penetration, and consequently the active transport of herbicides into plant cells. Many authors showed that more important is indirect effects of mineral adjuvants consisting of overcome any antagonism of Ca^{+2}, Mg^{+2}, Na^{+1} and other, often present in high concentrations in natural waters used for the preparation of liquid spray (Woznica, 1990; Nalewaja and Matysiak, 1993; Thelen at al., 1995).

According to Nandula et al. (1995) nicosulfuron applied with adjuvants has a positive effect on the grain yield of corn. In our study the effect of herbicide treatment on corn yield was variable. Adjuvants and nitrogen fertilizers, especially AMN and UAN, had rather consistent effect on grain yield increase. Grain yield was higher on plots treated with nicosulfuron with oil adjuvants and AMN or UAN as compared with U. Grain yield was always much higher in plots treated with herbicide and adjuvants than in untreated plots.

CONCLUSIONS

Effective weed control is guaranteed by nicosulfuron applied alone at the recommended rate, but also at reduced rate but with edition of properly chosen adjuvant or adjuvants. The presence of adjuvant in the spray liquid with different mechanisms of action, first of all MSO, especially with nitrogen fertilizer, preferably improve nicosulfuron activity. Higher weed control generally positively impact on the growth and yield of maize.

LITERATURE CITED

Auskalniene, O., A. Auskalnis. 2006. Effect of sulfonylurea herbicides on weeds and maize. Agron. Res. 4: 129-132.

Anderson, D.M., C.J. Swanton, J.C. Hall, B.G. Mersey. 1993. The influence of temperature and relative humidity on the efficacy of glufosinate-ammonium. Weed Res. 33: 139-147.

Caseley, J.C. 1989. Variations in foliar pesticide performance attributable to humidity, dew and rain effects. Aspect Appl. Biol. 21: 215–224.

EFSA. 2012. Reasoned opinion on the review of the existing maximum residue levels (MRLs) for nicosulfuron according to Article 12 of Regulation (EC) No 396/2005. EFSA J. 2012, 10(12): 1-27.

Evans, S.P., S.Z. Knezevic, J.L. Lindquist, C.A. Shapiro, E.E. Blankenship. 2003. Nitrogen application influences the critical period for weed control in corn. Weed Sci. 51: 408-417.

EXTOXNET. 2010. Extension Toxicology Network. Pesticide Information Profile. Nicosulfuron. Available at: htpp://pmep.cce.cornell.edu/profiles/extoxnet/metiram-propxur/nicosulfuron-ext.html (Accessed 25 October 2010).

Gosheh, H.Z., D.L. Holshouser, J.M. Chandler. 1996. The critical period of johnsongrass (*Sorghum halepense*) control in field corn (*Zea mays*). Weed Sci. 44: 944-947.

Green, J.M., T. Hale. 2005. Increasing and decreasing pH to enhance the biological activity of nicosufluron. Weed Technol. 19: 468-475.

Gronwald, J.W., S.W. Jourdan, D.L. Wyse, D.A. Somer, M.U. Magnusson. 1993. Effect of ammonium sulfate on absorption of imazethapyr by quackgrass (*Elytriga repens*) and maize (*Zea mays*) cell suspension cultures. Weed Sci. 41: 325-334.

GWC. 2008. Corn (Field, Seed and Sweet). In: Publication 75 Guide to Weed Control 2012-2013, Ontario Ministry of Agriculture, Food and Rural Affairs, pp. 119-164.

Hazen, J.L. 2000. Adjuvants – terminology, classification and chemistry. Weed Technol. 14: 773-784.

HRAC. 2012. Herbicide Resistance Action Committee. Classification of Herbicides According to Site of Action. Available at: http://www.hracglobal.com (Accessed 07 January 2013).

Johnson, B.C., B.G. Young. 2002. Influence of temperature and relative humidity on the foliar activity of mesotrion. Weed Sci. 50: 157-161.

Joost, R.E. 1998. Benefits and more of action o nitrogen fertilizers as adjuvants. Proc. Fifth International Symposium on Adjuvants for Agrochemicals 1, Memphis, USA, p. 259-266.

Bruce, J.A., J.B. Carey, D. Penner, J.J. Kells. 1996. Effect of GrowthS tage and Environmento n FoliarA bsorption,T ranslocation, Metabolism, and Activityo f Nicosulfuroni n Quackgrass(*Elytrigia repens*). Weed Sci. 44: 447-454.

Kells, J.J., W.F. Meggitt, D. Penner. 1984. Absorption, translocation, and activity of fluazifop-butyl as influnced by plant growth stage and environment. Weed Sci. 32: 143-149.

Martins, J.M.F., A. Mermoud. 1999. Transport of rimsulfuron and its metabolites in soil columns. Chemosphere 38: 601-616.

Nalewaja, J.D., R. Matysiak. 1993. Influence of diammonium sulfate and Rother salts on glyphosate phytotoxicity. Pestic. Sci. 38: 77-84.

Nalewaja, J.D., T. Praczyk, R. Matysiak. 1995. Surfactants and oil adjuvants with nicosulfuron. Weed Technol. 9: 689-695.

Nalewaja, J.D., T. Praczyk, R. Matysiak. 1998. Nitrogen fertilizer, oil, and surfactant adjuvants with nicosulfuron. Weed Technol. 12: 585-589.

NDWCG. 2010. Corn, corn resistant. In North Dakota Weed Control Guide: pp. 18-23.

Penner, D. 2000. Acitvator Adjuvants. Weed Technol. 14: 785-790.

Praczyk, T., K. Adamczewski. 1996. The importance of adjuvants in chemical plant protection. Prog. Plant Protection 36 (1): 117-121.

Regitano, J.B., W.C. Koskinen. 2008. Characterization of nicosulfuron availability in aged soils. J. Agric. Chem. 56 (14): 5801-5805.

Rittel, R.L., H.D. Coble. 1981. Influence of temperature and relative humidity on the activity of acifluorfen. Weed Sci. 29: 480-485.

Russell, M.H., J.L. Saladini, F. Lichtner. 2002. Sulfonylurea herbicides. Pestic. Outlook 13: 166-173.

Sanyal, D., P.C. Bhowmik, K.N. Reddy. 2006. Influence of leaf surface micromorphology, wax content, and surfactant on primisulfuron droplet spread on barnyardgrass (*Echinochloa crus-galli*) and green foxtail (*Setaria viridis*). Weed Sci. 54: 627-633.

Thelen, K.D., E.P. Jackson, D. Penner. 1995. The basis for the hard-water antagonism of glyphosyte activity. Weed Sci. 43: 541-548.

Wilson, R.G., P. Westra. 1991. Wild proso millet (*Panicum miliaceum*) interference in corn (*Zea mays*). Weed Sci. 39: 217-220.

Woznica, Z. 1990. Influence of mineral compounds present in water on salt phytotoxicity of 2,4-dichlorophenoxyacetic acid. Rocz. AR Poznan 203: 1-43.

Woznica, Z. 2003. Adjuvant interaction and efficacy of herbicides. Prog. Plant Protection 43 (1): 473-480.

PERFORMANCES OF SOME PERENNIAL WARM SEASON GRASSES ALFALFA (*Medicago sativa* L.) MIXTURES UNDER MEDITERRANEAN CONDITIONS

Selahattin CINAR[*1], *Rustu HATIPOGLU*[2], *Feyza Dondu GUNDEL*[3], *Arif AKTAS*[3], *Mustafa AVCI*[3]

[1] *Gaziosmanpasa University, Faculty of Agriculture, Deparment of Field Crops, Tokat, TURKEY*
[2] *Cukurova University, Faculty of Agriculture, Deparment of Field Crops, Adana, TURKEY*
[3]*East Mediterranean Agricultural Research Institue, Adana, TURKEY*
[*]*Corresponding author: scinar01@hotmail.com*

ABSTRACT

This research was conducted to determine forage yield and botanical composition of duo (alfalfa + one grass) and trio (alfalfa + two grasses) mixtures of some warm season perennial grass species such as dallis grass (*Paspalum dilatatum* Poir.), Rhodes grass (*Chloris gayana* L.), Bermuda grass (*Cynodon dactylon* (L.) Pers.) Guineae grass (*Panicum maximum* Jacq.), blue couch grass (*Digitaria didactyla* Willd) and finger grass (*Digitaria milanjiana* (Rendle) Stapf) with alfalfa (*Medicago sativa* L.) as well as their pure sowings under irrigated conditions of Mediterranean during the years of 2011 and 2012. In the study, green herbage yield, dry matter yield, alfalfa proportion in dry matter yields and land equivalent ratio (LER) were determined. The experimental design was completely randomized block design with three replications. According to the results, it was determined that the ratios of warm season grasses in the botanical composition of the mixtures declined with the establishment age while rate of alfalfa in the botanical composition of the mixtures increased. Increasing the rate of alfalfa in the botanical composition of the mixtures also increased hay quality of the mixtures. The mixtures gave higher dry matter yield than pure sowings of the species. The highest green herbage yield (98.14 t ha^{1}) was obtained from the mixture of Rhodes grass + alfalfa. The highest dry matter yield (22.46 t ha^{1}) was obtained from the mixture of dallis grass + Guinea grass + alfalfa. The highest value of land equivalent ratio (LER) (1.49) was obtained from the mixture of bermuda grass +Rhodes grass + alfalfa. From the results of the study, it was concluded that Guinea grass and finger grass did not show good adaptaion to the Cukurova conditions due to low temperatures during the winter season. Bermuda grass, Rhodes grass, blue couch grass and dallis grass could be used to establish pasture mixtures with alfalfa, having long grazing season and giving high hay yield and quality. Before the establishment of such pasture mixtures, it was needed to research proper mixture ratios and proper management techniques of the mixtures.

Key words: Warm season perennial grasses, alfalfa, mixture, yield, LER

INTRODUCTION

Grass-legume mixtures are preferred over pure-grass forage stands throughout the world because they often increase the total yields of herbage and protein and offer balanced nutrition (Albayrak and Ekiz, 2005). Mixtures offer several potential advantages over pure grasses or pure legumes. Some other advantages of mixtures include the control of erosion, weed control and prolonged stand longevity (Casler, 1988). Alfalfa is one of the most commonly used legumes for both hay and pasture in Turkey because of its high yield, high nutritional quality, ability to fix nitrogen and vigorous fall regrowth (Acikgoz, 2001).

Grass-legume mixtures tend to provide a superior nutrient balance and produce higher forage yields. However, grass-legume mixtures are more difficult to manage than monoculture pastures because of competition among the mixture components for light, water, and nutrients (Charles and Lehmann, 1989; Serin et al., 1998; Berdahl et al., 2001; Albayrak and Ekiz, 2005; Cinar and Hatipoglu, 2014). The benefit of including legumes into mixture with grasses has long been recognized. Introducing legumes in pastures and hay fields improves forage quality and lowers the cost of production (Brown and Munsell, 1943).

The advantages of cool season legumes in cool season pastures and hay lands have been well documented, but there is little information on mixing legumes with warm season grasses. Some researcher reported that growing warm-season grasses with legumes resulted in higher dry matter yields than pure sowings (Aganoglu 1985; Blanchet et al., 1995; Larbi et al., 1995; Gettle et al., 1996; Tessema and Baars 2006; Cinar and Hatipoglu, 2014).

Mediterranean region of the Turkey experiences long periods of drought, especially in mid-summer. High temperatures and low precipitation are detrimental to production of cool season forages. Lack of productivity of cool season pastures during the summer months often results in overgrazing. Perennial warm season grasses produce their yields in midsummer, which could provide forage production during that period (Jung et al., 1978).

Methods to effectively establish and manage some warm season grasses for yield have been characterized for Mediterranean conditions (Cinar and Hatipoglu 2014). Using the potential of these species to provide a dependable pasture forage supply during summer months warrants investigation of options to enhance their quality.

Rhodes grass is one of the main sub-tropical grasses and is widely grown in Africa, Australia, Japan and South America as well as in the Middle East under irrigation for both forage and soil conservation purposes. Rhodes grass is a morphologically variable out-crossing species, which is native to east, central and southern Africa where it occurs in open grasslands.

Dallis grass is a summer-active perennial grass native to the humid sub-tropics of southern Brazil, Argentina and Uruguay. It is now widespread in many areas of the world, and is capable of very high production under suitable conditions.

Bermuda grass is native to southern Africa and southeast Asia. It has been widely used in tropical and warm temperate regions as a pasture grass and is one of the major turf grasses in the world. Bermuda grass is widely sown in the south-eastern United States, where its hybrid cultivars with improved productivity and forage quality have been developed Blue couch grass similar to bermuda grass. It differs from Bermuda grass in its shorter, broader leaf and its distinctive bluish colour (Skerman and Riveros, 1990).

Guinea grass is widely distributed in the tropical and sub-tropical areas of the world. It can produce over 20 tons DM/ha but suffers from drought when dry matter production can be reduced to as low as 40% of its yield potential. Guinea grass responds well to cutting and grazing, and is fairly pest resistant (Nung and Binh, 2014).

Finger grass is a native of tropical Eastern and Southern Africa, from Ethiopia down to South Africa. It is found in semiarid to wet equatorial areas, with average annual rainfall between 450 to 1700 mm. It grows in grasslands or sandy loam soils and in open woodlands on heavy black or sandy soils. An annual dry matter yield of up to 15 t/ha has been achieved from well fertilized. İt is very palatable to all types of stock as green feed, dry feed or as hay. It can be used in mixed pastures or as a hay crop. It should not be grazed in the wet season of establishment (Cameron, 2010).

This research was conducted to determine forage yield and botanical composition of duo (one grass + alfalfa) and trio (two grass + alfalfa) mixtures of some perennial warm season grass species such as dallis grass (*Paspalum dilatatum* Poir.), Rhodes grass (*Chloris gayana* L.), Bermuda grass (*Cynodon dactylon* (L.) Pers.) Guineae grass (*Panicum maximum* Jacq.), blue couch grass (*Digitaria didactyla* Willd) and finger grass (*Digitaria milanjiana* (Rendle) Stapf) with alfalfa (*Medicago sativa* L.) as well as their pure sowings under irrigated conditions of Mediterranean

MATERIALS AND METHODS

Materials

The research was carried out during 2011 and 2012 years in Adana province (36°51'N, 35°20'E, elevation 12 m), located in the Mediterranean region of Turkey. The experimental area has typically natural Mediterranean climate with hot and dry summer and heavy precipitation during winter. According to the average of long years, the coolest month is January with a monthly mean temperature of 9.9 ºC and the hottest month is July with 29.7 ºC. The total precipitation was higher in 2011 (687.7 mm) and in 2012 (1048.8 mm) compared with the long-term precipitation mean (647.1 mm). The means of the temperature and relative humidity during the experimental period were close to the long-term means (Anonymous, 2013). The research area has a flat land; its soil texture is classified as silty clay with slightly alkaline, which contained medium organic matter, poor phosphorus (P), zinc (Zn) and calcium (Ca) contents (Anonymous, 2011).

Dallis grass (*Paspalum dilatatum* Poir.), Rhodes grass (*Chloris gayana* L.), Bermuda grass (*Cynodon dactylon* (L.) Pers.) Guineae grass (*Panicum maximum* Jacq.), blue couch grass (*Digitaria didactyla* Willd), finger grass (*Digitaria milanjiana* (Rendle) Stapf) and alfalfa (*Medicago sativa* L.) were used as experimental material. In the research, duo and trio mixtures of the mentioned grass species with alfalfa as well as their pure sowings were studied.

Methods

The experiment was established according to the randomized complete block design with three replications. There were 28 treatment plots in each block. Each plot consisted of 6 rows each 5 m in length. The row spacing was 25 cm. The seeding rates for dallis grass, Rhodes grass , Bermuda grass, Guineae grass, blue couch grass and alfalfa in pure sowing were 11 kg ha^{-1}, 4 kg ha^{-1},10 kg ha^{-1} , 6 kg ha^{-1} , 8 kg ha^{-1} , 8 kg ha^{-1} and 10 kg ha^{-1} , respectively. Seed mixtures consisted of 30% alfalfa and 70% grasses. Seed ratio of each grass species in trio mixtures was 35%. The seeds of alfalfa and grass species were sown in the same rows. Sowing was done by hand on May 28th 2011. Plots were fertilized with 100 kg ha^{-1} N and 100 kg ha^{-1} P at sowing (Avci, 2000). The plots were irrigated once after each harvest.

The harvest time was based on the 10% flowering stage of alfalfa (Serin et al., 1998). The plots were harvested at 5 cm cutting height. The plots were harvested 3 and 5 times during the growing season of 2011 and 2012, respectively. The botanical composition and yield were determined from the 3 randomly located quadrats each having 0.5 m^2 area in each plot. Green herbage from each quadrat was hand-separated to the mixture components. All samples were dried at 70 °C for 48 h and weighed. Dry matter content was determined from ground samples dried at 105 °C for 24 h.

The land equivalent ratio (LER) was defined as the relative area of monocrop plant required for the same yield obtained from its mixture. The LER was calculated using the formula given below (Ta and Faris, 1987):

$$LER: \frac{\text{yield of alfalfa in mixture}}{\text{yield of alfalfa alone}} + \frac{\text{yield of grass in mixture}}{\text{yield of grass alone}}$$

When LER is greater than 1, the mixed growing favors the growth and yield of the mixture species. In contrast when LER is lower than 1, the mixed growing negatively affects the growth and yield of plants grown in mixture (Caballero et al., 1995; Dhima et al., 2007)

Statistical analyses

Data were analyzed according to the completely randomized design by using MSTATC (V.1.2, Michigan State University, USA). The differences between means were separated by Duncan multiple range test (P ≤0.05) (Steel and Torrie, 1960).

RESULTS AND DISCUSSION

Green herbage yield

The results of the variance analysis showed that effects of treatments, years on the green herbage yield were as well as year x treatment interactions statistically significant (Table 1). Average green herbage yield in the first year were significantly lower than that in the second year. Higher averaged green herbage yield in the second year was due to higher yield of the alfalfa in that year. The lower yields of the grasses in the second year may be due to harmful effect of the winter cold on the warm season grasses (Anonymous, 2013).

Table 1. Averaged green herbage yields (t ha^{-1}) of pure sowings and mixtures of some perennial warm season grasses with alfalfa

Mixture or Pure Sowing	Years 2011	Years 2012	Average
Bermuda grass (BG)	34.64 j*	19.27 l	26.96 j
Dallis grass (DG)	71.25 e-h	28.24 k	49.74 hı
Rhodes grass (RG)	89.17 c-f	40.45 j	64.81 fg
Guinea grass (GG)	124.89 a	10.49 m	67.69 e-g
Blue couch grass (BCG)	36.35 j	12.17 lm	24.26 j
Finger grass	81.97 d-g	13.21 lm	47.59 ı
Alfalfa (A)	30.87 j	89.42 f	60.15 gh
BG+A	43.86 ıj	115.37 ab	79.61 b-e
DG+A	57.19 g-j	115.69 ab	86.44 ab
RG+A	96.45 b-e	99.85 de	98.15 a
GG+A	95.99 b-e	48.45 hı	72.22 c-g
BCG+A	32.96 j	91.37 f	62.16 g
FG+A	57.06 g-j	102.22 cd	79.64 b-e
BG+DG+A	38.74 ıj	90.37 f	64.56 fg
BG+RG+A	57.31 g-j	70.84 g	64.08 fg
BG+GG+A	112.24 a-c	49.87 hı	81.05 b-d
BG+BCG+A	40.17 ıj	92.92 ef	66.55 fg
BG+FG+A	47.43 h-j	117.72 a	82.57 bc
DG+RG+A	84.74 a-d	108.61 bc	96.67 a
DG+GG+A	120.62 ab	54.21 hı	87.41 ab
DG+BCG+A	39.36 ıj	112.27 ab	75.81 b-f
DG+FG+A	49.49 h-j	102.03 cd	75.76 b-f
RG+GG+A	108.04 a-d	56.15 h	82.09 b-d
RG+BCG+A	65.60 f-ı	76.68 g	71.14 c-g
RG+FG+A	88.23 c-f	86.10 f	87.17 ab
GG+BCG+A	80.83 e-g	49.73 hı	65.28 fg
GG+FG+A	88.45 c-f	46.17 ıj	67.31 e-g
FG+BCG+A	51.74 h-j	86.79 f	69.26 d-g
Average	**68.77 b+**	**70.95 a**	**69.86**

* Means with the same letter in a column are not statistically different from each other according to the Duncan test at P≤0.05
+) Means with the same letters in the row not statistically different from each other at P≤0.05

Pure sowings and mixtures in the first year showed statistically significant differences in green herbage yield. Pure sowing of Guinea grass gave significantly higher green herbage yield (124.89 t ha^{-1}) than all of the other mixtures and pure growings with the exceptions of mixtures of DG+GG+A (120.62 t ha^{-1}), BG+GG+A (112.24 t ha^{-1}) and RG+GG+A (108.04 t ha^{-1}).

In the second year, the highest green herbage yield was obtained from mixture of BG+FG+A (117.72 t ha^{-1}). The lowest green herbage yield was obtained from pure sowing of Guinea grass (10.49 t ha^{-1}). Warm season grasses in general have been damaged by the cold winter period 2011-2012. Especially Guinea grass and finger grass has losted greatly from winter damage in period 2011-2012.

According to the values averaged over two years, the mixtures of RG+A gave significantly higher green herbage yield than the pure sowings and other mixtures with the exceptions of the mixtures of DG+RG+A, DG+GG+A, RG+FG+A and DG+A.

As a result of the study, it appeared that some of the mixtures studied were superior to the single grass or legume stands. The similar results were also reported by Albayrak and Ekiz (2005), and Gökkus et al. (1999) and Cinar and Hatipoglu (2014). The higher performance of the mixtures may be due to utilization of symbiotically fixed nitrogen by the grasses in the mixture and more enhanced interception of light (Hay and Walker, 1995) as well as due to synergetic and some other effects. Sleugh et al. (2000) stated that alfalfa and mixtures of alfalfa produced greater yields because the deep root system of alfalfa plants was able to tap deeper soil water.

Dry matter yield

The results of the variance analysis showed that effects of treatments, years, and year x treatment interactions on the dry matter yield were statistically significant (Table 2). The averaged dry matter yield in the first year was statistically significant lower than those in the second year. In the first year, grasses in the mixture were dominant but alfalfa was the dominant species in the second year. Therefore, dry matter yields of the mixtures in the first year were lower than those in the second year.

Table 2. Averaged dry matter yields (t ha^{-1}) of pure sowings and mixtures of some perennial warm season grasses with alfalfa

Mixture or Pure Sowing	Years		Average
	2011	2012	
Bermuda grass (BG)	10.68 ı-k*	6.09 ıj	8.38 m
Dallis grass (DG)	13.87 d-k	8.22 hı	11.05 k-m
Rhodes grass (RG)	17.99 b-g	10.62 gh	14.31 h-k
Guinea grass (GG)	24.10 ab	2.58 j	13.34 j-l
Blue couch grass (BCG)	11.61 g-k	5.40 ıj	8.51 m
Finger grass	15.98 c-ı	3.94 ıj	9.96 lm
Alfalfa (A)	7.88 k	21.80 c-e	14.84 g-j
BG+A	12.19 f-k	28.40 ab	20.29 a-e
DG+A	12.16 f-k	29.34 a	20.75 a-d
RG+A	19.97 a-d	24.22 b-d	22.09 ab
GG+A	19.26 a-e	15.14 fg	17.20 d-j
BCG+A	8.77 k	23.17 cd	15.97 f-j
FG+A	12.99 e-k	24.80 a-d	18.89 a-g
BG+DG+A	10.21 ı-k	22.94 cd	16.58 e-j
BG+RG+A	13.49 d-k	26.35 a-c	19.92 a-f
BG+GG+A	19.83 a-d	12.98 g	16.41 e-j
BG+BCG+A	10.93 ı-k	23.81 b-d	17.37 d-j
BG+FG+A	10.58 ı-k	25.42 a-d	18.00 c-ı
DG+RG+A	18.62 a-f	25.00 a-d	21.81 a-c
DG+GG+A	24.57 a	20.35 de	22.46 a
DG+BCG+A	8.95 jk	26.45 a-c	17.70 d-ı
DG+FG+A	11.11 h-k	22.77 c-e	16.94 d-j
RG+GG+A	22.05 a-c	14.74 fg	18.39 b-h
RG+BCG+A	14.65 d-k	20.38 de	17.51 d-ı
RG+FG+A	17.83 b-h	17.99 ef	17.91 c-ı
GG+BCG+A	16.58 c-ı	13.55 fg	15.07 g-j
GG+FG+A	15.67 c-j	12.47 gh	14.07 ı-k
FG+BCG+A	11.28 g-k	21.21 de	16.24 e-j
Average	**14.78 b+**	**18.22 a**	**16.50**

* Means with the same letter in a column are not statistically different from each other according to the Duncan test at $P \leq 0.05$
+) Means with the same letters in the row not statistically different from each other at $P \leq 0.05$

In the first year, mixture of DG+GG+A gave statistically significant higher dry matter yield than all of the other mixtures and pure sowings with the exception of pure Guinea grass and mixtures of RG+GG+A, RG+A, BG+GG+A and GG+A.

In the second year, mixture of DG+A gave significantly higher dry matter yield than all of the pure sowings and other mixtures with exceptions of mixtures of BG+A, DG+BCG+A, BG+RG+A, BG+FG+A, DG+RG+A and FG+A.

Based on the averaged dry matter yields over two years, the mixture of DG+GG+A was the best performed mixture. This mixture gave significantly higher dry matter yield than all of the pure sowings and mixtures with the exceptions of mixtures of RG+A, DG+RG+A, DG+A, BG+A, BG+RG+A and FG+A.

Several researches reported that mixtures had higher dry matter yields than monocultures of grasses and legumes (Gokkus et al., 1999; Sleugh et al., 2000; Albayrak et al., 2011, Cinar and Hatipoglu, 2014), consistent with our results.

Botanical composition

Contribution of alfalfa to the dry matter yields of the mixtures significantly changed depending on the years (Table 3). In the first year of the experiment, averaged ratio of the alfalfa (24.4 %) was significantly lower than that in the second year (87.5 %). This result was an expected result because alfalfa could not show its potential yield in the establishment year while growing of the warm season grasses in the first year were much better than the alfalfa. This result coincides with the result of the study conducted by El Hadj (2000), Cinar and Hatipoglu (2014).

Table 3. Alfalfa proportions (%) in dry matter yields of different alfalfa-warm season grass mixtures

Mixture	Years		Average
	2011	2012	
BG+A	47.7 a*	95.7 bc	71.7 ab
DG+A	40.4 a	88.7 f-h	64.5 a-e
RG+A	7.6 d	95.7 bc	51.6 f-j
GG+A	1.8 d	100.0 a	50.9 g-j
BCG+A	51.2 a	92.0 c-f	71.6 ab
FG+A	48.6 a	99.3 ab	73.9 a
BG+DG+A	44.7 a	90.5 d-g	67.6 abc
BG+RG+A	12.2 d	78.5 j	45.4 ı-k
BG+GG+A	2.4 d	70.2 k	36.3 kl
BG+BCG+A	42.7 a	89.9 e-g	66.3 a-d
BG+FG+A	34.8 a-c	91.1 d-g	62.9 a-f
DG+RG+A	11.2 d	85.4 hı	48.3 hıj
DG+GG+A	2.5 d	62.0 l	32.1 l
DG+BCG+A	45.2 a	81.7 ıj	63.4 a-e
DG+FG+A	38.8 ab	79.9 j	59.3 c-h
RG+GG+A	1.8 d	95.0 b-d	48.4 hıj
RG+BCG+A	19.0 b-d	89.3 e-h	54.2 e-ı
RG+FG+A	16.0 cd	93.5 c-e	54.8 d-ı
GG+BCG+A	3.8 d	79.0 j	41.4 jkl
GG+FG+A	2.8 d	92.9 c-f	47.9 hıj
FG+BCG+A	37.1 ab	86.7 gh	61.9 b-g
Average	**24.4 b+**	**87.5 a**	**56.0**

* Means with the same letter in a column are not statistically different from each other according to the Duncan test at P≤0.05
+) Means with the same letters in the row not statistically different from each other at P≤0.05

The analysis of variance showed that contribution of the alfalfa to the dry matter yield of the mixtures in the first and second years of the experiment was significantly influenced by the mixtures. In the first year, ratio of the alfalfa in the mixture containing DG+GG+A was significantly higher than the other mixtures with the exceptions of RG+A, GG+A, BG+GG+A and RG+GG+A. In the second year, ratio of the alfalfa in the mixture containing DG+GG+A was significantly lower than the other mixtures. From this results, it could be said that alfalfa can better compete with dallis grass, finger grass, blue couch grass, Guinea grass and bermuda grass than with Rhodes grass.

Based on the averaged values of the two years, proportion of alfalfa in the dry matter yield of DG+GG+A mixture was significantly lower than those in the dry matter yields of the mixtures.

Land Equivalent Ratio (LER)

LER verifies the effectiveness of mixtures for using available resources of environment compared to pure cropping. For this purpose, LER value is being used frequently to compare effectiveness of the mixtures and pure growing (Yilmaz et al., 2005, Dhima et al., 2007, Atis et al., 2012, Cinar and Hatipoglu, 2014). According to the average values over two years, LER values were greater than 1 in all mixtures with the exceptions of the mixtures of GG+A, BG+GG+A, RG+GG+A, GG+BCG+A, GG+FG+A (Table 4).

Table 4. Land equivalent ratios for different alfalfa-warm season grass mixtures

Mixture	Years 2011	2012	Average
BG+A	1.32 a*	1.47 d	1.40 b
DG+A	1.16 c	1.56 c	1.36 b
RG+A	1.23 b	1.18 jk	1.21 d
GG+A	0.85 gh	0.99 l	0.92 ı
BCG+A	0.95 f	1.32 h	1.13 ef
FG+A	1.23 b	1.16 jk	1.20 d
BG+DG+A	1.06 de	1.24 ı	1.15 e
BG+RG+A	1.15 c	1.83 a	1.49 a
BG+GG+A	0.89 g	1.03 l	0.96 h
BG+BCG+A	1.18 bc	1.39 fg	1.29 c
BG+FG+A	1.01 e	1.46 de	1.24 d
DG+RG+A	1.22 b	1.34 gh	1.28 c
DG+GG+A	1.06 de	1.14 k	1.10 d
DG+BCG+A	0.76 j	1.64 b	1.20 d
DG+FG+A	1.02 e	1.41 ef	1.22 d
RG+GG+A	1.23 b	0.71 o	0.97 h
RG+BCG+A	0.82 hı	1.22 ıj	1.02 g
RG+FG+A	1.19 bc	0.89 m	1.04 g
GG+BCG+A	0.78 ıj	1.02 l	0.90 ı
GG+FG+A	0.75 j	0.77 n	0.76 j
FG+BCG+A	1.09 d	1.38 fg	1.23 d
Average	**1.04 c+**	**1.25 b**	**1.15**

* Means with the same letter in a column are not statistically different from each other according to the Duncan test at P≤0.05
+) Means with the same letters in the row not statistically different from each at P≤0.05

LER value was significantly influenced by the years and mixtures. The interaction of year x mixture was also statistically significant. The average LER value in the first year was significantly lower than that in the second year. Reasons of the higher LER values in the second year compared to the first year could be increasing the contribution of the alfalfa to the dry matter yield of the mixtures as well as better benefiting of the grasses from the fixed nitrogen through the alfalfa in the second year of the experiment. Erkovan (2005) reported that nitrogen transfer from the fixed nitrogen by legumes to the grasses in the legume-grass mixtures was higher in the second year than in the first year.

CONCLUSIONS

From the results of the study, it was concluded that Guinea grass and finger grass did not show good adaptation to the Cukurova conditions due to low temperatures during the winter season. Bermuda grass, Rhodes grass, blue couch grass and dallis grass could be used to establish pasture mixtures with alfalfa, having long grazing season and giving high hay yield and quality. Before the establishment of such pasture mixtures, it was needed to research proper mixture ratios and proper management techniques of the mixtures.

ACKNOWLEDGEMENTS

Thanks to TUBITAK (Project of 110O695) for their financial support.

LITERATURE CITED

Acikgoz, E. 2001. Forage crops. Uludag University Faculty of Agriculture Publications, Bursa

Aganoglu, V. 1985. A Research on the possibility of growing Rhodes grass (*Chloris gayana* Kunth) and alfalfa (*M. sativa* L.) in mixture under Cukurova conditions. Msc thesis, Field Crops Department of Institute for Natural and Applied Sciences, Cukurova University, Adana.

Albayrak, S. and H. Ekiz, H. 2005. An investigation on the establishment of artificial pasture under Ankara's ecological conditions. T. Jo. of Agri. Forest. 29:69-74

Albayrak, S., M. Turk., O. Yuksel and M. Yilmaz. 2011. Forage yield and the quality of perennial legume-grass mixtures under rainfed conditions. Notulae Botanicae Horti Agrobotanici Cluj. 39(1):114-118.

Anonymous. 2011. Chemical analysis results, University of Cukurova Laboratory of Soil Science, Faculty of Agriculture, Adana, Turkey.

Anonymous. 2013. Meteorological data for Adana. http://www.cukurovataem.gov.tr/iklim.html, (Accessed February 15, 2013)

Atis, A., K.Kokten., R.Hatipoglu., S. Yilmaz., M. Atak and E.Can. 2012. Plant density and mixture ratio effects on the competition between common vetch and wheat. Australian Jo. of Crop Science. 6 (3): 498-505

Avci, M. 2000. Determining suitable mixtures of cool season perennial grasses and legumes for establishment of short term pastures in Cukurova Region. PhD Thesis, Cukurova University, Institue of Natural and Applied Sciences, Department of Field Crops, 113 p. Adana

Berdahl, D. J., J.K.Karn and J.R.Hendrickson. 2001. Dry Matter yields of cool-season grass monocultures and grass-alfalfa binary mixtures. Agronomy Journals, 93: 463-467.

Blanchet, K.M., J.R. George., R.M. Gettle., D.R. Buxton and K.J. Moore. 1995. Establishment and persistence of legumes interseeded into switchgrass. Agron. Jo. 87: 935-941

Brown, B.A. and R.I. Munsell. 1943. Grasses fertilized with nitrogen compared with legumes for hay and pasture. Agron. Jo. 35: 811-816

Caballero, A.R., E.L. Goicoechea-Oicoechea and P.J.Hernaiz-Ernaiz. 1995. Forage yields and quality of common vetch and oat sown at varying seeding ratios and seeding rates of vetch.Field Crops Research. 41:135-140

Cameron, A. G. and B. Lemcke, 2010. Pasture species sowing guide for the top end. http://www.nt.gov.au/d/Content/File/p/Pasture/544.pdf. (Accessed April, 20, 2014)

Casler, M.D. 1988. Performance of orchardgrass, smooth bromrgrass and ryegrass in binary mixtures with alfalfa. Agron. Jo. 80:509-514

Charles, J.P. and J. Lehman. 1989. The İmportance Of Grass/Legume Mixtures İn Forage Production İn Switzerland. Fourages. 119:311-320

Cinar, S. and Hatipoglu, R, 2014. Forage yield and botanişcal composition of mixtures of some perennial warm season grasses with alfalfa (*Medicago sativa* L.) under Mediterranean conditions. T. Journal of Field Crops. 19(1): 13-18

Dhima, K.V., A.S. Lithourgidis, I.B.Vasilakoglou and C.A.Dordas. 2007. Competition indices of common vetch and cereal intercops in two seeding ratio. Field Crops Res. 100: 249-256

El Hadj, M. 2000. Compatibility, yield, and quality of perennial warm-season grass-legume mixtures. Master of Science In Crop and Soil Environmental Sciences, Submitted to the Faculty of the Virginia Polytechnic Institute and State University. P: 86

Erkovan, H. 2005. Some of legumes and grasses perennial forage crops in pure and mixtures pure determination of yield and transfer of nitrogen fixation. PhD Thesis, Ataturk University, Institue of Natural and Applied Sciences, Department of Field Crops, Erzurum.

Gettle, R.M., J.R. George, K.M. Blanchet, D.R. Buxton and K.J. Moore. 1996. Frost-seeding legumes into established switchgrass. Establishment, density, persistence, and sward composition. Agrono Jo. 88: 98-103

Gokkus, A., A. Koc, Y.Serin, B.Comakli, M.Tan and F.Kantar. 1999. Hay yield and nitrogen harvest in smooth bromegrass mixtures with alfalfa and red clover in relation to nitrohen application. Eur Journal Agron. 10: 145-151.

Hay, R.K.M. and A.J. Walker. 1989. An introduction to the physiology of crop yield. Longman Scientific and Technical, p.292

Jung, G.A., C.F.Gross. R.E. Kocher, L.A.Burdette and W.C.Sharp. 1978. Warm season range grasses extend beef cattle forage. Penn.Agric.Exp.Stn.Sci.Agric. 64:211-219

Larbi, A., Lazier, J., Ochang, J and Addie, A., 1995. Dry Matter Production of Thirteen Tropical Legumes İn Association with Rhodes Grass (*Chloris gayana* cv. Callide) on an Acid Soil in Etiopia. Trop. Grass. Volume 29 p: 88-91.

Nung, H.V and Binh, L.H, 2014. Forage productivity of guinea grass in different farming systems. http://www.fao.org/ag/agp/agpc/doc/publicat/VIET95/V95_97.PDF (Accessed April, 15, 2014)

Serin, Y., A.Gokkus, M.Tan, A.Koc and B. Comakli. 1998. Determination of appropriate mixtures artificial grass and forage crops, Tr. J. of Agri. and Forest. 22: 13-20

Skerman, P.J. and F.Riveros. 1990. Tropical Grasses. FAO Plant Production and Protection Series 23, Roma. 832 p.

Sleugh, B., K.J.Moore, L.R.George and C.E.Brummer. 2000. Binary legume grass mixtures improve forage yield, quality and seasonal distrubition. Agron Jo. 92: 24-29.

Steel, R.G.D. and J.H.Torrie. 1960. Principles and Procedures of Statistics With Special Reference to the Biological Sciences. McGraw-Hill Book Co., Inc., London.

Ta, T.C. and M.A.Faris. 1987. Effects of alfalfa proportions and clipping frequencies on timothy-alfalfa mixtures. I. Competition and yield advantages. Agron Jo. 79:817-820

Tessema, Z. and Baars, R.M.T. 2006. Chemical Composition Dry Matter Production and Yield Dynamics of Tropical Grasses Mixed With Perennial Forage Legumes, Tropical Grasslands. Volume 40 p: 150-156.

Yilmaz, S. 2005. Determining intra and inter specific competition in vetch (*Vicia sativa* L.) and barley (*Hordeum vulgare* L.) mixtures. MSc Thesis, Institute of Naturel and Applied Sciences. University of Cukurova. Adana/Turkey

MOLECULAR CLONING AND SEQUENCE ANALYSIS OF RIBOSOMAL PROTEIN GENES IN MAIZE

*Jun LI, Xiaohong LIU **

College of Life Science, China West Normal University, NanChong City, CHINA
**Corresponding author: liuxiaohong0817@sina.com*

ABSTRACT

Ribosomal proteins are the essential components of ribosome, playing an important role in cell, involved protein synthesis and some other enzyme activities. In the current study, five ribosomal protein genes were successfully cloned from the maize (*Zea mays* L.) inbred line Southern 202 with reverse transcription polymerase chain reaction (RT-PCR) strategy. All of them were sequenced and analyzed preliminarily. The results showed that the complete coding regions of *RPS9*, *RPS10*, *RPS14*, RPS16 and *RPS18* genes encode mature proteins with 193, 179, 150, 148 and 152 amino acids (AA), respectively. Bioinformatics analysis using the related data within NCBI/GenBank indicated that these sequences share high similarities at both DNA and protein levels. But same gene not only can encode different AA in the different variety in same species, but the number of encoded AA residues is probably different. Whereas in animals especially mammalian, the DNA and protein sequences of ribosomal protein genes have higher coincidence, and the number of AA residues encoded is very stable. The knowledge stated here could be beneficial for maize breeding programs and studying the functions of ribosomal proteins.

Key words: Maize (*Zea mays* L.), Ribosomal protein gene, Cloning, Bioinformatics analysis

Abbreviations: AA – amino acids; EB - ethidium bromide; ORF - open reading frame ; PCR - polymerase chain reaction; pI - isoelectric point; RT-PCR - reverse transcription polymerase chain reaction; SNP single-nucleotide polymorphism

INTRODUCTION

Selecting an inbred line and utilizing its heterosis can greatly improve maize yield, which is also the basic principles for corn breeders in maize (*Zea mays* L.) breeding project. The maize breeding targets with high-stable yield, resistant diseases, high combining ability and good quality are controlled by the traits of parental inbred lines (When, 2002). Southern 202, bred by Nanchong Academy of Agricultural Sciences, Sichuan Province, China, is an elite inbred line with high general combining ability and special combining ability, comprehensively good agronomic traits, high disease and stress resistance. Currently, the inbred line has been widely used as a key donor line in maize breeding programs in China and several superior hybrids had been developed from the parental line. For example, in 2001, the hybrid Zhengtian 1 was obtained by the hybridization between inbred lines Southern 202 and Southern 637, which has highly resistant to *Exserohilum turcicum*, and moderate resistant to *Rhizoctonia solani* and *Bipolaris maydis*, high yielding potential, and in 2007 it has been validated through Sichuan Province and widely planted in China today.

Ribosome, the organelle that catalyzes protein synthesis, consists of a small 40S subunit and a large 60S subunit. The two subunits are composed of 4 rRNA and approximately 80 structurally distinct proteins. According to the order of binding to rRNA, the ribosomal proteins are divided into three groups, Primary binding protein, Subprime binding protein, and Late binding protein (Arnold and Reilly, 1999). In 30S small subunit assembling of *E. coli*, S4、S7、S8、S15、S17 and S20 that can directly bind to 16S rRNA are called Primary binding protein, the second groups protein including S5、S6、S9、S12、S13、S16、S18 and S19 that can bind to 16S rRNA are called Subprime binding protein, and the others including S2、S3、S10、S11、S14 and S21 are called Late binding protein (Hamacher et al., 2006).

Ribosomal proteins not only participate in balancing the synthesis of the RNA and protein components of the ribosome itself, but also perform other extraribosomal functions such as DNA replication, transcription and repair, RNA splicing and modification, cell growth and proliferation, regulation of apoptosis and development, and cellular transformation, et al. (Wool, 1996; Lai and Xu, 2007; Warner and McIntosh, 2009). The previous studies on ribosomal proteins were focused mainly on *Escherichia coli*, many ribosomal proteins have been identified, and some of their functions have already been clarified. However, the reports on ribosomal proteins and their genes in plant are limited (Arnold and Reilly, 1999; Hamacher et al., 2006).

Therefore, five ribosomal proteins of 40S small subunit from maize inbred line Southern 202 were cloned in this present study, and their sequence characteristics are analyzed in detail. The knowledge stated here is beneficial for maize breeding and studding the function of ribosomal proteins.

MATERIALS AND METHODS

The young and fresh leaves of maize inbred line Southern 202 were friendly provided by Nanchong Academy of Agricultural Sciences, Sichuan Province, China. The tissues were frozen in liquid nitrogen and stored at -80 °C before processing.

Total RNA extraction

Total RNA was isolated from the leaves of maize inbred line Southern 202 with RNAiso Plus reagent (TAKARA Biotechnology (Dalian) Co., LTD., China). The purity and yield of total RNA were determined by monitoring the A260/A280 absorbance ratio with NanoDrop 2000 instrument. The sample integrality was tested by 1.0% agarose gel electrophoresis stained with ethidium bromide (EB). Finally, the RNA samples were diluted to 300 ng/μl.

Cloning of the cDNA encoding region

According to the manufacturer's protocol, cDNA was synthesized from total RNA with a reverse transcription kit named TaKaRa RNA PCR Kit (AMV) 3.0 (TAKARA Biotechnology (Dalian) Co., LTD., China). The 10 μl reaction system included AMV Reverse Transcriptase (5 U/μl) 1 μl, dNTP Mixture (10 mM) 1 μl, $MgCl_2$ (25 mM) 2 μl, 10 × RT Buffer 1 μl, RNase Inhibitor 0.25 μl, Oligo dT-Adaptor Primer (2.5 pmol/ μl) 0.5 μl, RNA sample (500 ng/μl) 1 μl, and finally added RNase Free dH_2O up to 10 μl. The reverse transcriptase reaction was carried out at 42 °C for 30 min, 99 °C for 5 min and 5 °C for 5 min.

The PCR (polymerase chain reaction) primers were designed by Primer Premier 5.0 software, according to the open reading frame (ORF) sequence of reported ribosomal protein genes. The sequences of the forward and reverse primes and other information were listed in Table 1.

The synthesized first-strand cDNA was used as a template. The total reaction volume for DNA amplification was 50 μl. Reaction mixtures contained 5 × PCR Buffer 10 μl, *TaKaRa Ex Taq* HS (5 U/ μl) 0.25 μl, Reverse transcriptase mixture 10 μl, the special primers (20 μM) (Table 1) each 0.5 μl, and added dH_2O up to 50 μl. The amplification procedure was as follows: 94 °C for 2 min, followed by 30 cycles of 94 °C for 30s, 46 °C for 30s, 72 °C for 1 min, and a final elongation step of 72 °C for 7 min.

Table 1. Sequences of primes used for RT-PCR amplification

Gene	Primer sequence (5'→3')		Reference sequences and GenBank accession number
RPS9	forward	ATGGTGCATGTTAACTTCTACC	*Zea mays*, EU964329.1
	reverse	TTACTCCTCGTCCTCCTCA	
RPS10	forward	ATGATCATCTCCAAGAAGAAC	*Zea mays*, EU977087.1
	reverse	TCACTCCATGGAAGATCCACTG	
RPS14	forward	ATGTCGAGGAGGAAGACCA	*Zea mays*, EU960087.1
	reverse	CTACAGCCTCCTTCCCCT	
RPS16	forward	ATGACCGTGCTGAGCC	*Zea mays*, EU963976.1
	reverse	TCAACGGTACGACTTCTGG	
RPS18	forward	ATGTCGCTGATCGCC	*Zea mays*, EU952700.1
	reverse	TTATCGCTTCTTGGAGACACC	

After amplification, PCR products were separated by electrophoresis on 1.5% agarose gel with 1 × TAE (Tris-acetate-EDTA) buffer, stained with EB and visualized under UV light. The expected fragments of PCR products were harvested and purified from gel using a DNA extraction kit (Sangon Biotech (Shanghai) Co., Ltd., China), and then ligated into a pUC18-T vector at 16°C for 12 h. The recombinant molecules were transformed into *E. coli* complete cells (JM109), and then spread onto the LB-plate containing 50 μg/ml ampicillin, 200 mg/ml IPTG, and 20 mg/ml X-gal. After 13 h, pick 3-5 single colonies as template for PCR identification. Finally, the positive cloning was sequenced by Sanger method at Sangon Biotech (Shanghai) Co., Ltd., China.

Sequence analysis

The sequences were analyzed using ORF Finder (http://www.ncbi.nlm.nih.gov/gorf/orfig.cgi); BLAST algorithms (http://blast.ncbi.nlm.nih.gov/Blast.cgi) were used to search for sequence homology; multiple sequence alignment was performed by DNAMAN 6.0 software; theoretical pI and molecular weight of protein were computed online (http://web.expasy.org/compute_pi/); the analysis of protein functional sites was performed online (https://www.predictprotein.org/).

RESULTS

Gene cloning

The target DNA fragments were obtained from the RT-PCR products of electrophoresis, and their sizes were approximately 700 bp, 700 bp, 500 bp, 500 bp, and 500 bp in length. The positive clones were screened out and sequenced, and the results indicated that all the five cloned genes contain a complete ORF, with 582 bp, 540 bp, 453 bp, 447 bp and 459 bp in length, respectively. According to the identity comparison results between our nucleotide sequences and the data in GenBank, it was concluded that the five cloned genes here should be *RPS9*, *RPS10*, *RPS14*, *RPS16* and *RPS18* of maize, and the DNA and encoded amino acids (AA) sequences were shown in Figure 1, Figure 2, Figure 3, Figure 4 and Figure 5, respectively.

```
1     ATGGTGCATGTTAACTTCTACCGCAACTATGGGAAGACTTTCAAGAAGCCAAGGCGGCCG
1      M  V  H  V  N  F  Y  R  N  Y  G  K  T  F  K  K  P  R  R  P
61    TATGAGAAGGAGCGCCTAGATGCTGAGCTGAAGCTGGTTGGTGAGTATGGCCTGCGGTGC
21     Y  E  K  E  R  L  D  A  E  L  K  L  V  G  E  Y  G  L  R  C
121   AAGCGTGAGCTGTGGCGCGTGCAGTATGCCCTGAGCCGTATCAGGAATGCAGCCAGGGAG
41     K  R  E  L  W  R  V  Q  Y  A  L  S  R  I  R  N  A  A  R  E
181   TTGCTCACCCTGGATGAGAAGAACCCACGCCGTATCTTTGAGGGCGAGGCGCTCCTCCGT
61     L  L  T  L  D  E  K  N  P  R  R  I  F  E  G  E  A  L  L  R
241   CGCATGAACAGATATGGTCTTCTTGGCGAGGGACAGAACAAGCTTGATTACGTGCTTGCC
81     R  M  N  R  Y  G  L  L  G  E  G  Q  N  K  L  D  Y  V  L  A
301   CTCACTGTTGAGAACTTCCTCCAGCGCCGCCTCCAGACCATCGTCTTCAAGAATGGCATG
101    L  T  V  E  N  F  L  Q  R  R  L  Q  T  I  V  F  K  N  G  M
361   GCCAAGTCCATCCACCATGCTCGTGTCCTGATCAGGCAGCGCCACATCAGGGTGGGAAGG
121    A  K  S  I  H  H  A  R  V  L  I  R  Q  R  H  I  R  V  G  R
421   CAGCTCGTCAACATCCCGTCGTTCATGGTCAGGGTCGAATCAGAGAAGCACATCGACTTC
141    Q  L  V  N  I  P  S  F  M  V  R  V  E  S  E  K  H  I  D  F
481   TCCCTCACCAGCCCTCTGGGTGGTGGCCCCGCCGGAAGGGTGAAGCGGAAGAACCAGAAG
161    S  L  T  S  P  L  G  G  G  P  A  G  R  V  K  R  K  N  Q  K
541   AAGGCCTCAGGGGGCGGCGACGCTGAGGAGGACGAGGAGTAA
181    K  A  S  G  G  G  D  A  E  E  D  E  E  *
```

Figure 1. The nucleotide and deduced AA sequences of *RPS9* gene cloned in maize. With an ORF of 582 bp encoding 193 AA.

```
1     ATGATCATCTCCAAGAAGAACCGCCGCGAGATCTGCAAGTACATCTTCCATGAGGGGGTT
1      M  I  I  S  K  K  N  R  R  E  I  C  K  Y  I  F  H  E  G  V
61    CTATATGCCAAGAAGGACTACAACCTGGCCAAGCACCCCAAGCTTGACGTGCCCAACCTG
21     L  Y  A  K  K  D  Y  N  L  A  K  H  P  K  L  D  V  P  N  L
121   GAGGTGATTAAGCTCATGCAGAGCTTCAAGTCCAAGGAGTATGTCAGGGAGACCTTCTCC
41     E  V  I  K  L  M  Q  S  F  K  S  K  E  Y  V  R  E  T  F  S
181   TGGCAGTACTACTACTGGTACCTCACCAACGATGGCATTGAGCACCTCCGCAGCTTCCTC
61     W  Q  Y  Y  Y  W  Y  L  T  N  D  G  I  E  H  L  R  S  F  L
241   AACCTGCCGTCAGAGGTTGTGCCCAACACCCTCAAGAAGTCCTCCAAGCCCCCGTCCCGT
81     N  L  P  S  E  V  V  P  N  T  L  K  K  S  S  K  P  P  S  R
301   CCCTTTGGCTCTGGCCCACCGGGTGACCGCCCCAGGGGTCCCCCTCGCTTTGGGGAAGAC
101    P  F  G  S  G  P  P  G  D  R  P  R  G  P  P  R  F  G  E  D
361   AGACCTAGGTTTGGGGATAGGGATGGTTACAGAGGAGGTCCACGAGGTGCAATGGGTGAT
121    R  P  R  F  G  D  R  D  G  Y  R  G  G  P  R  G  A  M  G  D
421   TTTGGTGGTGAGAAGGGTAGTGCTCCTGCGGATTTCCAGCCATCTTTTAGGGGTAGCAGA
141    F  G  G  E  K  G  S  A  P  A  D  F  Q  P  S  F  R  G  S  R
481   CCTGGCTTCGGCCGTGGTGGCGGCAGCGCTTTTGGTGCCAGTGGATCTTCCATGGAGTGA
161    P  G  F  G  R  G  G  G  S  A  F  G  A  S  G  S  S  M  E  *
```

Figure 2. The nucleotide and deduced AA sequences of *RPS10* gene cloned in maize. With an ORF of 540 bp encoding 179 AA.

```
1     ATGTCGAGGAGGAAGACCAGGGAGCCCAAGGAGGAGAACGTCACCCTTGGACCCACTGTC
1      M  S  R  R  K  T  R  E  P  K  E  E  N  V  T  L  G  P  T  V
61    CGTGAAGGAGAGTATGTCTTTGGTGTCGCTCACATCTTTGCATCCTTCAATGACACCTTC
21     R  E  G  E  Y  V  F  G  V  A  H  I  F  A  S  F  N  D  T  F
121   ATTCATATCACTGATTTGTCTGGGAGGGAAACTCTGGTTCGGATCACCGGTGGCATGAAG
41     I  H  I  T  D  L  S  G  R  E  T  L  V  R  I  T  G  G  M  K
181   GTGAAGGCTGACCGTGACGAGTCGTCACCTTACGCTGCTATGCTTGCTGCTCAAGACGTC
61     V  K  A  D  R  D  E  S  S  P  Y  A  A  M  L  A  A  Q  D  V
241   GCACAGCGCTGCAAGGAGCTTGGCATTACTGCACTGCACATTAAGCTTCGTGCCACCGGA
81     A  Q  R  C  K  E  L  G  I  T  A  L  H  I  K  L  R  A  T  G
301   GGCAACAAGACCAAGACCCCTGGACCTGGTGCCCAGTCTGCCCTCAGGGCGCTTGCTCGT
101    G  N  K  T  K  T  P  G  P  G  A  Q  S  A  L  R  A  L  A  R
361   TCCGGGATGAAAATCGGACGCATTGAGGACGTTACCCCGGTCCCCACGGACAGCACTCGC
121    S  G  M  K  I  G  R  I  E  D  V  T  P  V  P  T  D  S  T  R
421   AGAAAGGGCGGTAGGAGGGGAAGGAGGCTGTAG
141    R  K  G  G  R  R  G  R  R  L  *
```

Figure 3. The nucleotide and deduced AA sequences of *RPS14* gene cloned in maize. With an ORF of 453 bp encoding 150 AA.

```
1     ATGACCGTGCTGAGCCGCCCTACCCCGGGCACGGCCCAGTGCTTCGGGCGGAAGAAGACC
1      M  T  V  L  S  R  P  T  P  G  T  A  Q  C  F  G  R  K  K  T
61    GCCGTCGCGGTCGCGTACACGAAGCCGGGGCGCGGCCTGATCAAGGTGAACGGCGTCCCG
21     A  V  A  V  A  Y  T  K  P  G  R  G  L  I  K  V  N  G  V  P
121   ATTGAGCTGATCCGACCGGAGATGCTCCGCCTCAAGGCCTACGAGCCCATCCTGCTGGCG
41     I  E  L  I  R  P  E  M  L  R  L  K  A  Y  E  P  I  L  L  A
181   GGGCGGTCCCGGTTCAAGGACATCGACATGCGGATCCGCGTCCGCGGCGGCGGGAAGACG
61     G  R  S  R  F  K  D  I  D  M  R  I  R  V  R  G  G  G  K  T
241   TCGCAGATCTACGCCATCCGCCAGGCCGTCGCCAAGGGGCTCGTCGCCTACTACCAGAAG
81     S  Q  I  Y  A  I  R  Q  A  V  A  K  G  L  V  A  Y  Y  Q  K
301   TACGTCGACGAGGCCGCCAAGAAGGAGATCAAGGACATCTTTACCCGCTACGATCGCACC
101    Y  V  D  E  A  A  K  K  E  I  K  D  I  F  T  R  Y  D  R  T
361   CTCCTCGTCGCTGACCCCCGGCGCTGCGAGCCGAAGAAGTTCGGCGGACGTGGTGCCCGC
121    L  L  V  A  D  P  R  R  C  E  P  K  K  F  G  G  R  G  A  R
421   GCCAGGTTCCAGAAGTCGTACCGTTGA
141    A  R  F  Q  K  S  Y  R  *
```

Figure 4. The nucleotide and deduced AA sequences of *RPS16* gene cloned in maize. With an ORF of 447 bp encoding 148 AA.

```
1     ATGTCGCTGATCGCCGGGGAGGACTTCCAGCATATCCTGCGTTTGCTGAACACCAACGTG
1      M  S  L  I  A  G  E  D  F  Q  H  I  L  R  L  L  N  T  N  V
61    GATGGGAAGCAGAAGATCATGTTCGCCATGACCTCAATCAAGGGTGTCGGGCGCCGCTTC
21     D  G  K  Q  K  I  M  F  A  M  T  S  I  K  G  V  G  R  R  F
121   TCCAACATCGTCTGCAAGAAGGCCGACATCGACATGAACAAGAGGGCCGGCGAGCTGACG
41     S  N  I  V  C  K  K  A  D  I  D  M  N  K  R  A  G  E  L  T
181   CCTGATGAGCTGGAGCGCCTGATGACGGTCGTGGCCAACCCTAGGCAATTCAAGGTGCCG
61     P  D  E  L  E  R  L  M  T  V  V  A  N  P  R  Q  F  K  V  P
241   GACTGGTTCCTCAACAGGAAGAAGGATTACAAGGACGGCAGGTTCTCGCAGGTCGTCTCC
81     D  W  F  L  N  R  K  K  D  Y  K  D  G  R  F  S  Q  V  V  S
301   AACGCCCTTGATATGAAGCTCAGGGACGACCTTGAGAGGCTCAAGAAGATCAGGAACCAC
101    N  A  L  D  M  K  L  R  D  D  L  E  R  L  K  K  I  R  N  H
361   CGTGGTCTGCGTCACTACTGGGGCCTCCGTGTCCGTGGCCAGCACACCAAGACTACTGGC
121    R  G  L  R  H  Y  W  G  L  R  V  R  G  Q  H  T  K  T  T  G
421   AGGCGTGGAAAGACCGTTGGTGTCTCCAAGAAGCGATAA
141    R  R  G  K  T  V  G  V  S  K  K  R  *
```

Figure 5. The nucleotide and deduced AA sequences of *RPS18* gene cloned in maize. With an ORF of 459 bp encoding 152 AA.

These genes have been submitted to GenBank, and the access number was also allocated. The statistic information on length of ORF, number of encoded AA, pI value, molecular weight and GenBank accession number for the five cloned genes was listed in Table 2. The longest gene is *RPS9*, with 582 bp coding 193 AA. Followed by *RPS10*, *RPS18* and *RPS14*, and the shortest gene is RPS16, only 447 bp in length, including an ORF of 148 AA. The pI values of the five protein deduced from the cloned genes ranged from 9.81 (*RPS10*) to 10.74 (*RPS18*). The calculated molecular weight of the five proteins are also different, *RPS9* had the largest value, up to 22.52 kD, followed by *RPS10*, *RPS18* and RPS16 in turn, the least molecular weight was found in *RPS14*, only 16.39 kD.

Table 2. The cloned ribosomal protein genes in this study

Genes	Length of ORF (bp)	Number of encoded AA	pI value of encoded protein	Molecular weight of encoded protein (kD)	GenBank accession number
RPS9	582	193	10.29	22.52	JX232263.1
RPS10	540	179	9.81	20.06	JX232264.1
RPS14	453	150	10.56	16.39	JX232265.1
RPS16	447	148	10.59	16.77	JX232266.1
RPS18	459	152	10.74	17.69	JX232267.1

Functional sites analysis of cloned genes

The cloned genes were further analyzed functional sites using their encoded proteins online, and the results indicated that *RPS9*, *RPS10*, *RPS14*, *RPS16* and *RPS18* genes had 4, 5, 7, 6 and 5 patterns, respectively (Table 3). There were total ten different patters found in the five proteins deduced from the cloned genes, including cAMP- and cGMP-dependent protein kinase phosphorylation site, protein kinase C phosphorylation site, casein kinase II phosphorylation site, N-myristoylation site, tyrosine kinase phosphorylation site, N-glycosylation site, amidation site, pibosomal protein S11 signature, ribosomal protein S9 signature and Ribosomal protein S13 signature. All the five encoded protein had cAMP- and cGMP-dependent protein kinase phosphorylation site, protein kinase C phosphorylation site and casein kinase II phosphorylation site, and N-myristoylation site was found in all of them but for *RPS10* gene. Whereas, Ribosomal protein S11, S9 and S13 signatures only occurred in *RPS14*, RPS16 and *RPS18* genes, respectively. The results revealed that all the sites were very important for their biological functions in cell.

DISCUSSION

As a part of ribosome, ribosomal proteins play a crucial role in catalyzing protein synthesis and other enzyme activities. In this paper, five ribosomal protein genes (*RPS9*, *RPS10*, *RPS14*, *RPS16*, *RPS18*) of 40S small subunit were successfully cloned from maize inbred line Southern 202. By analyzing these genes with the NCBI/GenBank data, these sequences share high similarity both at the nucleotide sequence and the protein levels. For the five genes cloned in this study, their experimental results were discussed one by one as follows.

Table 3. The functional sites of proteins deduced from cloned genes in this study

Functional patterns found	No.	Position
***RPS9* gene**		
cAMP- and cGMP-dependent protein kinase phosphorylation site	1	180 to 183 KKAS
Protein kinase C phosphorylation site	2	13 to 15 TFK; 154 to 156 SEK
Casein kinase II phosphorylation site	1	63 to 66 TLDE
N-myristoylation site	3	119 to 124 GMAKSI; 167 to 172 GGGPAG; 184 to 189 GGGDAE
***RPS10* gene**		
cAMP- and cGMP-dependent protein kinase phosphorylation site	1	92 to 95 KKSS
Protein kinase C phosphorylation site	5	4 to 6 SKK; 48 to 50 SFK; 90 to 92 TLK; 94 to 96 SSK; 155 to 157 SFR
Casein kinase II phosphorylation site	1	176 to 179 SSME
Tyrosine kinase phosphorylation site	2	6 to 14 KNRREICKY; 123 to 130 RFGDRDGY
N-myristoylation site	6	19 to 24 GVLYAK; 132 to 137 GGPRGA; 142 to 147 GGEKGS; 158 to 163 GSRPGF; 166 to 171 GGGSAF; 172 to 177 GASGSS
***RPS14* gene**		
N-glycosylation site	3	13 to 16 NVTL; 37 to 40 NDTF; 102 to 105 NKTK
cAMP- and cGMP-dependent protein kinase phosphorylation site	1	3 to 6 RRKT
Protein kinase C phosphorylation site	4	2 to 4 SRR; 19 to 21 TVR; 47 to 9 SGR; 138 to 140 STR
Casein kinase II phosphorylation site	3	19 to 22 TVRE; 35 to 38 SFND; 47 to 50 SGRE
N-myristoylation site	4	100 to 105 GGNKTK: 110 to 115 GAQSAL; 122 to 127 GMKIGR; 143 to 148 GGRRGR
Amidation site	1	143 to 146 GGRR
Ribosomal protein S11 signature	1	115 to 137 LRALARSGMKIGRIEDVTPVPTD
***RPS16* gene**		
cAMP- and cGMP-dependent protein kinase phosphorylation site	1	17 to 20 RKKT
Protein kinase C phosphorylation site	1	146 to 148 SYR
Casein kinase II phosphorylation site	1	115 to 118 TRYD
N-myristoylation site	4	10 to 15 GTAQCF; 76 to 81 GGGKTS; 135 to 140 GGRGAR
Amidation site	1	15 to 18 FGRK
Ribosomal protein S9 signature	1	76 to 94 GGGKTSQIYAIRQAVAKGL
***RPS18* gene**		
cAMP- and cGMP-dependent protein kinase phosphorylation site	1	38 to 41 RRFS
Protein kinase C phosphorylation site	3	32 to 34 SIK; 139 to 141 TGR; 149 to 151 SKK
Casein kinase II phosphorylation site	2	18 to 21 TNVD; 60 to 63 TPDE
Amidation site	2	36 to 39 VGRR; 139 to 142 TGRR
Ribosomal protein S13 signature	1	121 to 134 RGLRHYWGLRVRGQ

RPS9 gene

Sequence comparison through Blast Search (http://www.ncbi.nih.gov) using Genbank database showed that the *RPS9* gene cloned by us shares high homology with many other gramineous plants at both nucleic acid and protein levels. The protein sequence encoded by the gene shares 100% similarity with reported *RPS9* gene of maize (EU964329.1). At nucleic acid level, only eight SNPs were found between the *RPS9* sequence of inbred line South 202 and EU964329, one of them is transversion (G-C), others are T-C transition.

Previous studies displayed that altering the C terminus of yeast *RPS9* protein will lead to a significant change in ribosomal association with many mRNAs, *RPS9* may take part in the translation of many mRNAs (Pnueli and Arava, 2007). Lindström and Zhang (2008) thought *RPS9* is a novel B23/NPM-binding protein, which is required at the stage of normal cell proliferation. Subsequently, Lindström and Nistér (2010) found functional silence of ribosomal protein S9 will cause induce senescence or apoptosis of cancer cells. But the functional difference of *RPS9* protein in plants has not been studied. In our results, *RPS9* gene was successfully cloned from the maize inbred line Southern 202, encoding 193AA. According to the search results in GenBank database, there may be two different *RPS9* genes in different maize variety: one gene encodes 193 AA including our results; the other encodes 194 AA (NM_001155768.1). Comparison of functional sites indicated that N-myristoylation site is also different between the gene cloned by us (JX232263.1) and NM_001155768.1. On the contrary, in the animal kingdom, *RPS9* gene of fruit fly and other insects encodes 195 AA residues, and all vertebrates also encode 195 AA residues. These results indicated the *RPS9* gene shares high conservative, but unlike all vertebrates, encoded AA is stable in number. So the evolution process of *RPS9* gene is worth studying.

RPS10 gene

The sequence of cloned *RPS10* gene from maize inbred line South 202 shares high similarity with other reported genes in gramineous plants. In different varieties of maize, the sequence presents difference. At protein level, only the 125th AA residue (V-G) is different between the sequence we cloned and NP_001152734.1, which is caused by the 374th site base (T-G) change. Another single-nucleotide polymorphism (SNP) sit is found in the 487th site base, the change located in the third nucleotide of a codon, belonging to synonymous mutation.

For *RPS10* protein, it is general encoded in the mitochondrion, whereas in several other species it is encoded in nuclear and thus must be imported into the mitochondrial matrix to function (Knoop et al., 1995). Interestingly, in Lactuca sativa and Daucus carota, *RPS10* independently gained different N-terminal extensions from other genes, which are essential for mitochondrial import, and following transfer to the nucleus. However, in maize, *RPS10* has not acquired an extension upon transfer, but it can be readily imported into mitochondria (Murcha et al., 2005). In fact, the *RPS10* gene is absent from the mitochondrial genome of rice and has been transferred to the nucleus (Kubo et al., 2000). There are two *RPS10* genes in the rice nuclear genome and that their transcripts differ in abundance, the two *RPS10* genes were mapped on chromosomes 6 and 12 by RFLP (restriction fragment length polymorphism) markers. Majewski et al. (2009) found that *RPS10* takes part in photoperiod regulation, its express level gradually increase in vegetative phase in Arabidopsis. However, about the report on its function is fewer. Recently, Yazaki (2012) found some DBA patients with mutations in *RPS10*. In maize, there is at least another *RPS10* gene, which encodes 182 AA (NM_001155450) and shares high similarity with our gene encoding 179 AA (JX232263.1). In animal kingdom, the AA number encoded by *RPS10* gene is also different. The *RPS10* proteins of fruit fly and zebra fish encode 163, 166 AA, respectively; while mammalia it encodes 165 AA residues, and thus, the AA number of *RPS10* protein is quite different between animal and plant.

RPS14 gene

RPS14 protein is widespread in the lower and higher organisms. Previous research displayed that *RPS14* have transferred from the mitochondrion to the nucleus in grasses, and maize mitochondrial genome does not contain *RPS14*, nevertheless, in wheat there is a nonfunctional *RPS14* pseudogene in mitochondrial DNA of wheat, the functional *RPS14* gene is located in the nucleus (Sandoval, et al., 2004; Figueroa et al., 2000). As for its function, Zhou et al. (2013) thought *RPS14* takes part in regulation of MDM2-p53 feedback loop in response to ribosomal stress. It binds to the central acidic domain of MDM2 and inhibits MDM2 activity, elevating p53 level and activity. Oliva et al. (2010) found that its loss is a potential causal factor of 5q- syndrome in humans. In this study, we successfully cloned the coding region of *RPS14* gene, the length of cDNA fragment cloned is 453 bp in size, encoding 150 AA.

Alignment analysis indicated the nucleotide sequence and the deduced AA sequence showed high similarity with other species or variety studied. However, both the nucleotide sequence and the deduced AA sequence are different even in the same species. The *RPS14* gene in maize with accession number NM_001158971.1 encodes 160 AA residues, while the others encode 150 AA residues, including EU960087.1, NM_001158971, NM_001137768 and ours (JX232265.1). Additionally, the *RPS14* genes from sorphum (XM_002437295) and rice (NM_001052517) also encode 150 AA, but the *RPS14* genes encoding 151 AA were already found from fruit fly to higher mammalian. To be noticed, compared to ours (JX232265.1), the *RPS14* gene of NM_001158971 in maize has an additional Casein kinase II phosphorylation site, but missing an N-glycosylation site and Protein kinase C phosphorylation site. So, the function of *RPS14* gene needs to be further confirmed.

RPS16 gene

The study on *RPS16* gene is rather less, especially in plant. Wood et al. (2000) found that *RPS14* and *Rps16* performed high expression level in tortula ruralis gametophytes during a desiccation-rehydration cycle. Ajuh et al. (2000) found RPS16 protein has interaction with CDC5L, which act as a positive regulator of cell cycle G2/M progression. Here, we reported the RPS16 gene of the maize inbred line South 202, which contains an ORF of 447 bp encoding 148 AA.

The RPS16 proteins derived from maize and sorghum encode 148 and 149 AA, respectively, but they share over 92% similarity in AA sequence and the difference of N-terminal between them is obvious. To be noticed that RPS16 genes coding 148 and 146 AA was found in fruit fly and vertebrate, respectively. The knowledge here will be helpful for further study the function of *RPS16* gene.

RPS18 gene

RPS18 protein could be a novel substrate for CaMK II, providing a potential link between Ca^{2+}-mobilizing agents and protein translation (Mishra-Goruret et al., 2002), but subsequently, *RPS18* was identified as a cofilin-binding protein (Kusuiet et al., 2004). In *E. coli*, *RPS13* protein, the ortholog product of *RPS18* gene, is involved in the binding of fMet-tRNA and in the initiation of translation (Kenmochi et al., 1998). In this study, we successfully cloned the *RPS18* gene, which encode 152 AA, its nucleic acids sequences and protein sequence is highly conservative in evolution according to the data in GenBank.

The sequence cloned here (JX232267.1) is full accord with NM_001147782.1. Compared with NM_001158140.1 and EU952700.1, only an AA residue (the 62 site) was found, but at the nucleic acid level, Many SNPs sites are displayed between them. Unlike the *RPS9*, *RPS10*, *RPS14* and *RPS16* genes, the AA number of *RPS18* protein is identical between plant and animal, which suggested that the *RPS18* shares high conservation. For its function, Maroniche et al. (2011) found *RPS18* gene exhibits a stable expression levels in virus-infected plant hoppers and can be used as a reference gene of real-time quantitative PCR.

CONCLUSION

We have successfully isolated five ribosomal protein genes from maize inbred line Southern 202, including *RPS9*, *RPS10*, *RPS14*, RPS16 and *RPS18* genes, these genes have high similarities to other related genes at both nucleic acids and protein sequences levels. In addition, the functional sites of the five genes were analyzed online. Compared to previous studies on ribosomal protein genes with out results, it was found that the number of AA residues encoded presents differences between same ribosomal protein genes in different species or varieties. These results indicated the conservation is lower in plant even the same species than higher animal. Comparatively, the ribosomal genes in animal, especially mammalian, are more conservative than those in plant. The knowledge stated here are helpful for future studying the physiological function of ribosomal proteins in plant.

ACKNOWLEDGEMENTS

The study was financially supported by Scientific Research Fund of Sichuan Provincial Education Department (13ZA0012).

LITERATURE CITED

Ajuh, P., B. Kuster, K. Panov, J.C. Zomerdijk, M. Mann and A.I. Lamond, 2000. Functional analysis of the human CDC5L complex and identification of its components by mass spectrometry. EMBO J. 19: 6569-6581.

Arnold, R.J. and J.P. Reilly, 1999. Observation of *Escherichia coli* ribosomal proteins and their posttranslational modifications by mass spectrometry. Anal. Biochem. 269: 105-112.

Figueroa, P., L. Holuigue, A. Araya and X. Jordana, 2000. The nuclear-encoded SDH2-RPS14 precursor is proteolytically processed between SDH2 and RPS14 to generate maize mitochondrial RPS14. Biochem. Biophys. Res. Commun. 271: 380-385.

Hamacher, K., J. Trylska and J.A. McCammon, 2006. Dependency map of proteins in the small ribosomal subunit. PLoS Comput. Biol. 2: e10.

Kenmochi, N., T. Kawaguchi and S. Rozen, 1998. A map of 75 human ribosomal protein genes. Genome Res. 8: 509-523.

Knoop, V., T. Ehrhardt, K. Lättig and A. Brennicke, 1995. The gene for ribosomal protein S10 is present in mitochondria of pea and potato but absent from those of Arabidopsis and Oenothera. Curr. Genet. 27: 559-564.

Kubo, N., X. Jordana, K. Ozawa, S. Zanlungo, K. Harada, T. Sasaki and K. Kadowaki, 2000. Transfer of the mitochondrial rps10 gene to the nucleus in rice: acquisition of the 5' untranslated region followed by gene duplication. Mol. Gen. Genet. 263: 733-739.

Kusui, K., H, Sasaki, R. Adachi, S. Matsui, K. Yamamoto, T. Yamaguchi, T. Kasahara and K. Suzuki, 2004. Ribosomal protein S18 identified as a cofilin-binding protein by using phage display library. Mol. Cell. Biochem. 262: 187-193.

Lai, M.D. and J. Xu, 2007. Ribosomal proteins and colorectal cancer. Curr. Genomics 8: 43-49.

Lindström, M.S. and M. Nistér, 2010. Silencing of ribosomal protein S9 elicits a multitude of cellular responses inhibiting the growth of cancer cells subsequent to p53 activation. PLoS One 5: e9578.

Lindström, M.S. and Y. Zhang, 2008. Ribosomal protein S9 is a novel B23/NPM-binding protein required for normal cell proliferation. J. Biol. Chem. 283: 15568-15576.

Majewski, P., M. Wołoszyńska and H. Jańska, 2009. Developmentally early and late onset of Rps10 silencing in Arabidopsis thaliana: genetic and environmental regulation. J. Exp. Bot. 60: 1163-1178.

Maroniche, G.A., M. Sagadín, V.C. Mongelli, G.A. Truol and M. del Vas, 2011. Reference gene selection for gene expression studies using RT-qPCR in virus-infected planthoppers. Virol. J. 8: e308.

Mishra-Gorur, K., H.A. Singer and J.J.Jr. Castellot, 2002. The S18 ribosomal protein is a putative substrate for Ca^{2+}/calmodulin-activated protein kinase II. J. Biol. Chem. 277: 33537-33540.

Murcha, M.W., C. Rudhe, D. Elhafez, K.L. Adams, D.O. Daley and J. Whelan, 2005. Adaptations required for mitochondrial import following mitochondrial to nucleus gene transfer of ribosomal protein S10. Plant Physiol. 138: 2134-2144.

Oliva, E.N., M. Cuzzola, F. Nobile, F. Ronco, M.G. D'Errigo, C. Laganà, F. Morabito, S. Galimberti, A. Cortelezzi, M.A. Aloe Spiriti, G. Specchia, A. Poloni, M. Breccia, R. Ghio, C. Finelli, P. Iacopino, G. Alimena and R. Latagliata, 2010. Changes in RPS14 expression levels during lenalidomide treatment in Low- and Intermediate-1-risk myelodysplastic syndromes with chromosome 5q deletion. Eur. J. Haematol. 85: 231-235.

Pnueli, L. and Y. Arava, 2007. Genome-wide polysomal analysis of a yeast strain with mutated ribosomal protein S9. BMC Genomics 8: e285.

Sandoval, P., G. León, I. Gómez, R. Carmona, P. Figueroa, L. Holuigue, A. Araya and X. Jordana, 2004. Transfer of RPS14 and RPL5 from the mitochondrion to the nucleus in grasses. Gene 324: 139-147.

Warner, J.R. and K.B. McIntosh, 2009. How common are extraribosomal functions of ribosomal proteins? Mol. Cell 34: 3-11.

Wen, Z.M., 2002. Recurrent selection in maize strain population. Journal of Maize Sciences 10(2): 11-13.

Wood, A.J., R. Joel Duff and M.J. Oliver, 2000. The translational apparatus of Tortula ruralis: polysomal retention of transcripts encoding the ribosomal proteins RPS14, RPS16 and RPL23 in desiccated and rehydrated gametophytes. J. Exp. Bot. 51: 1655-1662.

Wool, I.G., 1996. Extraribosomal functions of ribosomal proteins. Trends Biochem. Sci. 21: 164-165.

Yazaki, M., M. Kamei, Y. Ito, Y. Konno, R. Wang, T. Toki and E. Ito, 2012. A novel mutation of ribosomal protein s10 gene in a Japanese patient with diamond-blackfan anemia. J. Pediatr. Hematol. Oncol. 34: 293-295.

Zhou, X., Q. Hao, J. Liao, Q. Zhang and H. Lu, 2013. Ribosomal protein S14 unties the MDM2-p53 loop upon ribosomal stress. Oncogene 32: 388-396.

14

THE SHORT BREEDING CYCLE PROTOCOL EFFECTIVE ON DIVERSE GENOTYPES OF SUNFLOWER (*Helianthus annuus* L.)

Nazan DAGUSTU [*], *Gamze BAYRAM*, *Mehmet SINCIK*, *Melek BAYRAKTAROGLU*

Uludag University Agricultural Faculty, Department of Field Crops, Bursa, TURKEY
**Corresponding author: ndagustu@uludag.edu.tr*

ABSTRACT

Immature embryo culture of sunflower (*Helianthus annuus* L.) was studied for shortening the generation time in breeding programs. The seed development from pollination to maturity in sunflower takes 50-60% (60 days) of the life cycle duration (120-150 days). This technique allows the production of fertile plants from immature embryos of 11 sunflower genotypes. Immature embryos of 10-12 days after pollination were dissected from seed grown plants (SGP), were transferred into MS medium allowing shoot and root development for 5-10 days. Young plantlets were transferred to soil, developed to maturity and were then self pollinated and set seed. The first cycle of immature embryo raised plants (IERP) was obtained. The four cycles of IERP were obtained from immature embryo culture technique in contrast to one generation per year with conventional breeding. The majority of cultured embryos developed into vigorous plantlets with 3-6 leaves. Out of 1320 immature embryos, the average response of the explants were 92.1% (1216). The 75% of the developed plantlets had vigorous roots and were transplanted into viol containing 1:1:2 peat: perlite: soil mixture (v/v) at 24 ± 2°C in 16h/8 h (light/dark) in the growth chamber. The only 70.3% of them was grown to maturity, self-pollinated and set seed. The overall result was average 40-45 regenerated and matured plants per 100 immature zygotic embryos. The regenerated plants also showed no morphological changes. The analysis of variance for all agronomic characters (plant height, head diameter, number of leaves, stem diameter, number of branches and seed number per head) taken from the mean of four generations *in vitro* grown plants resulted in significant differences among genotypes at 5% level. All the agronomic characters examined at *in vitro* regenerated plants decreased compared to field grown plants.

Key words: fertile plant regeneration, immature embryo, shortening the breeding cycle, sunflower.

INTRODUCTION

Sunflower (*Helianthus annuus* L.) is one of the most important oil seed crops in both Turkey and the world. Development of hybrids which have valuable agronomic performances is an important target in sunflower breeding program. Sunflower is a highly cross-pollinated crop and a breeding program of these crops involves several breeding cycles. Obtaining homozygous parental lines for hybrid seed production in short period of time are of great importance for commercial hybrid production. In general sunflower plants were grown from seeds once a growing season in the field conditions in many countries. Therefore, only one generation a year is taken during classical breeding program. The homozygous lines and conversion of an inbred to a cytoplasmic male sterile (CMS) line can be obtained within minimum 6 generations by conventional methods (Jambhulkar, 1995; Hu et al., 2008). On the other hand, it is also known that sunflower seeds keep germination capacity about 6 days after pollination, becomes dormant 16 days after pollination (Maiti et al., 2006), remain dormant for 45-60 days (Jambhulkar, 1995) and seed maturation in sunflower takes 50-60% of the life cycle duration (Serieys, 1992).

A field grown sunflower crop requires at least 3-4 months from sowing to maturity for reseeding the next generation. Of this time, seed development from pollination to maturity takes 2 months of the life cycle duration (Hahne, 2002). *In vitro* immature embryo culture reduces the flowering and seed maturation, shortening the entire lifecycle, thus increasing the number of generation that can be produced annually. The length of one cycle is quite often limiting factor for performing a breeding aim in a reasonable period of time. This system can also reduces the cycle of backcross and selfing of lines, thus breeding process speed up and overcome seed dormancy problem. Consequently, reducing the duration of one cycle should have a great importance for the breeder to obtain seed propagated cultivars, lines and new source of maintainers (B), restorers (R) and cytoplasmic male sterile (CMS) lines.

Application of new technologies such as *in vitro* tissue culture methods in sunflower (anther culture, microspore culture, ovule culture, immature embryo culture etc.) accelerates the breeding process. Embryo culture has been used successfully by plant breeders to solve the problems of seed set, seed dormancy, slow seed germination, inducing embryo growth in the absence of symbiotic partner, shortening the breeding cycle, rapid seed viability test, obtaining rare hybrids and homozygous lines, haploid production (Yeung et al., 1981; Chandler and Beard, 1983; Gürel et al., 1991; Zhong et al., 1995; Bhojwani and Razdan, 1996; Torresàn et al., 1996; Saji and Sujatha, 1998; Raghavan, 2003). By using embryo culture, breeding cycle of sunflower (Jambhulkar, 1995; Torresàn et al., 1996), artichoke (Cravero and Cointry, 2007), wheat (Sharma and Gill, 1982), soybean (Roumet and Morin, 1997), orchids, roses and banana (Yeung et al., 1981), iris (Randolph, 1945) azalea (Michishita et al., 2001) were shortened from 1 to 3 years to a few months depended on the plants. The sunflower embryo culture system was first developed for interspecific hybridization by Chandler and Beard (1983). Jambhulkar (1995) developed a rapid embryo-raised plant system for sunflower production from immature embryos, which allow five cycles in 316 days. The purpose of the present study is to increase the generation time *via* immature embryo culture. This technique was developed for rapid propagation of parental lines (B, R and CMS) and hybrids by speeding up the breeding process in sunflower industry.

MATERIALS AND METHODS

The 11 sunflower genotypes (Record 109/Sanay 3-5 B, Record 109/Sera B, BGC0565, IMIN3, RHA 04, RHA 07, RHA 14, RHA 16, IMICMS6 X IMIN6, T0910131-2, T091241-3) were obtained from 3 different sources (sunflower collections of Field Crops Department, Uludağ University, Bursa; Trakya Agricultural Research Institute, Edirne; TMT Ltd. Şti., Tekirdağ, Turkey). These genotypes were chosen for their early flowering and maturing capacity and the shortest plant height within population. Sunflower collections were planted into 5 m lines with 0.65 m row spacing and 0.3 m plant-plant spacing at Uludağ University, Agricultural Research and Experimental Station in 2010.

The plants were hand-pollinated, and maintained until the seeds were set. The first regeneration cycle has started from seed grown plants (SGP). Ten to 12 days after pollination the heads were brought into the laboratory. The achenes placed at the outer 3-4 rows of the inflorescence were separated. For homogenize the embryos, achenes with dark color had been selected. The embryos with pericarps were surface sterilized in 20% commercial bleach with a few drops of detergent for 15 minutes. They were then rinsed three times in sterile distilled water. The immature embryos were excised from the pericarps, endosperm and embryo sac for producing the first cycle of plants and placed into Murashige and Skoog (MS, 1962) medium with 2% sucrose and 0.8% agar at pH 5.6-5.7 as described by Jambhulkar (1995). All studies were carried out under sterile conditions. The petri dishes were sealed with parafilm to prevent moisture loss and placed in the light (16h/8h day/night) at 25 ± 2 °C and cultured until plantlets production. Within 1-2 weeks of culture in MS, developing embryos grew plantlets with strong roots (2-6 cm and branching) and stem (3-5 cm) with 6-8 leaves (~1 cm). Plantlets washed agar debris were then placed in 31 by 51 cm plastic multipot trays of 48 pots per tray containing sterile 1:1:2 peat:perlite:soil mixture (v:v). The trays were covered with a plastic bag and incubated for a period of 7-10 days in the growth chamber at 25 ± 2 °C in the light (16 h/ 8 h day/night), after which the plantlets were transferred to 320 X 270 mm (16 lt) pots with regular soil, keeping 5 plants per pot at the same growth chamber. Three pots were set up for each genotype. The plantlets were irrigated with sterile water 3 times weekly to maintain a high humidity environment and fertilized every 10 days. Plantlets had flowers approximately 7-8 weeks after sowing and set seed.

The regeneration response of immature embryos was scored after 2 weeks, evaluating the growth of the shoot, root and their intensity. The number of plantlets regenerated and mature plants obtained were recorded. Each genotype consisted of 10-15 petri dishes (9 cm) with 10 embryos in each. Observations on efficiency of the immature embryos (%), embryos developing plantlets (number) were evaluated.

A number of growth parameters (head production, 50% of flowering dates, pollination dates, embryo emasculation dates, transfer to soil dates) and agronomic traits (plant height, head diameter, stem diameter, number of leaves per plant, number of branches, number of seeds per head) were determined for each of the individual plants in the field and plants grown in pots. Five plants were randomly selected from each genotype grown from original seeds in the field conditions. Mean data were presented with standard deviations. For *in vitro* grown plants, 3 pots with 5 plants each were subjected to variance analysis in the completely randomized design with JUMP statistical program. The agronomic performances of *in vitro* grown plants were determined by the mean of four generations. The significance of genotypes was calculated at the 0.05 probability level by the Student's t-test. The experiments were carried out for 2 years to obtain reliable results.

RESULTS AND DISCUSSION

The first cycle of immature embryo raised plants (IERP) had taken 85-90 days after sowing. The 2nd, 3rd and then 4th cycle of IERP were obtained from *in vitro* grown plants. Immature seeds removed as early as 10-12 days after pollination were successfully cultured *in vitro*. All of the cultured genotypes developed and produced plantlets at the first stage in MS medium. Within two weeks, developing embryos mainly grew plantlets with strong roots (2-6 cm and branching) and stem with 6-8 leaves (~1 cm) as described by Freyssinet and Freyssinet (1988).

The well developed plantlets were selected to transfer into viol containing peat:perlite:soil mixture for acclimatization. The plants were kept for two weeks in growth chamber. Almost all of the plantlets grew healthy in the viols in green house conditions and were phenotypically similar to original seeds. They were generally short, often branched with a flower on each branch. The only 15 of them placed into pots (5 plantlets /

pot). They survived to flower and set seed after 8 weeks. The premature flowering did not occur *in vitro*. Mature plants exhibited normal phenotype and were fertile. When the unshaped and small plantlets were transplanted to soil, they did not survive. Days from culturing to harvesting of mature seeds were almost 80-95 days depended on the genotype.

Out of 1320 immature embryos, the average response of the explants was 92.1% (1216) showing morphogenesis with a range of 90-100%. The majority of cultured embryos developed into plantlets like Jeannin and Hahne (1991). Of the embryos that placed into embryo growth medium, 85% was in peat:perlite:soil mixture and the only 77.5% of them was grown to maturity, self-pollinated and set seeds. This study reports the results of improvement of the growing sunflower plants from *in vitro* immature embryo culture in the growth chamber.

The mean values for agronomic performances of SGP and IERP sunflower genotypes are given in Table 1 and Table 2. Although there were no big morphological changes in *in vitro* regenerated plants, they showed variations for plant height, head diameter, stem diameter, number of branches, number of leaves per plant and number of seed per head.

Table 1. The agronomic performances (plant height, head diameter and number of leaves) of seed grown and *in vitro* grown plants in sunflower genotypes.

Genotypes	Plant height (cm)		Head diameter (cm)		Number of leaves (no)	
	SGP	IERP*	SGP	IERP*	SGP	IERP*
RHA04	93.0 ± 2.6	40.8ef	7.5 ± 0.2	2.06de	21.0 ± 1.1	16.4bc
RHA 07	97.4 ± 3.8	41.2ef	9.0 ± 0.3	2.14de	22.4 ± 0.5	17.0bc
RHA 14	84.8 ± 3.4	60.6b	6.1 ± 0.3	4.50a	21.2 ± 0.4	17.3bc
RHA 16	58.8 ± 6.0	51.04c	5.2 ± 1.0	2.06de	18.6 ± 1.9	18.3ab
T0910131-2	70.4 ± 1.9	41.8ef	11.0 ± 0.6	2.32cd	24.6 ± 2.4	20.2a
T091241-3	120.0 ± 8.4	61.2b	10.6 ± 1.4	2.82bc	18.4 ± 1.6	18.3ab
IMIN3	75.8 ± 5.8	41.9ef	7.5 ± 1.1	1.54f	22.4 ± 1.4	16.2c
IMICMS6 X IMIN6	78.8 ± 4.1	37.9f	7.5 ± 0.8	1.77ef	22.8 ± 0.9	15.6c
Record 109/Isera	70.2 ± 3.1	45.2de	5.3 ± 0.1	2.37cd	19.0 ± 0.5	17.0bc
Record 109/Sanay 3-5(B)	81.4 ± 7.9	69.1a	7.8 ±0.9	2.67bc	20.8 ± 0.8	16.8bc
BGC0505	68.0 ± 2.4	48.4cd	4.5 ± 0.2	3.08b	18.6 ± 0.9	16.6bc

*the mean of 4 generations were taken from *in vitro* grown plants

Table 2. The agronomic performances (stem diameter, number of branches and number of seed per head) of seed grown and *in vitro* grown plants in sunflower genotypes.

Genotypes	Stem diameter (cm)		Number of branches (no)		Number of seed per head (no)	
	SGP	IERP*	SGP	IERP*	SGP	IERP*
RHA04	0.46 ± 0.0	0.38c-e	14.6 ± 0.7	3.12bc	558.3 ±33.5	16.3e
RHA 07	0.54 ± 0.0	0.47bc	8.6 ± 0.5	3.31b	579.3 ±148.9	41.5ab
RHA 14	1.86± 0.0	1.28a	7.4 ± 0.4	4.30a	646.3 ±167.8	72.7a
RHA 16	0.86± 0.0	0.36de	8.0 ± 0.7	1.79ef	172.7 ±58.0	22.1c-e
T0910131-2	2.04 ± 0.2	0.43b-d	2.8 ± 1.4	1.04g	841.7 ±160.9	17.7de
T091241-3	1.64 ± 0.2	0.39b-e	3.6 ± 0.4	1.50fg	787.0 ±133.4	30.1b-e
IMIN3	0.56± 0.2	0.47bc	10.4 ± 0.5	2.56cd	382.0 ±58.9	32.2a-c
IMICMS6 X IMIN6	0.88± 0.0	0.41b-e	10.6 ± 0.9	2.44c-e	535.3±21.5	18.8de
Record 109/Isera	0.60± 0.0	0.49b	7.6 ± 0.7	3.11bc	415.7±84.0	39.3a-c
Record 109/Sanay 3-5(B)	0.40± 0.0	0.31e	10.4 ± 1.2	2.13d-f	494.7±27.1	35.8a-c
BGC0505	0.72± 0.0	0.43b-d	7.0 ± 0.3	3.80ab	508.7±81.1	38.9a-c

*the mean of 4 generations were taken from *in vitro* grown plants

The seed grown plants had generally higher agronomic performances compared to *in vitro* grown plants as reported by Jeannin and Hahne (1991). In general, the plant height, stem diameter, number of seed per head, head diameter decreased when *in vitro* technique was used as indicated by Encheva et al., (2003) and Gopalkrishnan (1993). The possible reason for decreasing of the agronomic performances is growing conditions of the plants. The plant height of the individual sunflower plants obtained from *in vitro* culture varied between 37.9 and 69.1 cm as it was revealed earlier by Hahne (2002). Record 109/Sanay A 3-5 had the highest plant height (69.1 cm) while IMICMS6 X IMIN6 had the shortest plant height (37.9 cm). The plant height of seed grown plants varied between 58.8 and 120.0 cm. It was lower than 100 cm except T091241-3 (120.0 cm). The regenerated plants had minimum 1.54 and maximum 4.5 cm head diameter at harvest stage. The similar results with reduced head diameter of sunflower after *in vitro* grown conditions were published by Encheva et al., (2003). The size of the individual flowers chanced from a few ray flowers surrounding a small number of disk flowers to sunflower heads of reasonable size with more than 70 seeds in this study. The seed production of SGP varied from 173 to 842 while *in vitro* grown plants had 16-73 seeds (average of 24 seeds) under our growth conditions.

All the plants transferred to soil produced flower, contained both fertile and sterile seeds. When the plants were harvested, all the heads produced seeds changing from 16.3 to 72.7 in total depended on the genotype. Similarly, Gopalkrishnan et al., (1993) observed that the vegetative phase of embryo derived plants was very short resulting in small plants (12-25 cm), flower buds and early flowering compared to field-grown plants. The possible reason for decreasing of the agronomic performances is growing conditions of the plants.

One of our study conducted at field conditions in Marmara region showed that the agronomic features of sunflower in the field conditions was very high (Göksoy et al., 2003). Our results indicated that it is possible to produce viable seeds from *in vitro* grown plants and to obtain four generations a year.

The overall result was average 54 regenerated and matured plants per 563 excised immature embryos. Although agronomic characters of *in vitro* regenerated plants decreased, more than 16 seeds were harvested for each genotype and majority of them was viable. Therefore, it was sufficient for continuing to the next generation. This method will be useful for breeders to advance sunflower breeding. The four generations in a year advanced the breeding of genetic population. Therefore, homozygosity may be developed in 2 years if the population is derived from a cross between two inbred lines.

It is also known that when the mature seeds of most sunflower lines are freshly harvested, they do not germinate due to dormancy (Gopalkrishnan et al., 1993; Hahne, 2002). By using immature embryo culture dormancy problem can be solved easily. This technique will shorten the breeding cycle for sunflower, permitting the production of plantlets almost one third of a year earlier than (waiting for mature seed collection, sowing and germination) seed grown in field. In conclusion, embryo culture techniques avoid embryo abortion and seed dormancy. This technique can be useful in breeding schemes where small numbers of plants are needed per generation.

The rapid cycling genotypes of sunflower selected in our conditions were completed their life cycles in about 106-112 days in the field conditions compared to the 150- 160 days of late flowered genotypes (Data were not shown). In general, the genotypes selected above needed 85-95 days for completing one generation by *in vitro* immature embryo culture.

Consequently, both immature embryo culture and selection of possible rapid cycling lines of sunflower produced 4 generations per year. However, the minimum requirement for completing the seed to seed cycle was 86 days by using dwarf and early maturing genotype Morden and, the embryo raised plants completed their 1st cycle at 65 days. Thus, the researcher managed to obtain 5 generations in a year (Jambhulkar, 1995).

Based on the results from the present study, the time required to recover immature embryos averaged 3 months. The length of time includes the following steps: 1) 2 weeks in *in vitro* conditions for production of plantlets 2) 2 weeks acclimatization period from *in vitro* to growth room chamber conditions, 3) 4-5 weeks of plant growth in the green house to flowering and pollen shed, 4) 2 weeks to allow the seeds to fill and discarded immature embryo.

This study reports the results of improvement growing sunflower plants from *in vitro* immature embryo culture in the growth chamber and green house conditions. This system is available for use by sunflower breeders and researches to create hybrids, parental lines in a very short time at breeding studies.

ACKNOWLEDGMENT

This research was supported by the Scientific Research Projects Unit of Uludag University (Project No: 2009/28). The author is grateful to the Scientific Research Projects Unit of Uludag University for financial support.

LITERATURE CITED

Bhojwani, S. S., M. K. Razdan,1996. Chapter 11: Zygotic Embryo Culture. In: Plant Tissue Culture: Theory and Practice, A Revised Edition. Elsevier, Science B.V. The Netherlands. pp: 297-335.

Chandler, J. M., B. H. Beard,1983. Embryo culture of *Helianthus* hybrids. Crop Science, 23: 1004–1007.

Cravero, V., E. Cointry, 2007. Short communication. Shortening the seed-to-seed cycle in artichoke breeding by embryo culture. Plant Breeding, 126: 222-224.

Encheva, J., H. Köhler, W. Friedt, F. Tsvetkova, P. Ivanov, V. Encheva, P. Shindrova, 2003. Field evaluation of somaclonal variation in sunflower (*Helianthus annuus* L.) and its application for crop improvement. Euphytica, 130: 167-175.

Freyssinet M., G. Freyssinet, 1988. Fertile plant regeneration from sunflower (*Helianthus-annuus* L.) immature embryos. Plant Science, 56: 177-181.

Gopalkrishnan K., M. R. Naidu,D. Sreedhar, 1993. Shortening breeding cycle through immature embryo culture in sunflower (*Helianthus annuus* L.). Helia, 16: 61-68.

Göksoy, A. T., A. O., Demir, Z. M. Turan, N. Dağüstü, 2003. Responses of sunflower (*Helianthus annuus* L.) to full and limited irrigation at different growth stages. Field Crops Research, 87: 167-178.

Gürel, A., K. Nichterlein, W. Friedt, 1991. Shoot regeneration from anther culture of sunflower (*Helianthus annuus*) and some interspecific hybrids as affected by genotype and culture procedure. Plant Breeding, 106: 68-76.

Hahne, G., 2002. 54 Sunflower seed. In: Transgenic Plants and Crops. Eds., G.G. Khachatourians, A. McHughen, R. Scorza, W-K. Nip, Y. H. Hui, pgs: 813-830.

Hu, J., B. Yue, W. Yuan, B. A. Vick, 2008. Growing sunflower plants from seed to seed in small pots in greenhouse. Helia, 31(48): 119-126.

Jambhulkar, S. J., 1995. Rapid cycling through immature embryo culture in sunflower (*Helianthus annuus* L). Helia, 18(22): 45-50.

Jeannin, G., G Hahne, 1991. Donor plant growth conditions and regeneration of fertile plants from somatic embryos induced on immature zygotic embryos of sunflower (*Helianthus annuus* L.). Plant Breeding, 107: 280-287.

Maiti, R. K., P. Vidyasagar, S. C. Shahapur, S. K. Ghosh, G. J. Seiler, 2006. Development and standardization of a simple technique for breaking seed dormancy in sunflower (*Helianthus annuus* L.). Helia, 29(45): 117-126.

Michishita A., K. Ureshino, I. Miyajima, 2001. Shortening the period from crossing to the seedling stage through ovule culture of interspecific crosses of azelea (*Rhododendron* spp.). J. Japan. Soc. Hort. Sci., 70(1): 54-59.

Murashige, T., F. Skoog, 1962. A revised medium for rapid growth and bioassays with tobacco tissue cultures. Physiol. Plant., 15(3): 473-497.

Raghavan, V., 2003. One hundred years of zygotic embryo culture investigations. In vitro Cell. Dev. Biol.-Plant., 39: 437-442.

Randolph , L. F., 1945. Embryo culture of *Iris* seeds. Bulletin of the American Iris Society. 97: 33-45.

Roumet, P., F. Morin, 1997. Germination of immature soybean seeds to shorten reproductive cycle duration. Crop Science, 37: 521-525.

Saji, K. V., M. Sujatha, 1998. Embryogenesis and plant regeneration in anther culture of sunflower (*Helianthus annuus* L.). Euphytica, 103(1): 1-7.

Serieys, H., 1992. Cytoplasmic effects on some agronomical characters in sunflower. Proceedings of the 13th International Sunflower Conference, 2: 1245-1250.

Sharma, H. C., B. S. Gill, 1982. Effect of embryo age and culture media on plant growth and vernalization response in winter wheat. Euphytica, 31: 629-634.

Torresán, A., J. Kesteloot, F. Castano, R. Rodrígues, M. Colabelli, 1996. Use of immature seed germination technique as an alternative to *in vitro* culture of sunflower (*Helianthus annuus* L.) embryos. Euphytica, 91: 1-3.

Yeung, E. C., T. A. Thorpe, C. I. Jensen, 1981. İn vitro fertilization and embryo rescue. In Plant Tissue Culture: Methods and Applications in Agriculture, ed. Thorpe, T.A. pp. 253-271. New York, Academic Press.

Zhong, D., N. Michaux-Ferriere, M. Coumans, 1995. Assay for doubled haploid sunflower (*Helianthus annuus*) plant production by androgenesis: fact or artifact? Part 1. *In vitro* anther culture. Plant Cell, Tissue and Organ Culture, 41: 91–97.

SPAD GREENNESS TO ESTIMATE GENOTYPIC VARIATION IN FLAG LEAF CHLOROPHYLL IN SPRING WHEAT UNDER MEDITERRANEAN CONDITIONS

Celaleddin BARUTCULAR[1], Irem TOPTAS[1], Hatice TURKTEN[2], Mehmet YILDIRIM[3], Mujde KOC[1]*

[1]Çukurova University, Faculty of Agriculture, Department of Field Crops, Adana, TURKEY
[2]Ondokuz Mayıs University, Faculty of Agriculture, Department of Agricultural Economics, Samsun, TURKEY
[3]Dicle University, Faculty of Agriculture, Department of Field Crops, Diyarbakır, TURKEY
[*]Corresponding author: cebar@cu.edu.tr

ABSTRACT

Leaf chlorophyll (Chl) is emphasized as an indicator for photosynthesis in wheat (*Triticum aestivum* L.). SPAD greenness meters are used to predict extractable Chl, but few studies have evaluated relationships between flag leaf greenness and Chl among wheat genotypes. Sixteen spring wheat genotypes with similar development patterns were studied in eight environments (2 years, 2 irrigation treatments and 2 sowing times) to investigate the precision of the SPAD-502 meter to predict Chl content/concentration. Relationships of Chl with SPAD greenness generally best fit linear and quadratic models. Relationships of SPAD greenness with Chl concentration were weak and inconsistent. Strongest associations were observed with Chl_a content (highest R^2 = 0.71 under late sowing), whereas associations with Chl_b content were weak (highest R^2 = 0.46) or insignificant. Relationships between SPAD greenness and total Chl content ranged from low ($R^2 = 0.24$, $p = 0.054$) to moderate ($R^2 = 0.64$, $p < 0.001$) under rainfed conventional-sowing and late-sowing conditions in the second year, respectively. SPAD greenness can be used to diagnose spring wheat genotypes differing in flag leaf Chl_a and total Chl content under warm Mediterranean conditions, but may not always applicable in cooler Mediterranean conditions, where genetic variability, especially in Chl_b, was not expressed adequately.

Key words: Chlorophyll meter, Chlorophyll content, Chlorophyll concentration, Flag leaf, Spring wheat, drought

INTRODUCTION

Chlorophyll *a* and chlorophyll *b* are essential pigments of the plant photosystems (Richardson et al., 2002). In wheat (*Triticum aestivum* L.), chlorophyll content is positively correlated with photosynthetic rate (Evans, 1983). Leaf chlorophyll (measured by extraction and quantification of pigment content or with optical chlorophyll meters) is increasingly emphasized as an indicator for photosynthetic capacity and its stability during senescence (the 'stay green' phenotype) under both stress and non-stress conditions, and therefore for yield in wheat (Reynolds et al., 1994; Araus et al., 2008; Parry et al., 2011; Xiao et al., 2012).

Traditional extraction-based methods of measuring chlorophyll *a* (Chl_a), chlorophyll *b* (Chl_b) and total chlorophyll (Chl_{a+b}) in leaves are destructive and time consuming (Richardson et al., 2002). Therefore, handheld optical chlorophyll meters, such as SPAD (soil plant analysis development) meters, based on the absorbance, reflectance and/or transmittance of radiation by the intact leaf, have been developed. SPAD chlorophyll meters determine the relative amount of total leaf chlorophyll in intact leaves and produce values that express relative chlorophyll content (SPAD greenness), but not of absolute chlorophyll amount per unit area (content) nor per mass (concentration) of leaf tissue. Taking SPAD readings is easy, rapid and non-destructive; therefore, use of SPAD greenness meters has increased dramatically in recent years.

SPAD readings have been used to predict chlorophyll content in a large number of crop species, such as corn (Markwell et al., 1995), sorghum and pigeonpea (Yamamoto et al., 2002), soybean (Fritschi and Ray, 2007), wheat and potato (Uddling et al., 2007) and basil (Ruiz-Espinoza et al., 2010). A number of studies, restricted to one or a small number of genotypes, have reported a curvilinear relationship between extractable chlorophyll and SPAD values in wheat (Monje and Bugbee, 1992; Castelli et al., 1996; Richardson et al., 2002; Uddling et al., 2007). However, Castelli et al.

(1996) reported that monocots (wheat and maize) and dicots (tobacco and soybean) show distinct patterns in terms of the relationship between non-destructive readings and analytical results obtained by solvent extraction.

Indeed, the SPAD value depends not only on extractable chlorophyll concentrations, but also on other aspects of leaf optics, which may be affected by a variety of environmental and biological factors (Palta, 1990; Hoel and Solhaug, 1998; Markwell et al., 1995; Martínez and Guiamet, 2004). The non-homogeneous distribution of chlorophyll molecules within the leaf and/or differential scattering and reflection of radiation can alter the relationship between extractable chlorophyll and SPAD greenness (Monje and Bugbee, 1992; Markwell et al., 1995; Uddling et al., 2007).

The lack of a consistent relationship between chlorophyll estimated with an extraction method and by the greenness meter for different genotypes may limit the potential use of SPAD as a complementary selection trait. The relationship between chlorophyll content/ concentration determined with an extraction method and by a SPAD chlorophyll meter have been verified depending on sowing time and irrigation threatment among wheat genotypes

In synthetic hexaploid wheat the variation in SPAD greenness is not completely explained by changes in chlorophyll concentration (Del Blanco et al., 2000). Thus it is necessary to verify the relationship between extractable chlorophyll and SPAD greenness in different genotypes under different environmental conditions to enable breeders to select more easily for genotypes differing in chlorophyll content or concentration, when the source capacity is to be improved. This objective is particularly important if not only chlorophyll content/concentration, but also chlorophyll components (Chl_a, Chl_b, Chl_{a+b} and ratio of Chl_a to Chl_b [$Chl_{a/b}$]), are to be considered.

The objective of this study was to examine if SPAD greenness could be used as a rapid screening method to predict genotypic variation in wheat for extractable leaf chlorophyll content/concentration (Chl_a, Chl_b and Chl_{a+b}) and $Chl_{a/b}$, and to assess the stability of these relationships in different environments. To eliminate possible phenotypic differences in chlorophyll content, genotypes with similar or known development patterns were used.

MATERIALS AND METHODS

Plant materials and cultural practices

Field experiments were conducted during the 2011/2012 and 2012/2013 growing seasons at the experimental field of the Faculty of Agriculture, University of Cukurova, Balcalı, Adana (37°00′ N, 35°21′ E, 29 m above sea level), Turkey. The soil type is a fine loamy, montmorillonitic typic xerofluvent, low in organic matter and slightly alkaline (pH 7.1–7.6).

Sixteen unrelated spring bread wheat (*Triticum aestivum*) genotypes with a similar developmental pattern selected for diversity in leaf traits were used. The widely adapted genotypes comprised commercial cultivars from Turkey (Adana-99, Balattila, Cemre, Cumakalesi, Genç-99, Karacadağ-98, Meta-2002 and Özkan), ICARDA (Cham-6 and Siete Cerros), Israel (Dariel and Galil), Italy (Colfiorito), Spain (Mané-Nick) and Pakistan (Inqilab-91 and V-3010).

In each growing season (2011/2012 and 2012/2013) in addition to conventional (optimal) sowing time (CS) and late sowing (LS) times were used to expose the cultivars to lower and higher temperatures. Each replication of the sowing times (main plots) was nested on the same field with two irrigation treatments (rainfed, RF; and irrigated, IR) as subplots and genotypes as mini-plots in a split-split plot randomised design of four replications.

Except for sowing time and irrigation treatment, the growing conditions were maintained as similar as possible through fertilization and pest control. The first-year sowings were on 25 November 2011 and 8 March 2012, and the second-year sowings were on 29 November 2012 and 8 March 2013, for CS and LS, respectively. Phosphorus (40 kg ha^{-1} P_2O_5) was applied before sowing in the form of triple super phosphate. Nitrogen was applied as ammonium nitrate in three split doses (40 + 80 + 40 kg N ha^{-1}) at Zadok's (Zadoks et al., 1974) growth stages (ZGS) 00, 20 and 30. The sowing density was 450 viable seeds m^{-2}. Plots consisted of 8 rows, each of which was 6.0 m long with a row spacing of 0.15 m.

The irrigation applications were initiated on 6 April 2012 with CS and 12 May 2012 with LS in the first year. The respective dates in the second year were 19 March 2013 and 1 May 2013. Water supplies between irrigated and rainfed conditions in first year were similar. First-year water supply (rainfall + irrigation) until sampling time was 671 and 682 mm for RF and IR, repsectively, and 255 and 255 mm in CS and LS, respectively. The respective water supply in the second year was 459, 501, 169 and 199 mm.

Measurement of SPAD greenness and extractable leaf chlorophyll

SPAD greenness and extractable leaf chlorophyll were measured on flag leaves. Sampling was performed before the onset of leaf senescence at the end of anthesis (2012 CS), at the beginning of anthesis (2012 LS and 2013 CS), and of heading (2013 LS) 1 week (in LS) or 2 weeks (in CS) after initiation of the irrigation regimes. Five healthy flag leaf blades were sampled at random from the central rows of each plot and transported in cool isolation boxes to the laboratory. The leaves were kept in a refrigerator before processing. Samples from one replication were processed within approximately 2 h from collection in the field.

In the laboratory, at first SPAD chlorophyll was measured on sampled fresh leaves using a hand–held SPAD-502 Chlorophyll Meter (Minolta Camera Co., Osaka, Japan), which generates a value predictive of chlorophyll concentration (Minolta Camera Co. 1989).

The SPAD-502 meter measures transmittance at 650 nm (red light is strongly absorbed by chlorophyll) and 940 nm (near-infrared light is a 'reference wavelength' that is used to adjust for differences in leaf structure) of radiation through the leaf, and calculates a relative SPAD meter value that is termed SPAD greenness. The average of three readings from the center of each leaf blade was used as a SPAD greenness score per leaf (5 × 3 readings per plot). The midpoint of the leaf is the best position in winter wheat plants on which to take chlorophyll meter readings (Hoel and Solhaug, 1998). The adaxial side of the leaves was always placed toward the emitting window of the instrument and major veins were avoided.

Following each SPAD reading, the amount of chlorophylls (Chl_a and Chl_b) was determined in two leaf disks of 8 mm diameter taken from an identical position of the same leaves. After the disks were weighed, pigments were thoroughly extracted in 80% (v/v) acetone using a glass mortar, and the homogenate was then filtered. Absorbance was measured at the relevant wavelength with a spectrophotometer (Shimadzu, UV-1208, Kyoto, Japan). Chlorophyll content/concentration was calculated on a leaf area/fresh weight basis using the equations of Porra et al. (1989). These equations are an improvement over the widely used equations of Arnon (1949).

Data analysis

Although irrigation was initiated a short time prior to sampling for SPAD greenness and extractable leaf chlorophyll measurement, and differences in water supply between irrigation treatments were small, analysis of variance (ANOVA) of the traits was carried out first on the basis of the experimental design used (a randomized split-split plot) with sowing date as the main factor, irrigation as a sub-plot, and genotype as a mini-plot with four replications. Subsequently, analyses of variance for each growing condition were performed separately to obtain a clearer picture for flag leaf traits of the genotypes. Least significant difference (LSD) values were calculated at the 0.05 level of probability. A regression was used when appropriate. The SPSS (1999) package was used for all analyses.

RESULTS AND DISCUSSION

Development of genotypes

As a result of the small pre-anthesis differences in water supply between the irrigation treatments, no effect of irrigation (and its interaction with genotype) on development of the wheat genotypes was observed (data not shown). Under CS, anthesis (ZGS 60, averaged over genotypes) occurred 138 and 123 days after sowing in 2012 and 2013, respectively. The respective values under LS were 69 and 66 days. When the relative thermal duration (determined as cumulative growing degree-days from sowing to sampling date, base temperature 0 °C) was considered, acceleration of anthesis by LS was in a narrow range from 1376 to 1149 °C days (16%) and from 1439 to 1151°C days (20%) in 2012 and 2013, respectively. These results indicated the overwhelming effect of growing temperature on development.

Despite the obvious genotypic trends, the development of most genotypes was very similar. Only the local Turkish cultivar Cumakalesi was consistently the latest to flower and Karacadağ-98 tended to be the earliest to anthesis. Development of the remaining genotypes was tightly clustered in the first year, especially with LS. Genotypic differences in the second year (particularly with LS) were more pronounced, but ZGS values changed only within a narrow range of development such as from heading (ZGS 57) to completion of anthesis (ZGS 67).

It is well known that the growth stage also affects chlorophyll pigments, and therefore SPAD readings (Lopes and Reynolds, 2012). The similar (or known) developmental pattern of each genotype in the present study allowed valid comparison of SPAD greenness and extractable chlorophylls among the genotypes.

Flag leaf SPAD greenness and extractable chlorophylls

The results of ANOVA over years showed consistently highly significant genotypic effects, whereas other effects were inconsistent (sowing time), negligible (irrigation) or extremely variable (interactions) (data not shown). To obtain a clearer picture, the maximum, minimum and mean values for flag leaf traits of each genotype under each conditions (two sowing times and two irrigation treatments in the 2 study years) and LSD values calculated based on the results of ANOVA for individual conditions (hereafter termed 'environments') are presented separately in Table 1. With a few exceptions for chlorophyll concentration and content under certain conditions, genotypic differences for SPAD greenness and extractable chlorophylls were significant (Table 1) and more or less consistent (data not shown). Here emphasis is placed more on the extent of genotypic changes rather than on genotypic differences *per se*.

Senescing leaves in wheat suffer the loss of chlorophylls, especially Chl_a (Reynolds et al., 2000; Zhang et al., 2006). Thus the leaf age affects chlorophyll pigments, and therefore SPAD readings (Lopes and Reynolds, 2012). Except only one cultivar, the similar developmental pattern and similar leaf age of the genotypes in the present study allowed valid comparison of SPAD greenness and extractable chlorophylls among the genotypes.

SPAD greenness

Although highly significant differences among genotypes were observed in all environments, SPAD greenness values changed within rather narrow ranges (mean change 23%) (Table 1). The highest change of 40% occurred from 38.5 to 53.9 in the first year in the LS rainfed environment (2012 LSRF), and the lowest changes (17%) from 44.6 to 52.0 and 44.5 to 52.2 were observed in 2013 in the CS rainfed and CS irrigated environments (2013 CSRF and CSIR), respectively. Changes in the other environments were more or less similar.

Table 1. Maximum, minimum and mean flag leaf SPAD greenness, chlorophyll *a* (Chl_a) and chlorophyll *b* (Chl_b) concentration (mg g^{-1}), and $Chl_{a/b}$ ratio of 16 spring wheat genotypes grown under two sowing times (CS, conventional; LS, late) and irrigation treatments (RF, rainfed; IR, irrigated) during the 2011/2012 (2012) and 2012/2013 (2013) crop growing seasons at Adana, Turkey.

Values	2012					2013				
	SPAD	Chl_a	Chl_b	Chl_{a+b}	Chl a/b	SPAD	Chl_a	Chl_b	Chl_{a+b}	Chl a/b
			CSRF					CSRF		
Maximum	51.6	2.53	0.75	3.28	3.70	52.0	2.36	0.67	3.01	3.95
Minimum	42.6	1.85	0.55	2.39	2.90	44.6	1.95	0.53	2.48	3.35
Mean	45.8	2.16	0.67	2.82	3.26	46.5	2.13	0.59	2.72	3.66
$LSD_{0.05}$	4.04	0.255	0.101	0.337	0.343	3.41	ns	ns	ns	ns
			CSIR					CSIR		
Maximum	47.7	2.20	0.69	2.85	3.90	52.2	2.25	0.68	2.93	3.75
Minimum	39.1	1.65	0.45	2.10	3.08	44.5	1.87	0.51	2.38	3.30
Mean	45.8	2.03	0.59	2.62	3.48	47.0	2.09	0.60	2.69	3.51
$LSD_{0.05}$	2.90	ns	0.090	0.393	0.398	2.83	ns	ns	ns	ns
			LSRF					LSRF		
Maximum	53.9	2.54	0.85	3.39	3.80	56.2	2.32	0.81	3.06	3.30
Minimum	38.5	1.78	0.54	2.42	2.35	46.4	1.67	0.55	2.26	2.68
Mean	45.8	2.12	0.70	2.81	3.10	50.3	1.95	0.66	2.62	2.98
$LSD_{0.05}$	2.99	0.247	0.110	0.315	0.448	3.73	0.331	0.127	0.439	0.355
			LSIR					LSIR		
Maximum	51.8	2.66	1.04	3.63	3.65	53.8	1.97	0.62	2.59	3.38
Minimum	42.2	1.95	0.57	2.55	2.53	44.4	1.50	0.49	2.01	2.95
Mean	46.9	2.28	0.74	3.02	3.13	48.7	1.79	0.58	2.37	3.16
$LSD_{0.05}$	3.01	0.346	0.135	0.446	0.439	4.97	0.243	ns	ns	ns

ns, not significant at the 0.05 probability level.

SPAD values measured in this study were comparable to those observed in previous studies of spring wheat grown under irrigated and non-limited nitrogen conditions (Del Blanco et al., 2000; Gutiérrez-Rodríguez et al., 2000; Babar et al., 2006) and were near the upper limit of calibration studies in wheat (Monje and Bugbee, 1992; Castelli et al., 1996; Richardson et al., 2002; Uddling et al., 2007). Despite the narrow range within higher values of SPAD, significant and consistent genotypic differences were found. This futher promised analysis of relationships SPAD with extracted chlorophylls.

Extractable chlorophylls

Except for occasional cases, such as for both pigments in 2013 CSRF and CSIR, for Chl_a in 2012 CSIR, and for Chl_b in 2013 LSIR, genotypic differences for extractable concentration of flag leaf Chl_a, Chl_b, and Chl_{a+b} were generally significant (Table 1). Differences were not evident for both pigments (Chl_a, Chl_b, and therefore Chl_{a+b}) in the 2013 CSRF and for Chl_b in the 2013 LSIR environments.

The highest genotypic range in chlorophyll content was observed in 2012 in the LSIR environment, in which Chl content ranged from 318 to 481 mg m^{-2} for Chl_a, from 96 to 193 mg m^{-2} for Chl_b, and from 427 to 674 mg m^{-2} for Chl_{a+b} (data not shown, but can be seen in Figure 2). Genotypic differences in $Chl_{a/b}$ were not evident in three environmets (CSRF, CSIR and LSIR) in 2013. The extent of significant genotypic differences in the other environments changed in a wide range from 23% (2013 LSRF) to 62% (2012 LSRF). Changes shown in Table 1 were largely similar when the two genotypes that exhibit early (Karacadağ-98) and late (Cumakalesi) development were not considered.

One of the most remarkable results in the present study was the trend for declining $Chl_{a/b}$ under LS warm conditions in both years due to increased Chl_b. The ratio of Chl_a to Chl_b is known to change with environmental factors such as light. Lichtenthaler and Wellburn (1983) reported that plants exposed to high irradiance show high $Chl_{a/b}$ ratios, whereas plants growing in environments with reduced irradiance have a lower $Chl_{a/b}$ ratios. The variation in $Chl_{a/b}$ ratio in the current study seem not to be due to reduced radiation, as light intensity above the canopy constantly increases during the spring months, when late-sown crops are grown. The reduction $Chl_{a/b}$ under LS conditions may be caused by increased growing temperatures during the flag leaf expansion. For a precise interpretation, temperature and irradiance measurements at the flag leaf level in the canopy during flag leaf expansion are needed.

Relationship of SPAD greenness with chlorophyll concentration and $Chl_{a/b}$

The linear, quadratic and exponential (data not shown) relationships among chlorophyll concentration and $Chl_{a/b}$ with SPAD greenness were generally similar. The linear and quadratic regression coefficients of determination (R^2) for flag leaf chlorophyll concentration (Chl_a, Chl_b, and Chl_{a+b} per unit fresh weight) and $Chl_{a/b}$ with SPAD greenness are illustrated in Table 2. The degree of quadratic fit was improved only in two environments, namely in 2012 LSRF and 2013 CSRF for Chl_a and Chl_b, respectively.

Table 2. Relationships of flag leaf extracted chlorophyll (Chl) concentration and $Chl_{a/b}$ ratio with SPAD greenness for 16 spring wheat genotypes grown under two sowing times (CS,conventional; LS, late) and irrigation treatments (RF, rainfed; IR, irrigated) during the 2011/2012 (2012) and 2012/2013 (2013) crop growing seasons at Adana, Turkey.

Regression		2012				2013			
		Chl_a	Chl_b	Chl_{a+b}	$Chl_{a/b}$	Chl_a	Chl_b	Chl_{a+b}	$Chl_{a/b}$
		CSRF				CSRF			
Linear	R^2	0.17	0.15	0.18	0.02	0.03	0.25*	0.07	0.19
Quadratic	R^2	0.17	0.15	0.18	0.22	0.04	0.49*	0.14	0.20
		CSIR				CSIR			
Linear	R^2	0.10	0.31*	0.18	0.38*	0.11	0.17	0.13	0.08
Quadratic	R^2	0.18	0.35 (p:0.06)	0.22	0.47*	0.11	0.17	0.13	0.08
		LSRF				LSRF			
Linear	R^2	0.22	0.04	0.18	0.04	0.40**	0.31*	0.39*	0.00
Quadratic	R^2	0.41*	0.07	0.25	0.15	0.41*	0.37 (p:0.052)	0.39*	0.00
		LSIR				LSIR			
Linear	R^2	0.20	0.14	0.20	0.02	0.49**	0.30*	0.47**	0.00
Quadratic	R^2	0.32	0.14	0.21	0.02	0.51**	0.32	0.49*	0.02

*,** significance at 0.05 and 0.01 probability levels, respectively.

Linear relationships between SPAD greenness and chlorophyll concentration (Chl_a, Chl_b, and Chl_{a+b} per unit fresh weight) were consistently positive (Table 2). However, the relationships were significant only in the 2013 LS, RF and IR environments for Chl_a, Chl_b, and Chl_{a+b} and in the 2012 CSIR and 2013 CSRF environments for Chl_b. Except a weak negative correlation under CSIR in the first year ($r = -0.38$*), relationships between SPAD greenness and $Chl_{a/b}$ were not significant.

Although many studies have reported a nonlinear (curvilinear) relationships between extracted chlorophylls and SPAD greenness (Monje and Bugbee, 1992; Markwell et al., 1995; Castelli et al., 1996; Bindi et al., 2002; Richardson et al., 2002; Jifon et al., 2005; Uddling et al., 2007), a linear model was also used in the present study, as chlorophyll amounts recorded were quite high and ranged within very narrow limits.

Relationship of SPAD greenness with chlorophyll content

Only linear relationships were considered, as the relationship of flag leaf SPAD greenness and extractable chlorophyll content estimated by linear and quadratic relationships did not show large differences. Exponential relationships (data not shown) also did not improve prediction of chlorophyll content.

The equations ($Chl_a = 8.6093x - 29.496$, $Chl_b = 3.0695x - 28.085$, $Chl_{a+b} = 11.673x - 57.265$) quantifying the linear relationship between SPAD greennes and chlorophyll content for the 16 genotypes averaged over the eight environments demonstrated a moderate association with R^2 values of 0.65***, 0.550** and 0.64*** for Chl_a, Chl_b, and Chl_{a+b}, respectively. However, the equations were not always valid if the individual environments were considered seperately, as shown in Figure 1 and Figure 2 for CS and LS, respectively. The relationship of SPAD greenness with Chl_a, Chl_b, and Chl_{a+b} was positive; hence, in all environments increased flag leaf greenness was associated with increased chlorophyll content. Nevertheless, the distribution of the data around the regression line (R^2) varied with the environment.

A generally weak or moderate relationship was observed between total extractable chlorophyll content and SPAD greenness (R^2 values ranged from 0.24 [p = 0.054] in 2013 CSRF to 0.64*** in 2013 LSRF) (Figure 1 and Figure 2). Most relationship between flag leaf Chl_{a+b} and SPAD greenness were significant but of limited predictive value, especially under CS (Figure 1). However, the association was improved under LS (Figure 2).

The relationships between Chl_a content and SPAD greenness were similar to those between total extractable chlorophyll content and SPAD greenness (Figure 1 and Figure 2). The associations were to a certain degree stronger, with highest R^2 values of 0.71*** in 2013 LSRF and 0.61*** in 2012 LSIR. The relationship between SPAD greenness and Chl_a was insignificant in 2013 CSRF, consistent with the significance of the relationship between SPAD greenness and total chlorophyll content.

The association of SPAD greenness with Chl_b content was weaker than those with total chlorophyll and Chl_a, as also reported by Richardson et al. (2002) for paper birch leaves. The strongest relationship was observed in 2012 CSIR (R^2 = 0.46**) followed by 2013 CSIR (R^2 = 0.43**) (Campbell et al., 1990; Monje and Bugbee, 1992; Jifon et al., 2005; Neufeld et al., 2006). The associations in CSRF and LSRF in 2012 could not be proven. The coefficients of determination in the remaining environments ranged between 0.33* and 0.40**.

The association of SPAD greenness with chlorophyll content (when chlorophyll was expressed on a leaf area basis) observed in this study was stronger than the association with chlorophyll concentration (when chlorophyll was expressed on a leaf fresh weight basis), as has been reported previously (Marquard and Tipton, 1987; Uddling et al., 2007).

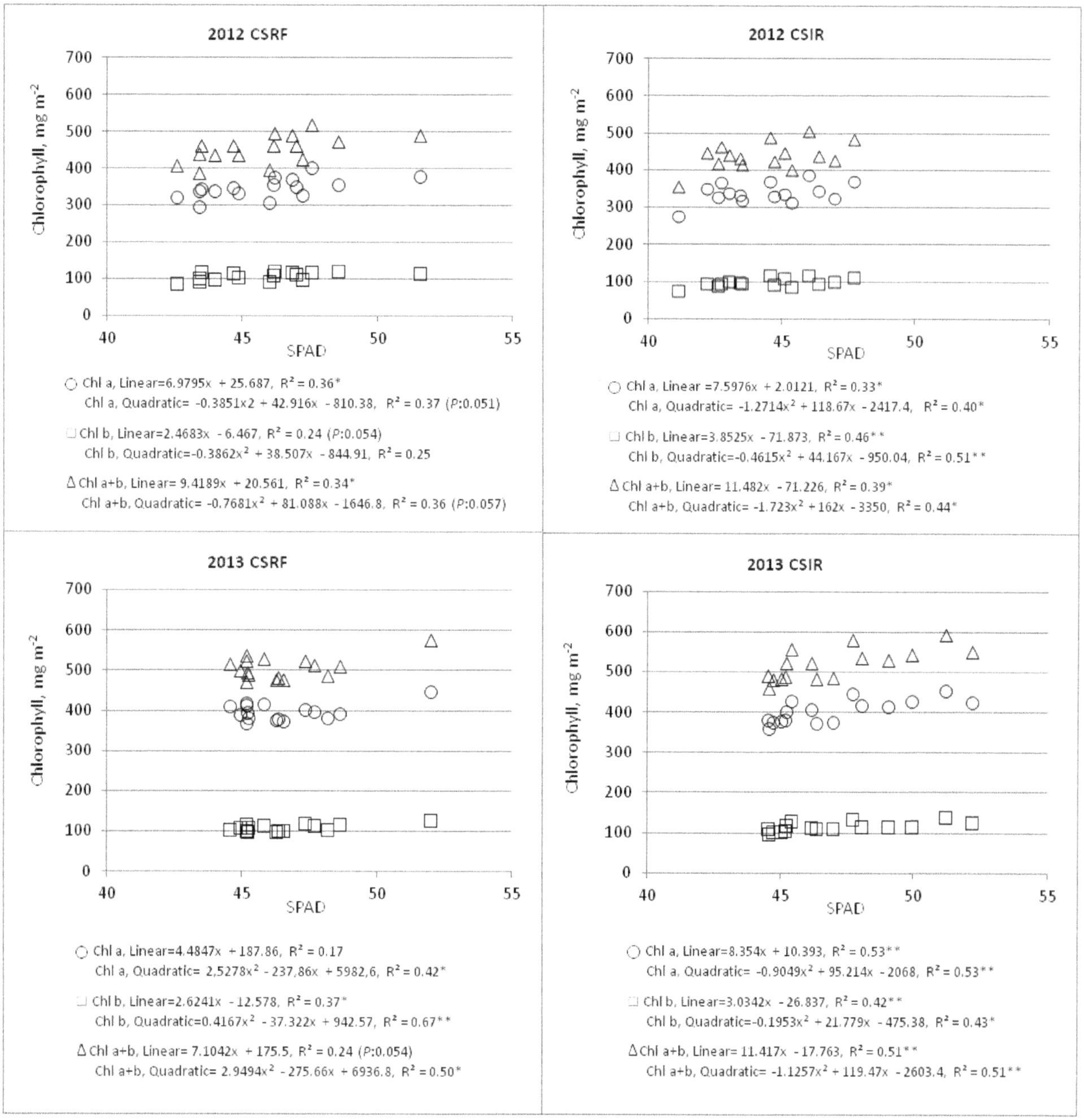

Figure 1. Relationship between flag leaf extracted chlorophyll (Chl) content and SPAD greenness for 16 spring wheat genotypes grown under a conventional sowing (CS) time and two irrigation treatments (RF, rainfed; IR, irrigated) during the 2011/2012 (2012) and 2012/2013 (2013) crop growing seasons at Adana, Turkey. *, ** significance at 0.05 and 0.01 probability levels, respectively.

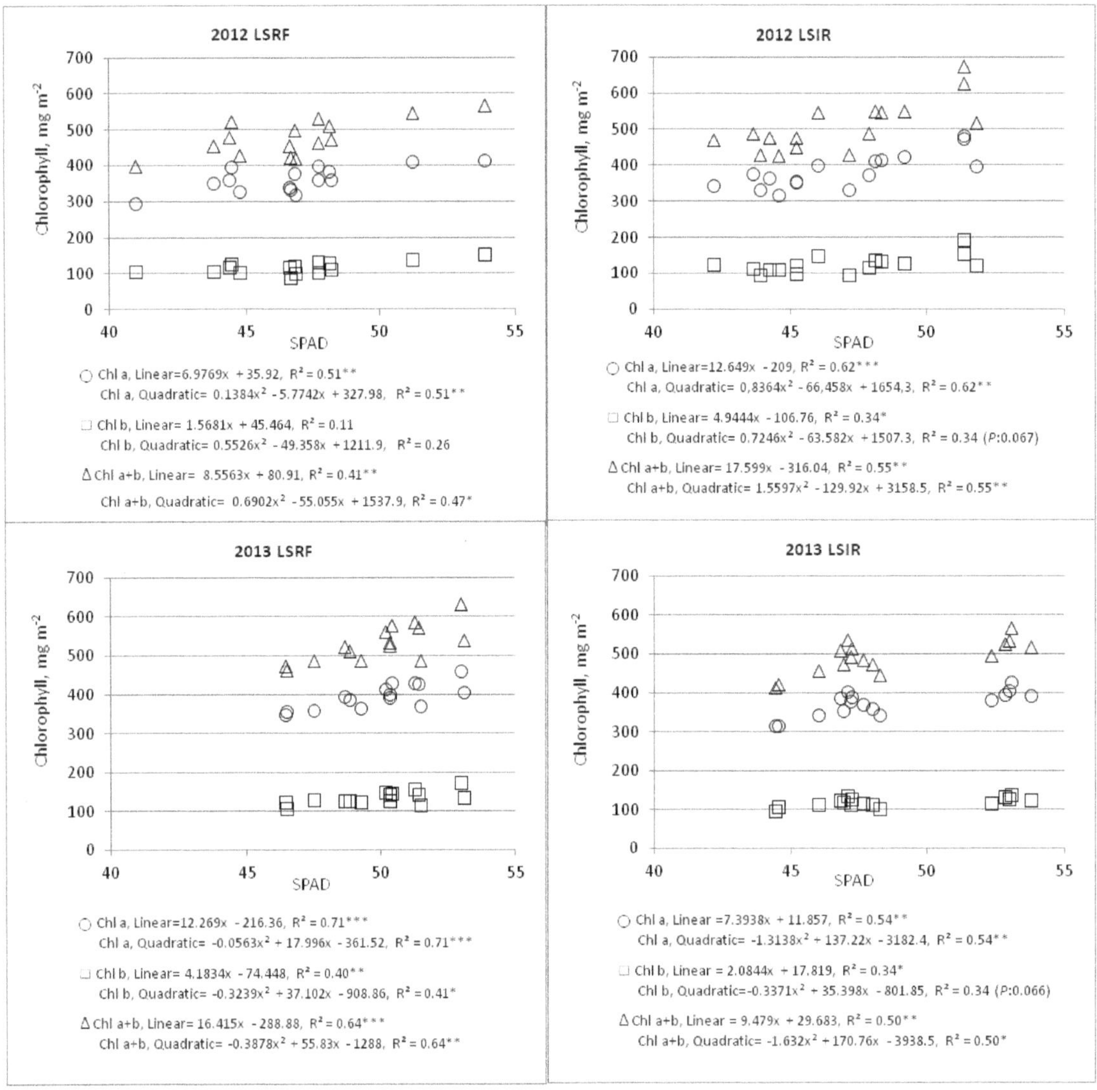

Figure 2. Relationship between flag leaf extracted chlorophyll (Chl) content and SPAD greenness for 16 spring wheat genotypes grown under a late sowing (LS) time and two irrigation treatments (RF, rainfed; IR, irrigated) during the 2011/2012 (2012) and 2012/2013 (2013) crop growing seasons at Adana, Turkey. *,** , *** significance at 0.05, 0.01 and 0.001 probability levels, respectively.

The results of this study further suggest that the usefulness of SPAD readings in typical Mediterranean wheat growing conditions is questionable and, for prediction purposes, growing conditions are an important consideration and the date (growing conditions) cannot be pooled. Complex optical properties due to factors other than the chlorophyll amount in the leaf, which change with growing environment, seem problematic for application of SPAD greenness.

CONCLUSIONS

Genotypic differences in flag leaf SPAD greenness and extractable chlorophylls among spring wheat genotypes with similar development provided the opportunity to assess the accuracy and precision of SPAD readings to estimate genotypic variation in flag leaf chlorophyll amounts independent of phenology (leaf age).

The ability of the SPAD-502 meter to estimate genotypic differences in chlorophyll content per unit flag leaf area was satisfactory as long as the leaves developed under warm (and irrigated) conditions. It might be due to the fact that cell under warm condition perform well having more chlorophyll contents of leaf morphology. Secondly the genetic variation could be the main reason of high chlorophyll contents of some varieties performing well under Mediterranean conditions. However, the SPAD meter was less reliable for estimation of genotypic variation in leaf chlorophyll under the typical cooler wheat-growing conditions of the Mediterranean, where genetic variability, especially in Chl_b, was not expressed adequately. Applicability of the SPAD-502 meter also might be restricted when the objective is to assess the capacity of the photosynthetic antenna system, as SPAD greenness was generally a better indicator of Chl_a and Chl_{a+b} than Chl_b. This is the one of the main drawback of SPAD meter in determining the photosynthetic antenna system of plants. Therefore instrument must be updated to determine the both chlorophyl contents as well as light harvesting capacity of leaf. Further research is also indispensible at cellular level to understand the efficiency of SPAD for putting a theoratical conclusions. In spite of these limitations, chlorophyll meters are valuable instruments if they are properly calibrated with extractable chlorophyll measurements for a given set of growing conditions.

ACKNOWLEDGMENTS

This research was supported by the Scientific and Technological Research Council of Turkey (TUBITAK) Project no. 110O345.

LITERATURE CITED

Araus, J.L., G.A. Slafer, C. Royo, M.D. Serret. 2008. Breeding for Yield Potential and Stress Adaptation in Cereals. Critical Reviews in Plant Sci. 27: 377–412.

Arnon, D.I.. 1949. Copper enzymes in isolated chloroplasts; Polyphenoloxidases in Beta vulgaris. Plant Physiol. 24: 1–15.

Babar, M.A, M.P. Reynolds, M. Van Ginkel, A.R. Klatt, W.R. Raun, M.L. Stone. 2006. Spectral Reflectance to Estimate Genetic Variation for In-Season Biomass, Leaf Chlorophyll, and Canopy Temperature in Wheat. Crop Sci. 46: 1046–1057.

Bindi, M., A. Hacour, K. Vandermeiren, J. Craigon, K. Ojanpera, G. Sellden, P. Hogy, J. Finnan, L. Fibbi. 2002. Chlorophyll concentration of potatoes grown under elevated carbon dioxide and/or ozone concentrations. European Journal of Agron. 17: 319 -335.

Campbell, R.J., K.N. Mobley, R.P. Marini, D.G. Pfeiffer. 1990. Growing conditions alter the relationship between SPAD-501 values and apple leaf chlorophyll. Hortsci. 25: 330–331.

Castelli, F., R. Contillo, F. Miceli. 1996. Non-destructive Determination of Leaf Chlorophyll Content in Four Crop Species. Journal of Agronomy and Crop Sci. 177: 275–283.

Del Blanco, I.A., S. Rajaram, W.E. Kronstad, M.P. Reynolds. 2000. Physiological performance of synthetic hexaploid wheat-derived populations. Crop Sci. 40: 1257-1263.

Evans, J.R.. 1983. Nitrogen and photosynthesis in the flag leaf of wheat (*Triticum aestivum* L.). Plant Physiol. 72: 297 - 302.

Fritschi, F.B., J.D. Ray. 2007. Soybean leaf nitrogen, chlorophyll content, and chlorophyll a/b ratio. Photosynthetica 45 (1): 92-98.

Gutiérrez-Rodríguez, M., M.P. Reynolds, A. Larqué-Saavedra. 2000. Photosynthesis of wheat in a warm, irrigated environment II. Traits associated with genetic gains in yield. Field Crops Research 66: 51-62.

Hoel, B.O., K.A. Solhaug. 1998. Company Effect of Irradiance on Chlorophyll Estimation with The Minolta SPAD-502 Leaf Chlorophyll Meter. Annals of Bot. 82: 389-392.

Jifon, J.L., J.P. Syvertsen, E. Whaley. 2005. Growth environment and leaf anatomy affect non-destructive estimates of chlorophyll and nitrogen in Citrus sp. leaves. Journal of American Society of Horticultural Sci., 130: 152 – 158.

Lichtenthaler, H.K., A.R.Wellburn. 1983. Determinations of total carotenoids and chlorophylls a and b of leaf extracts in different solvents. Biochemical Society Transactions, 11: 591–592.

Lopes, M.S., M.P. Reynolds. 2012. Stay-green in spring wheat can be determined by spectral reflectance measurements (normalized difference vegetation index) independently from phenology. Journal of Experimental Bot. 63: 3789-3798.

Markwell, J., J.C. Osterman, J.L. Mitchell. 1995. Calibration of the Minolta SPAD-502 leaf chlorophyll meter. Photosynthesis Res. 46: 467-472.

Marquard, R.D., J.L. Tipton. 1987. Relationship between extractable chlorophyll and an in situ method to estimate leaf greenness. HortSci. 22: 1327

Martínez, D.E., J.J. Guiamet. 2004. Distortion of the SPAD 502 chlorophyll meter readings by changes in irradiance and leaf water status. Agronomie, 24, 41-46.

Monje, O.A., Bugbee B.. 1992. Inherent limitations of nondestructive chlorophyll meters: A comparison of two types of meters. HortSci. 27: 69-71.

Neufeld, H.S., A.H. Chappelka, G.L. Somers, K.O. Burkey, A.W. Davison, P.L. Finkelstein. 2006. Visible foliar injury caused by ozone alters the relationship between SPAD meter readings and chlorophyll concentrations in cutleaf coneflower. Photosynthesis Res. 87, 281–286.

Palta, J.P.. 1990. Leaf chlorophyll content. Remote Sensing Reviews, 5: 207-213.

Parry, M.A.J., M. Reynolds, M.E. Salvucci, C. Raines, P.J. Andralojc, X-G. Zhu, G.D. Price, A.G. Condon, R. Furbank. 2011. Raising Yield Potential of Wheat: (II) Increasing photosynthetic capacity and efficiency. Journal of Experimental Bot. 62: 453-467.

Porra, R.J., W.A. Thompson, P.E. Kreidemann. 1989. Determination of accurate extinctions coefficients and simultaneous equations for assaying chlorophylls a and b extracted with four different solvents: verification of the concentration of chlorophyll standards by atomic absorption spectroscopy. Biochimica et Biophysica Acta (BBA) – Bioenergetics, 975: 384–394.

Reynolds, M.P., M. Balota, M.I.B. Delgado, I. Amani, R.A.Fischer. 1994. Physiological and morphological traits associated with spring wheat yield under hot, dry irrigated conditions. Australian Journal of Plant Physiol. 21: 717-730.

Reynolds, M.P., M.I.B. Delgado, M. Gutierrez-Rodriguez, A. Larque-Saavedra. 2000. Photosynthesis of wheat in a warm, irrigated environment. I. Genetic diversity and crop productivity. Field Crops Res. 66: 37–50.

Richardson, A.D., S.P. Duigan, G.P. Berlyn. 2002. An evaluation of noninvasive methods to estimate foliar chlorophyll content. New Phytologist, 153: 185-194.

Ruiz-Espinoza, F. H., B. Murillo-Amador, J. L. García-Hernández, L. Fenech-Larios, E. O. Rueda-Puente, E. Troyo-Diéguez , C. Kaya, A. Beltrán-Morales. 2010. Field evaluation of the relationship between chlorophyll content in basil leaves and a portable chlorophyll meter (SPAD-502) readings. Journal of Plant Nutrition 33:423 - 438.

SPSS, 1999. SYSTAT User's Guide: Statistics. Version 9.0. SPSS, Inc., Chicago, IL

Uddling, J., J. Gelang-Alfredsson, K. Piikki, H. Pleijel. 2007. Evaluating the relationship between leaf chlorophyll concentration and SPAD-502 chlorophyll meter readings. Photosynthesis Res. 91: 37–46.

Xiao, Y.G., Z.G. Qian, K. Wu, J.J. Liu, X.C. Xia, W.Q. Ji, Z.H. He. 2012. Genetic gains in grain yield and physiological traits of winter wheat in Shandong Province, China, from 1969 to 2006. Crop Sci. 52: 44–56.

Yamamoto, A., T. Nakamura, J.J. Adu-Gyamfi, M. Saigusa. 2002. relationship between chlorophyll content in leaves of sorghum and pigeonpea determined by extraction method and by chlorophyll meter (SPAD-502). Journal of Plant Nutrition 25(10): 2295 – 2301.

Zadoks, J.C., T.T. Chang, C.F. Konzak. 1974. A decimal code for the growth stages of cereals. Weed Res. 14: 415–421.

Zhang, C.J., G.X. Chen, X.X. Gao, C.J. Chu. 2006. Photosynthetic decline in flag leaves of two field-grown spring wheat cultivars with different senescence properties. South African Journal of Bot. 72: 15 -23.

INHERITANCE OF FEMALE STERILITY IN INDUCED *Cicer* SPECIES

*Tolga YILDIRIM[3], Hüseyin CANCI[1], Nisa ERTOY INCI[1], Fatma Oncu CEYLAN BALOGLU[1], Cengiz IKTEN[2], Cengiz TOKER[*1]*

[1]Akdeniz University, Faculty of Agriculture, Department of Field Crops, Antalya, TURKEY.
[2]Akdeniz University, Faculty of Agriculture, Department of Plant Protection ,Antalya, TURKEY.
[3]Akdeniz University, Faculty of Science, Department of Biology, Antalya, TURKEY.
**Corresponding author: toker@akdeniz.edu.tr*

ABSTRACT

Mutations play an important role to induce new genes, creating variation. The present study deals with inheritance of female sterilities in the induced mutants of the cultivated chickpea (*Cicer arietinum* L.) and its progenitor (*Cicer reticulatum* Ladiz.). Air-dried seeds of *Cicer* species were irradiated with 200, 300 and 400 Gy gamma rays. Two induced mutations conferring open flower and determinate growth habit with small leaf characteristics in the cultivated chickpea and its progenitor were discovered in M_2 generation. These mutant chickpeas were female sterile. Inheritance study showed that the female sterility in the induced mutants was controlled by a single recessive gene (*fs*). The female sterility, determinate growth habit, open flower and small leaf characteristics were first reported for *C. reticulatum* Ladiz. Results indicated that a joint segregation of the female sterility, open flower, determinate growth habit and small leaf characteristics might be linked in both of induced mutants of the cultivated chickpea and its progenitor. The new gene discovered on the sterility in the induced mutants may be useful for gene mapping in *Cicer* species.

Key words: Chickpea, *Cicer arietinum, Cicer reticulatum,* mutagenesis, mutant, sterility

INTRODUCTION

The number of species in the genus *Cicer* L. has reached up to 45 taxa with a new endemic perennial species, *C. uluderensis* Donmez (van der Maesen *et al.*, 2007; Donmez, 2011), recently found in Southeast Turkey. In the genus *Cicer* L., *Cicer arietinum* L. is only cultivated species, with 2n = 16 chromosomes, and self-pollinated species due to its cleistogamic flowers (Cubero, 1987). The cultivated chickpea is a derivative from its wild progenitor species, *C. reticulatum* Ladizinsky (Zohary and Hopf, 2000; Toker, 2009). *C. reticulatum* Ladiz. is originated from South-Eastern Turkey (Ladizinsky and Adler, 1976), and it can easily be crossed with the cultivated chickpea (Abbo *et al.*, 2007). The morphological variations in *C. reticulatum* Ladiz. are narrower than those of the cultivated chickpeas (Robertson *et al.*, 1997).

Induced mutations are useful for genetic exploitation of crop plants when the genetic variation is narrow. So far, several morphological mutants have been found and utilized in chickpea improvement as well as in linkage studies (Dahiya *et al.*, 1984; Pundir and Reddy, 1998; Gaur and Gour, 2002; McNeil *et al.*, 2007; Rajesh *et al.*, 2007; Salimath *et al.*, 2007; Srinivasan *et al.*, 2006; Wani and Anis, 2008; Ali *et al.*, 2010; Kharkwal *et al.*, 2010; Si *et al.*, 2010; Wani, 2011). The aim of the present study is to report inheritance of induced male and female sterilities in the cultivated chickpea and its progenitor.

MATERIALS AND METHODS

Air-dried seeds of two *Cicer* species including the cultivated chickpea genotype namely '*macrosperma*' chickpeas, 'CA 2969' (Rubio *et al.*, 2003), and a genotype of *C. reticulatum* Ladiz 'AWC 612' (Toker, 2005; Toker *et al.*, 2007; Canci and Toker, 2009) were irradiated with 200, 300 and 400 Gy of gamma rays from a ^{60}Co source in Turkish Atomic Energy Agency (TAEK), Ankara, Turkey (Toker *et al.*, 2005). There were approximately 500 seeds in each treatment level, after which they were stored at 4°C. The field was fertilized with N, P and K at a rate of 20 kg per ha prior to sowing. The irradiated seeds were sown in plots with 45 cm row and 5 cm plant spacing at Antalya location (30° 38' E, 36° 53' N, 32 m from sea level) in the spring of 2005. Weed control was done by hand during seedling stage. Additional irrigation was not practiced because of sufficient rainfall.

The seeds from M_1 generation were individually harvested in the summer of 2005. M_2 generation was grown as single plant progenies in separate rows at the same location in 2006. After germination, M_2 plants were carefully screened for all morphological mutations until harvest. Putative mutants in M_2 generation were isolated and individually harvested. Furthermore, the sibs of selected putative mutants were also harvested as single plants. For observation of mutant and segregation ratios, the selected mutants and their sibs were grown as M_3 generation at the same location in the following year. The normal and female sterile plants were counted in segregating rows for M_3 generation. The mode of inheritance was performed using chi-square (χ^2) test according to the formula: $\chi2 = \Sigma(O\text{-}E)^2/E$, where O and E are observed and expected values, respectively (Toker *et al*., 2012b).

RESULTS AND DISCUSSION

The mutants selected from 'CA 2969' and 'AWC 612' was similar to each other for some respects. They had lesser number of stamens than those of parents, and both were female sterile bearing no stigma. Also, flowers in these mutants were open as shown in Figure 1. Furthermore, these mutants were determinate growth habit, while their parents were indeterminate growth habit (Table 1). The sterile mutant selected from 'CA 2969' and 'AWC 612' had smaller leaf with 4-7 leaflets than those of parents (Table 1).

Table 1. Characteristics of the mutants and their parent in *Cicer* species

Characteristics	***C. arietinum* L.**		***C. reticulatum* Ladiz.**	
	CA 2969	Mutant	AWC 612	Mutant
Flower color	White	White	Pink	Pink
Flower shape	Normal	Open	Normal	Open
Flower per peduncle	2	3-6	1	1
Stamens	9+1	<10	9+1	<10
Stigma	Yes	Absent	Yes	Absent
Leaflet per leaf	16-18	4-7	14-16	4-7
Leaf shape	Normal	Normal	Normal	Normal
Growth habit	Indeterminate	Determinate	Indeterminate	Determinate

Figure 1. Female sterile mutants with open flower, small leaf and determinate growth habit in *C. arietinum* L. (a and c) and in *C. reticulatum* Ladiz. (b and d).

Although flowers of the mutants were pollinated by hand, no pods were obtained from the crosses between the induced female sterile mutants (♂) and their parents (♀). Apparently pollen fertilities were deficient confirmed by with negative pollen staining (data not shown).

In M_1 generation, some morphological mutants were observed in plots. In M_2 generation, there were two female sterile mutants in plots. The ratio of the fertile sibs to the sterile mutants were 20:4 in 'CA 2969' and 17:3 in 'AWC 612' in M_2 generation, respectively (Table 2). The mutant plants in M_2 rows had no pods and seeds due to female sterility, whereas normal sibs were fertile. Therefore, only fertile sibs were sown in M_3 generation. In M_3 generation, the ratios of the fertile sibs to the mutants in 'CA 2969' and 'AWC 612' 'were recorded as 121:35 and 110:39, respectively (Table 2). Chi-square analysis of data from these mutants indicated that segregation ratios fit well to the ratio of 3:1 (i.e. 3 fertile sibs: 1 sterile mutant). Therefore, all these mutations described here were governed by a single recessive gene.

Table 2. Inheritance of induced sterility in *Cicer* species

Parents	M_2		M_3		$\chi 2$	*P*
	No. of filials	F: S*	Observed F: S*	Expected F: S*		
CA 2969	24	20:4	121:35	3:1	0.5	0.90-0.75
AWC 612	20	17:3	110:39	3:1	0.1	0.90-0.95

* F and S are the induced fertile sibs and the induced sterile mutant, respectively.

The nomenclature guidelines proposed by Muehlbauer and Singh (1987) were used for gene symbols. The gene symbol *fs* was designed for the induced female sterility in 'CA 2969' and 'AWC 612'. A similar study on inheritance of the induced sterility was reported by Ashri (1968) in peanut (*Arachis hypogeae* L.). van Rheenen *et al.* (1994) found a sterile mutant in the cultivated chickpea. Although male sterility in the cultivated chickpea has been reported in different studies (Sethi 1979; Muehlbauer and Singh, 1987), the sterility has not been used as sustainable so far.

The induced mutants and their sibs segregating in M_2, M_3, and latter generations can be used for gene mapping. These sterile mutations indicated that useful male sterile chickpea could be induced if chemical mutagen could have been used instead of gamma rays. Large chromosomal aberrations are possible in the present study because physical mutagens are one of the reasons for chromosome aberrations (Salimath *et al.*, 2007).

After Muehlbauer and Singh (1987), many induced and some spontaneous genes in the cultivated chickpea and its progenitor were identified (Dahiya *et al.*, 1984; Pundir and Reddy, 1998; Gaur and Gour, 2002; McNeil *et al.*, 2007; Rajesh *et al.*, 2007; Salimath *et al.*, 2007; Srinivasan *et al.*, 2006; Wani and Anis, 2008; Ali *et al.*, 2010; Kharkwal *et al.*, 2010; Si *et al.*, 2010; Wani, 2011; Toker *et al.*, 2012a). Pundir and Reddy (1998) isolated mutant having few leaflets. Open flower was also reported as a useful trait in the cultivated chickpea because it will reduce the time needed for cross pollinations (Pundir ve Reddy, 1998). van Rheenen *et al.* (1994) isolated an induced mutant chickpea having indeterminate growth habit. The symbol *cd* was designed for the allele conditioning for determinancy and *Dt* for the allele expressing the determinate trait (van Rheenen *et al.*, 1994). To our knowledge, the mutant having open flower and determinate growth habit with small leaf is the first report in *C. reticulatum* Ladiz. (Figure 1 and Table 1). The mutant selected in 'CA 2969' was identical with the mutants reported by van Rheenen *et al.* (1994) and Pundir and Reddy (1998). These results suggested that the same loci could be induced again and also it was more inducible than the others.

A joint segregation of the induced sterility and open flower in 'CA 29169' and 'AWC 612' indicated that these characteristics may be linked to each other. Gaur and Gour (1999) reported similar results on linked genes in the cultivated chickpea.

The results indicated that there was a parallelism for the induced sterility between the cultivated chickpeas and *C. reticulatum* Ladiz., since both had the same mutation. The known genes related to determinate growth habit and small leaf was also induced in the cultivated chickpea and *C. reticulatum* Ladiz. Although the induced female sterility in the cultivated chickpeas and *C. reticulatum* Ladiz. is undesirable, this new gene can be useful for gene mapping.

ACKNOWLEDGEMENTS

We are grateful to Prof. J. Gil (Universidad de Córdoba, Córdoba, Spain) for kindly supplying materials. The work was supported by the Scientific and Technical Research Council of Turkey (TUBITAK) and Akdeniz University Scientific Research Projects Coordination Unit. We are also grateful to Dr. Z. Sagel (TAEK, Ankara, Turkey) for irradiation of seeds of chickpea. The gene discovered in the present study was dedicated to the last authors' mother (Dudu Toker) and father (Azmi Toker).

LITERATURE CITED

Abbo, S., R.J. Redden, S.S. Yadav, 2007. Utilization of wild relatives. In: Yadav, S.S., Redden, R., Chen, W., Sharma, B. (eds), Chickpea Breeding and Management. CAB Int. pp. 338–354.

Ali, H., T.M. Shah, N. Iqbal, B.M. Atta, M.A. Haq, 2010. Mutagenic induction of double-podding trait in different genotypes of chickpea and their characterization by STMS marker. Plant Breed. 129: 116–119.

Ashri, A., 1968. Morphology and inheritance of sterile brachytic dwarfs in peanuts, *Arachis hypogaea*. Crop Sci. 8: 413–415.

Canci, H., C. Toker, 2009. Evaluation of annual wild *Cicer* species for drought and heat resistance under field conditions. Genetic Resour. and Crop Evol. 56: 1-6.

Cubero, J.I., 1987. Morphology of chickpea. In: Saxena, M.C., Singh, K.B. (eds), The Chickpea. CAB Int. pp. 35–66.

Dahiya, B.S., V.S. Lather, I.S. Solanki, R. Kumar, 1984. Useful spontaneous mutants in chickpea. Int. Chickpea Newslet. 11: 4–7.

Donmez, A.A., 2011. *Cicer uludereensis* Donmez: a new species of *Cicer* (Chickpea) (Fabaceae) from around the Fertile Crescent, SE Turkey. Turk. J. Bot. 35: 71–76.

Gaur, P.M., V.K. Gour, 1999. An induced fasciated mutant of chickpea (*Cicer arietinum* L.). Indian J. Genet.Pl. Br. 59: 325–330.

Gaur, P.M., V.K. Gour, 2002. A gene producing one to nine flowers per flowering node in chickpea. Euphytica 128: 231–235.

Kharkwal, M.C., M.I. Cagirgan, C. Toker, T. Shah, M.M. Islam, H. Nakagawa, X. Xu, P. Si, 2010. Legume mutant varieties for food, feed and environmental benefits. In: 5th International Food Legumes Research Conference (IFLRC V) and 7th European Conference on Grain Legumes (AEP VII), 26-30 April 2010, Antalya, Turkey, pp: 196.

Ladizinsky, G., A. Adler, 1976. The origin of chickpea *Cicer arietinum* L. Euphytica 25: 211–217.

McNeil, D., F. Ahmad, S. Abbo, P.N. Bahl, 2007. Genetics and cytogenetics. In: Yadav, S.S., Redden, R., Chen, W., Sharma, B. (eds), Chickpea Breeding and Management. CAB Int. pp. 321–337.

Muehlbauer, F.J., K.B. Singh, 1987. Genetics of chickpea. In: Saxena, M.C., Singh, K.B. (eds), The Chickpea. CAB Int. pp. 99–125.

Pundir, R.P.S., G.V. Reddy, 1998. Two new traits – open flower and small leaf in chickpea (*Cicer arietinum* L.). Euphytica 142: 357–361.

Rajesh, P.N., K.E. McPhee, R. Ford, C. Pittock, J. Kumar, F.J. Muehlbauer, 2007. Ciceromics: Advancement in genomics and recent molecular techniques. In: Yadav, S.S., Redden, R., Chen, W., Sharma, B. (eds), Chickpea Breeding and Management. CAB Int. pp. 445–457.

Robertson, L.D., B. Ocampo, K.B. Singh, 1997. Morphological variation in wild annual *Cicer* species in comparison to the cultigen. Euphytica 95:309–319.

Rubio, J., M.T. Moreno, C. Martinez, J. Gil, 2003. Registration of CA2969, an ascochyta blight resistant and double-podded chickpea germplasm. Crop Sci. 43: 1567–1568.

Salimath, P.M., C. Toker, J.S. Sandhu, J. Kumar, B. Suma, S.S. Yadav, P.N. Bahl, 2007. Conventional breeding methods. In: Yadav, S.S., Redden, R., Chen, W., Sharma, B. (eds), Chickpea Breeding and Management. CAB Int. pp. 369–390.

Sethi, S.C., 1979. Male sterility. Int. Chickpea Newslet. 1: 4.

Si, P., Y. Chen, S. Weerakoon, J. Quealy, S. Powles, W. Erskine, 2010. Chickpea (*Cicer arietinum* L.) breeding lines tolerant to metribuzin applied post-emergence. In: 5th International Food Legumes Research Conference (IFLRC V) and 7th European Conference on Grain Legumes (AEP VII), 26-30 April 2010, Antalya, Turkey, pp: 216.

Srinivasan, S., P.M. Gaur, S.K. Chaturvedi, B.V. Rao, 2006. Allelic relationships of genes controlling number of flowers per axis in chickpea. Euphytica 152: 331–337.

Toker, C., 2005. Preliminary screening and selection for cold tolerance in annual wild *Cicer* species. Genet. Resour. and Crop Evol. 52: 1–5.

Toker, C., B. Uzun, H. Canci, F.O. Ceylan, 2005. Effects of gamma irradiation on the shoot length of *Cicer* seeds. Radiat. Phys. Chem. 73: 365–367.

Toker, C., H. Canci & T. Yildirim, 2007. Evaluation of perennial wild *Cicer* species for drought resistance. Genetic Resour. and Crop Evol. 54: 1781–1786.

Toker, C., 2009. A note on the evolution of kabuli chickpeas as shown by induced mutations in *Cicer reticulatum* Ladizinsky. Genet. Resour. and Crop Evol. 56: 7–12.

Toker, C., H. Canci, N.E. Inci, F.O. Ceylan, 2012a. Improvement in imidazolinone resistance in *Cicer* species by induced mutation. Plant Breeding 131: 535-539.

Toker, C., F.O. Ceylan, N.E. Inci, T. Yildirim, M.I. Cagirgan, 2012b. Inheritance of leaf shape in the cultivated chickpea (*Cicer arietinum* L.). Turkish J. of Field Crops 17: 16–18.

van Rheenen, H.A., R.P.S. Pundir, J.H. Miranda, 1994. Induction and inheritance of determinate growth habit in chickpea (*Cicer arietinum* L.). Euphytica 78: 137–141.

van der Maesen, L.J.G., N. Maxted, F. Javadi, S. Coles, A.M.R. Davies, 2007. Taxonomy of the genus *Cicer* revisited. In: Yadav, S.S., Redden, R., Chen, W., Sharma, B. (eds), Chickpea Breeding and Management. CAB Int. pp. 14–46.

Wani, A.A., 2011. Spectrum and frequency of macromutations induced in chickpea (*Cicer arietinum* L.). Turkish. J. of Biol. 35: 221–231.

Wani, A.A., M. Anis, 2008. Gamma ray- and ems-induced bold-seeded high-yielding mutants in chickpea (*Cicer arietinum* L.). Turkish J. of Biol. 32: 161–166.

Zohary, D., M. Hopf, 2000. Domestication of Plants in the Old World: the origin and spread of cultivated plants in West Africa, Europe, and the Nile Valley, 3. ed. Oxford: Clarendon Press, 316 p.

PHENOLOGICAL RESPONSE OF RICE PLANTS TO DIFFERENT MICRONUTRIENTS APPLICATION UNDER WATER SAVING PADDY FIELDS ON CALCAREOUS SOIL

[1]Naeem Sarwar*, [1]Hakoomat Ali, [2]Muhammad Maqsood, [2]Ehsan Ullah , [2]Muhammad Shahzad, [3]Khuram Mubeen, [1]Ahmad Naeem Shahzad, [4]Muhammad Asghar Shahid and [1]Shakeel Ahmad

[1]Department of Agronomy, Bahauddin Zakariya University Multan, PAKISTAN
[2]Department of Agronomy, University of Agriculture Faisalabad, PAKISTAN
[3]Department of Agronomy, University of Poonch Rawalakot, PAKISTAN
[4]Department of Pest Warning and Quality of Pesticides Punjab, PAKISTAN
*Corresponding author E-mail: bajwa834@gmail.com

ABSTRACT

Due to a reduction of world water resources, rice cultivation has shifted from flooded to aerobic condition. This shift may alter soil physico-chemical conditions and cause reduced availability and uptake of nutrients, especially boron and zinc in calcareous soils, and may ultimately reduce yield. Thus a field experiment in a calcareous soil was conducted in which basal application of boron & zinc alone, or in combination, was undertaken in rice grown aerobically (T_1), flooded for the entire growth period (T_2) or flooded for two weeks after transplanting, given supplemental irrigation up to panicle initiation and then flooded from panicle initiation to physiological maturity (T_3). The study showed that rice crop was significantly reduced growth in term of Leaf Area Index (LAI), Leaf Area Duration (LAD), Crop Growth Rate (CGR), Total Dry Matter accumulation (TDM) and Net Assimilation Rate (NAR) when it was grown in aerobic condition. However, the crop was not affected significantly different from normal flooded rice (T_2) when it was grown in modified rice culture (T_3). Application of zinc alone or in combination with boron enhanced chlorophyll contents in T_2 similar to T_3 while the crop grown in aerobic condition (T_1) had significantly lower chlorophyll content.

Key words: Rice, growth, boron, zinc and rice cultures

INTRODUCTION

Water shortage raised another question for rice growers all over the world and most of the farmers has shifted cultivation from conventional flooded method to new system of aerobic rice. Lot of water saving technologies are under study like aerobic rice , alternative wetting drying and system of rice intensification (Bouman et al., 2005; Stoop et al, 2002; Tabbal et al., 2002; Borell et al., 1997) to save the water input. In Pakistan, basmati rice is grown which have good national and international demand because of its specific fragrance. This system of aerobic rice and other water saving techniques are also under consideration in Pakistan without having specific aerobic cultivars. This cultivation shifting for water saving may broader the deficiencies of micronutrients such as boron and zinc in rice growing areas due to which high yield could not be realized. In Pakistan, soil is calcareous in nature with low organic matter and due to intensive cultivation deficiency of nutrients is increasing day by day especially zinc and boron (Rafique et al., 2006, 2008). Plant grown in aerobic field takes less zinc than plants grown in flooded fields (Gao et al., 2006). In 1969, zinc deficiency was first time detected when scientists explored its role for "hadda" disease of rice (Yoshida and Tanaka, 1969). Zinc deficiency is found to be major constraint for annual crops as well as in rice area in developing as well as in developed countries like India, China, USA and Australia (Cakmak, 2002; Karak et al., 2006; Gao et al., 2006). Almost 30 percent soil is zinc deficient which cover 49 developed and developing countries of the world (Alloway, 2004). Crop production and quality is desperately affecting with emerging deficient micronutrients (B & Zn) all over the world (Rashid and Ryan, 2004; Rafique et al., 2008). Numerous studies have indicated that zinc deficiency is a serious nutritional problem for upland crops. Zinc is essential for several biochemical processes in the rice plant, such as cytochrome and nucleotide synthesis, auxin metabolism, chlorophyll content, enzyme activation and membrane integrity (IRRI, 2000). Zinc is not only imperative for rice crop but it also plays a significant role in other crops. Application of zinc resulted in improved yield as well grain zinc contents not only in rice but also in another crops like wheat, maize, sorghum and alfalfa (Sahrawat et al., 2008; Khan et al., 2009; Ceylan et al., 2009). Boron is emerging deficient nutrient which is affecting crop impressively in rice tract. Rashid et al. (2004) perceived the effect of boron on rice cultivars Super Basmati, Basmati-385 and KS-282 and reported 14-25% increase in paddy yield as compared with control. They further explained that rice crop was affected positively with optimum boron dose at 0.75 kg/ha (Rashid et al., 2004). Studies on boron fertilization showed that paddy yield consistently increased with boron application (Rakshit et al., 2002; PARC, 2002; Tandon, 1999; Ali et al., 1996). Stimulating effects of boron may be linked with greater availability of sugars, increased enzymatic activity and respiration which favored the pollen viability and leads to improved yield or number of rice grains (Garge et al., 1979). In the case of severe boron deficiency, the root cap, quiescent centre and protoderm of root tips disappear and root growth ceases, leading to the death of root tips. Inhibited leaf expansion by low boron also indirectly decreases the photosynthetic ability of plants (Bernie and Longbin, 1997). Paddy yield was significantly higher with the application of micronutrients (Zn, B and Mo) alone or in combination with each other (Hossain et al., 2001). Keeping in view the shift of rice cultivation on micronutrients deficient soil, a field experiment was designed to see the uptake of different micronutrients (B & Zn) under various rice cultures and its resultantly effect on rice plant growth.

MATERIALS AND METHODS

The proposed field study was conducted at Agronomic Research Farm, Department of Agronomy, University of Agriculture, Faisalabad (310-25'N, 73o-09'E) during 2008 and 2009. Experiment was laid out in randomized complete block design with split plot arrangement using three replications, having a net plot size of 2.20 m × 4 m. Rice cultures were kept in main plots while micronutrients were in subplots of experimental layout. The soil texture was sandy clay loam (50.16% sand, 21.75% Silt, 25.40% Clay) with an average pH of 7.9, total soluble salts 0.21%, organic matter 0.74, 0.65ppm Zn and 0.38ppm B. Nursery was raised in 3rd week of June with recommended seed rate (2 kg/marla) and aerobic rice sowing was done at the same date using seed rate 75 kg per hectare with 20 cm apart rows by using hand drill. Basal application of boron (F1) in the form of borax (7.5 kg ha-1 and zinc (F2) in the form of zinc sulphate (12.5 kg ha-1 alone and in combination (F3) along with control treatment (F0) was done in aerobic rice (T1), flooded rice (T2) and "Flooding for two weeks after transplanting and will be maintained at field capacity up to panicle initiation and again will be kept flooded starting from panicle initiation up to physiological maturity" (T3). Manual transplanting was done with 20 cm apart rows. _ Nitrogen was applied at the rate of 140 kg N ha-1, 80 kg P ha-1 and 60 kg K ha-1 and 10 kg Zn ha-1 were applied in the form of urea, diammonium phosphate, sulfate of potash and zinc sulfate (ZnSO4 35%) respectively. Whole of P and K and 1/3 of N was incorporated into the soil at the time of seed bed preparation while remaining N was top dressed in 2 equal splits at the time of booting and panicle initiations. Water depth was maintained at 3-4 cm during transplanting. Measured amount of water was applied as per treatments for different rice cultures by using cut throat flume (90 cm x 20cm). Following formulae was used for calculating the time of application for required amount of water.

$$T = \frac{AD}{Q}$$

Where T represents application time (hours), A for field area (m-2), D shows the depth of irrigation (m) and Q signed for flow rate (m3/sec). Before one week of harvesting, irrigation was stopped when the sign of physiological maturity was appeared. For weed control Butachlor 60 % EC @ 800 mL ha-1 was applied after 7-d of transplanting in standing water (Reddy, 2004). Pre-emergence herbicide (Penoxolan @75ml/ha) was used in aerobic rice for controlling weeds. Carbofuran 10% GR was broadcasted @ 25 kg ha-1 at 55-d after transplanting to protect the crop from insects attack. Harvesting was done manually at harvest maturity and threshing was done separately for each plot.

Plant samples were collected fortnightly for observing growth behavior of the crop. Four plants were cut out from each treatment and then weighed for its fresh weight. In aerobic rice, row of 30cm was cut out from the base of the plants and fresh weight was calculated. Leaves were separated from each sample and then weighed separately for recording leaf and stem weight. Leaf sample (5g) from each treatment was used for recording leaf area. Growth parameters LAI, LAD and CGR were calculated by following procedure of Hunt, 1978.

$$LAI = \frac{\text{Leaf Area}}{\text{Land area}}$$

$$LAD_1 = \frac{(LAI_1 + LAI_1) \times (T_2 - T_1)}{2}$$

$$CGR = \frac{(W_2 - W_1)}{(T_2 - T_1)}$$

$$NAR = \frac{TDM}{LAD}$$

Sample (10g) was air dried and then placed into oven at 70 C0 for recording dry weight. Data of leaf area index (LAI) was used for further calculation of leaf area duration (LAD). Similarly the dry weight from different harvest was used for recording total dry matter production (TDM) which was further used for recording crop growth rate (CGR). Final yield was calculated at harvest following the standard procedure. Chlorophyll content was measured at different stages of the rice crop (seedling, tillering, panicle initiation and maturity) by using the procedure of Yoshida et al. (1976). Leaf samples were collected at respective stage and were weighed separately.

Leaf sample of 0.05 g was measured on digital scale balance and was grinded in mortar by using 80% acetone. After dissolving all green portion of leaf final volume of 5ml was made with acetone (80%). The absorbance of chlorophyll solution was recorded at 645 and 663 wavelengths with a spectrophotometer. Chlorophyll content (expressed as mg/g fw of a sample) was estimated as follows:

$$\text{Total chlorophyll} = (20.2 \times D645 + 8.02 \times D663) \times \frac{5}{1000 \times 0.05} mg/gfw$$

$$\text{Chlorophyll a} = (12.7 \times D663 - 2.69 \times D645) \times \frac{5}{1000 \times 0.05} mg/gfw$$

$$\text{Chlorophyll b} = (22.9 \times D645 - 4.68 \times D663) \times \frac{5}{1000 \times 0.05} mg/gfw$$

The meteorological data for the growing period of the crop were provided from the Department of Crop Physiology, Agronomic Research Institute Faisalabad, Pakistan. Meteorological data showed (Table 3) that the year 2008 was cooler and rainy while the year 2009 was hot and dry. All the collected data was analyzed by using the Fisher's analysis of variance function of MSTAT statistical computer package and LSD at 5% probability was used to compare the treatment's means (Steel et al., 1997).

RESULTS

Growth Performance

Leaf area Index (LAI) progressively increased in all treatments during the growth period and was highest at 3rd harvest in both years (Figure 2, 3).

Combined effect of rice culture and nutrient application was significant on leaf area index in both the years (2008-09). Highest leaf area index (LAI) was calculated in T_2 & T_3 (8.06 & 7.86) rice cultures with the combined application of boron and zinc, while the minimum value was found in T_1F_0 where these nutrients were missed. Flooded rice (T_2) with combined application of boron and zinc produced maximum LAD (205.4) which was similar with T_3 (196.83) while the minimum value was recorded in T_1F_0 (73.92) which was similar with T_1F_1 (83.72) during 2008. Almost similar trend was observed for leaf area duration in second year of experimentation. Maximum average crop growth rate in 2008 occurred in T_2 (29.0) and T_3 (27.4) rice cultures with the combined application of boron and zinc while minimum value was observed in T_1F_0 (6.5). Maximum TDM was recorded in treatment combination of T_2F_3 (1642.57) & T_3F_3 (1564.90) while the minimum value (398.67) was found in T_1F_0 treatment combination. Similar trend was observed in second year of experimentation. Net assimilation rate is the physiological potential for converting the total dry matter into yield. In 2008, maximum value for NAR was noted in T_2F_3, T_3F_3 and T_2F_2 treatment combinations while the minimum value was found in T_1F_0 combination. Similar trend was observed in second year (2009) (Table 1). All treatments progressively increased chlorophyll contents up to tillering stage in both of the years (2008-09). The maximum chlorophyll contents were recorded in T_2 (1.55) and T_3 (1.55) respectively, while crop grown in T_1 rice culture had lower chlorophyll contents (1.46). As regarding nutrients application, comparison of mean values of two years shows that the maximum chlorophyll contents were produced with F_3 & F_2 (1.63 & 1.62) respectively, while the minimum contents were recorded in F_0 (1.39) which was similar with F_1 (1.43) (Table 2).

Table 1. Growth parameters as affected by boron and zinc in different rice cultures during 2008 and 2009

Treat	LAI (m^{-2})		LAD (days)		CGR ($g\ m^{-2}d^{-1}$)		TDM ($g\ m^{-2}$)		NAR ($g\ m^{-2}d^{-1}$)	
	2008	2009	2008	2009	2008	2009	2008	2009	2008	2009
T_1F_0	3.7e	3.3f	73.9f	64.1e	6.5f	6.08f	398g	338f	5.3e	5.2e
T_1F_1	4.2cd	4.4de	83.7ef	77.2d	9.3e	8.9e	541f	481e	6.4d	6.2d
T_1F_2	5.0c	4.7d	122.4d	114.1c	15.3d	15.1c	854d	794c	7.0cd	7.0cd
T_1F_3	6.0b	4.7d	146.3c	146.7b	18.2c	19.0b	1018c	1033b	7.0cd	7.1bcd
T_2F_0	4.23cd	4.1de	91.4e	85.3d	11.0e	10.7de	655e	595d	7.2bcd	7.0cd
T_2F_1	4.6cd	4.4de	96.3e	90.1d	11.0e	10.8de	674e	614d	7.0cd	6.8cd
T_2F_2	6.4b	6.0b	160.5b	146.3b	21.0b	20.9b	1183b	1123b	7.4abc	7.7abc
T_2F_3	8.1a	7.7a	205.4a	196.4a	29.0a	27.4a	1642a	1582a	8.0a	8.1a
T_3F_0	4.3d	4.1e	89.3e	83.4d	10.3e	10.0de	622ef	562de	7.0cd	6.7d
T_3F_1	4.6cd	4.3de	95.3e	86.8d	11.1e	10.9d	673e	613d	7.1cd	7.1bcd
T_3F_2	6.2b	5.3c	153.3bc	145.2b	18.8c	19.1b	1077bc	1031b	7.0cd	7.1bcd
T_3F_3	7.9a	7.5a	196.8a	187.3a	27.4a	25.7a	1564a	1486a	7.9ab	8.0ab
LSD	0.58	0.61	13.32	12.96	2.2	1.99	113	97.97	0.79	0.93

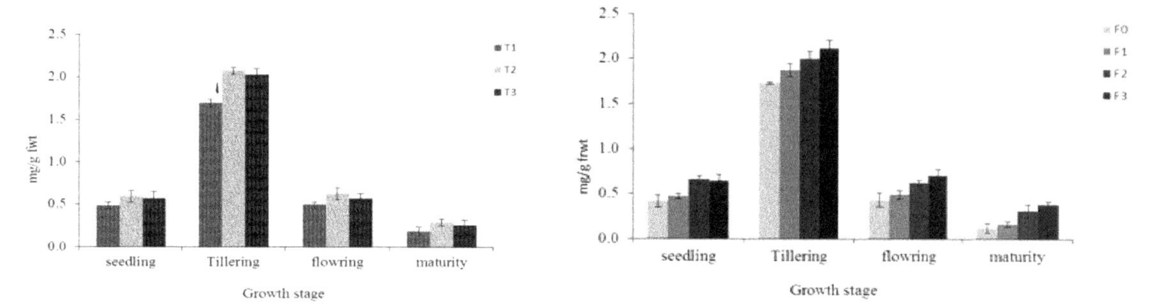

Figure 1. Total chlorophyll (mg/g fw) contents as affected by micronutrients application in different rice cultures

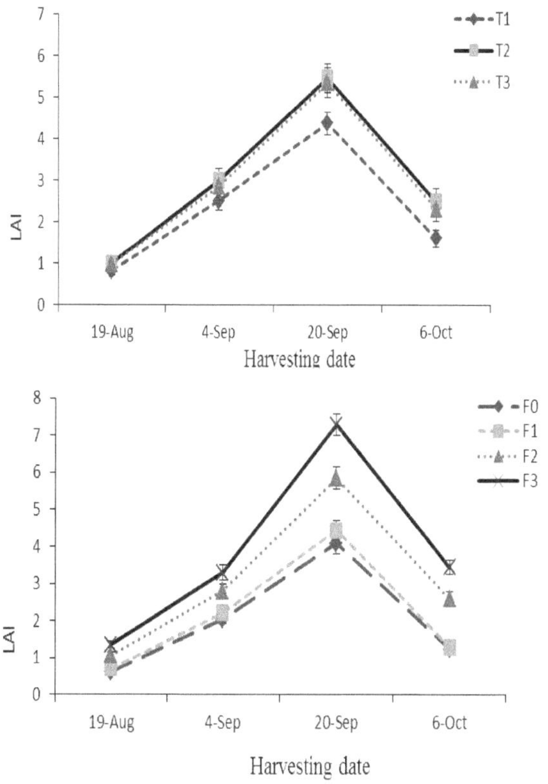

Figure 2. LAI as affected with basal application of boron and zinc (b) in different rice cultures (a) during 2008

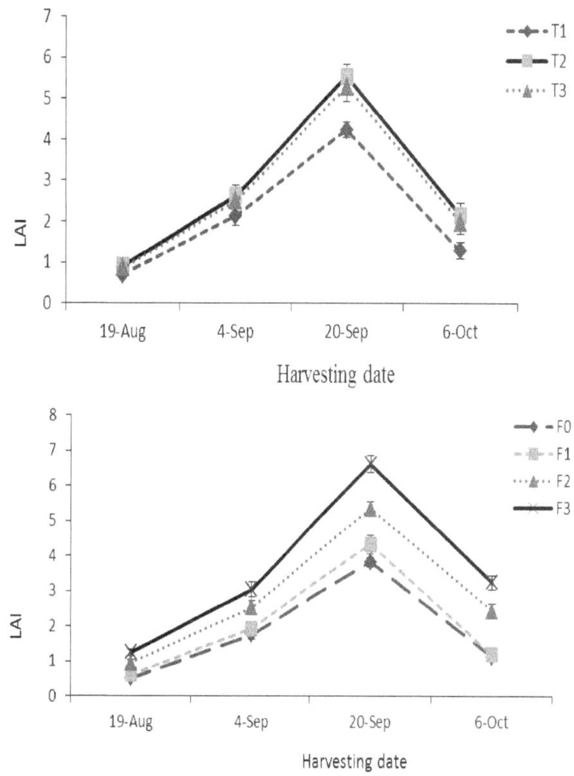

Figure 3. LAI as affected with basal application of boron and zinc (b) in different rice cultures (a) during 2009

Table 2. Maximum chlorophyll contents (mg/g fw) as affected by boron and zinc in different rice cultures during 2008 and 2009

Treatments	Chlorophyll-a (mg/g fw)	Chlorophyll-b (mg/g fw)	Total chlorophyll a+b (mg/g fw)
Rice cultures			
1	0.94b	0.52b	1.46b
2	0.99a	0.56a	1.55a
3	0.98a	0.56a	1.55a
$LSD_{0.05}$	0.014	0.013	0.027
Fertilizers			
0	0.90b	0.49b	1.39b
1	0.92b	0.51b	1.43b
2	1.02a	0.60a	1.62a
3	1.03a	0.60a	1.63a
$LSD_{0.05}$	0.022	0.016	0.036

(T_1) Aerobic rice, (T_2) Flooded rice, (T_3) Flooding for two weeks after transplanting and then maintained at field capacity up to panicle initiation and again kept flooded starting from panicle initiation up to physiological maturity, (F_0) Control, (F_1) Boron, (F_2) Zinc and (F_3) Boron + Zinc

DISCUSSION

This study showed that aerobic cultivation for fine rice caused significant reductions in growth rate and other growth parameters like LAI, LAD, CGR, TDM and NAR. However, chlorophyll contents were similar between flooded rice and rice that was grown under partial flooding with supplemental irrigation indicating that flooded condition may not be necessary for whole crop growth period. The significant decline in growth rate for the crop grown in aerobic condition might be due to the higher moisture and nutrients stress at different growth stages of plant. Chlorophyll contents is an imperative indicator of crop growth which were higher in flooded rice although it was similar with modified rice culture (T_3) indicating that flooded condition is also not necessary for whole crop growth period. Almost similar results were reported by Viraktamath (2006) who found similar rice plant yield while grown in flooded and alternate wetting and drying conditions. Water use efficiency was decreased in full irrigation treatment as compared to deficit irrigation in some other crops like wheat, safflower (Erekul et al., 2012; Omidi et al., 2012). Chlorophyll contents were highest at tillering stage while the minimum chlorophyll contents were recorded at maturity due to the leaf senescence. Chlorophyll concentration showed that significant decline in growth rate was recorded when crop was grown in aerobic condition. This might be due to the higher moisture and nutrients stress at different growth stages of plant. Previous studies support this argument (Peng et al., 2006; Zulkarnain et al., 2009) and also show that above ground biomass significantly increases in flooded soil. Many factors are expected to change in aerobic condition like redox potential (Gao et al., 2002) which effect the availability of nutrients. Soil pH may increase or decrease according to soil nature which will ultimately effect the availability of nutrients (Liu, 1996). Reduction of water contents in soil solution can also restrict the movement of the nutrient toward the root of the plant (Yoshida, 1981) and organic matter will be oxidized which will also cause the boron & zinc availability. As the availability of the zinc in aerobic rice declined which exhibited a suppressing effect on rice growth. All growth parameters were improved in flooded as well as in intermittent flooding and drying condition with the combined application of both micronutrients. This might be due to the higher availability of nutrients in these rice cultures as compared to aerobic rice. Rajesh and Thanunathan (2003) found that alternate wetting and drying uptake more nutrients as compared to aerobic rice culture due to improved root activity. Furthermore our results are also supported by Gao et al. (2006) who recorded inferior zinc contents in rice crop grown under aerobic condition. Crop performance was higher in year 2008 as compared to year 2009 which might be due the variation in weather condition. In the growing season of year 2008, crop received more rainfall with low temperature which favoured the rice growth and yield as compared to next growing season (Table-3). Aerobic rice cultivation is a best option for water saving rice production but a successful transition requires special rice genotypes. Thus plant breeder should play a role in developing aerobic varieties which can adopt that environment without harming growth and yield.

Table 3. Meteorological data recorded during rice growing seasons

Month	Mean Temperature (C^0)		Relative Humidity (%)		Rainfall (mm)	
	2008	2009	2008	2009	2008	2009
June	32.805	33.85	32	33.6	41.7	9.6
July	32.755	32.95	53	59	81.6	43.5
August	30.95	32.1	65	65.8	204.5	116
Sept	29.05	30.35	59.3	61	28.8	20.6
October	26.65	24.9	57.6	57.9	0	17.5
November	19.75	18.25	58.9	64.7	0	0.7

Due to the intense cropping and continuous removal of nutrients resulted in decreased availability of essential micronutrients like boron and zinc. Farmers are just concentrating on macronutrients as they are well aware about these nutrients but they are not conscious about micronutrients. It is the need of the time to grow crops with timely and balanced use of these nutrients. Deficiency of these nutrients can become a bigger problem which will affect the crop severely as farmers are shifting cultivation from flooded to aerobic condition. In this study, application of micronutrients in T_2 & T_3 recorded similar and maximum yield as compared with T_1 rice culture indicating that decreasing moisture contents decreased the availability of nutrients to plants.

CONCLUSIONS

Results of the present study showed that cultivation of rice under fully aerobic conditions significantly reduced the crop growth as compared to that of fully-flooded rice. However partial flooding did not significantly affect the growth attributes. It is therefore concluded that a shift in rice cultivation from fully flooded to fully aerobic is likely to change the growth behavior of rice crop.

ACKNOWLEDGEMENTS

We gratefully acknowledge the Higher Education Commission (HEC) for the support provided to Naeem Sarwar for whole study period under Indigenous Scholarship Programme.

LITERATURE CITED

Ali, A., M.S. Zia, F. Hussain , M. B. Khan, 1996. Boron requirement of rice and its management for rice production. Pakistan J. Soil Sci. 11: 68-71.

Alloway, B. J., 2004. Zinc in Soils and Crop Nutrition. International Zinc Association, Brussels, Belgium.

Bernie, D., H. Longbin, 1997. Physiological response of plants to low boron. Plant Soil. 193:03-120.

Borell, A., A. Garside, F. K. Shu, 1997. Improving efficiency of water for irrigation rice in semi-arid tropical environment. Field Crop Res.52: 231-248.

Bouman, B.A.M., S. Peng, A. R.Castaneda, R. M. Visperas, 2005. Yield and water use of irrigated tropical aerobic rice systems. Agric. Water Manage. 74: 87–105

Cakmak, I., 2002. Plant nutrition research: Priorities to meet human needs for food in sustainable ways. Plant Soil. 247: 3-24.

Ceylan, S., S. Hikmet, B. Bulent, A. Huseyin, E. C. Bihter, 2009. Effect of zinc on yield and some related traits of alfalfa. Turkish Journal of Field Crops, 14(2):136-143.

Erekul, O., G. Klaus-Peter, G. Talih, 2012. Effect of supplemental irrigation on yield and bread-making quality of wheat (*Triticum aestivum* L.) varieties under the mediterranean climatical conditions. Turkish Journal of Field Crops, 17(1):78-86.

Gao, S., K. K. Tanji, S. C. Scardaci, A. T. Chow, 2002. Comparison of redox indicators in a paddy soil during rice growing season. Soil Sci. Soc. Am. J. 66: 805–817.

Gao, X., C. Zou, X. Fan, F. Zhang, E. Hoffland, 2006. From flooded to aerobic conditions in rice cultivation: Consequences for zinc uptake. Plant Soil. 280: 41-47.

Garge, O. K., A. N. Sharma, G. R. S. Konica, 1979. Effect of boron on pollen vitality and yield of rice plant (Oryza Sativa. Var. Java). Plant Soil. 52: 591-594.

Hunt, R. 1978. Plant growth analysis. Edward Arnold, UK: 26-28.

Hossain, M. B., T. N. Kumar, S. Ahmed, 2001. Effect of zinc, boron and Molybdenum application on the yield and nutrient uptake by BRRI Dhan 30. J. Bio. Sci. 1: 698-700.

IRRI, 2000. Nutritional disorders and nutrient management in rice. Inter. Rice Res. Ins. Manila, Philippines.

Karak T., D. K. Dad , D. Maiti, 2006. Yield and zinc uptake in rice (*Oryza sativa*) as influenced by sources and times of zinc application. Indian J. Agri. Sci. 76: 346-348.

Khan, R., A. R. Gurmani, M. S. Khan, A. H. Gurmani, 2009. Residual, direct and cumulative effect of zinc application and rice yield under rice-wheat system. Soil and Environ. 28: 24-28.

Liu, Z., 1996. Microelements in Soils of China. Jiangsu Science and Technology Publishing House, Nanjing, China. 188 pp.

Omidi, A.H, K. Hamid, M. Philippe, S. Frederick, 2012. Effect of cultivar and water regime on yield and yield components in safflower (*Carthamus tinctorius* L.). Turkish Journal of Field Crops, 17(1):10-15.

PARC, 2002. Annual Report, 2002. Effect of boron on rice productivity. Pak. Agri. Res. Coun. Islamabad: 97.

Peng, S., B. Bouman, R. M. Visperas, A. Castaneda, A. Nie, H. Park, 2006. Comparison between aerobic and flooded rice in the tropics: Agronomic performance in an eight-season experiment. Field crop Res. 96: 252-259.

Rafique, E., A. Rashid, J. Ryan, A. U. Bhatti, 2006. Zinc deficiency in rainfed wheat in pakistan: Magnitude, spatial variability, management and plant analysis diagonistic norms. Commun. Soil Sci. and Pl. Anal. 37: 181-197.

Rafique, E., M. Mahmood-ul-hassan, K.M. Khokar, G. Nabi, T. Tabassam, 2008. Zinc nutrition of onion: Proposed diagonistic criteria. J. Pl. Nutr. 31: 307-316.

Rajesh, V., K., Thanunathan. 2003. Effect of seedling age, number and spacing on yield and nutrient uptake of traditional Kambam chamba rice. Madras Agric. J. 90: 47-49.

Rakshit, A, P. B. S. Bhadoria, D. Ghosh, 2002. Influence of boron on NPK uptake of rice (Oryza sativa L) in acid alluvial soils of Coochbehar, West Bengal amd. Ecol. 20:188-190.

Rashid, A., J. Ryan. 2004. Micronutrient constraint to crop production in soil with mediterranen type characteristics: A review. J. Pl. Nutr. 27: 959-975.

Reddy, S., 2004. Agronomy of Field Crops. Kalyani Publishers, New Delhi, India.

Sahrawat, K. L., T. J. Rego, S. P. Wani and G. Pardhasaradhi. 2008. Sulfur, Boron and Zinc Fertilization Effect on Grain and Straw Quality of Maize and Sorghum Grown in Semi-Arid Tropical Region of India. J. Pl. Nutr. 32: 1578-1584.

Steel, R. G. D., J. H. Torrie, D. A. Deckey, 1997. Principle and Procedures of Statistics: A Biometrical approach. 3rd ed. Megraw Hill Book Co.Inc. New York.

Stoop, W. A, N. Uphoff, A. Kassam, 2002. A review of agricultural research issue raised by the System of Rice Intensification (SRI) from Madagascar: opportunities for improving system for resource poor farmers. Agric. Syst. 71:249-274.

Tabbal, D. F., B. A. M. Bouman, S. I. Bhuiyan, E. B. Sibayan, M. A. Sattar, 2002. On-farm stratigies for reducing water input in irrigated rice; case studies in the Philppine. Agric. Water Mange. 56: 93-112.

Tandon, H. L. S., 1999. Micronutrients in Soils, Crops and Fertilizer. Fer. Dev. Con. Org. New Delhi, India: 41-50.

Viraktamath, B .C. 2006. Evaluation of System of Rice Intensification (SRI) under All India Coordinated Rice Improvement Project. In 'Abstracts of National Symposium on System of Rice Intensification (SRI) – Present status and future prospects' November 17-18, pp.11-13.

Yoshida, S., 1981. Fundamentals of Rice Crop Science. International Rice Research Institute, Los Banos, Philippines.

Yoshida, S., A. Tanaka, 1969. Zinc deficiency of the rice plant in calcareous soils. Soil Sci. Plant Nutr. 15: 75-80.

Yoshida, S., D. A. Forno, D. H. Cock, K. A. Gomez, 1976. Labortary Manual for Physiological Studies of Rice, 3rd edition. IRRI, Los Banos, laguna, Philippines.

Zulkarnain, W. M., M. R. Ismail, M. Ashrafuzzaman, H. M. Saud, I. C. Haroun, 2009. Rice growth and yield under rain shelter house as influenced by different water regimes. Int. J. of Agri. Bio. 11: 566-570.

GENETIC CONTROL OF PURPLE PLANT COLOR IN SESAME

Engin YOL[1], *Seymus FURAT*[1], *Bulent UZUN*[1*]

[1]*Department of Field Crops, Faculty of Agriculture, Akdeniz University, Antalya, TURKEY*
[*]*Corresponding author:bulentuzun@akdeniz.edu.tr*

ABSTRACT

The exploitation of exotic genetic resources in plant breeding is necessary to enhance diversity of cultivars. Purple color is an exceptional character in sesame and monitored rarely in stems, capsules and leaves. The purple sesame is suitable for commercial production with high antioxidant capacity. For understanding of its genetic behaviour, inheritance study was carried out for four years (2008-2011) by crossing Muganli-57 (♀) and ACS 70 (♂). Muganli-57 parent had green color in canopy while ACS 70 was purple. All the plants in the F_1 generation were purple colored. In F_2 population, 3:1 segregation ratio showed that purple color character was controlled by a single dominant gene. F_2 progenies were sown to single rows separately, either purple or green plants. The green plants in F_2 were also green colored in F_3 with no segregation while the purple plants obtained from F_2 indicated purple and green colors in F_3 with a segregation ratio of 3:1. These segregations demonstrated that purple color in sesame was under the control of a single gene and this unique color was dominant over traditional green color. This accession (ACS 70) is a valuable genetic resource by providing a unique color character in sesame. The result presenting in this study would be of importance for further improvement of high antioxidant capacity in sesame.

Keywords: Antioxidants, genetic control, purple capsule, *Sesamum indicum* L.

INTRODUCTION

Sesame (*Sesamum indicum* L.) is one of the world's important oilseed crops and has long been used as ingredient in human foods and animal feeding in the form of seed, seed oil and meal (Hahm et al., 2009). Sesame has also medicinal and pharmaceutical value owing to high quality oil (Uzun et al., 2008) and antioxidant lignans such as sesamin and sesamolin (Peschel et al., 2007; Erbas et al., 2009). Recently, its antioxidants have been used exclusively for anti-aging processes.

Sesame is a plant breeder's dream because of its great variability (Langham and Wiemers, 2002). Hundreds of characters have been studied with respect to range of variability and many of them were classified by several researchers (Bisht et al., 1998; Furat and Uzun, 2010; Yol et al., 2010; Yol and Uzun, 2012). Color establishment in sesame also contributes to this variability with different cotyledon, petiole, corolla, capsule, leaf, stem, and seed coat color. Seed coat color has been studied intensively as it was related with amino acid composition (Lee et al., 1990), oil content (Namiki, 1995; Were et al., 2001; Zhang et al., 2004) and antioxidant capacity (Lee et al., 1999; Lee et al., 2002; Moazzami and Kamal-Eldin, 2006; Shahidi et al., 2006) whilst there has been limited investigation about plant color because of low variability.

In sesame, during the vegetative and reproductive phases the color is usually a shade of green, and then as the plants mature and begin to drop their leaves, the color will turn to many shades of yellow/green, and rarely purple (Langham, 2007). When purple lines reach late bloom stage until the drying phase, plants indicated purple color especially in stems and capsules on the contrary of usual sesame plants. Previously, purple plants were observed in Indian genotypes by Bedigian et al. (1986) who stated that purple form had dominancy however inheritance mechanism was not exhibited. Erbas et al. (2009) studied chemical composition of purple plant in sesame and they observed high antioxidant capacity in purple sesame compared with normal sesame plants.Purple forms should be assessed as commercial types due to their being a source of high antioxidant capacity. The aim of the present study is to determine genetic behaviour of purple sesame accession for understanding the gene system operating in the expression of purple color in sesame.

MATERIALS AND METHODS

In 2008 growing season, the accession with purple color, ACS 70 (♂) was crossed with Muganli-57 (♀) which is registered cultivar (green), at the West Mediterranean Agricultural Research Institute's fields of

Antalya (36°52'N. 30°50'E., 15 m elevation). F_1s were selfed in the growing season of 2009 and F_2 populations were developed in 2010. All the plants in F_2 were scored either purple or green colored. F_2 progenies showing both colors were sown in single rows separately in 2011. Purple and green plants were also scored either purple or green in F_3 generation.

The crosses between Muganlı-57 (green color) and ACS 70 (purple color) were made using flower buds emasculated just before anthesis and pollinated the second day with pollen grains from freshly dehisced anthers of the male parents (Falusi and Salako, 2003). Parents, F_1s, F_2s and F_3s were grown in 70 cm row and 10 cm plant spacing in all the experiments. Standard cultural practices like regular weeding, irrigation and fertilization were followed. Fertilizer was applied at rates of 60 kg N, 60 kg P_2O_5 and 60 kg K_2O per hectare just prior to sowing. F_1, F_2 and F_3 generations were grown at the same location.

A chi-square ($\chi 2$) goodness of fit test was performed on the segregating populations against a possible theoretical segregation ratio using the formula:

$\chi 2 = \Sigma\ (O - E)^2 / E$, where O and E are the observed and expected values, respectively (Steel and Torrie, 1980).

RESULTS

Following to Muganli-57 (♀) x ACS 70 (♂) cross, all the F_1 filials had purple color on their stems, capsules and leaves. In F_2 progeny, the ratio of purple color to green color was 65 to 22 (Table 1). Phenotypically, 65 plants indicated purple form (Figure 1a) and 22 plants indicated green form (Figure 1b). Chi-square values obtained from this cross showed a good fit for a monogenic inheritance with the F_2 phenotypic ratio of 3:1 (Table 1).

Figure 1. Green and purple sesame plants a - green stem, capsule and leaf, respectively. b - purple stem, capsule and leaf, respectively.

Table 1. Chi-square values of the cross between Muganlı-57 and ACS 70

Cross	Experimental		Theoretical		χ^2	p	Ratio
	Purple Plants	Normal Plants	Purple Plants	Normal Plants			
Muganli-57 x ACS 70	65	22	65.25	21.75	0.0038	0.95	3:1

The plants with purple and green color in F_2 were separately advanced to F_3 generation in single rows. Totally, 106 offsprings in F_3 were obtained from four F_2 plants with green color. All these offsprings had green color and showed no segregation (Table 2).

Table 2. Segregation ratios of purple and green colored sesame in F_3

Cross	Number of plants			
Muganli-57 x ACS 70	**Purple plants**	**Normal (green) plants**	**Ratio**	**Symbol**
F_2 progeny with purple color				
Offspring 1	43	0	1:0	*P/P*
Offspring 2	37	0	1:0	*P/P*
Offspring 3	39	0	1:0	*P/P*
Offspring 4	32	0	1:0	*P/P*
Offspring 5	45	0	1:0	*P/P*
Offspring 6	41	0	1:0	*P/P*
Offspring 7	30	8	3:1 (*P=0.57*)	*P/p*
Offspring 8	51	0	1:0	*P/P*
Offspring 9	12	5	3:1 (*P=0.67*)	*P/p*
F_2 progeny with normal (green) color				
Offspring 1	0	45	0:1	*p/p*
Offspring 2	0	12	0:1	*p/p*
Offspring 3	0	30	0:1	*p/p*
Offspring 4	0	19	0:1	*p/p*

Totally, 343 offsprings in F_3 were obtained from nine F_2 plants with purple color. Those of 288 offsprings coming from seven F_2 plants had purple color with no segregation. The other two F_2 progeny (offspring 7 and 9) consisted of 42 purple and 13 green color plants. This result supported that purple and green color plants fitted the expected 3:1 ratio (Table 2). These segregations proved that purple color in sesame was under the control of a single gene and it was dominant over traditional green color.

DISCUSSION

Monogenic inheritance for different colors in sesame was identified for petiole, nectar (Van Rheenen, 1970) and flower (Khidir, 1973) however there was no previous information about the inheritance of stem, leaf and capsule color. Same color characteristics were investigated for different crop plants. Red/purple color had dominancy over green stem color in *Ricinus communis* L. (Lavanya and Gopinath, 2008), in groundnut (Jadhav and Shinde, 1979), in pigeonpea (Narkhada et al., 1980) and purple plant color had dominancy over green plant color in *Vigna radiate* (L.) Wilczek (Khattak et al., 2000).

More knowledge on the inheritance mechanisms of desirable traits will indeed contribute much to varietal improvement in sesame (Ashri, 2007). Recently, antioxidant capacity has been taken into consideration as desirable character in sesame. Researchers use plant genetic resources to produce new crop varieties with specific characteristics like disease resistance, drought tolerance, or color; develop pharmaceutical or medical products; and determine the origins of a particular species (Delheimer, 2013). In particular, an adequate knowledge of genetic structure of plant genetic resources provides basic science and applied aspects like the efficient management of crop genetic diversity (Mondini et al., 2009). From this point of view, identifying inheritance mechanism of new traits would be beneficial for trait utilization. Purple sesame as a valuable genetic resource offers exploiting antioxidant capacity. Genetic behaviour of this unique character was explored and monogenic inheritance was established in this investigation.

Breeding objectives in sesame have focused many characters such as seed size, shape, flavor, coat color and oil content. Especially for physiological needs, color is an important argument because color production is an integral part of the development of various plant parts; type of color may adapt the plant part for a specific function (Padi, 2003). Therefore, color is one of the most important attributes of foods, being considered as a quality indicator and determining frequently their acceptance (Azerado, 2009). Particularly purple colored plants hold many healthy and agricultural advantages. Purple color can be due to anthocyanins which are phenolic compounds with high antioxidant activity that give many fruits and vegetables red or purple color (Wolfe and Liu, 2002). Purple sweet potato (*Ipomoea batatas*) has been concerned as a rich source of stable anthocyanins (Wu et al., 2008). Purple tomatillo (*Physalis ixocarpa*) genotypes appear to be good and safe source of antioxidants (Mendoza et al., 2010). Purple-pigmented rice bran includes high amounts of hydrophilic phenolic compounds and lipophilic antioxidants (Hu et al., 2003). Purple corn kernels are affluent in anthocyanins with well-established antioxidants (Adom and Liu, 2002). With this respect, the introduction of purple color to sesame plants will satisfy new demands and markets as in other cultivated plants with purple color. The results of the study showed that purple color character is dominant over green color and information about inheritance of purple color character would be useful for the genetic improvement of sesame and improve new varieties with higher antioxidant capacity.

ACKNOWLEDGEMENTS

The authors are thankful to the Akdeniz University, The Scientific Research Projects Coordination Unit for continuous support and the Scientific and Technological Research Council of Turkey (TUBITAK) under the project number of TOVAG 107O912 for the financial support.

LITERATURE CITED

Adom, K.K. and R.H. Liu. 2002. Antioxidant activity of grains. J. Agr. Food Chem.50:6182-6187.

Ashri, A. 2007. Sesame (*Sesamum indicum* L.). In: Genetics Resources, Chromosome Engineering, and Crop improvement, Oilseed Crops,ed. Singh, R.J., 231-289, CRC Press, Florida.

Azeredo, H.M.C. 2009. Betalains: properties, sources, applications, and stability – a review. Int. J. Food Sci. Tech. 44:2365-2376.

Bedigian, D., C.A. Smyth. and J.R. Harlan. 1986. Patterns of morphological variation in sesame. Econ. Bot. 40:353-365.

Bisht, I.S., R.K., Mahajan, T.R. Loknothan and R.C. Agrawal. 1998. Diversity in Indian sesame collection and stratification of germplasm accessions in different diversity groups. Genet. Resour. Crop Evol. 45:325-335.

Delheimer, S. 2013. Conserving plant genetic resources. Fact sheet of the plant genetic resources conservation unit.

Erbas, M., H. Sekerci, S. Gül, S. Furat, E. Yol, B. Uzun. 2009. Changes in total antioxidant capacity of sesame (*Sesamum* sp.) by variety. Asian J. Chem. 21:5549-5555.

Falusi, O.A. and E.A. Salako. 2003. Inheritance studies in wild and cultivated 'Sesamum' L. species in Nigeria. J. Sustain. Agr. 22:75-80.

Furat, S. and B. Uzun. 2010. The use of agro-morphological characters for the assessment of genetic diversity in sesame (*Sesamum indicum* L). Plant Omics J. 3:85-91.

Hahm, T.S., T.S.J. Park and Y.M. Lo. 2009. Effects of germination on chemical composition and functional properties of sesame (*Sesamum indicum* L.) seeds. Bioresource Technol. 100:1643-1647.

Hu, C., J. Zawistowski, W. Ling and D.D. Kitts. 2003. Black rice (*Oryza sativa* L. indica) pigmented fraction suppresses both reactive oxygen species and nitric oxide in chemical and biological model systems. J. Agric. Food Chem. 51:5271-5277.

Jadhav, G.D. and N.N. Shinde. 1979. Genetic studies in groundnut (*Arachis hypogaea*). Indian J. Agric. Res. 13:93-96.

Khattak, G.S.S., M.A. Haq, M. Ashraf and M.A. Awan. 2000. Inheritance of plant and flower bud color in mungbean (*Vigna radiata* (L.) Wilczek). Songklanakarin J. Sci. Technol. 22:249-251.

Khidir, M.O. 1973. Genetic studies in sesame. II. Inheritance of flower color and number of locules per pod.Exp. Agr. 9:361-364.

Langham, D.R. and T. Wiemers. 2002. Progress in mechanizing sesame in the US through breeding. In: Trends in New Crops and New Uses, ed.Janickand, J. and Whipkey, A., 157-173, ASHS Press, Virginia.

Langham, D.R. 2007. Phenology of sesame. In:),Issues in New Crops and New Uses, ed. Janickand, J. and Whipkey, A., 144-182, ASHS Press, Alexandria, Virginia.

Lavanya, C. and V. Gopinath. 2008. Inheritance studies for morphological characters and sex expression in pistillate lines of castor (*Ricinus communis* L.).Indian J. Genet. 68:275-282.

Lee, J.I., B.H. Lee and J.K. Bang. 1990. Breeding of sesame (*Sesamum indicum* L.) for oil quality improvement. V. Varietal differences of amino-acid composition in protein of sesame seed. Kor. J. Breed. 21:300-307.

Lee, S.W., C.W. Kang, D.H. Kim, Y. Satoko and K. Masumi. 1999. Varietal variation of sesamin, sesamolin and oil contents according to seed coat colors in sesame. Kor. J. Breed. 31:286-292.

Lee, W.C., J.Y. Wen, C.H. Shiow and D.D. Pin. 2002. Antioxidant activity of sesame coat. Food Chem. 78:347-354.

Mendoza, D.G., O.G. Juárez, R.S. Ortiz, F.E.Garcia andJ.F.S. Hernández. 2010. Evaluation of total phenolics, anthocyanins and antioxidant capacity in purple tomatillo (*Physalis ixocarpa*) genotypes. Afr. J. Biotechnol. 9:5173-5176.

Moazzami, A.A. and A. Kamal-Eldin. 2006. Sesame seed is a rich source of dietary lignans. J. Am. Oil Chem. Soc. 83:719-723.

Mondini, L., A. Noorani and M.A. Pagnotta. 2009. Assessing plant genetic diversity by molecular tools. Diversity1:19-35.

Namiki, M. 1995. The chemistry and physiological functions of sesame. Food Rev. Int. 11:281-329.

Narkhada, B.N., A.B. Daokar and R. Cruz. 1980. Inheritance of some characters in a cross of pigonpea, *Cajanus cajan* (L.) Millsp. Journal of Maharashtra Agricultural Universities.5:205-208.

Padi, K.F. 2003. Genetic analyses of pigmentation in Cowpea. Pakistan J. Biol. Sci. 6:1655-1659.

Peschel, W., W. Dieckmann, M. Sonnenschein and A. Plescher. 2007. High antioxidant potential of pressing residues from evening primrose in comparison to other oilseed cakes and plant antioxidants. Ind. Crop. Prod. 25:44-54.

Shahidi, F., C.M.L. Pathirana and D.S. Wall. 2006. Antioxidant activity of white and black sesame seeds and their hull fractions. Food Chem. 99:478-483.

Steel, R.G.D. and J.H. Torrie. 1980. Principles and Procedures of Statistics: A Biometrical Approach. 2. ed. New York: McGraw-Hill, 631p.

Uzun, B., C. Arslan and S. Furat. 2008. Variation in fatty acid compositions, oil content and oil yield in a germplasm collection of sesame (*Sesamum indicum* L.). J. Am. Oil Chem. Soc. 85:1135-1142.

Van Rheenen, H.A. 1970. Intergeneric hybridization between *Ceratotheca sesamoides*, Endl. and *Sesamun indicum* L. Nigerian J. Sci. 4:251-254.

Were, B.A., M. Lee,and S. Stymne.2001. Variation in seed oil content and fatty acid composition of *Sesamum indicum* L., and its wild relatives in Kenya. Sveriges Utsadesforenings Tidskrift.111:178-193.

Wolfe, K. and R.H. Liu. 2002. Apple peels are rich in phytochemicals and have high antioxidant activity. New York Fruit Quarterly. 10:9-11.

Wu, D.M., J. Lu, Y.L. Zheng, Z. Zhou, Q. Shan and D.F. Ma. 2008. Purple sweet potato color repairs D-galactose-induced spatial learning and memory impairment by regulating the expression of synaptic proteins. Neurobiol. Learn. Mem. 90:19-27.

Yol, E., E. Karaman, S. Furat and B. Uzun. 2010. Assessment of selection criteria in sesame by using correlation coefficients, path and factor analyses. Aust. J. Crop Sci. 4:598-602.

Yol, E. and B. Uzun. 2012. Geographical patterns of sesame (*Sesamum indicum* L.) accessions grown under Mediterranean environmental conditions, and establishment of a core collection. Crop Sci. 52:2206-2214.

Zhang, J.H., H. Guo, Y. Wang and D.C. Chen. 2004. Study on method of determination of the content of sesame oil by second derivative ultraviolet spectrophotometry. J. Hunan. Agric. Univ. 30:367-370.

INHERITANCE OF RESISTANCE TO RUSSIAN WHEAT APHID (*Diuraphis noxia* Kurdjumov) IN BREAD WHEAT (*Triticum aestivum* L.)

Ferit TURANLI[1*], *Emre ILKER*[2], *Firdevs ERSIN DOGAN*[1], *Lale ASKAN*[1], *Deniz ISTIPLILER*[2]

[1] *Ege Üniversitesi, Ziraat Fakültesi, Bitki Koruma Bölümü, Izmir, TURKEY*
[2] *Ege Üniversitesi, Ziraat Fakültesi,Tarla Bitkileri Bölümü, Izmir, TURKEY*
[*] *Corresponding author: ferit.turanli@ege.edu.tr*

ABSTRACT

Diuraphis noxia (Kurdjumov) (Hemiptera; Aphididae), Russian wheat aphid (RWA) causes death at heavily infested bread wheat (_Triticum aestivum_ L.) plants. The pest is living in rolling leaves so its chemical control is difficult. The most effective and economical means of controlling Russian wheat aphid is through the use of resistant cultivars. Two resistant lines, 15 (FL302//BUC/PVN/3/RSK/CA8055//CHAM6) and 27 (BJN C 79/F96PYN3-1828) were crossed with susceptible cultivar Basribey in 2010 and the F_1 plants were selfed to produce the F_2 progeny to determine the inheritance mode after exposing to pest in 2011. Chi-square test showed that the segregation ratio fits a one gene segregation ratio of 3:1 resistant:susceptible phenotyping segregation ratio at the $P \leq 0.75$-0.50 and $P \leq 0.50$-0.25 level of probability in combinations respectively. In both combinations, chi-square test results indicated that a single dominant gene with minor genes governed resistance to RWA. Differential set from USDA (United State Department of Agriculture) with different _Dn_ resistance genes used to find out the single dominant gene carrying the resistance. Variance analyses showed that the differences between reactions of resistant genes were important for chlorosis and leaf rolling. The lines carried _Dn7_ resistant gene exhibited high level of resistance to the pest while moderate resistance conferred by the plants containing _Dn6_ gene. Reactions of other _Dn_ genes were close to susceptible control cultivars.

Key words: Resistance, Inheritance, resistant cultivars, _Dn7_ gene, Russian wheat aphid

INTRODUCTION

As a wheat pest Russian wheat aphid (RWA), is one of the most important and widely distributed insect species in the world. Pest causes leaf rolling and streaking, head trapping and even death heavily infested plants. Russian wheat aphid after the first record as a pest in Crimea in 1901, is one of the most devastating pest of wheat. After occurring in South Africa in 1978 it is a limiting factor in wheat production. It is also known as the key pest of wheat in the USA and in the other significant grain producing countries of the world. It was found that the pest caused the economic yield losses over one billion dollar in the USA, including not only losses in the wheat yield but also insecticide usage (Stoetzel, 1987; Elmali, 1998; Morrison and Peairs, 1998; Haley et al., 2004).

In Turkey, *D. noxia* was first recorded in Bitlis province in 1959. A few years later, it was recorded in Isparta, Ankara and wider areas of Central Anatolia, and in some areas of Adiyaman and Malatya provinces (Tuatay and Remaudiere, 1964). An epidemic occurred in 1962 and caused 25-60% crop losses in Konya province (Duran and Koyuncu, 1974; Altinayar, 1981). A survey was carried out to determine the current status of the *D. noxia* in major wheat-producing areas of Turkey in May 2010. In the survey conducted in the important wheat growing areas in Turkey (Izmir, Manisa, Usak, Kutahya, Eskisehir, Aksehir, Ankara, Konya, Aksaray, Nevsehir, Yozgat and Erzurum provinces) were checked in view of distribution, damage and population level of the pest. The survey showed that almost all areas were infested with the pest (Turanli et al., 2012).

Several effective management approaches have been proposed to reduce economic losses from *D. noxia*. The pest is living in rolling leaves so its chemical control is difficult. Although cultural and biological controls may have some effectiveness, the most effective and economical means of control growing resistant cultivars with one or more resistance genes. Resistant varieties could be the source of fundamental breeding studies. For almost two decades, there has been a worldwide effort by wheat breeders to identify and use new source of genetic resistance to the *D. noxia* (Du Toit, 1987, 1989). The first sources of resistance to Russian wheat aphid were identified from wheat that originated from countries where the pest is endemic, namely the former Soviet Union, the Balkans, Iran, Turkey and the rest of the Middle East (Harvey and Martin, 1990; Do Toit, 1992). In the United States, winter wheat resistance to the Russian wheat aphid was first identified in 1987 in the PI 372129 genotype (Quick et al., 1991). Inheritance of this type of resistance was governed by a single dominant gene, *Dn4* (Nkongolo et al., 1991a; Saidi and Quick, 1996). The first North American Russian wheat aphid-resistant hard red winter wheat cultivar, 'Halt', contained *Dn4*, was released to growers by the Colorado Agricultural Experiment Station in 1994 (Quick et al., 1996). Since the release of Halt, other cultivars containing *Dn4* have been released and are currently available to growers (Quick et al., 2001a, b, c, d; Haley et al., 2004). Roughly 25% of Colorado wheat acreage is planted with this Russian wheat aphid resistant variety. Many researchers have categorized resistance to Russian wheat aphid-resistant in wheat and have found varying degrees of antibiosis, antixenosis, and tolerance. Hawley et al. (2003) and Miller et al. (2003) reported that wheat containing *Dn4* exhibited antixenosis, antibiosis, and tolerance. Tolerance is the ability of the plant to grow when infested with aphids. Antixenosis refers to non-preference of a host by aphids and is measured by the number of adult aphids per plant (Kogan and Ortman, 1978). Antibiosis is measured by a significantly reduced fecundity of aphids (i.e. the number of nymphs per aphid) grown on plant and can occur together with tolerance or antixenosis. The incorporation of different types of resistance would provide plants with wider range of responses to the aphid.

In the previous studies of resistance, sources of resistance genes have been identified in wheat germplasm accessions including *Dn1* in the PI 137739, *Dn2* in the PI 262660 (Du Toit, 1989), *Dn4* in the PI 372129 (Nkongolo et al., 1991b), *Dn6* in the PI 243781 (Saidi and Quick, 1996), *Dn8 and Dn9* in the PI 294994, and *Dnx* in the PI 220127 (Liu et al., 2001). The PI 292994 was first hypothesized to have a single dominant gene called *Dn5* (Marais and Du Toit, 1993), but later was found to contain more than one gene in some genotypes. Three types of inheritance were observed in the PI292994: a single dominant gene, two dominant independent genes, or one dominant and one recessive gene carrying resistance to the RWA (Zhang et al., 1998).

The identification of a new, *Dn4*-virulent Russian wheat aphid biotype in the southeastern Colorado necessitated a rapid usage of new resistance sources to reduce economic losses from infestations with the new biotype. Information on the effectiveness of the *Dn7* gene or other resistance sources under field conditions would be helpful to evaluate the value of the resistance in breeding wheat cultivars with Russian wheat aphid-resistance (Collins et al., 2005).

Dn7 is a rye gene located on chromosome 1RS that carries resistance to the Russian wheat aphid. This gene was previously transferred from rye to a wheat background via a 1 RS/1BL translocation. The resistant line "94M370" was crossed with a susceptible wheat cultivar which also contains a pair of 1 RS/1BL translocation chromosomes. The F_2 progeny from this cross segregated for resistance in a ratio of 3 resistant 1 susceptible indicating a single dominant gene. The *Dn7* gene confers antixenosis and provides a higher level of resistance than that provided by the *Dn4* (Garret et al., 2003)

The aim of this study was to determine the inheritance mode by using the F_2 segregating generations and identify the gene which carries the resistance to the pest. Thus this study will be one of the first breeding attempt to determine the mode of inheritance to this pest and important problem of wheat production.

MATERIALS AND METHODS

Genetic Materials

Seeds of resistant lines 15 (FL302//BUC/PVN/3/RSK/CA8055//CHAM6) and 27 (BJN C 79/F96PYN3-1828) were obtained from the ICARDA (International Center for Agricultural Research in The Dry Areas) and the CIMMYT (International Maize and Wheat Improvement Centre). Seeds of susceptible cultivar (Basribey) were obtained from the Aegean Agricultural Research Institute. The lines number 15 and 27 were crossed with Basribey in Department of Field Crops, Faculty of Agriculture of the Ege (Aegean) University. The F_1 plants were selfed to produce the F_2 progeny in 2010. The plants of the F_2 generations were infested with the Russian wheat aphid and the injury evaluations were done on individual plants in 2011.

The differential set which have different resistant genes to the RWA, was obtained from the ICARDA (Originated from USDA, United States Department of Agriculture) and was presented in Table 1.

Table 1. The list of differential set from the United State Department of Agriculture (USDA)

USDA Small Grain	Plant/Variety/ID No	Origin	Resistant Gene
	Yama	USA	*Dn0*
PI 262660 TR05ID	Turtsikum	Azerbaijan, Naxcıvan	*Dn2*
PI 372129	Yamar (Colorado)	Turkmenistan	*Dn4*
PI 294994	CO950043 (Colorado)	Bulgaria	*Dn5*
PI 47545 TR05ID	CI 6501	Iran	*Dn6*
RWA-MATRIX	CI 6501	Iran	*Dn6*
Tukey 77 (Rye)	94M370 (South Africa)	Turkey	*Dn7*
Gamatoos R	94M370 (South Africa)	Turkey	*Dn7*

Insect Culture

The insect materials were three different populations (İzmir, Kütahya and Ankara) of *Diuraphis noxia*. The aphid populations which had been taken as alive from wheat fields and reared as purely in the Department of Plant Protection, Faculty of Agriculture, Ege University were used in the studies. Aphid cultures were established and maintained on barley plants in the greenhouse with 20±1 ^{O}C temperature, 60±5% humidity and 16:8 photoperiods.

RWA Resistance Phenotyping

The F_2 individual plants were infested with Izmir population to test their reaction to RWA. Test plants were also reared in viols under same conditions. One wheat seed was sown in the mixture of soil and peat (1/1 ratio) in viol from each combinations or line. Irrigation was done every two days.

Tests with the differential set were done by hill plots with four replications. Each hill plots consisted of 5 plants. When the plants reached to 2-4 leaf stage, they were infested with 5 individuals of the pest per plants. The RWA damage was rated at 28 days after infestation. In 2011, leaf chlorosis and leaf rolling were evaluated based on 1-6 and 1-3 scales, respectively (Ennahli et al., 2009) (Table 2). According to leaf chlorosis scale, plants with 1, 2 and 3 values were accepted as resistant and with 4, 5 and 6 values as susceptible. For the leaf rolling scale 1 and 2 values were referred to resistant and 3 referred to susceptible. For this purpose, differential set including different *Dn* resistant genes were exposed to three different RWA populations.

Statistical Analysis

A chi-square test analysis was performed on the F_2 segregation data for the RWA resistance accepting a single dominant–gene model (Toker et al., 2012). All the data obtained from differential set were analyzed by the ANOVA. The average value of three different populations were used in the analyses variance of using TARIST (Acikgoz et al., 2004) and the differences between the means were compared by the LSD multiple range test according to Steel et al. (1997).

Table 2. Descriptors and rating scales used for evaluation of the F_2 population for resistance to Russian wheat aphid (Ennahli et al., 2009).

Scale	Description
Scale for leaf chlorosis (1-6)	
1	no chlorosis
2	<1/3 of leaf area chlorotic
3	1/3- 2/3 leaf area chlorotic
4	>2/3 of leaf area chlorotic
5	necrosis in at least one leaf
6	plant death
Scale for leafrolling (1-3)	
1	no rolling
2	trapping or curling in one or more leaves.
3	rolling in one or more leaves

RESULTS AND DISCUSSION

The Chi-square values of the plants which is belongs to F_2 generations of two crosses [Basribey (♀) x 15 (♂) and Basribey (♀) x 27 (♂)] were shown in Table 3. The 199 F_2 individual plants from the cross between Basribey (Susceptible) and line number 15 (Resistant) were segregated for RWA resistance. In the F_2, 144 plants were found as resistant, while 55 plants were identified as susceptible. Chi-square test showed that this segregation ratio fits to 3 (Resistant) :1 (Susceptible) ratio. The 200 F_2 individuals from the cross between Basribey (Susceptible) and line number 27 (Resistant), segregated for resistance to RWA, were shown in Table 3. Out of 200 individuals in the F_2, 159 plants were scored as resistant, whereas 41 plants were scored as susceptible. Chi-square test results of this combination were also similar with the previous segregation the ratio 3:1 resistant:susceptible.

Beside the distribution of the plants in the F_2 generation in two combinations according to leaf chlorosis scale showed that resistant and susceptible plants were not divided only in two groups (Figure 1). Results of the chi-square test and distribution of the plants in the F_2 generation indicated that a single dominant gene possibly with minor genes controls the resistance to the RWA (Table 3 and Figure 1). These results also implied that resistance in the parents (15 and 27) was also governed by a single dominant gene.

Table 3. Number of the F_2 plants, chi-square value and segregation ratio from two crosses

F_2 combinatons	Number of resistant plants	Number of susceptible plants	Total plants	Chi-square value	P value
Basribey X 15	144	55	199	0.735	0.75-0.50
Basribey X 27	159	41	200	2.16	0.50-0.25

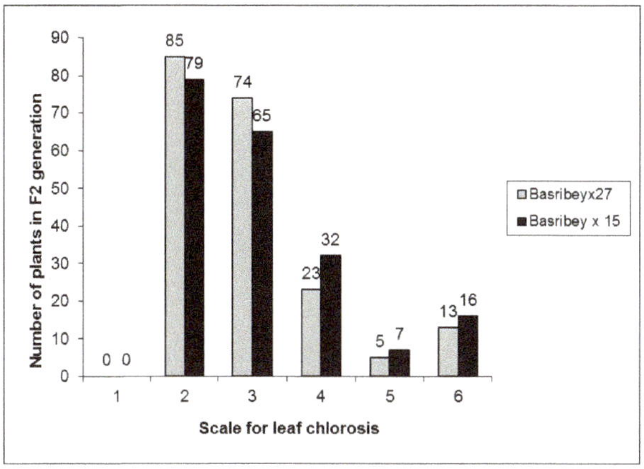

Figure 1. The classification of plants in the F_2 generation for leaf chlorosis scale.

According to Smith et al. (2004), 11 RWA resistance genes have been reported in wheat, rye and goat grass (*Aegilops* sp.). The locations of some of these genes on the chromosome are known, and the *Dn1*, *Dn2*, *Dn5*, *Dn6*, *Dn8* and *Dnx* genes were located on chromosome arm 7DS (Liu et al., 2005). Also, the *Dn7* gene is derived from rye, located on the chromosome arm 1RS, and it carries resistance to different RWA populations. In another study, Peng et al. (2007), the ratio of [H (Heterozygotes Dominance) + B (Homozygotes Dominance)] to A (Homozygotes Recessive) by using the molecular markers fits the expected ratio of 3:1 a single dominant gene model, indicating that the RWA resistance in the line 2414-11 is governed by a single dominant gene. Our findings studies were parallel with previous studies (Peng et al., 2007; Valdez et al., 2012).

In second part of this study, the resistant gene responsible for resistant to RWA was determined by using the differential set. Based on leaf chlorosis and leaf rolling reactions of all *Dn* genes to three RWA populations were similar. Chlorosis and leaf rolling scale values were shown separately in Figure 2 and 3 to demonstrate the reactions to each resistant gene.

The analyses of variance indicated that the differences between reactions of resistant genes were statistically significant for chlorosis and leaf rolling. The lines carried *Dn7* resistance gene exhibited high level of resistance to the pest, while moderate resistance conferred by the plants containing *Dn6* gene. Reactions of other *Dn* genes were found to be close to susceptible control cultivars (Figure 2 ve 3). In breeding studies against to RWA in the USA *Dn4* used successfully. In 2003, a new biotype of the RWA which is virulent to *Dn4* were identify in Canada (Haley et al., 2004). It was understood that *Dn7* were resistant to the new virulent biotype of RWA and it is appear that this gene could be used in resistant breeding programs against to this pest. In this study, the lines carrying the *Dn4* gene were found susceptible to the RWA populations in Turkey. These findings seemed to be parallel to the studies in the USA indicating usage of the *Dn7* resistance genes in a breeding program for resistance.

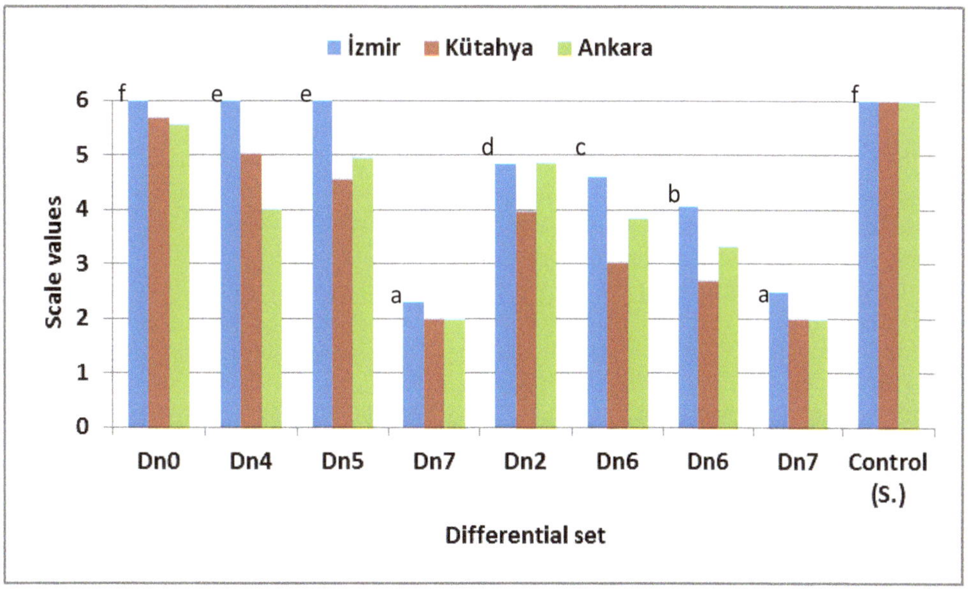

Figure 2. Scale values for chlorosis of three different populations of RWA and their groups based on variance analyses. LSD P≤ 0.05: 0.382.

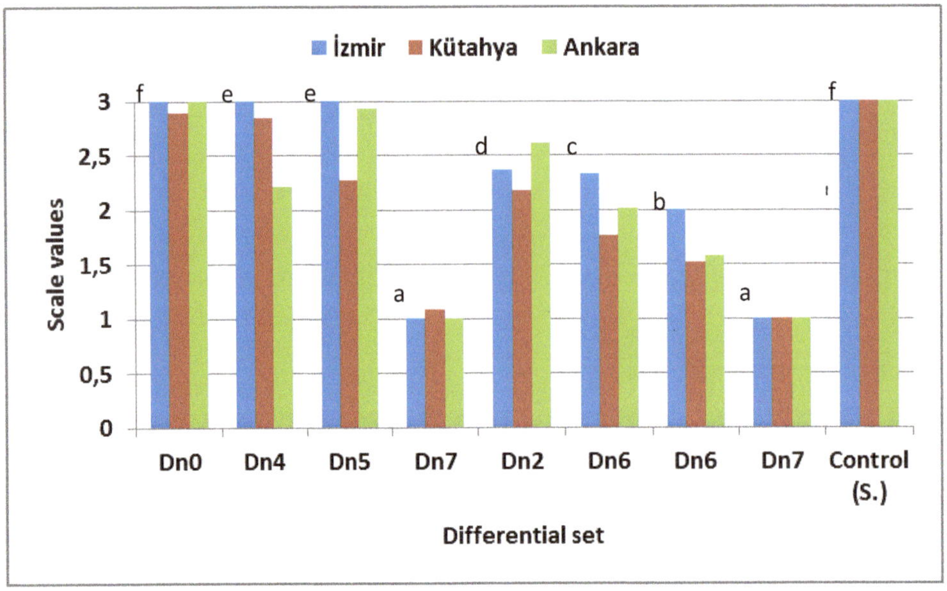

Figure 3. Scale values for leaf rolling of three different populations of RWA and their groups based on variance analyses. LSD P≤ 0.05: 0.157.

CONCLUSION

The usage of host plant resistance at the low cost and environmentally safe is an ideal method to control the Russian wheat aphid. The genetic bases of the resistance to Russian wheat aphid would be crucial step in the future of resistance breeding studies. Continuous efforts are necessary to identify and introduce additional resistance genes into the commercial wheat cultivars in Turkey. The results of this study would be beneficial to the researchers in conducting resistance breeding programs to the RWA. Continuous efforts are necessary to identify and introduce additional resistance genes into commercially acceptable cultivars in Turkey.

ACKNOWLEDGEMENTS

We would like to thank TUBITAK (The Scientific and Technological Research Council of Turkey) (Project Number: 110 O 636) for the financial support. We are also thankful to ICARDA (International Center for Agricultural Research in The Dry Areas Aleppo-Syria) and the Turkish office of CIMMYT (International Maize and Wheat Improvement Centre, Mexico) for providing seeds.

LITERATURE CITED

Acikgoz, N., E. Ilker, A. Gokcol, 2004. TARIST-Computer Based Analyze of Biological Researches (in Turkish). ISBN: 973-483-607-8 E.U. Seed Technology Center, Publication No:2 Bornova-İzmir, TURKEY.

Altinayar, G., 1981. Studies on the insect fauna in cereal fields of Middle Anatolian Region. Plant Protection Bull. 21: 53-89.

Anderson, G.R. , D. Papa, J. Peng, M. Tahir, N.L.V. Lapitan, 2003. Genetic mapping of *Dn7*, a rye gene conferring resistance to the Russian wheat aphid in wheat. Theor. and Appl. Genet. 107(7): 1297-1303.

Collins, M.B., S.D. Haley, F.B. Peairs, J.B. Rudolph, 2005. Biotype 2 Russian wheat aphid resistance among wheat germplasm accessions. Crop Sci. 45:1877-1880.

Du Toit, F., 1987. Resistance in wheat (*Triticum aestivum*) to *Diuraphis noxia* (Homoptera: Aphididae). Cereal Res. Commun. 15: 175–179.

Du Toit, F., 1989. Inheritance of resistance in two *Triticum aestivum* lines to Russian wheat aphid (Homoptera: Aphididae). J. Econ. Entomol. 82: 1251–1253.

Du Toit, F., 1992. Russian wheat aphid resistance in a wheat line from the Caspian sea area. Cereal Res. Commun. 20: 56–61.

Duran, M., N. Koyuncu, 1974. Preliminary studies on damage rate and control of wheat aphid (*Diuraphis* (= *Brachycolus*) *noxius* Mordv.) in wheat field in Middle Anatolian Region. Ankara Central Plant Protection Research Institute. Grain Pest Lab. Project no: 104.653 (In Turkish.).

Elmali, M., 1998. Russian Wheat Aphid (*Diuraphis noxia* (Kurdjumov) (Hom.: Aphididae)) in Konya Province. Euphytica. 100: 69-76.

Ennahli, S., M. El Bouhssini, S. Grando, R. Anathakrishnan, T. Niide, L. Starkus, S. Starkey, C.M. Smith, 2009. Comparison of categories of resistance in wheat and barley genotypes against biotype 2 of the Russian wheat aphid, *Diuraphis noxia* (Kurdjumov). Arthropod Plant Interactions. 3: 45-53.

Haley, S.D., F.B. Pearis, C.B. Walker, J.B. Rudolph, T.L. Randolph, 2004. Occurrence of a new Russian wheat aphid biotype in Colorado. Crop Sci. 44: 1589-92.

Harvey, T.L., T.J. Martin, 1990. Resistance to Russian wheat aphid, *Diuraphis noxia*, in wheat (*Triticum aestivum*). Cereal Res. Commun. 18: 127-129.

Hawley, C.J., F.B. Peairs, T.L. Randolph, 2003. Categories of resistance at different growth stages in Halt, a winter wheat resistant to the Russian wheat aphid (Homoptera: Aphididae). J. Econ. Entomol. 96: 214-219.

Kogan, M., E.E. Ortman, 1978. Antixenosis-a new term proposed to define Painter's "nonpreference" modality of resistance. Bull. Entomol. Soc. Am. 24: 175-176.

Liu, X.M., C.M. Smith, B.R. Friebe, B.S. Gill, 2005. Molecular mapping and allelic relationships of Russian wheat aphid-resistance genes. Crop Sci. 45: 2273-2280.

Liu, X.M., C.M. Smith, B.S. Gill, V. Tolmay, 2001. Microsatellite markers linked to six Russian wheat aphid resistance genes in wheat. Theor. Appl. Genet. 102: 504-510.

Marais, G.F., F. Du Toit, 1993. A monosomic analysis of Russian wheat aphid resistance in the common wheat PI 292994. Plant Breed. 111: 246-248.

Miller, H.R., T.L. Randolph, F.B. Peairs, 2003. Categories of resistance at four growth stages in three wheats resistant to the Russian wheat aphid (Homoptera: Aphididae). J. Econ. Entomol. 96: 673-679.

Morrison, W.P., F.B. Peairs, 1998. Response model concept and economic impact. In: Quisenberry, S.S., Peairs, F.B. (eds), A Response Model for an Introduced Pest, The Russian Wheat Aphid. Entomol. Soc. Amer., London, Maryland, pp. 1-11.

Nkongolo, K.K., J.S. Quick, F.B. Peairs, W.L. Meyer, 1991b. Inheritance of resistance of PI 372129 wheat to the Russian wheat aphid. Crop Sci. 31: 905-907.

Nkongolo, K.K., J.S. Quick, A.E. Limin, D.B. Fowler, 1991a. Sources and inheritance of resistance to Russian wheat aphid in *Triticum* species amphiploids and *Triticum tauschii*. Can. J. Plant. Sci. 71: 703-708.

Peng, J., H. Wang, S.D. Haley, F.B. Peairs, N.L.V. Lapitan, 2007. Molecular mapping of the Russian wheat aphid resistance gene *Dn2414* in wheat. Crop Sci. 47: 2418-2429.

Quick, J.S., G.E. Ellis, R.M. Normann, J.A. Stromberger, J.F. Shanahan, F.B. Peairs, J.B. Rudolph, K. Lorenz, 1996. Registration of 'Halt' wheat. Crop Sci. 36: 210.

Quick, J.S., S.D. Haley, J.A. Stromberger, S. Clayshulte, B. Clifford, J.J. Johnson, F.B. Peairs, J.B. Rudolph, K. Lorenz, 2001a. Registration of 'Prowers 99' wheat. Crop Sci. 41: 929.

Quick, J.S., J.A. Stromberger, S. Clayshulte, B. Clifford, J.J. Johnson, F.B. Peairs, J.B. Rudolph, K. Lorenz, 2001b. Registration of 'Prairie Red' wheat. Crop Sci. 41: 1362-1363.

Quick, J.S., J.A. Stromberger, S. Clayshulte, B. Clifford, J.J. Johnson, F.B. Peairs, J.B. Rudolph, K. Lorenz, 2001c. Registration of 'Prowers' wheat. Crop Sci. 41: 928-929.

Quick, J.S., J.A. Stromberger, S. Clayshulte, B. Clifford, J.J. Johnson, F.B. Peairs, J.B. Rudolph, K. Lorenz, 2001d. Registration of 'Yumar' wheat. Crop Sci. 41: 1363-1364.

Quick, J.S., K.K. Nkongolo, W. Meyer, F.B. Peairs, B. Weaver, 1991. Russian wheat aphid reaction and agronomic and quality traits of a resistant wheat. Crop Sci. 31: 50-53.

Quisenberry, S.S., F.B. Peairs, 1998. A response model for an introduced pest-The Russian wheat aphid. in Entomol., Thomas Say Publ. Entomol. Soc. Amer., Lanham, MD.

Saidi, A., J.S. Quick, 1996. Inheritance and allelic relationships among Russian wheat aphid resistance genes in winter wheat. Crop Sci. 36: 256-258.

Smith, C.M., T. Belay, C. Stauffer, P. Stary, I. Kubeckova, S. Starkey, 2004. Identification of Russian wheat aphid (Homoptera: Aphididae) populations virulent to the *Dn*4 resistance gene. J. Econ. Entomol. 97: 1112-1117.

Steel, R.G.D., J.H. Torrie, D.A. Dickey, 1997. Principles and Procedures of Statistics. A Biometrical Approach. 3rd Ed., New York, USA: McGraw Hill Book Co., 666p.

Stoetzel, M.B., 1987. Information on and identification of *Diuraphis noxia* (Homoptera: Aphididae) and other aphid species colonizing leaves of wheat and barley in the United States. J. Econ. Entomol. 80: 696-704.

Toker, C., F.O. Ceylan, N.E. Inci, M.I. Cagirgan, 2012. Inheritance of leaf shape in the cultivated chickpea (*Cicer arietinum* L.). Turkish Journal of Field Crops 17: 16-18.

Tuatay, N., G. Remaudiere, 1964. Premiere contribution au catalogue des Aphididae (Hom.) de la Turquie. Rev. Path. Veg. Entomol. Agric. Fr. 43: 243-278.

Turanli, F., A. Jankielsohn, A. Morgounov, M. Cakir, 2012. The distribution of Russian wheat aphid, *Diuraphis noxia* (Kurdjumov) (Hemiptera: Aphididae) in Turkey. African Journal of Agricultural Research 7(39): 5396-5404.

Valdez, V.A., P.F. Byrne, N.L.V. Lapitan, F.B. Peairs, A. Bernardo, G. Bai, S.D. Haley, 2012. Inheritance and genetic mapping of Russian wheat aphid resistance in Iranian wheat landrace accession PI 626580. Crop Sci. 52: 676-682.

Zhang, Y., J.S. Quick, S. Liu, 1998. Genetic variation in PI 294994 wheat for resistance to Russian wheat aphid. Crop Sci. 38: 527-530.

EFFECT OF BIO-ACTIVATORS ON THE TUBER YIELD AND TUBER SIZE OF POTATOES

*Gulsum OZTURK, Zihin YILDIRIM**

Ege University, Faculty of Agriculture, Department of Field Crops, Bornova-Izmir, TURKEY
**Corresponding Author: zihin.yildirim@ege.edu.tr*

ABSTRACT

Two bio-activators, the Crop-set and the ISR-2000 were sprayed on the potato plants before flowering in order to increase tuber set and tuber size in a field trial conducted at Bornova, Izmir-Turkey in 2009 and 2010. The Crop-set increased the number of tubers per plants, plot yield, tuber width over the control in 2009 and 2010. The ISR-2000 had the highest single tuber weight such as 36.7 g in 2009 and 35.9 g in 2010. It could be concluded that bio-activators could be used in the minituber propagation stage in order to increase tuber number and tuber size in a seed potato program.

Key words: Bio-activator, number of tubers, potato, tuber size, tuber yield

INTRODUCTION

In the seed potato production technology, increasing minitubers in the field is an important stage. In this practice, micro and minitubers obtained from the micropropagation of meristem plantlets are grown in the greenhouse in small pots. Then minitubers obtained are grown under the field conditions in order to obtain large amount of tuber to distribute to the seed potato growers. There are several difficulties during this stage. One of the main constraints is the size and amount of minitubers produced in the greenhouse as well as in the field (Struik, 2007). Minitubers produced in the pots under the greenhouse conditions are not uniform in size so they are classified in two classes such as suitable and not suitable to field production (Struik and Lommen,1999). As a result, significant amount of the minitubers should be discarded (Melching et al., 1993).

In Turkey, seed potato technology has been developing slowly and supported by the government in order to establish a dependable seed production and distribution system. Therefore a national research project has been completed by the Ministry of Food, Agriculture and Livestock in cooperation with the research institutions and the universities and the productions of virus free seed stocks were accomplished (Yildirim, et al., 2009). Minituber production in the field is the first stage of certified seed potato production, since the minitubers are advanced to the potato seed growers. But there were some difficulties and deficiencies in the field propagation stage of minitubers, such as the size and amount of minitubers produced in the field (Lommen, 1994).

Bio-activators have been proposed and used in plant production for a long time and their positive effects on increasing yield and yield components have been reported (Plissey, 2003). On the other hand bio-activators have also been proposed to enhance the induced systemic resistance (ISR) in plant protection. Tosun (2003) gave information about the ISR-2000^{TM} and the Crop-set^{TM} and reported the effects of ISR-2000 on the control of late blight in tomatoes and potatoes in the western Turkey. Since they have organic ingredients in nature bio-activators have been accepted to use in plant production (Karavas, 2000; Ozturk et al., 2006; Akbudak and Tezcan, 2006; Unlu et. al., 2010). Turkan et al., (2004) speculated that plant activators could enhance plant growth and development so yield and quality might increase. Bower (2004), stressed the importance of the Crop-set and the ISR-2000 to enhance growth and yield in plant production. Koca, (2003) reported the positive effect of bio-activators on tuber set and tuber size in potato production. Koca and Yildirim, (2003) discussed the effects of the Crop-set and the ISR-2000 on tuber numberper plant, plot yield, single tuber weight and tuber size in potatoes. The positive effects of bio-activators on yield and tuber size of the minitubers were reported so the seed tuber producing farmers have used them in practice (Plissey, 2003). Therefore the purpose of this study was to test the effect of two bio-activators, the Crop-set and the ISR-2000, on the yield and tuber size of minitubers grown under the field conditions in the western Turkey.

MATERIALS AND METHODS

This study was carried out in the tissue culture laboratory, greenhouse, and the experimental field of the Department of the Field Crops of the Agricultural Faculty of the Ege University located at Bornova, Izmir, Turkey during the 2009 and 2010 growing seasons.

The potato plantlets originated from the seed stocks were planted in the plastic containers (19cm x 12cm x 9 cm in size) in the greenhouse and the minitubers obtained were selected for suitable size to increase in the field. The remaining minitubers, 20-30 mm in size, were used as starting material of this study. The potato cultivars grown, were Hermes, Agria and Granola. The characteristics of these cultivars could be found in Ozturk and Yildirim, (2010). The minitubers obtained for these genotypes in the greenhouse in October, 2008 were stored in the storage room at +4 ^{0}C. They were planted in the field in 2009 and the minitubers obtained in the field trial run in 2009 were planted in the field trial of 2010.

Production of Minitubers in the Greenhouse

The starting materials of the research were from the meristem cultures of 3 potato cultivars in the laboratory, *in vitro,* conditions. Micropropagation procedure and nodal cuttings of meristem plantlets were practiced as described by Yildirim, (1995), Ozturk and Yildirim (2011). Virus free *in vitro* plantlets were transferred in to plastics containers (19cm x 12cm x 9 cm in size) having the mixture of 1:2:1 turf, sand and manure. Minitubers not suitable to regular minituber production in the field were used to be grown in the greenhouse. The harvested tubers in 2008 were advanced to the field trial run in 2009.

Field Trial

The minitubers obtained in the greenhouse 20-30 mm in size were grown in the field trial during the 2009 growing season. The tubers obtained in the field trial of 2009 were planted in the field trial of 2010. The meteorological data of two years are given in Table 1.

Table 1. The temperature and the precipitation during the potato growing revised of 2009 and 2010*

Months	Temperature (^{0}C)		Rainfall (kg/m^2)	
	2009	2010	2009	2010
March	11.7	13.3	175.7	16.1
April	16.0	17.4	83.8	20.4
May	21.4	21.8	44.3	27.1
June	26.2	25.5	9.2	76.3
July	29.0	28.8	0	0
August	27.9	30.2	0	0

*: based on the Guzelyali - Izmir Meteorological Station records

The minitubers were grown in the experimental field of the Department of the Field Crops of the Ege University located at Bornova, Izmir-Turkey. The experimental design was the Randomized Complete Block Design arranged in Split Plots with 3 replications. One plot consisted of one single row 1.5 m long and 50 cm apart. Spacing on the row was 30 cm, so each plot had 5 plants. The plantings were done by hand in the second week of March in 2009 and in the first week of April in 2010. Standard agronomical practises were applied during the growing seasons.

The bio-activators, the ISR-2000 (600 ml/ha) and the Crop-Set (300 ml/ha) described by Tosun, (2003), were sprayed before the flowering on 20 May, 2009 and on 4 June, 2010.

The field trials were harvested on 30 July in 2009 and on 9 August in 2010 and the number of tuber per plant and plot yields were recorded and the sizes of tubers were measured.

The data obtained for number of tuber per plant, plot yield, single tuber yield and tuber width and length were analyzed by using the Standard Analysis of Variance techniques. The means were compared by using the Least Significant Difference Test described by Steel et al., (1997).

RESULTS AND DISCUSSION

The F values of the yield characteristics based on the analysis of variance combined over 2009 and 2010 are given in Table 2.

Table 2. The F values of the yield characteristics based on the ANOVA combined over 2009 and 2010.

Source of Variation	F Values				
	Tuber Number Per Plant	Single Tuber Weight(g)	Plot Yield (kg)	Tuber Width (cm)	Tuber Legth (cm)
Bioactivator	13.946*	31.541**	84.800**	9.471*	4.163^{ns}
Genotype	2.269^{ns}	0.232^{ns}	6.399**	0.087^{ns}	1.950^{ns}
Bioactvator x Genotype Year	1.187^{ns}	1.513^{ns}	1.631^{ns}	0.622^{ns}	0.139^{ns}
Bioactivator xYear	23.676**	0.337^{ns}	12.548**	11.193**	5.952*
Genotype x Year	3.748*	0.425^{ns}	2.546^{ns}	0.686^{ns}	1.237^{ns}
Bioactivaor x	2.478^{ns}	0.735^{ns}	0.848^{ns}	0.646^{ns}	0.715^{ns}
Genotype xYear	1.605^{ns}	0.542^{ns}	0.468^{ns}	0.651^{ns}	0.274^{ns}

*: significant at the p≤0.05 level
**: significant at the p≤0.01 level
^{ns}: non significant

It could be seen in Table 2 that bio-activator treatment had significant F values for number of tuber per plant, single tuber weight (g), plot yield (kg), and tuber width (cm). Genotype source of variation had only one significant F value for plot yield. Years had significant variation for number of tuber per plant, plot yield, tuber width and tuber length. Bio-activator x year interaction was significant for only number of tuber per plant.

The non significant F values for the interactions between bio-activator x genotype, genotype x year and bio-activator x genotype x year indicated that bio-activators could be discussed separately by using the means over the genotypes.

The Effect of Bio-activators

The mean of the yield related traits and tuber size obtained for the bio-activator applications are shown in Table 3.

Table 3. The means of tuber number per plant, single tuber weight (g), plot yield (kg) and tuber size obtained in the field trial run in 2009 and 2010.

Traits	Year							
	2009				**2010**			
	Control	Crop-set	ISR-2000	Mean	Control	Crop-set	ISR-2000	Mean
Tuber Number Per Plant	2.7c	**7.5a**	5.6b	**5.26a**	2.2b	**4.4a**	3.6ab	3.4b
Single Tuber Weight(g)	23.8b	**34.2a**	**36.7a**	**31.5a**	24.7b	30.7ab	**35.9a**	**30.4a**
Plot Yield (kg)	0.232b	**1.102a**	**0.896a**	**0.773a**	0.268b	**0.635a**	**0.624a**	0.509b
Tuber Width (cm)	2.9b	**3.5a**	3.2ab	**3.2a**	2.4b	**3.1a**	**3.0a**	2.8b
Tuber Length (cm)	3.5b	**4.5a**	4.0ab	**3.9a**	3.2b	**3.8a**	**3.9a**	3.6b

$LSD(p\leq0.05)_{tuber\ number}$: 1.4 LSD for means of 2009 and 2010 $(p\leq0.05)_{tuber\ number}$: 0.8
$LSD(p\leq0.05)_{single\ tuber\ weight}$: 6.92 LSD for means of 2009 and 2010 $(p\leq0.05)_{single\ tuber\ weight}$: 3.9
$LSD(p\leq0.05)_{plot\ yield}$: 0.265 LSD for means of 2009 and 2010 $(p\leq0.05)_{plot\ yield}$: 0.153
$LSD(p\leq0.05)_{tuber\ width}$: 0.4 LSD for means of 2009 and 2010 $(p\leq0.05)_{tuber\ width}$: 0.23
$LSD(p\leq0.05)_{tuber\ length}$: 0.5 LSD for means of 2009 and 2010 $(p\leq0.05)_{tuber\ length}$: 0.3

It could be seen in Table 3 that the Crop-set application increased tuber number per plant (7.5 versus 2.7), plot yield (1.1 kg versus 0.3 kg), tuber width (3.5 cm versus 2.9 cm) over the control in 2009. Although the over all yield was lower as compared to 2009, the same trend could also be observed in 2010. In general, control plots had lower means that those of the two type bio-activators applied. Among the bio-activators, the Crop-set, appeared to have higher means for plot yield (1.102 kg), tuber number per plant (7.5) and tuber width (3.5 cm). The untreated control plot had 0.23 kg and 0.27 kg overall mean in 2009 and in 2010, The Crop-set had 1.1 kg and 0.6 kg plot yield in 2009 and 2010. If the plot yields are converted to t/ha by using the conversion factor 1.3 the control plot had 3.9 t/ha as compared to the Crop-set yield of 13.4 t/ha. The significant point is that the net effects of bio-activator on the increase of tuber number from 2.7 to 7.5 for Crop-set and from 2.7 to 5.6 for the ISR-2000. These result, were in good agreement with the earlier reports (Koca and Yildirim, 2003; Plissey, 2003).

The Crop-set also had higher means for tuber length (control: 3.5 cm Crop-set: 4.5 cm) in 2009. The ISR-2000 had the highest mean for single tuber weight (36.7 g) as compared to 23.8 g for the control and 34.2 g for Crop-set in 2009.

Since there was no significant interaction between the bio-activator and the genotype, it could be concluded that bio-activator had significant effect on the yield related traits of potato. The bio-activators positively increased the tuber number per plant, single tuber weight, plot yield, tuber width of the potato. It could also be observed in Table 3 that the means of the traits studied in 2009 were significantly higher than those of in 2010. One significant result of the trial was the effect of bio-activators on potato yield in two years had the same trend.

Years might have different means due to the adverse weather differences and late planting done in 2010 as compared to early planting done in 2009. Plant development could be fastened due to late planting done in 2010 so the yield related traits could not properly express themselves. On the other hand the effect of bio-activators on the traits expressed a similar trend in two years.

The low yields obtained in 2010 as compared to 2009 might also be due to the usage of harvested field grown seed tubers of 2009. Although the same size tubers were planted in 2010 and they were stored in the storage room until March 2010, the physiological disorders coming from the 2009 harvested could effect the age of tubers. In some minituber propagation practises used in Turkey the tuber harvested are also, used as seed tubers in the coming year, although they have yield losses.

There were statistically significant differences between the means of genotypes tested for plot yield, as shown in Figure 1.

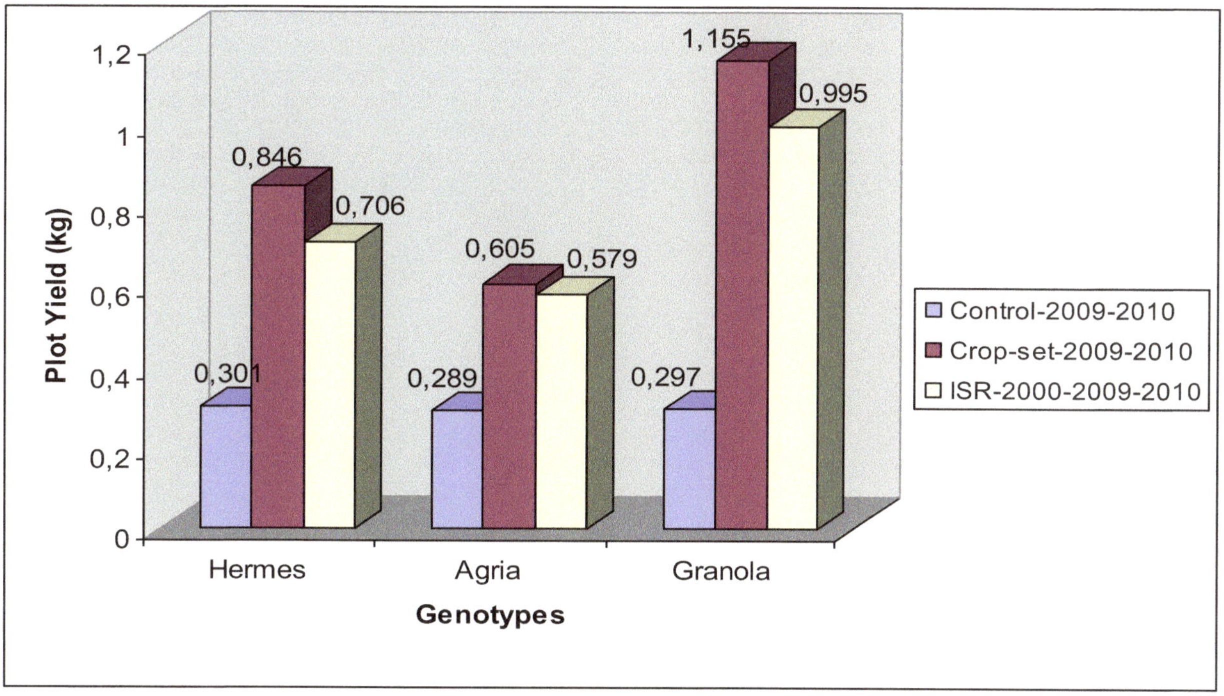

Figure 1. Histogram of the genotypes for plot yield.

The positive effects of the two bio-activators used on the yield related traits were in good agreement with earlier reports given by Koca, (2003); Koca and Yildirim, (2003) and Plissey, (2003).

Based on the results and the discussion given above it could be concluded that the bio-activators could be used to increase tuber set and tuber size in the minituber propagation stage of a seed potato program.

ACKNOWLEDGEMENTS

The authors would like to thank to Professors Metin B. YILDIRIM, Necip TOSUN and Sevgi CALISKAN for their kind and valuable helps during the preparation and revision of the manuscript.

LITERATURE CITED

Akbudak, N. and H. Tezcan. 2006. A new active chemical: Harpin in plant production and plant protection. Journal of Uludag University Faculty of Agricultural. 21:39-42 (in Turkish).

Bower, J.P., 2004. The pre-and postharvest application potential for Crop-set TM and ISR-2000TM on citrus. In: Lyons, T.P, Jacques, K.A. (eds) Nutrional Biotechnology in the Feed and Food Industries, Proceedings of Alltech's 20th Annual Symposium Nottingham University Pres, U.K, pp.361-367.

Melching, J.B., S.A. Slack, E.D. Jones, 1993. Field performance of peat-lite mix

encapsulated small minitubers. American Potato Journal. 70: 285-290.

Karavas, B. 2000. the effects of fungicide plant activator and plant stimulant on the anatomic and morphological structure of paper (Capsicum Annual) Master Thesis. Institute of Natural and Applied Sciences of Ege University. Bornova-Izmir, Turkey, (in Turkish).

Koca, Y.O. 2003. The effect of two plant activators on the certain agronomical traits in potatoes (*Solanum tuberosum* L.). Master Thesis. Institute of Natural and Applied Sciences of Ege University. Bornova-Izmir, Turkey, (in Turkish).

Koca, Y.O. and M.B. Yildirim. 2003. The effect of two plant bioactivators on some agronomical traits in potato (*Solanum tuberosum* L.) 5th Congress of Field Crops of Turkey. 13-17 October, 2003. Diyarbakır, Turkey (in Turkish). 207-212.

Lommen, W.J.M. 1994. Effect of weight of potato minitubers of sprout growth, emergence of plant characteristics at emergence. Potato Research.37:315-327.

Ozturk, I. N.Tort and N.Tosun. 2006. The effect of metalaxyl application on the anatomical structure of tomatoes (*Lycopersicon esculentum*).Tarim Bilimleri Dergisi. 12: 14-22 (in Turkish).

Ozturk, G. and Z. Yildirim. 2010. A comparison of field performances of minitubers and microtubers used in seed potato production. Turkish Journal of Field Crops. 15: 141-147.

Ozturk, G. and Z. Yildirim. 2011. Uniformity of potato minitubers derived from meristem cultures of nuclear seed stocks. Turkish Journal of Field Crops. 16: 149-152.

Plissey, E.S. 2003. Applying crop stimulant technology to the potato crop. In: Lyons, T.P, Jacques, K.A. (eds) Nutrional Biotechnology in the Feed and Food Industries, Proceedings of Alltech's 19th Annual Symposium. Nottingham University Pres, U.K, pp.415-421.

Steel, R.G.D., J.A. Torrie and D.A. Dickey. 1997. Principles and procedures of statistics. A. Biometrical Approach 3rd Edi. Mc Graw Hill Book.INC.N.Y.

Struik, P.C and W.J.M. Lommen. 1999. Improving the field performance of micro-and minitubers. Potato Research. 49: 558-568.

Struik, P.C. 2007. The canon of potato science: 25. minitubers.Potato Research. 50:305-308.

Tosun, N. 2003. Disease control with ISR-2000 TM (ecelitor) in conjuction with fungicides. In: Lyons, T.P, Jacques, K.A. (eds) Nutrional Biotechnology in the Feed and Food Industries, Proceedings of Alltech's 19th Annual Symposium. Nottingham University Pres, U.K, pp.423-426.

Turkan, I.,T. Demiral, A.H. Sekmen, N. and Tosun. 2004. Abiotic stresses and plant activators. In: Lyons, T.P, Jacques, K.A. (eds) Nutrional Biotechnology in the Feed and Food Industries, Proceedings of Alltech's 20th Annual Symposium. Nottingham University Pres, U.K, pp.387-390.

Unlu, H.,H. Ozdamar, Y. Karadavut. and H. Padem. 2010. Organic and conventional production systems, microbial fertilization and plant activators affect tomato quality during storage. Afr.Jour. of Biotechnology. 9(46): 7909-7914.

Yildirim, Z. 1995. Microtuber production in potato (*Solanum tuberosum* L.).The Journal of Agricultural Faculty of Ege University (in Turkish). 32: 73-77.

Yildirim,Z.,S. Erkan, M. Gumus.,2009. Development of Seed Potato Production Programmes. Final report of the project: 105G117-TUBİTAK-KAMAG (in Turkish).

RELATIONSHIP BETWEEN WATER USE EFFICIENCY AND $\delta^{13}C$ ISOTOPE DISCRIMINATION OF SAFFLOWER (*Carthamustinctorius* L.)UNDER DROUGHT STRESS

Öner CANAVAR[1,2],*Klaus-Peter GÖTZ*[3],*Yakup Onur KOCA*[1],*Frank ELLMER*[2]

[1]*Adnan Menderes University, Faculty of Agriculture, Department of Field Crops, Aydın, TURKEY*
[2]*Humboldt University, Agriculture and Horticulture Faculty, Crop Science Department, Berlin, GERMANY*
[3]*Humboldt Universität, Agriculture and Horticulture Faculty, Agroclimatology, Department, Berlin, GERMANY*
**Correspondingauthor: oner.canavar@hotmail.com*

ABSTRACT

Drought stress is one of the most limiting factors in agricultural productivity because of its highly negative effect on photosynthesis and growth of plants. The main objectives of this study were to determine the performance of four selected safflower genotypes (Remzibey, Dinçer, Balcı and TRE-ASL09/14) against drought stress. The relationship between water use efficiency (WUE) and $\delta^{13}C$ (isotope discrimination) was investigated under well watered (60%) and drought stress (30%) irrigation in controlled conditions in a green house. The result showed that drought stress clearly reduced plant biomass, leaf area, leaf number, relative water content (RWC), specific leaf weight (SLW), WUE and $\delta^{13}C$ in all genotypes, while chlorophyll increased under drought stress. There were significant differences between performances of all safflower genotypes in terms of response to drought stress. High WUE and low $\delta^{13}C$ discrimination under drought stress were inversely associated with a strong regression relationship (R^2=0.75). The analysis of $\delta^{13}C$ revealed a substantial variation in water use efficiency among the genotypes under drought stress. It was revealed that low $\delta^{13}C$ discrimination types had high WUE, RWC and total biomass under drought stress. Thus, the ability of the low $\delta^{13}C$ genotypes (high water use efficiency, WUE) to maintain higher RWC may provide a good indication of the differences in drought tolerance of safflower genotypes differing in $\delta^{13}C$. Overall, on the basis of the consistent percentage reductions in plant heights, total dry weight, leaf area, RWC, WUE and low $\delta^{13}C$, the TRE-ASL09/14 new breeding line was found to be more drought tolerant when compared with the other safflower hybrids under drought stress. As a result of our study it is suggested that there is a strong relationship between WUE and lower $\delta^{13}C$ under drought stress, indicating that it may be used as a selection criterion for developing safflower genotypes with drought tolerance.

Key words: Safflower, drought stress, WUE, $\delta^{13}C$.

Abbreviation:P.H: Plant Height, ChyII: Chlorophyll, TLA: Total Leaf Area, RWC: Relative Water Content, LDW: Leaf Dry Weight, SDW: Stem Dry Weight, TDW: Total Dry Weight, LN: Leaf Number, SLW: Specific Leaf Weight, $\delta^{13}C$isotope: Carbon Isotope Discrimination, WUE: Water Use Efficiency, FC: Field Capacity, WW: Well Watered, DS: Drought Stress.

INTRODUCTION

With regard to economic benefits, researchers have been impelled to search for alternative edible oilseed crops which have tolerant genotypes against changing environmentalconditions in a new crop rotation system for every region. This is owing to the increase in global warming over the years, limited irrigation water and also the increasing lack of vegetable crude oil and oilseed crops in the world. Although safflower (*Carthamustinctorius* L.) is one of the oldest crops, it is traditionally grown for its seeds and used for coloring and flavoring foods and for making red and yellow dyes (Arslan, 2003; Zohary and Hopf, 2000). It is not a popular oilseed crop compared to other oilseed crops, such as soybean, sunflower, peanut, cotton seed, and rapeseed, since it is not commonly grown globally. Much of the literature has denoted that safflower is a C3 plant tolerant to drought and salinity stress because of having deep-rooting ability (Dordas and Sioulas, 2008), water uptake from soil (Majidiet al., 2011), and different osmolyte accumulation (Bhatia et al. 1994).

Drought, which is low water availability or random and unpredictable changes in weather conditions during the period of plant growth, is considered one of the most effective abiotic stress factors limiting production from plants. All field crops respond differently at different phenological stages to changing water status of the soil under drought stress, which means that plants are more sensitive to drought stress at some stages. For example, Blum (2005) explained that drought resistance in seedlings grown in a pot has nothing to do with drought resistance during grain filling in the field. Although the drought-resistant ideotype is still not well defined, drought resistance in its physiological context is defined according to Levitt (1972) as being determined by 'dehydration avoidance' (maintenance of water potential in tissue) and/or 'dehydration tolerance' (Levitt, 1980; Price et al., 2002). Dehydration avoidance or osmotic adjustment is defined as the plant's capacity to sustain high plant water status or cellular hydration under drought stress (Blum, 2005; Cushman, 2001).

There is no consistent relationship between plant production and water use efficiency (WUE). However, Munoz et al., (1998) pointed out that high yield potential of plants under water-limited conditions is generally associated with reduced WUE mainly because of high water use. In contrast, other researchers have explained that high WUE is largely a function of reduced water use rather than a net improvement in plant production or the biochemistry of assimilation (Blum, 2005). WUE is generally equated with drought resistance and the improvement of crop yield under stress, due to variations in water use. In addition, as explained by Farquhar et al., (1989), carbon isotope distribution can reveal information about the physical, chemical, and metabolic processes involved in carbon transformations. This is because carbon isotope discrimination occurs during photosynthetic CO_2 uptake leading to a ^{13}C-depletion of plant organic matter. Therefore, it is not surprising that selection of high WUE using carbon isotope discrimination has resulted in earlier flowering plants that use less water over the growing season. These plants were found to be very suitable for conditions where moderated use of the given amount of stored soil moisture is crucial (Condon et al. 2002).

Maintenance of leaf turgor in the face of decreasing soil moisture has been emphasized as an important adaptation trait that contributes to drought tolerance (Hsiao et al., 1976). Tolerance to internal water deficit has been characterized by turgor loss at lower relative water content (RWC), promoting the maintenance of chloroplast functioning during dehydration (Gupta and Berkowitz, 1987; Ranneyet al., 1991). The studies of dehydration tolerance in crop plants have revealed genotypic variation in plant recovery from dehydration as a measure of tolerance to be positively correlated with the plant water status (RWC) (Chaves et al., 2002). Anyia and Herzog (2004) pointed out that the high relative water content (RWC) of cowpea leaves was maintained in some of the genotypes by stomata closure and a reduction in leaf area. Many techniques and parameters such as leaf water potential, leaf osmotic potential, and canopy temperature have been used to screen drought tolerant plants in different crops (Askahniet al., 2007; David and Duniway, 1997). Therefore, drought resistance and its components are almost constantly being redefined to express the inventive capacity for terminology. However, none of the previous experiments attempted to identify whether $\Delta^{13}C$ isotope is activated in safflower genotypes of any safflower cultivars under drought stress. As a result of these statements, the main aims of our study were 1) to determine selected cultivars which may be tolerant against drought stress, and 2) to investigate the relationship between plant biomass, RWC, WUE and $\delta^{13}C$ and other traits in safflower cultivars.

MATERIALS and METHODS

Plant materials and experimental conditions

Remzibey,Balcı (brambly) andDincer (unbrambly) safflower hybrid cultivars and the new safflower breeding line TRE-ALS09/14 (brambly), all of Turkish origin, were used to determine variation under controlled drought stress and well-watered environmental conditions. The plants were grown under light/dark regime 12/12 h, at 25/15 ±3 °C and relative humidity of approximately 30-50%at the research greenhouse of the Crop Science department of the Agriculture and Horticulture Faculty in Humboldt University, Germany in 2012. The safflower cultivars were planted in Mitscherlich cylindrical pots (30-cm deep 25 cm dia.) in the greenhouse with only natural sunlight of the summer months. Clay loam soil was used to fill pots and all cultivars were arranged completely in a randomized block design with five replications. Required amounts of chemical fertilizers were applied and seeds were sown according to the recommendations for field conditionsin the literature (1 g nitrogen from 3.70 g KAS fertilizer). Watering began immediately after sowing and once the seedlings had emerged, thinning was carried out and the plant populations were maintained (4 plants in a pot). Changes in the soil water potential of each pot were measured and checked daily by weighing each pot at the beginning and end. The soil water factor included two irrigation regimes including irrigation at 30% (drought stress) and 60% (well-watered) of field capacity.

Determination of water holding capacity of soil

Field soil which had already been taken from the field experiment area was air-dried and ground to pass through a 2-3 mm sieve at room temperature. Water holding capacity was determined using a gravimetric method with five replicates as the amount of moisture (percentage). Firstly the bottoms of five 100 cm^3 cylindrical tubes were covered with paper and a plastic strap as a filter, they were tared without soil and then completely filled with soil (by compression). Each cylindrical tube with soil was weighed and settled in a big tray which was approximately as deep as the height of the cylindrical tubes. The tray was fully filled with water up to the top of the cylindrical tubes and left for 3 h (saturation). Then, all cylindrical tubes were left on quartz soil for 2 h (for drainage and filtering). After that, all the saturated cylindrical tubes were cleaned and weighed again (wet weight). Then all the tubes were oven-dried at 105 °C for 24 h and the weight of the oven-dry soil samples was measured (dry weight). The field capacity of undisturbed soil was calculated according to the following formula;

F.C. % = wet soil weight (saturated) – dry weight / dry weight x 100. The mean of five replicates was 38.89%.

Drought stress treatment

To determine the amount of irrigation required to produce irrigation regimes at 30% and 60% of field capacity, the soil was firstly weighed to exactly 6091 g for each pot and then the net weight of each pot with filled soil was measured. Pots were initially watered with 1330 and 619 g per pot which corresponds to 60% and 30% of field capacity. The soil water content was continuously monitored and maintained at 30% and 60% of field capacity by daily irrigation during the experiment.

Plants were harvested 50 days after sowing, when the plants were at the heading stage (head visible).

Determination of relative water content

RWC (*Relative Water Content*); the youngest fully expanded leaves were collected from each pot in the morning. The leaves were weighed immediately to obtain the fresh weight. Afterwards the leaves were rehydrated by floating for 12 h in a covered water bath cap at approximately 23 °C under dark conditions. All leaves were dried for 72 h at 70 °C in an oven and RWC was calculated by dividing the amount of water in the fresh leaf tissue by the water in the leaf tissue after rehydration multiplied by 100.

RWC = Fresh weight – Dry weight / Turgid weight – Dry weight x 100

Measurement of Chlorophyll

Chlorophyll content was assessed using a chlorophyll meter (SPAD-502, Minolta) and measurements were taken at two points on both sides of young fully developed sunlit leaves (upper, middle and lower parts) two times during the experiment (Days 30. and 40. of the experiment after sowing). Forty readings were averaged per genotype (twelve readings of two fully developed leaves per plant five replicates) for each treatment. The average of these forty readings was considered to be the SPAD value for one cultivar under one condition.

Measurement of the growth parameters

Plant height (PH cm) was measured from the soil surface to the top of plant 3 times between emergence and harvest time. Total leaf area (cm^2) was measured immediately for all leaves on a plant using a leaf area meter (Li-Cor 3000, Lambda Instruments Co., USA) after the plant was removed. Leaf and stem dry weights were obtained after all parts of the plant were separately dried at 70 °C for 72 h. The total dry matter production per plant (g $plant^{-1}$) was obtained with the summation of dry weight of all plant parts and was expressed on a per plant basis. All leaves on a plant were numbered before the leaf area was measured. After that the plant was removed. Specific leaf weight (SLW) was calculated by dividing the total leaf area by leaf dry weight (LDW/LA) (g m^{-2}). The water use efficiency (WUE) measurements were completed in order to note another important index to estimate the water productivity over time. WUE was given in terms of the dry fresh weight per water consumed by evapotranspiration and evaluated as (g/g plant);

WUE = Total biomass / Water consumption (amount of irrigation (g) during the experiment).

Analyses of carbon isotope discrimination ($\delta^{13}C$)

Carbon isotope discrimination was analyzed from the same leaves (young fully expanded leaves) which were kept at -20 °C. The leaves were dried at 60 °C for 72 h and ground through a 0.1 mm screen to produce a flour for carbon isotope analysis. ^{13}C analyses were performed by Prof. Dr. K.D. Wutzke, Research Laboratory, University of Rostock, Germany, measured by isotope ratio mass spectrometry with the Tracer mass 20-20, SerCon, Crewe, UK and calculated as: $\delta^{13}C$ (‰) = [(R sample/R reference – 1) × 1000], with R being the $^{13}C/^{12}C$ ratio. Carbon isotope discrimination (Δ) was calculated using the following formula (Farquhar *et al.* 1989): ^{13}C (‰) = [(δa – δp) / (1 + δp)], where δp is the $\delta^{13}C$ of the leaves and δa is the $\delta^{13}C$ of the atmospheric CO_2 (–8‰).

Statistical analysis

To determine the effect of drought stress on the four safflower hybrid cultivars, the samples were analyzed statistically as a randomized block design with five replications. ANOVA was applied to analyze the variance of drought stress on safflower hybrid cultivars and the interaction of drought and cultivars. The ANOVA (analyses of variance) of this study and correlation coefficients among the traits are given as the mean of genotypes under each condition. Significant differences between the means of replications were tested using Fisher's least squares difference (LSD) method. All differences referred to in the text were significant at 0.05. Regression analyses were computed using Microsoft Excel office program y diagram to assess the relationship between $\delta^{13}C$ and WUE only under drought stress conditions.

RESULTS

The variance analyses clearly showed the significant effects of drought stress statistically on the physiologicalandmorphological traits measured in the safflower genotypes during the experiment (Table 1). Drought stress x cultivar interaction was also statistically significant in terms of $\delta^{13}C$, WUE, plant height, leaf number and chlorophyll of safflowers in Table 1. The correlation coefficients obtained from our study shown in Table 2 indicate that drought stress was considerably negatively correlated with total leaf area, relative water content, total dry weight, specific leaf weight, $\delta^{13}C$ and water use efficiency, while it was positively correlated with chlorophyll. Also a significant negative correlation coefficient of -0.46** was found between RWC and $\delta^{13}C$. Total leaf area had a significantly positive relationship with WUE, $\delta^{13}C$ and leaf number under drought stress conditions (Table 2). Total dry weight had a significantly positive relationship with plant height, leaf number, total leaf areas, $\delta^{13}C$ and water use efficiency (WUE) (Table 2). It appears that SLW was positively correlated with plant height, total leaf area, $\delta^{13}C$, RWC and slightly less with WUE (Table 2).

Table 1. The result of variance analyses for all components measured of safflower under the well watered and drought stress conditions.

Variance Source	d.f	Calculated of Mean Square						
		PH1	**PH2**	**PH3**	**LDW**	**SDW**	**TDW**	**LN**
D.S	1	83.377**	152.881**	2061.809**	1.895**	7.043**	15.952**	172.225**
C	3	2.619**	18.614**	156.340**	0.102**	0.432**	0.888**	38.425**
D.S x C	3	3.004**	32.426**	18.074**	0.012ns	0.055ns	0.083ns	8.425**
		TLA	**SLW**	**ChyII1**	**ChyII2**	**WUE**	**RWC**	**$\delta^{13}C$**
D.S	1	34686.901**	1578.541**	577.600**	1037.852**	0.047**	0.014**	156.816**
C	3	2935.079ns	119.807*	15.479ns	13.974**	0.012**	0.003ns	7.381**
D.S x C	3	867.342ns	20.449ns	9.259ns	21.870**	0.001ns	0.001ns	3.637**

D.S: Drought Stress, C: Cultivars, d.f.: Degree of Freedom, ns: non-significant; $^{*}P< 0.05$; $^{**}P< 0.01$, P.H.: Plant Height, LDW: Leaf Dry Weight, SDW: Stem Dry Matter, TDW: Total Dry Weight, LN: Leaf Number, TLA: Total Leaf Area, SLW: Specific Leaf Weight, ChyII: Chlorophyll, WUE: Water Use Efficiency,RWC: Relative Water Content, $\delta^{13}C$ discrimination

Table 2. Correlation coefficientsbetween condition and traits of Safflower.

	Condition	LA	TDW	Chyll2	SLW	RWC	WUE	$\delta^{13}C$	PHU	LN
Condition	-	-0.87**	-0.70**	0.95**	-0.68**	-0.46**	-0.70**	-0.89**	-0.89**	-0.71**
LA	-	-	0.84**	-0.82**	-0.58**	-0.35*	0.69**	0.74**	0.81**	0.74**
TDW	-	-	-	-0.66**	ns	0.37*	0.80 **	0.55**	0.75**	0.69**
Chyll2	-	-	-	-	-0.57**	-0.46**	-0.69**	-0.77**	-0.87**	-0.81
SLW	-	-	-	-	-	ns	0.39*	0.67**	0.55**	ns
RWC	-	-	-	-	-	-	0.32*	-0.46**	0.48**	0.55**
WUE	-	-	-	-	-	-	-	-0.43**	0.86**	0.46**
$\Delta^{13}C$	-	-	-	-	-	-	-	-	0.66**	0.44**
PHU	-	-	-	-	-	-	-	-	-	0.75**
LN	-	-	-	-	-	-	-	-	-	-

ChyII: Chlorophyll, LA: Total Leaf Area, RWC: Relative Water Content, TDW: Total Dry Weight, SLW: Specific Leaf Weight, $\delta^{13}C$ discrimination, WUE: Water Use Efficiency, LN: Leaf Number, PHU. Plant Height3.

As all plants were exposed to drought stress, all safflower genotypes responded to the drought stress by decreasing the plant height measured three times (Figure 1). There were statistical differences among the genotypes in terms of plant height under both well-watered and drought stress conditions. For example, although the TRE-ASL09/14 genotype was not statistically in the first group at the 1st and 2nd measuring times, it was statistically in the first group with regard to plant height at the 3rd measuring time under both well-watered and drought stress conditions. Also, the percentage decrease in the plant height of TRE-ASL09/14 was generally the lowest when compared with the other cultivars at all measuring times under drought stress (Figure 1). This study identified differences among the plant heights of safflower genotypes in response to drought stress markedly change during the growth process (at 3rd measuring time), illustrated in Figure 1, because of the increasing percentage decrease in plant height of all safflowers during drought stress. The greatest reaction can be observed in the 17.8 cm decrease in the plant height of the Remzibey safflower genotype at the last measuring time under drought stress (Figure 1).

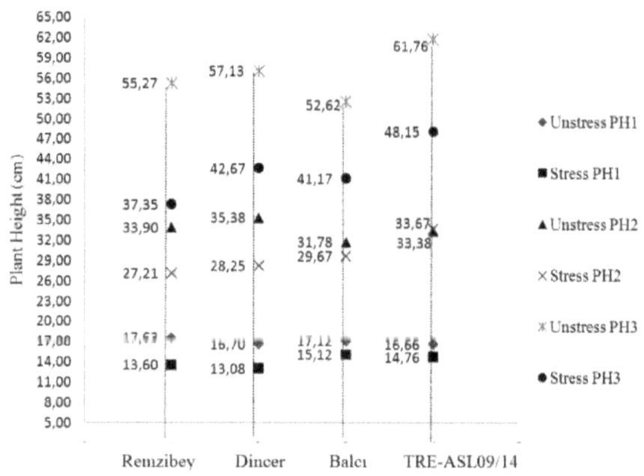

	PH1	PH2	PH3
$LSD_{0,05\ Drought}$	0.400**	0.693**	0.723**
$LSD_{0,05\ Cultivar}$	0.565**	0.981**	1.023**
$LSD_{0,05\ D x C}$	0.799**	1.387**	1.447**

Figure 1. Plant heights of safflower genotypes under the well watered(WW) and drought stress condition (DS).(PH1: first measuring; PH2: second measuring; PH3: third measuring time)

The results indicate that increasing drought, decreased leaf, stem and total dry weight were significant in all safflower genotypes (Figure 2). It was determined that the highest percentage of decrease in terms of total dry weight was determined in the TRE-ASL09/14 safflower genotype under drought stress conditions (Figure 2). The highest mean leaf, stem and total dry weight was observed in the TRE-ASL09/14 safflower genotype under both conditions (Figure 2). It was determined in this study that the greatest plant weight component was the plant stem, as it was 65% of the average of all safflower genotypes under both conditions in Figure 2 (not shown in Figure).

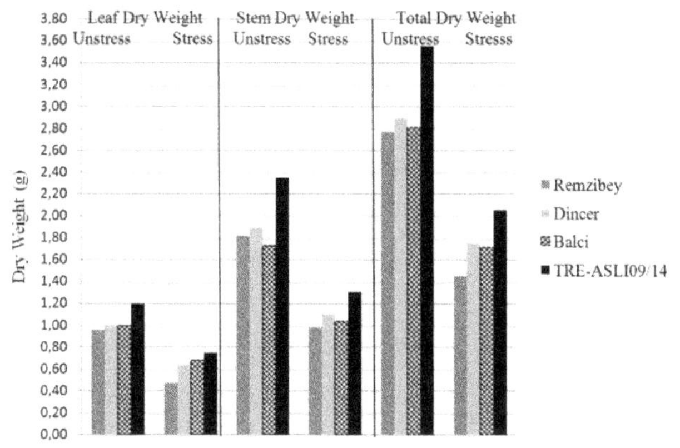

	Leaf Dry Weight	Stem Dry Weight	Total Dry Weight
$LSD_{0,05\ Drought}$	0.050**	0.236**	0.341**
$LSD_{0,05\ Cultivar}$	0.071**	0.334*	0.482*
$LSD_{0,05\ DxC}$	0.101	0.473	0.682

Figure 2. Leaf,stem and total dry weight of safflower genotypes under the well watered (WW) and drought stress conditions (DS).

Drought stress statistically reduced the leaf number of all safflower genotypes, shown in Figure 3. There were also significant differences in terms of leaf number among all the safflower genotypes under both the well-watered and drought stress conditions (Figure 3). However, it was revealed that although the Balcı genotype was in the highest statistical group under both conditions, the leaf number of TRE-ASL09/14 was more stable as there was a smaller reduction than in the other cultivars under drought stress (Figure 3).

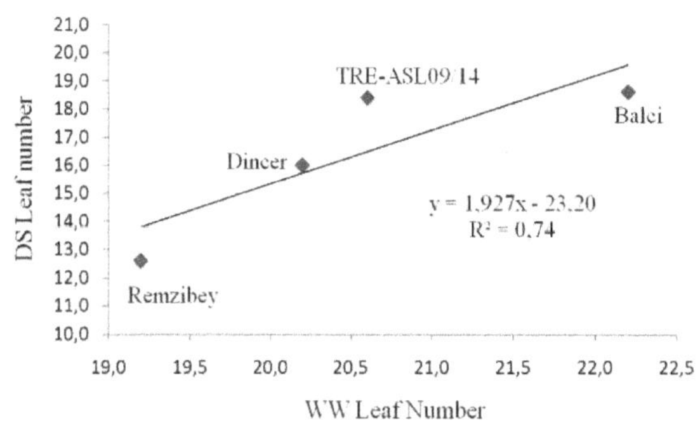

$LSD_{0,05Drought}$:0.564**, $LSD_{0,05\ Cultivar}$:0.797**, $LSD_{0,05}$ D_XC: 1.127**

Figure3. Relationship between total leaf number of safflower genotypes under drought stress (DS) and well watered conditions (WW) condition.

Our regression analyses, shown in Figure 3, indicate there was a significant relationship between the leaf number of unstressed plants and the leaf number of stressed plants with R^2=0.74. Drought stress significantly reduced the total leaf area of all safflower genotypes (Figure 4). Although all sunflower genotypes did not show statistically different reactions under well-watered and drought stress conditions, there was a difference of 53 cm^2 in terms of total leaf area between Dinçer and Remzibey cultivars under drought stress (Figure 4) (not shown in Figure). The largest leaf area was obtained from Dinçer and TRE-ASL09/14 under well-watered and drought stress conditions (Figure 4). Data obtained from the study shows that the percentage of reduction in terms of leaf area was lowest in the Dinçer safflower cultivar under the effects of drought stress (Figure 4). It was revealed that there was significant relationship, with R^2=0.68, between the total leaf area of stressed plants and the total leaf area of unstressed plants (Figure 4).

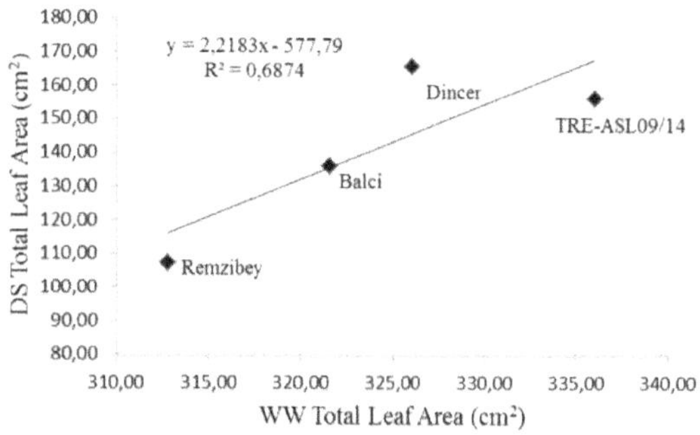

LSD0,05 D: 34.367**, LSD0,05 C: 48.603, LSD0,05 DXC: 68.735

Figure 4. Relationship between total leaf area of safflower genotypes under drought stress (DS) and well watered conditions (WW).

As shown in Figure 5a, the data obtained from both conditions indicated that $\delta^{13}C$ (‰) in all safflower genotypes significantly (*P*<0.01) decreased as they were exposed to drought stress. The highest reduction in $\delta^{13}C$ isotope percentage was determined from TRE-ASL09/14 with a 6.09% decrease under drought stress (Figure 5a). It was also observed that the lowest $\delta^{13}C$ was obtained from TRE-ASL09/14 under drought stress conditions. However, the highest WUE was determined in the TRE-ASL09/14 genotype under drought stress. There were significant differences in terms of $\delta^{13}C$ among all safflower genotypes in both well-watered and drought stress conditions (Table 1 and Figure 5a). In particular, significant differences in $\delta^{13}C$ among all safflowers could be observed under the effects of drought stress (Figure 5a). Drought stress reduced the WUE of all safflower genotypes (Figure 5b). There were also statistically significant differences in terms of WUE among all the safflower genotypes (Table 1 and Figure 5b). The highest reduction in WUE under drought stress conditions was observed in TRE-ASL09/14 and Remzibey with decreases of 0.5 and 0.6 g $g.plant^{-1}$ because of having high WUE under well watered conditions (Figure 5b). The highest WUE was determined in the TRE-ASL09/14 genotype in both well-watered and drought stress conditions (Figure 5b). Dinçer and Balcı safflower genotypes showed the lowest reduction in terms of WUE with decreases of 0.2 g g.plant^{-1} under drought stress (Figure 5b).

The decrease in the specific leaf weight (SLW) of all safflower genotypes under drought stress can be seen in Figure 5c. There were significant differences in terms of SLW among all safflower genotypes under both conditions. The SLW changes observed in all genotypes in response to drought stress indicated that the Balcı safflower genotype was the most sensitive genotype to drought stress (Figure 5c). The highest SLW under drought stress conditions was observed in the Balcı and TRE-ASL09/14 safflower genotypes (Figure 5c). It is evident from our study that the relative water content (RWC) of the four safflower genotypes decreased significantly under drought stress (Figure 5d). On average, the mean RWC of the genotypes showed a reduction of 3.62% under drought stress. The result indicates there were no statistical differences in the value of RWC among all safflower genotypes (Table 1 and Figure 5d). In particular, the RWC values of all genotypes were observed to be quite close under well-watered conditions, while the differences in these values among all safflower genotypes appeared to be more marked under drought stress (Figure 5d). Remzibey had the highest reduction of 0.6% in terms of RWC under drought stress. Figure 5d shows that although there were no statistically significant differences among the genotypes and drought x cultivar interactions, the lowest reductions in the mean RWC were observed in the Balcı and TRE-ASL09/14 genotypes under drought stress.

Our results show that there was a significant relationship (R^2=0.75) between water use efficiency and $\delta^{13}C$ under drought stress (Figure 6). The chlorophyll data obtained from SPAD twice showed that drought stress had significant effects on the chlorophyll of young fully developed leaves in all genotypes. Statistically the chlorophyll of all genotypes was increased by drought stress (Figure 7). On average, the increasingtendency of 44.08 obtained at the first measuring time and 47.13 obtained at the second measuring time under well-watered conditions were statistically different by 7.03 and 10.05 at $P < 0.01$ with respect to chlorophyll under drought stress (Figure 7). Our study also revealed that the chlorophyll of all genotypes showed changes depending on drought duration. There were significant differences between 1st and 2nd measuring times in all safflower genotypes under both conditions (Figure 7). In particular, at the second measuring time although the chlorophyll of TRE-ASL09/14 safflower under unstressed conditions was the highest when compared with the other genotypes, it was in the lowest group under stressed conditions.

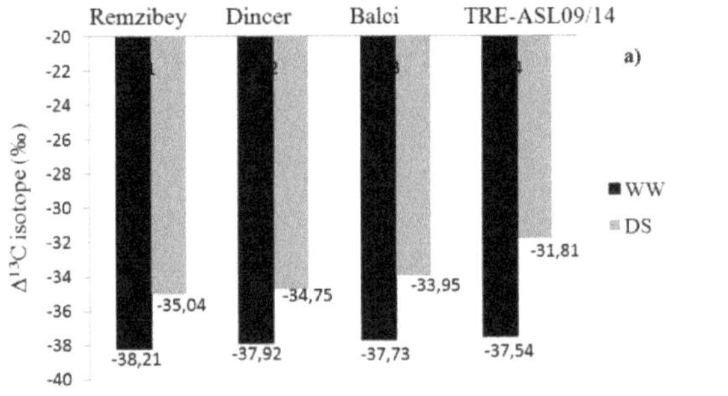

$LSD_{0,05\ Drought}$: 0.284 **, $LSD_{0,05\ Cultivar}$: 0.402 **, $LSD_{0,05}\ D_XC$: 0.568**

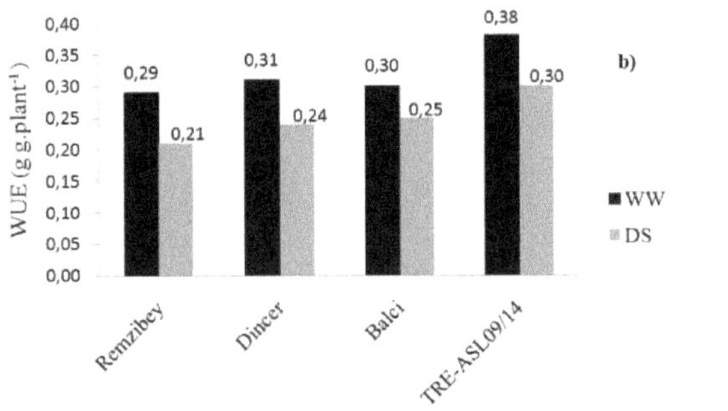

$LSD_{0,05\ Drought}$: 0.015**, $LSD_{0,05\ Cultivar}$: 0.021**, $LSD_{0,05}\ D_XC$: 0.029ns

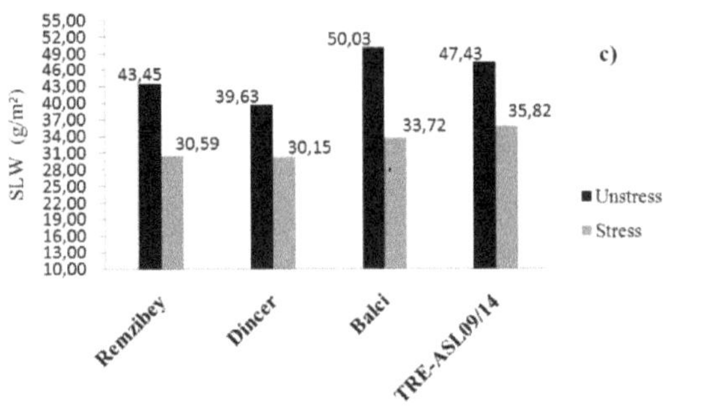

$LSD_{0,05\ Drought}$: 3.322**, $LSD_{0,05\ Cultivar}$: 4.698**, $LSD_{0,05}\ D_XC$: 6.644ns

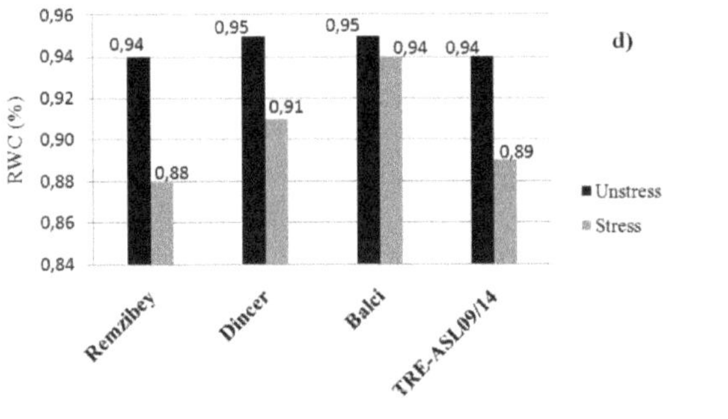

$LSD_{0,05Drought}$: 0.024**, $LSD_{0,05Cultivar}$: 0.033ns, $LSD_{0,05}\ D_XC$: 0.047ns

Figure 5. $\delta^{13}C$ isotope (‰) (a),water use efficiency (g g.plant^{-1}) (b), specific leaf weight (g/m^2) (c)and relative water content (%) (d) of safflower cultivars under the well watered (WW) and drought stress (DS).

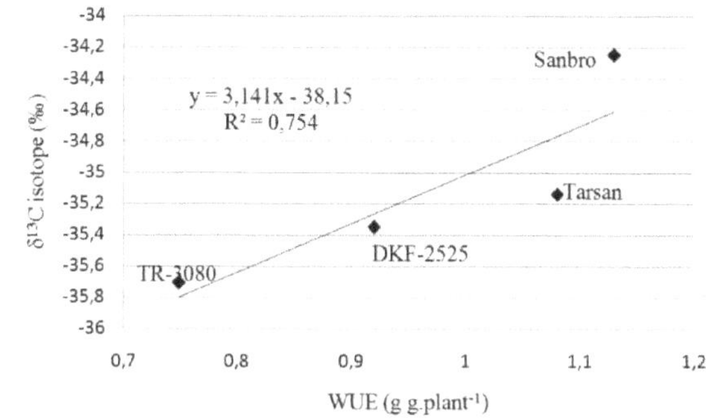

Figure 6. Relationship between plant water use efficiency and $\delta^{13}C$ isotope for four safflower genotypes under the drought stress condition.

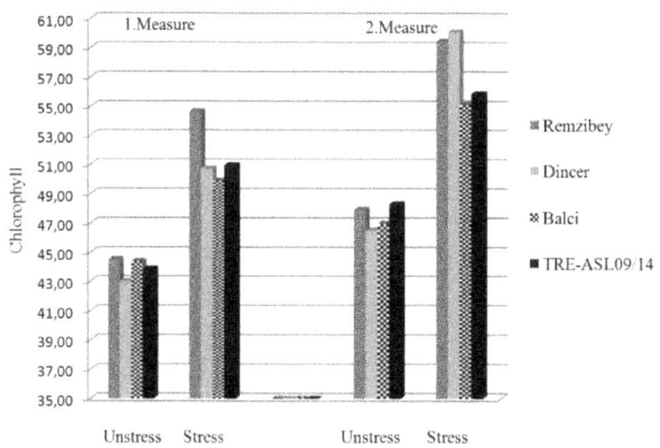

Figure 7. Chlorophyll content measured by SPAD in two times of safflower genotypes under well watered (WW) and drought stress conditions (DS).

DISCUSSION

Our data shows that the plant height, leaf, stem and total dry weight of all safflower genotypes decreased significantly under drought stress because plants respond to drought stress by improving root traits (Price et al., 2002) and, possibly diminishing cell expansion and cell division under drought stress (Munnset al., 2000; O'Neill, 1983). In parallel, the leaf area and leaf numbers of all genotypes also decreased to reduce the evaporative surface (leaf area) under drought stress, similar to the findings of Mitchell et al. (1998). This is because of reduced water loss through reduced epidermal conductance and reduced radiation absorption, as in the results of Farooq et al. (2009). The correlations found in our study, which were significantly positive between WUE and total dry matter (0.80**), further support this theory. Reduced leaf area results in reduced transpiration surface (Namirembeet al., 2008) and may be a drought avoidance strategy for plants. On the other hand, the reduction in leaf area limits photosynthesis and further decreases biomass production, consistent with the positive correlation between total leaf area and biomass production. Also Harbet al. (2010) pointed out in studies on *Arabidopsis thaliana* under drought stress that the reductions in dry matter accumulation and leaf expansion were dependent on the developmental stage of the plants and can be managed by many genes (explained in Figure 11 in the review of Harbet al., 2010).

The WUE values of all safflower genotypes markedly decreased under drought stress. The reason for the lower WUE under drought stress might be due to a greater effect on biomass production than on water use, or it might be that lower conductance of CO_2 and the activity of photosynthetic enzymes decreased photosynthesis and biomass, reducing WUE (Singh and Singh, 2003). The important relationship between WUE and $\delta^{13}C$ obtained from regression analyses, with R^2=0.75 under drought stress, showed that cultivars with high water-use efficiency can be selected using low carbon isotope discrimination under drought stress as presented by Condon et al. (2002). Thus, greater biomass production under stress has been associated with relatively greater WUE as pointed out by Condon et al. (2004), who also explained that the basic unit of production could be the moles of carbon gained by photosynthesis in exchange for water used in transpiration.

The RWC values of both unstressed and stressed plants in the present experiments were higher than those previously reported for safflowers in field experiments by Eslam (2011). One possible explanation for decreasing RWC in safflower leaves could be the limitation of carbohydrate supply caused by water stress.Lawlor and Cornic (2002) and Lawlor (2002) noted photosynthetic carbon assimilation and associated metabolism in relation to RWC in plants and distinguished two general types of relationship between photosynthetic potential and RWC. Our study determined that Balcı and TRE-ASL09/14 were sturdy safflower genotypes in terms of stable water retention in leaf tissue when compared with the other genotypes under drought stress. Françaet al. (2000) explained that genotypes characterized by better drought tolerance mechanisms and higher tissue water retention capacity lead to better growth in beans under water deficits. Thus, avoiding desiccation by maintaining leaf water status at a high level could be considered. Since drought stress decreased RWC, it seems that these indices could reflect the effect of stress that occurred during the vegetative stage (till head visible).

The results of this study indicate that the Balcı and TRE-ASL09/14 safflower genotypes were better than the other genotypes in terms of maintaining SLW under drought stress. According to O'Neill (1983), stressed leaves had a lower SLW, suggesting that these leaves were thicker or had more densely packed mesophyll cells with less intracellular air space. These alterations in leaf anatomy could also result from an inhibition of cell expansion. The possible relationship between specific leaf weight and WUE is based on the fact that SLW could be an indicator of leaf photosynthetic capacity. The increase in SLW could be due to carbohydrates and variation in mesophyll tissue density or leaf thickness, as proposed by Arauset al. (1986). Results of a study revealed that safflower genotypes with greater SLW provided more photosynthetic protein per unit ground area, as explained by Wells et al. (1986).

CONCLUSION

Our study suggested that using high relationship R^2: 0.75regression between WUE and $\delta^{13}C$ (carbon isotope discrimination) under drought stress can be used to selection criteriain response to drought stress for new developing safflower breeding programs. Also, TRE-ASL09/14 new safflower breeding line can be used for evaluation under drought stress, because of its high WUE, RWC, SLW, biomass and $\delta^{13}C$ against drought stress. Plants manifest a wide range of behaviors varying in response to drought stress. These results indicated that the observed variety of physiological and biochemical responses at cellular and whole-organism levels such as inducing the many biomass traits, $\delta^{13}C$, WUE, RWC and high SLW were triggered directly by the imposed drought stress treatment, thus making it a complex phenomenon. Hence, it is important to make comparisons among all the safflower genotypes against drought stress conditions using bilateral and multilateralrelations of traitsfor further breeding selection tests. Therefore, understanding the biological processes involved in the response of plants to drought is very useful for further study.

LITERATURE CITED

Anyia, A.O. andH. Herzog.2004. Water-use efficiency, leaf area and leaf gas exchange of cowpeas under mid-season drought. Eur. J.Agron.20: 327-339.

Araus, J.L., L.Alegre., l. Tapia andR. Calafell. 1986. Relationship between leaf structure and gas exchange in wheat leaves at different insertion levels. J.Exp. Bot.37:1323-1333.

Arslan, B., F. Altuner andM. Tunçtürk. 2003.A study onyield and yieldcharacteristics ofsomesafflower (Carthamustinctorius L.) cultivarsgrownin Van.Türkiye 5. Tarla Bitkileri Kongresi, 13-17 Ekim, Diyarbakır.

Askahni, J., H. Pakniyat and V.Ghotbi.2007. Genetic evaluation of severe physiological traits for screening of suitable spring safflower (*Carthamustinctorius* L.) genotypes under stress and non-stress irrigatipon regimes. Pakistan J. Bio. Sci. 10: 2320-2326.

Bhatia, P.K., A. Muktar andC.S. Vaidyanathan. 1994.Chloroplasticglutamine synthetase from normal and water stressedsafflower (*Carthamustinctorius* L.) leaves. Plant Sci.95: 153-164.

Blum, A.2005.Drought resistance, water-use efficiency, and yield potential—are they compatible, dissonant, or mutually exclusive?. Aust. J.Agr. Res.56:1159–1168.

Chaves, M.M., J.S. Pereira., J. Maroco., M.L. Rodrigues., C.P.P. Ricardo., M.L. Oserio., I. Carvalho., T. Faria andC. Pinheiro. 2002. How do plants cope with water stress in the field? Photosynthesis and growth.Ann. Bot-London.89: 907–916.

Condon, A.G., R.A. Richards., G.J. Rebetzke andG.D. Farquhar. 2002. Improving intrinsic water-use efficiency and crop yield. Crop Sci.42: 122-131.

Condon, A.G., R.A. Richards., G.J. Rebetzke andG.D. Farquhar. 2004.Breeding for high wateruse efficiency. J. Exp. Bot.55:2447–2460.

Cushman, J.C. 2001. Osmoregulation in plants: implications for agriculture. Am. Zool.41: 758-769.

David, R.B. andJ.M. Duniway. 2007. Effects of mycorhizal infection on drought tolerance and recovery in safflower and wheat. Plant Soil Environ.197:95-103.

Dordas, A.C. andC. Sioulas. 2008.Safflower yield, chlorophyll content,photosynthesis, and water use efficiency response tonitrogen fertilization under rainfed conditions. Ind. Crops Prod.27: 75-85.

Eslam, B.P. 2011.Evaluationof physiological indices for improving water deficit tolerance in spring safflower. J. Agric. Sci. Technol.13:327-338.

França, M.G.C., A.T.P. Thi., C. Pimentel., R.O.P. Rossiello., Y. Zuily-Fodil.andD. Lafrray. 2000.Differences in growth and water relations among *Phaseolus vulgaris* cultivars in response to induced drought stress. Environ. Exp. Bot.43: 227-237.

Farquhar, G.D., J.R. Ehleringer andK.T. Hubic. 1989.Carbon isotope discrimination and photosynthesis. Annu. Rev. Plant Phys.40:503-537.

Farooq, M., A. Wahid., N. Kobayashi., D. Fujita andS.M.A. Basra. 2009.Plant drought stress: Effects, mechanisms and management. Agron. Sustain. Dev.29:185–212.

Gates, D.M. 1965.Energy, plants and ecology.Ecology.46:1-13.

Gupta, S.A. andG.A. Berkowitz. 1987. Osmotic adjustment, symplast volume and nonstomatally mediated water stress inhibition of photosynthesis in wheat. Plant Physiol.87: 1040–1047.

Harbi, A.,A. Krishnan., M.M.R. Ambavaram andA. Pereira. 2010.Molecular and physiological analysis of drought stress in arabidopsis reveals early responses leading to acclimation in plant growth[1[C][W][OA]]. Plant Physiol.154:1254–1271.

Hsiao, T.C., R. Acevedo., E. Fereres andD.W. Henderson.1976.Stress, growth and osmotic adjustment. Phil. Trans. R. Soc.Lond. B.273: 479–500.

Lawlor, D.W. 2002.Limitation to photosynthesis in water stressed leaves: stomata versus metabolism and the role of ATP. Ann.Bot.89: 871-885.

Lawlor, D.W. andG. Cornic. 2002. Photosynthetic carbon assimilation and associated metabolism in relation to water deficits in higher plants. Plant Cell Environ.25:275–294.

Levitt, J. 1972.Responses of plants to environmental stresses.(Academic Press: New York).

Levitt, J. 1980.Responses of plants to environmental Stresses. Ed 2. 1: Chilling, Freezing and High Temperature Stresses. New York: Academic Press.

Majidi, M.M., V. Tavakoli.,A. Mirlohi andM.R. Sabzalian. 2011.Wild safflower species (*Carthamusoxyacanthus*Bieb.): Apossible source of drought tolerance for arid environments. Aus. J. Crop Sci. 5: 1055-1063.

Mitchell, J.H., D. Siamhan.,M.H. Wamala., J.B. Risimeri., E. Chinyamakobvu.,S.A. Henderson andS. Fukai. 1998. The use of seedling leaf death scorefor evaluation of drought resistance of rice. Field Crops Res. 55:129–139.

Munoz.P., J. Voltas., J.L. Araus., E. Igartua andI. Romagosa. 1998.Changes over time in the adaptation of barley releases in north-eastern Spain. Plant Breeding 117: 531–535

Munns, R., J.B. Passioura., J. Guo., O. Chazen andG.R. Cramer. 2000. Water relations and leaf expansion: importance of time scale. J. Exp. Bot. 51: 1495-1504.

Namirembe, S., R.M. Brook andC.K. Ong. 2008. Manipulating phenology and water relations in Senna spectabilis in waterlimited environment in Kenya. AgroforestSyst. 75: 197–210.

O'Neill, S.D. 1983.Role of osmotic potential gradients during water stress and leaf senescence in *Fragariavirginiana*. Plant Physiol.72:931-937.

Price, A.H., J.E. Cairns., P. Horton., H.G. Jones andH. Griffiths.2002. Linking drought-resistance mechanisms to drought avoidance in upland rice using a QTL approach: progress and new opportunities to integrate stomatal and mesophyll responses. J. Exp. Bot.53: 989–1004.

Ranney, T.G., R.E. Bir andW.A. Skroch. 1991. Comparative drought resistance among six species of birch (Betula): influence of mild water stress on water relations and leaf gas exchange. Tree Physiol. 8: 351–360.

Singh, B. andG. Singh. 2003. Biomass partitioning and gas exchange in Dalbergiasissoo seedlings under water stress. Photosynthetica.41: 407–414.

Zohary, D. andM. Hopf. 2000. Domestication of plants in the old world. Oxford University Press, Oxford.

Wells, R.,D.A. Ashley andH.R. Boerma. 1986.Physiological comparisons of two soybean cultivars differing in canopy photosynthesis. II. Variation in Specific Leaf Weight,Nitrogen, and Protein Components. Photosynth. Res. 9:295-304.

COMPARISON OF OLD AND NEW CLONES OF POTATO NUCLEAR SEED STOCKS FOR TUBER UNIFORMITY IN THE GREENHOUSE

Gulsum OZTURK[1*], *Zihin YILDIRIM*[1]

[1] *Ege University, Faculty of Agriculture, Department of Field Crops, Bornova-Izmir, TURKEY*
[*] *Corresponding author: gulsum.ozturk@ege.edu.tr*

ABSTRACT

The plantlets of old and new clones micropropagated from nuclear seed stocks were grown in the greenhouse to compare their tuber uniformity. The old and new clones nested in cultivars Agria, Hermes and Granola, were grown in Completely Randomized Plots design in the fall, 2012 and in the winter, 2013. The tuber characteristics were measured and the standard analyses of variance technique were applied and the regression coefficients of the means of new clones on the means of the old clones were estimated for tuber number, single tuber weight (g), single plant yield (g), plot yield (g), tuber width and tuber length. The F values for the clone type and clone were non significant indicating the uniformity of the means of the old and new clones. In addition, the b values of the regression of the means of the new clones on the means of the old clones were found to be equal to 1 supporting the hypothesis of the uniformity of the old clones and new clones for tuber characteristics.

Keywords: Basic seed, clone, potatoes, regression coefficient, uniformity

INTRODUCTION

The basic seed is the first step in a successful seed potato production programme to distribute healty seed tubers to the growers (Bus and Wustman, 2007). Meristem cultures have been used in obtaining virus free basic seed stocks in the seed potato programs (Yildirim and Yildirim, 1984; Ranalli et al., 1990; Regan et al., 1995; Yildirim et al., 1995; Ražukas, 2002; Bus and Wustman, 2007; Pruski, 2007; Struik, 2007; Sharma et al., 2013).

Since several factors such as fungi, bacteria and viruses are carried through the seed tubers, constructing nuclear seed stocks are the first step and requires *in vitro* techniques (Regan et al., 1995; Ahloowalia, 1999; Nielsen et al., 2007; Pruski, 2007). The seed stocks were obtained *in vitro* and they were maintained in the laboratory as nuclear seed stocks. There have been some arguments about keeping the nuclear seed stocks for several years in the laboratory and sampling meristem plants from them to produce mini tubers used in the elit seed production. Although some studies demonstrated that the basic seed stocks need to be renewed every year (Rosenberg et al., 2010), this practice has not been widely supported.

In this study, Nuclear Seed stocks obtained by meristem cultures in a National Seed Potato program in 2007 (Caliskan et al., 2010) and maintained in the laboratory for several years were compared with the new meristem clones constructed from their meristems. In accord with the purpose of this study, the uniformity of the old and new clones for the tuber characteristics could be studied (1) by evaluating the F values pertinent to the variation of the clones based on the ANOVA table. If the hypothesis of H_o: $m_{old}=m_{new}$ is valid the F values will be non significant for the tuber characteristics indicating their uniformity (2) by estimating the regression coefficient of the means of the new clones on the means of old clones. If the b values are equal to 1, there should be a close correspondence between the means of the old and new clones.

MATERIALS AND METHODS

The study was conducted in the Tissue Culture Laboratory and the Greenhouse of the Field Crops Department of the Agricultural Faculty of the Ege University, Izmir, Turkey in 2012 and 2013. The nuclear seed stocks of three potato cultivars, Agria, Hermes, Granola, had been established through the meristem cultures in 2007 and they were renewed by subculturing . The new clones originated from the tubers of the meristem samples taken from the nuclear seed stocks were obtained in 2012. The meristem plantlets were increased by nodal cuttings for the new and old clones. Then the new clones and the old clones of the original seed stocks were grown in the greenhouse to compare their tuber characteristics.

The old meristem stocks have been sub-cultured since 2007 and the plantlets propagated from these stocks were used for mini tuber production in the National Research Project supported by the Scientific and Technological

Research Council of Turkey. Sprouts of potato tubers were cultured in the MS (Murashige and Skoog, 1962) medium supplemented with Auxin, Cytokinin and Gibberelline as described as Ozturk and Yildirim (2011). The *in vitro* meristem plantlets were multiplied by nodal cuttings (Yildirim, 1995) and the meristems and cuttings were micro propagated in certain periods in the laboratory for their maintenance. The new meristem clones were composed of *in vitro* tubers from the meristem plants sampled from the seed stocks taken in 2012.

Production of mini tubers in the greenhouse

The old and new meristem clones described above obtained from *in vitro* plantlets were grown in two successive growing periods in the greenhouse. They were transferred to the plastic cups in September, 2012 and in February, 2013. Three clones nested in each of the three cultivars were grown in the Completely Randomized Design (CRD) with three replications. One plot consisted of 10 plantlets in plastic cups. Plants of two types of meristem clones were harvested in January, 2012 and in May, 2013. At the harvest, plot yield and tuber number were recorded. Following the harvest, tuber width (cm) and tuber length (cm) were measured on random samples of five mini tubers. Later the tuber number plant^{-1} and single plant yield were calculated.

Statistical analyses

Analyses of variance were run on the data of the traits measured by assuming clones nested in cultivars. The F values for the characteristics were obtained and the regression coefficients for the association between new clones (Y) and old clones (X) were estimated following the prosedures described by Steel et al. (1997).

RESULTS AND DISCUSSION

Mini tuber characteristics of the old and new meristem clones

Three old and three new clones nested in three cultivars were grown in the greenhouse in the fall, 2012 and in the spring, 2013. The mini tubers of 9 old and 9 new clones harvested were measured for tuber characteristics in order to compare the uniformity of the old and new meristem clones. The F values of the 6 tuber characteristics are shown in Table 1.

Table 1. The F values of the yield and tuber characteristics from the ANOVA combined over 2012 and 2013.

Source of Variation	Tuber Number Plant^{-1}	Single Tuber Weight(g)	Single Plant Yield (g)	Plot Yield (g)	Tuber Width (cm)	Tuber Length (cm)
Time	57.75**	16.09**	76.60**	76.60**	4.43**	15.33**
Cultivars	295.85**	80.29**	356.68**	356.68**	25.05**	28.00**
Type of clone*	0.726^{ns}	1.494^{ns}	1.453^{ns}	1.453^{ns}	0.143^{ns}	0.604^{ns}
Clone	1.70 ^{ns}	0.58 ^{ns}	0.94 ^{ns}	0.94 ^{ns}	0.25 ^{ns}	1.35 ^{ns}

*: old and new clones
**: significant at the $p \leq 0.01$
^{ns}: non significant

It could be seen in Table 1 that the variation for the type of clone was not significant for all the tuber characteristics studied. The variation of clones nested in cultivars had also non significant F values. On the other hand cultivars and time of growing had significant F values at the $p \leq 0.01$ probability level for all the traits.

Since the F values for the type of clone (old and new) and clone were non-significant we can accept the hypothesis that there was no difference between the means of the clones nested in cultivars for the 6 tuber characteristics. These results indicated the uniformity of the tuber characteristics of the clones originated from the old and new cultures.

Although the non-significant F values obtained for the old and new clones, the means of the old and new clones of 3 cultivars and the regression coefficients between them assuming the new clones as dependent variable are presented in Table 2.

Table 2 shows the means of the tuber characteristics of the old and new clones. The means of the 9 clones from Agria (3), Hermes (3) and Granola (3) had the uniformity confirmed by the non-significant F values mentioned above. The regression coefficients between the means of the old and new clones were found to be 1 indicating a one to one correspondence between the means of old and new clones. The uniformity of the old and new clones was also supported by the b values, so it could be accepted that under the greenhouse conditions, the mini tubers produced from old and new clones were uniform for the 6 tuber characteristics studied. The uniformity of the old and new clones for the tuber traits could also be observed in figures 1, 2, 3, 4, 5 and 6.

Since the callus formation was not observed due to the fast multiplication of the meristems, the genotypic variation among the meristem plants used in seed potato tuber production could not be expected. As a result, the uniformity of the tuber characteristics has continued for several steps of the multiplication. Ahloowalia (1999) reported that mini tubers of cultivars were similar to parents for uniformity of tuber and plant characteristics. Wright (1983) stated that variation observed for tuber traits within cultivars were due to other factors. Among them pathogens could be controlled, since their effect could be minimized due to strict control mechanism in the laboratory and also in the greenhouse.

Table 2. The means of the old and new clones for tuber characteristics

Clones	Tuber Number Plant^{-1}		Single Tuber Weight(g)		Single Plant Yield (g)		Plot Yield (g)		Tuber Width (cm)		Tuber Length (cm)	
	Old meristem	New meristem.	Old meristem	New meristem	Old meristem	New meristem	Old meristem	New meristem	Old meristem	New meristem	Old meristem	New meristem
Agria-1	2.8	2.7	2.8	3.0	7.9	8.2	79.1	81.7	1.8	1.8	2.5	2.5
Agria-4	2.7	2.7	2.9	2.9	7.7	7.8	77.1	77.9	1.8	1.8	2.4	2.5
Agria-5	2.7	2.8	3.0	2.9	8.0	8.0	79.9	80.4	1.8	1.8	2.5	2.5
Hermes-36	2.0	1.9	2.2	2.3	4.4	4.6	43.5	46.7	1.6	1.6	2.2	2.3
Hermes-27	1.8	1.9	2.2	2.2	4.2	4.3	42.1	42.5	1.6	1.6	2.3	2.2
Hermes-10	1.9	1.9	2.2	2.2	4.3	4.1	42.6	41.3	1.6	1.6	2.2	2.2
Granola-48	2.4	2.3	2.2	2.2	5.4	5.3	54.0	52.5	1.6	1.6	2.3	2.3
Granola-53	2.4	2.4	2.2	2.2	5.1	5.4	51.1	53.9	1.6	1.6	2.2	2.2
Granola-12	2.5	2.4	2.0	2.2	5.2	5.4	52.0	53.5	1.6	1.6	2.3	2.3
Regression coefficient (b)	0.96^{ns}		0.93^{ns}		0.99^{ns}		0.99^{ns}		1.0^{ns}		0.98^{ns}	

^{ns}: not significantly different from 1.

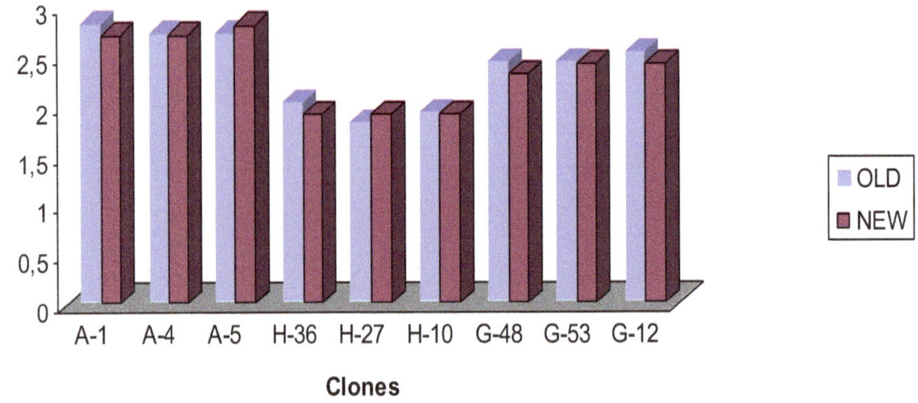

Figure 1. Histograms of tuber number of the mini tubers of old and new clones combined over 2012 and 2013. A: Agria, H: Hermes. G: Granola

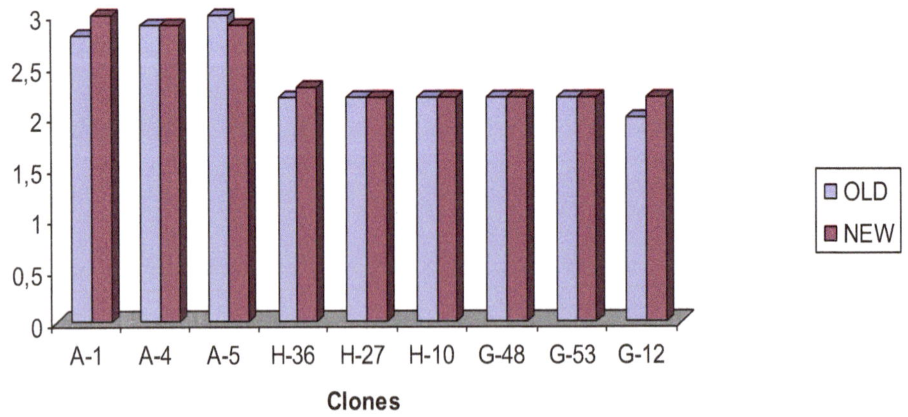

Figure 2. Histograms of single tuber weight (g) of the mini tubers of old and new clones combined over 2012 and 2013. A: Agria, H: Hermes. G: Granola

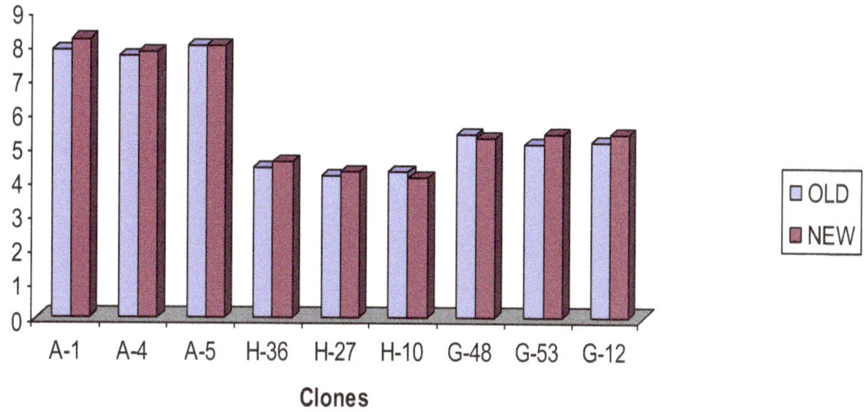

Figure 3.

Histograms of single plant yield (g) of the mini tubers of old and new meristem clones combined over 2012 and 2013. A: Agria, H: Hermes. G: Granola

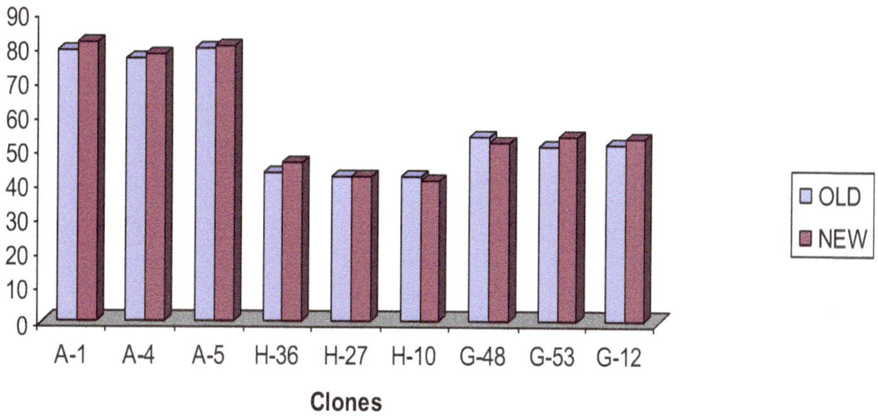

Figure 4.

Histograms of plot yield (g) of the mini tubers of old and new clones combined over 2012 and 2013. A: Agria, H: Hermes. G: Granola

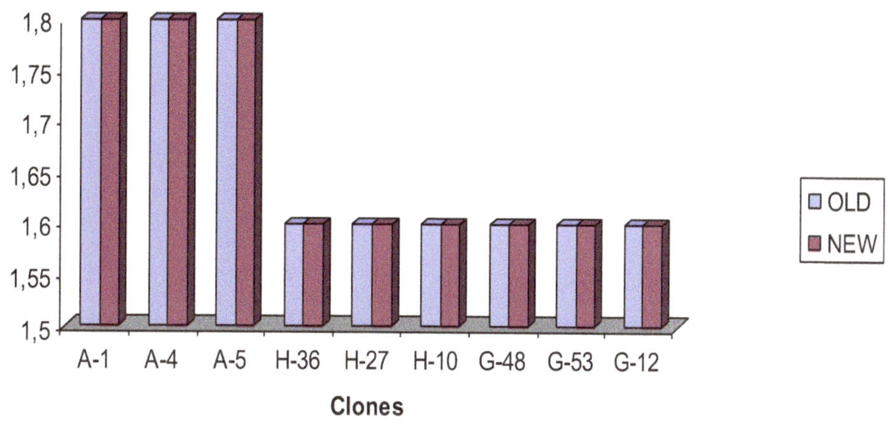

Figure 5.

Histograms of tuber width (cm) of the mini tubers of old and new clones combined over 2012 and 2013. A: Agria, H: Hermes. G: Granola

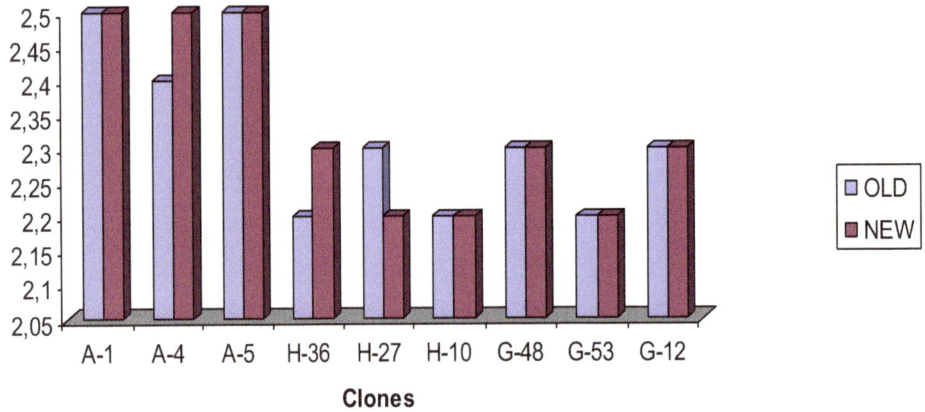

Figure 6. Histograms of tuber length (cm) of the mini tubers of old and new meristem clones combined over 2012 and 2013. A: Agria, H: Hermes. G: Granola

The tuber uniformity of the clones has been reported as consistent in the successive field multiplication stages of the mini tubers (Wright, 1983; Niemira et al., 1995; Kenneth and Charlton, 2004; Coltman, 2012). During the greenhouse multiplication of meristem plants, tuber number per plant was reported as 2 for Agria (Hassanpanah and Khodadadi, 2009) which was a little lower than the value obtained in this study.

Nielsen et al. (2002) reported the stability of Granola meristem clones tested several years in the field. Therefore the uniformity found in this study could be accepted in agreement with their result.

Balali et al. (2008) reported significant variation for planting time and genotype. The significant F values for planting time and for genotypes obtained in this study confirmed their report. The mini tuber yield of cultivars were found to be significantly different in this study conforming the reports of several workers (Powel et al., 1989, Venkatasalam et al., 2011 and Sharma et al., 2013).

In Turkey, there has been a steady problem in managing the mini tuber production which causes a significant impact on the seed potato growers. A major factor for this constraint was reported as sanitary problems and pathogen infection in the field (Caliskan et al., 2010). Since the strict controlled conditions were maintained in the greenhouse, these type of infections have not be observed. Therefore further problems to be encountered in the field multiplication stage should be resolved by selecting suitable field and management practices.

CONCLUSION

The uniformity of the seed stocks in the laboratory and the clones sampled from them were almost identical for the tuber characteristics. Therefor to achieve a successful mini tuber distribution in Turkey, the problems encountered during the field multiplication of mini tuber should be solved.

ACKNOWLEDGEMENT

The authors would like thank to Professor Rajesh K. Rana, Shimla, India, editor of Potato Journal for his kind and valuable help in evaluating the manuscript.

LITERATURE CITED

Ahloowalia, BS. 1999. Production of mini-seed tubers using a modular system of plant micropropagation. Potato Research. 42:569-575.

Balali, GR., MR. Hadi, P. Yavari, H. Bidran, AC. Nadeni, A. Eshami. 2008. Effect of pot size, planting date and genotype on mini tuber production of Marfona potato cultivar. Afri Jour Biyotechnology. 7: 1265-1270.

Bus, CB. and R. Wustman. 2007. Seed tubers. Potato Research. 50: 319-322.

Caliskan, MF., H. Onaran, H. Arioglu. 2010. Overview of the Turkish potato sector: Challenges, achievements and expectations. Potato Research. 53: 255-266.

Coltman, RR. 2012. Optimization of nuclear seed production: seed generations and potato crop productivity. Wisconsin Vegetable Production, vegetables.wisc.edu/ 11 Nov 2013.

Hassanpanah, D. and M. Khodadadi. 2009. Study the plantlet age effect and planting beds on Agria potato mini-tuber production under *in vivo* condition. Journal of Biological Sciences. 9(3): 243-248.

Kenneth, AR. and BA. Charlton. 2004. Effects of prenuclear mini tuber seed size on production of wallowa russet seed. The annual report Klamath Experiment Station, Klamath Falls, Oregon, U.S.A.

Murashige, T. and M. Skoog. 1962. A revised medium for rapid growth and bioassay with tobacco cultures. Physiol. Plant 15: 473-479.

Niemira, BA, GR. Safir, R. Hammerschmidt, GW. Bird. 1995. Production of prenuclear minitubers of potato with peat-based arbuscular mycorrhizal fungal inoculum. Agronomy Journal. 87(5):942-946.

Nielsen, SL, H. Bang, K. Kotkas, K. Kristensen, JP. Palohuhta, K. Tolstrup. 2002. Stability of potato meristem clones. TemaNord 2002:555 Nordic Council of Ministers Copenhagen pp 20-25.

Nielsen, SL, H. Bang, K. Kotkas, K. Kristensen, JP. Palohuhta, V. Rosenberg, K. Tolstrup. 2007. Variation of growth and disease characters between clones of potato (*Solanum tuberosum* L). Potato Research. 50: 97-114.

Ozturk, G. and Z. Yildirim. 2011. Uniformity of potato mini tubers derived from meristem cultures of nuclear seed stocks. Turk J Field Crops 16: 149-152.

Powel, W., J. Brown, PDS. Caligari, 1989. Variability in responce of potato cultivars to micro-propagation II. Subsequent field performance. Ann. Appl. Biol. 115: 123-128.

Pruski, K. 2007. *In vitro* multiplication through nodal cuttings. Potato Research. 50: 293-296.

Ranalli, PE. Forti, G. Mandolino, B. Casarini. 1990. Imrowing production and health of seed potato stocks in Italy. Potato Research. 33: 377-387.

Ražukas, A. 2002. Effect of genotype on distribution of diseases and viruses in potato. Biologija, ISSN: 1392-0146. Nr.1.

Regan, DR., V. Souza Machado, SMM. Alam, A. Alı. 1995. Greenhouse production of potato (*Solanum tuberosum* L. cv. Desire) seed tubers using *in vitro* plantlets and rooted cuttings in large propagation beds. Potato Research. 38: 61-68.

Rosenberg, V., M. Sarekanno, K. Kotkas, V. Vasar, A. Ojararnd. 2010. Variation of agronomic traits of potato somaclones produced by meristem culture. Agronomy Reasearch. 8, 697-704.

Sharma, A.K., E.P. Venkatasalam, V. Kumar. 2013. Potato mini-tuber production during main and off crop seasons in high hills of North-Western Himalaya.Potato J. 40(1): 29.37.

Steel, RGD., JA. Torrie, DA. Deckey. 1997. Principles and procedures of statistics. A. Biometrical Approach 3rd Edi, Mc Graw Hill Book.INC.N.Y.

Struik, PC. 2007. Mini tubers. Potato Research. 50: 305-308

Venkatasalam, E.P., J. Latawa, S. Sarma, S. Sarma, A.K.. Sarma, S. Sarma, R. Patial, S. Singh. 2011. *In vitro* and *in vivo* performance of potato cultivars for different seed production systems. Potato J. 38(2): 149-154.

Wright, NS. 1983. Uniformity among virus-free clones of ten potato cultivars. American Potato Journal. 60: 381-388.

Yildirim, MB. and Z. Yildirim. 1984. Virus free potato seed production by meristem culture. The Journal of Agricultural Faculty of Ege University. 21:45-50 (in Turkish).

Yildirim, MB., Z. Yildirim, C. Caliskan, O. Caylak. 1995. Production of healty potato seed through meristem culture. Final Report of Research of Project 91-ZRF-064 (in Turkish).

Yildirim, Z. 1995. Microtuber production in potato (*Solanum tuberosum* L). The Journal of Agricultural Faculty of Ege University. 32: 73-77 (in Turkish).

GENETIC DIVERSITY OF COMMON BEAN (*Phaseolus vulgaris* L.) LANDRACES GROWN IN NORTHEAST ANATOLIA OF TURKEY ASSESSED WITH SIMPLE SEQUENCE REPEAT MARKERS

Maya Izar KHAIDIZAR[1], Kamil HALILOGLU[2], Erdal ELKOCA[2], Murat AYDIN[2], Faik KANTAR[3]*

[1]*Strategic Resource Research Center Development Institute (MARDI), Kuala Lumpur, MALAYSIA*
[2]*Department of Agronomy, Faculty of Agriculture, Ataturk University, Erzurum, TURKEY*
[3]*Department of Agricultural Biotechnology, Faculty of Agriculture, Akdeniz University, Antalya, TURKEY*
**Corresponding author: eelkoca@atauni.edu.tr*

ABSTRACT

Simple sequence repeat (SSR) marker system was used to analyze the genetic diversity of 38 common bean landraces grown in Northern Anatolia Region and their relationship to twelve nationally registered cultivars. A total of 72 DNA fragments across all materials were scored using 30 primers pairs. More than 97% of the amplification products showed polymorphism, indicating high variation at the DNA level among these accessions. Pair-wise genetic similarity ranged from 0.218 to 0.759, with a mean of 0.585. Based on the SSR markers, two major clusters were formed of 50 genotypes investigated. A majority of the commercial cultivars (66.6%) and landrace accessions (84.2%) were clustered into second group, whereas, small portion of commercial cultivars (33.4%) and landrace accessions (15.7%) were clustered in to first group. A Mantel's test showed significant correlation between SSR distance matrix and cophenetic matrices. Considerable variation existed within the genotypes collected from Northeast Anatolia Region and registered cultivars that can be exploited for breeding programs.

Key words: Genetic relationships, molecular markers, UPGMA

INTRODUCTION

Crop plants rely on the broad genetic base of variation in order to adjust and adapt themselves for ever changing environments and pathogens. Characterization of germplasm using molecular marker techniques provides quantitative estimates of genetic diversity and the information which is essential for a rational utilization of germplasm in breeding programs. Common bean is a crucial food legume consumed worldwide. Genetic diversity in common bean have been studied using different molecular markers such as allozymes (Singh et al., 1991; Santalla et al., 2002), AFLP (Lioi et al., 2005; Svetleva et al., 2006), RAPD (Ocampo et al., 2005, Martins et al., 2006; Marotti et al., 2007), RFLP (Nodari et al., 1992), and ISSR (Svetleva et al., 2006; Marotti et al., 2007). Simple sequence repeats (SSRs), also termed microsatellites, are widely scattered at many different loci throughout the genome (Tautz and Renz, 1984; Kandemir et al., 2010). These small repetitive DNA sequences provide the basis for a PCR-based, multi-allelic, codominant genetic marker system (Saghai Maroof et al., 1994). Polymorphisms revealed by PCR amplification are due to the variation in the number of repeats in a defined region of the genome. The utility of SSR markers is due to their abundant distribution and high polymorphism in the whole genome and power to distinguish between closely related genotypes. Simple sequence repeats (SSRs) have been used to construct a PCR-based genetic map (Blair et al., 2003), to identify genetic variability (Gomez et al., 2005; Blair et al., 2006; Buso et al., 2006; Sarikamis et al., 2009), and to develop multiplex SSR-PCR in common bean (Masi et al., 2003).

In Turkey, dry beans are an important staple food and Turkey is an important producer in the world with annual production of 212,758 tons (FAO, 2010). There are limited molecular characterization studies that partially screened the common bean genotypes collected from the Black Sea region (Balkaya, 1999; Balkaya and Yanmaz, 2002) and Ercis – Gevas distric of Van province (Sarikamis et al., 2009). Northeast Anatolia of Turkey is one of main players of the common bean production in the country. However, genetic characterization of the common bean genotypes from this region has not been determined yet. Northeast Anatolia of Turkey differs significantly from the Black Sea region and Van province in terms of climate and elevation. Therefore, it is important to characterize the common bean germplasm from this region and determine the agronomically important genes that can potentially be integrated into future bean breeding programs.

The purpose of this study was to evaluate the level of genetic diversity present in the ecotypes from this region and compare them with twelve nationally registered cultivars. We believe that the information presented here will benefit in selection program and increase the efficiency of bean breeding programs.

MATERIALS AND METHODS

Plant material

The common bean genotypes used in this study were collected as part of The Scientific and Technological Research Council of Turkey (TUBITAK) supported project (Collection and Evaluation of Dry Bean Gene Resources from Northeastern Anatolia and Çoruh Valley TOVAG 107O400) from Northeast Anatolia of Turkey (Elkoca et al., 2010). A total of 38 common bean genotypes together with twelve nationally registered cultivars were used for SSR analysis (Table 1, 2 and Figure 1).

Table 1. List of common bean genotypes by collection and coordinates.

Accession No	Location	Latitude	Longitude	Altitude (m)
3	Kuruçalı Village Narman/Erzurum	40.393°	41.907°	1615
31	Yanıktaş Village Narman/Erzurum	40.274°	41.858°	1650
40	Kaleboğazı Village Oltu/Erzurum	40.523°	41.899°	1485
51	Burnaz Village Karaçoban/Erzurum	39.364°	41.994°	1580
58	Alacayar Village Narman/Erzurum	40.282°	41.895°	1690
63	Penek Village Şenkaya/Erzurum	40.662°	42.288°	1185
69	Gözalan Village Şenkaya/Erzurum	40.639°	42.262°	1489
77	Alabalık Village Narman/Erzurum	40.430°	41.936°	1560
91	Yeniköy Village Otlukbeli/Erzincan	39.999°	40.114°	1790
106	Şekerli Village Narman/Erzurum	40.313°	41.925°	1625
107	Çatalsöğüt Village Oltu/Erzurum	40.405°	42.005°	1595
115	Çatalsöğüt Village Oltu/Erzurum	40.405°	42.005°	1595
142	Kilimli Village Narman/Erzurum	40.327°	41.992°	1770
188	Petekli Village İspir/Erzurum	40.418°	40.914°	1480
240	Burnaz Village Karaçoban/Erzurum	39.364°	41.994°	1580
242	Hınıs/Erzurum	39.360°	41.700°	1720
244	Ortaköy Village Hınıs/Erzurum	39.403°	41.642°	1905
252	Konursu Village /Bayburt	40.403°	40.267°	1620
254	Konursu Village /Bayburt	40.403°	40.267°	1620
257	Konursu Village /Bayburt	40.403°	40.267°	1620
264	Ballıkaya Village /Bayburt	40.371°	40.389°	1480
268	Nişantaşı Village /Bayburt	40.356°	40.048°	1605
275	Cevizli Village Uzundere/Erzurum	40.627°	41.718°	1205
278	Çağlayan Village Uzundere/Erzurum	40.667°	41.689°	970
296	Arılı Village Tortum/Erzurum	40.369°	41.480°	1450
302	Doruklu Village Tortum/Erzurum	40.381°	41.443°	1749
303	Esendurak Village Tortum/Erzurum	40.326°	41.448°	1525
308	Kemerkaya Village Tortum/Erzurum	40.319°	41.442°	1715
318	Olur/Erzurum	40.820°	42.130°	1330
320	Olur/Erzurum	40.820°	42.130°	1330
328	Torul/Gümüşhane	40.550°	39.290°	933
332	Torul/Gümüşhane	40.550°	39.290°	933
335	Aşağıçayırlı Village Olur/Erzurum	40.876°	42.279°	1755
338	Aşağıçayırlı Village Olur/Erzurum	40.876°	42.279°	1755
351	Altunkaya Village Olur/Erzurum	40.877°	42.248°	1520
358	Altunkaya Village Olur/Erzurum	40.877°	42.248°	1520
361	Altunkaya Village Olur/Erzurum	40.877°	42.248°	1520
368	Çayırçukur Village Kürtün/Gümüşhane	40.665°	39.141°	1232

Table 2. List of nationally registered common bean cultivars used in this study.

Cultivars					
Sahin 90	Gungor	Sehirali 90	Onceler 98	Goynuk 98	Aras 98
Yakutiye 98	Tezibaba	Kantar 05	Akman 98	Elkoca 05	Eskisehir 855

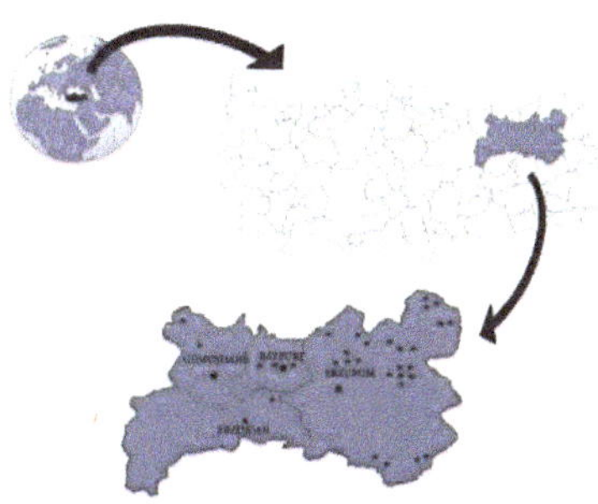

Figure 1. Locations of 38 common bean genotypes collected from Northeast Anatolia.

DNA Extraction

Bulk immature unifoliate leaves from each genotype were used for extracting DNA as described by Dellaporta et al. (1983).

SSR analysis

Thirty SSR primer pairs were selected from previous studies based on their reliable amplification patterns and high polymorphic information contents (Table 3). The reactions were performed in 0.2 µl tubes in Mastercycler personel apparatus (Eppendorf, Germany) programmed to cycle 45 times under the following conditions: for the first two cycles, denaturation for 30 s at 94°C, annealing for 60 s at 37°C, and elongation for 2 min at 72°C; second two cycles, denaturation for 30 s at 94°C, annealing for 60 s at 50°C, and elongation for 2 min at 72°C; the subsequent 41 cycles were run with the denaturation temperature reduced to 93°C, followed by a 4-min hold at 72°C. After amplification, the reaction products were separated by electrophoresis in 3% agarose gels, stained with ethidium bromide, and photographed under ultraviolet light with Nikon Coolpix5000. A total of 30 SSR primer pairs were used based on the band resolution and polymorphism they provided.

Table 3. Thirty SSR primers were used for diversity screening. The list of primer is as follows:

Marker Name	Motifs	Forward (5' – 3')	Backward (5' – 3')	Source; Year
BMd-1	(AT)9	CAAATCGCAACACCTCACAA	GTCGGAGCCATCATCTGTTT	Yu et al. 2000
BMd-15		TTGCCATCGTTGCTTAATTG	TTGGAGGAAGCCATGTATGC	Blair et al. 2003 & 2006
BMd18		AAAGTTGGACGCACTGTGATT	TCGTGAGGTAGGAGTTTGGTG	Blair et al. 2003 & 2006
BM053		TGCTGACCAAGGAAATTCAG	GGAGGAGGCTTAAGCACAAA	Blair et al. 2003 & 2006
BM114	(TA)8(GT)10	AGCCTGGTGAAATGCTCATAG	CATGCTTGTTGCCTAACTCTCT	Blair et al. 2003 & 2006
BM137	(CT)33	CGCTTACTCACTGTACGCACG	CCGTATCCGAGCACCGTAAC	Blair et al. 2003 & 2006
BM141	(GA)29	TGAGGAGGAACAATGGTGGC	CTCACAAACCACAACGCACC	Blair et al. 2003 & 2006
BM143	(GA)35	GGGAAATGAACAGAGGAAA	ATGTTGGGAACTTTTAGTGTG	Blair et al. 2003 & 2006
BM152	(GA)31	AAGAGGAGGTCGAAACCTTAAATCG	CCGGGACTTGCCAGAAGAAC	Blair et al. 2003 & 2006
BM153	(CA)5(TG)(CA)3CG(CA)10(TA)4	CCGTTAGGGAGTTGTTGAGG	TGACAAACCATGAATATGC	Blair et al. 2003 & 2006
BM154	(CT)17	TCTTGCGACCGAGCTTCTCC	CTGAATCTGAGGAACGATGACCAG	Blair et al. 2003 & 2006
BM156	$(CT)_{32}$	CTTGTTCCACCTCCCATCATAGC	TGCTTGCATCTCAGCCAGAATC	Blair et al. 2003 & 2006
BM160	(GA)15(GAA)5	CGTGCTTGGCGAATAGCTTTG	CGCGGTTCTGATCGTGACTTC	Blair et al. 2003 & 2006
BM161	(GA)7(GA)8	TGCAAAGGGTTGAAAGTTGAGAG	TTCCAATGCACCAGACATTCC	Blair et al. 2003 & 2006
BM167	(GA)19	TCCTCAATACTACATCGTGTGACC	CCTGGTGTAACCCTCGTAACAG	Blair et al. 2003 & 2006
BM175	(AT)5(GA)19	CAACAGTTAAAGGTCGTCAAATT	CCACTCTTAGCATCAACTGGA	Blair et al. 2003 & 2006
PV-CTT001	(CTT)3 (T)3 (CTT)6	GAGGGTGTTTCACTATTGTCACTGC	TTCATGGATGGTGGAGGAACAG	Yu et al. 1999
PV-AT001	(AT)22	GGGAGGGTAGGGAAGCAGTG	GCGAACCACGTTCATGAATGA	Yu et al. 1999
PV-AG004	(AG)8	TTGATGACGTGGATGCATTGC	AAAGGGCTAGGGAGAGTAAGTTGG	Blair et al. 2003 & 2006
PVBR14	(AG)23	ACGCCATCCACCATCCTT	TGAGAAAGTTGATGGGATTG	Buso et al.2006
BM183	(TC)14	CTCAAATCTATTCACTGGTCAGC	TCTTACAGCCTTGCAGACATC	Gaitan-Solis et al. 2002
BM187	(CT)10T(CT)14	TTTCTCCAACTCACTCCTTTCC	TGTGTTTGTGTTCCGAATTATGA	Gaitan-Solis et al. 2002
BM188	(CA)18(TA)7	TCGCCTTGAAACTTCTTGTATC	CCCTTCCAGTTAAATCAGTCG	Gaitan-Solis et al. 2002
BM199	(GA)15	AAGGAGAATCAGAGAAGCCAAAAG	TGAGGAATGGATGTAGCTCAGG	Gaitan-Solis et al. 2002
BM200	(AG)10	TGGTGGTTGTTATGGGAGAAG	ATTTGTCTCTGTCTATTCCTTCCAC	Gaitan-Solis et al. 2002
BM209	(TA)4(TG)16	CAACCAATGAATGCTGACAATG	CAATTTCTTGATTGAAAGGCAAT	Gaitan-Solis et al. 2002
BM210	(CT)15	ACCACTGCAATCCTCATCTTTG	CCCTCATCCTCCATTCTTATCG	Gaitan-Solis et al. 2002
BM211	(CT)16	ATACCCACATGCACAAGTTTGG	CCACCATGTGCTCATGAAGAT	Gaitan-Solis et al. 2002
BMd-42	(AT)5	TCATAGAAGATTTGTGGAAGCA	TGAGACACGTACGAGGCTGTAT	Blair et al. 2003 & 2006
GATS91	(GA)17	GAGTGCGGAAGCGAGTAGAG	TGTCACCTCTCTCCTCCAAT	Blair et al. 2003 & 2006

Data analysis

The DNA bands were scored as 0 (absence) or 1 (presence). Genetic similarity between two cultivars i and j was estimated following the formula of Nei and Li (1979). Based on the genetic similarity matrix (denoted GS), UPGMA cluster analysis were used to assess pattern of diversity among the bean entries. Dendrograms were created with the TREE program of NTSYS. All calculations were performed using the NTSYS-pc version 2.1 software (Rohlf, 2000).

RESULTS

Simple sequence repeat (SSR) marker system was used to analyze the genetic diversity of 38 common bean landraces grown in Northern Anatolia Region and their relationship to twelve nationally registered cultivars. A total of 72 alleles were detected. The number of alleles per locus ranged from one for Bmd-15 to 6 for BM-160 with an average number of 2.4 alleles per locus. A range of alleles of expected fragment sizes were obtained by different primer pairs with strong amplifications. Figure 2 shows SSR markers detecting different alleles. More than 97% of the amplification products showed polymorphism, indicating high variation at the DNA level among these accessions.

Pairwise genetic dissimilarity estimated among 50 genotypes was presented in Table 4. Pairwise genetic similarity ranged from 0.211 to 0.796. Based on dissimilarity index, A-320 and A-268 were the closest ones showing 79.6% similarity. On the other hand, genotype pairs A-351 and A-252 were the most dissimilar ones showing only 21.1% similarity.

Based on the SSR markers, two major clusters were formed of 50 genotypes investigated (Figure 3). A majority of the commercial cultivars (66.6%) and landrace accessions (84.2%) were clustered into second group, whereas, small portion of commercial cultivars (33.4%) and landrace accessions (15.7%) were clustered in to first group.

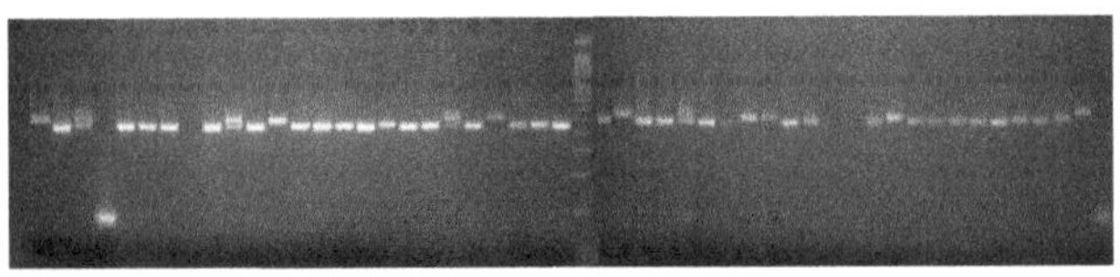

Figure 2. Example of amplification products of primer BM200 detecting different alleles in 50 bean genotypes

Table 4. Pairwise genetic dissimilarity index among 50 common bean genotypes based SSR data.

	A-338	A-278	A-358	A-31	A-296	A-240	A-106	A-51	A-188	A-77	A-242	A-252	A-332	A-302	A-361	A-351	A-142	A-115	A-264	A-58	A-320	A-275	A-63	A-268	A-107
A-338	0,00																								
A-278	0,74	0,00																							
A-358	0,67	0,54	0,00																						
A-31	0,69	0,39	0,55	0,00																					
A-296	0,72	0,44	0,59	0,39	0,00																				
A-240	0,69	0,42	0,60	0,44	0,42	0,00																			
A-106	0,71	0,39	0,58	0,41	0,35	0,33	0,00																		
A-51	0,51	0,73	0,63	0,68	0,71	0,74	0,72	0,00																	
A-188	0,73	0,39	0,55	0,37	0,26	0,41	0,29	0,72	0,00																
A-77	0,60	0,63	0,55	0,62	0,63	0,62	0,62	0,61	0,62	0,00															
A-242	0,74	0,44	0,59	0,46	0,41	0,39	0,39	0,69	0,39	0,66	0,00														
A-252	0,47	0,74	0,69	0,73	0,74	0,75	0,73	0,49	0,73	0,62	0,77	0,00													
A-332	0,73	0,42	0,58	0,37	0,35	0,37	0,29	0,70	0,29	0,67	0,31	0,75	0,00												
A-302	0,72	0,37	0,61	0,46	0,37	0,35	0,31	0,76	0,35	0,61	0,44	0,75	0,39	0,00											
A-361	0,72	0,41	0,61	0,49	0,41	0,46	0,39	0,76	0,35	0,63	0,47	0,72	0,42	0,33	0,00										
A-351	0,75	0,39	0,60	0,47	0,46	0,37	0,37	0,74	0,44	0,67	0,35	0,80	0,41	0,31	0,46	0,00									
A-142	0,68	0,53	0,68	0,54	0,53	0,51	0,49	0,71	0,54	0,68	0,50	0,70	0,49	0,50	0,50	0,54	0,00								
A-115	0,72	0,37	0,59	0,46	0,44	0,39	0,35	0,73	0,42	0,61	0,37	0,74	0,42	0,37	0,41	0,35	0,44	0,00							
A-264	0,69	0,59	0,71	0,62	0,61	0,62	0,58	0,61	0,60	0,67	0,59	0,67	0,58	0,61	0,61	0,62	0,57	0,54	0,00						
A-58	0,65	0,63	0,65	0,60	0,63	0,58	0,58	0,68	0,60	0,62	0,61	0,65	0,60	0,61	0,63	0,62	0,63	0,57	0,62	0,00					
A-320	0,71	0,46	0,60	0,44	0,39	0,37	0,33	0,77	0,37	0,62	0,51	0,73	0,41	0,31	0,39	0,41	0,51	0,39	0,60	0,58	0,00				
A-275	0,54	0,75	0,74	0,77	0,75	0,72	0,74	0,60	0,74	0,74	0,73	0,57	0,75	0,75	0,71	0,72	0,73	0,73	0,63	0,63	0,72	0,00			
A-63	0,74	0,41	0,59	0,42	0,37	0,46	0,31	0,73	0,31	0,61	0,41	0,74	0,31	0,33	0,37	0,39	0,47	0,33	0,54	0,59	0,31	0,76	0,00		
A-268	0,72	0,47	0,63	0,46	0,41	0,35	0,35	0,78	0,39	0,66	0,50	0,74	0,39	0,33	0,41	0,42	0,50	0,41	0,59	0,57	0,20	0,71	0,37	0,00	
A-107	0,70	0,50	0,61	0,51	0,50	0,42	0,42	0,73	0,46	0,66	0,44	0,70	0,42	0,47	0,47	0,46	0,53	0,41	0,57	0,57	0,39	0,67	0,41	0,37	0,00
A-318	0,67	0,54	0,65	0,58	0,57	0,41	0,47	0,77	0,53	0,69	0,49	0,75	0,47	0,54	0,57	0,50	0,54	0,49	0,62	0,53	0,50	0,70	0,54	0,49	0,46
A-91	0,44	0,75	0,71	0,73	0,75	0,76	0,76	0,54	0,75	0,71	0,75	0,50	0,75	0,75	0,72	0,76	0,74	0,75	0,69	0,67	0,75	0,54	0,74	0,75	0,74
Sahin-90	0,71	0,51	0,65	0,60	0,54	0,55	0,47	0,68	0,53	0,58	0,51	0,71	0,53	0,49	0,49	0,50	0,57	0,46	0,55	0,65	0,55	0,72	0,49	0,57	0,54
A-254	0,63	0,55	0,66	0,59	0,53	0,49	0,49	0,71	0,54	0,66	0,55	0,68	0,57	0,50	0,50	0,54	0,44	0,47	0,61	0,59	0,51	0,69	0,55	0,53	0,55
A-257	0,51	0,69	0,66	0,68	0,71	0,66	0,70	0,60	0,70	0,54	0,73	0,59	0,72	0,69	0,67	0,72	0,65	0,69	0,66	0,66	0,70	0,65	0,71	0,71	0,69
Yakutiye-98	0,66	0,53	0,61	0,59	0,58	0,49	0,49	0,73	0,54	0,68	0,47	0,72	0,51	0,47	0,47	0,46	0,55	0,47	0,59	0,61	0,49	0,65	0,50	0,53	0,53
A-40	0,73	0,46	0,62	0,53	0,51	0,41	0,47	0,75	0,50	0,62	0,46	0,78	0,50	0,46	0,51	0,41	0,59	0,42	0,67	0,62	0,50	0,74	0,51	0,51	0,51
Gungor	0,46	0,75	0,68	0,74	0,71	0,68	0,72	0,53	0,74	0,68	0,69	0,51	0,72	0,73	0,71	0,74	0,65	0,71	0,63	0,66	0,72	0,53	0,75	0,71	0,67
Terzibaba	0,51	0,73	0,77	0,72	0,73	0,68	0,70	0,62	0,74	0,72	0,73	0,57	0,72	0,67	0,67	0,70	0,67	0,71	0,70	0,63	0,68	0,62	0,71	0,69	0,69
A-368	0,68	0,53	0,70	0,59	0,58	0,42	0,51	0,75	0,57	0,70	0,50	0,74	0,57	0,47	0,53	0,46	0,60	0,44	0,61	0,68	0,51	0,69	0,55	0,50	0,50
Sehirali-90	0,67	0,51	0,60	0,53	0,42	0,44	0,37	0,63	0,44	0,58	0,46	0,67	0,47	0,46	0,51	0,47	0,57	0,46	0,60	0,62	0,47	0,68	0,46	0,51	0,51
Kantar-05	0,58	0,68	0,73	0,71	0,68	0,67	0,71	0,61	0,71	0,71	0,66	0,65	0,71	0,66	0,63	0,65	0,63	0,63	0,62	0,71	0,69	0,59	0,70	0,68	0,63
A-335	0,67	0,51	0,67	0,53	0,51	0,44	0,50	0,75	0,53	0,69	0,51	0,71	0,53	0,42	0,51	0,44	0,59	0,46	0,67	0,65	0,47	0,70	0,51	0,49	0,51
Onceler	0,69	0,51	0,65	0,55	0,54	0,41	0,47	0,70	0,53	0,69	0,49	0,73	0,50	0,46	0,54	0,47	0,51	0,49	0,58	0,58	0,50	0,66	0,54	0,46	0,46
Akman	0,55	0,63	0,71	0,69	0,72	0,69	0,71	0,61	0,73	0,71	0,68	0,62	0,71	0,68	0,68	0,65	0,66	0,68	0,71	0,73	0,71	0,66	0,70	0,72	0,70
E-5855	0,66	0,55	0,61	0,57	0,50	0,51	0,51	0,69	0,51	0,70	0,55	0,74	0,54	0,50	0,50	0,54	0,58	0,53	0,57	0,61	0,51	0,65	0,55	0,53	0,55
Goynuk	0,68	0,44	0,61	0,51	0,50	0,46	0,39	0,73	0,49	0,66	0,47	0,72	0,46	0,41	0,41	0,42	0,53	0,41	0,57	0,63	0,42	0,71	0,41	0,47	0,44
Elkoca-05	0,66	0,55	0,68	0,57	0,58	0,51	0,51	0,73	0,59	0,72	0,58	0,72	0,57	0,50	0,55	0,51	0,47	0,53	0,63	0,61	0,54	0,69	0,58	0,55	0,55
A-328	0,70	0,50	0,68	0,57	0,53	0,49	0,51	0,76	0,54	0,68	0,47	0,74	0,54	0,47	0,47	0,46	0,58	0,41	0,61	0,66	0,54	0,69	0,53	0,55	0,47
A-3	0,65	0,54	0,65	0,53	0,57	0,44	0,44	0,72	0,53	0,65	0,51	0,69	0,50	0,46	0,49	0,47	0,59	0,46	0,60	0,65	0,44	0,70	0,46	0,46	0,49
Aras-98	0,68	0,53	0,61	0,57	0,53	0,51	0,54	0,67	0,54	0,68	0,50	0,70	0,57	0,55	0,53	0,51	0,58	0,53	0,59	0,61	0,59	0,62	0,60	0,60	0,55
A-244	0,63	0,60	0,63	0,61	0,55	0,46	0,54	0,71	0,57	0,70	0,50	0,68	0,54	0,53	0,58	0,54	0,55	0,50	0,63	0,61	0,51	0,67	0,58	0,53	0,50
A-69	0,62	0,66	0,67	0,67	0,61	0,67	0,65	0,68	0,62	0,69	0,68	0,67	0,67	0,63	0,61	0,69	0,66	0,63	0,50	0,65	0,62	0,63	0,63	0,63	0,61
A-308	0,51	0,69	0,72	0,70	0,67	0,68	0,70	0,60	0,68	0,72	0,69	0,59	0,68	0,69	0,67	0,72	0,69	0,69	0,63	0,68	0,68	0,58	0,69	0,67	0,67
A-303	0,59	0,73	0,74	0,68	0,67	0,66	0,70	0,65	0,72	0,74	0,69	0,63	0,68	0,69	0,67	0,68	0,67	0,69	0,70	0,68	0,66	0,62	0,71	0,62	0,65

Table 4. Continue

	A-318	A-91	Sahin-90	A-254	A-257	Yakutiye-98	A-40	Gungor	Terzibaba	A-368	Sehirali-90	Kantar-05	A-335	Onceler	Akman	E-5855	Goynuk	Elkoca-05	A-328	A-3	Aras-	A-244	A-69	A-308	A-303
A-318	0,00																								
A-91	0,73	0,00																							
Sahin-90	0,60	0,75	0,00																						
A-254	0,51	0,70	0,51	0,00																					
A-257	0,70	0,61	0,61	0,53	0,00																				
Yakutiye	0,49	0,68	0,54	0,50	0,65	0,00																			
A-40	0,47	0,75	0,55	0,54	0,68	0,57	0,00																		
Gungor	0,68	0,54	0,70	0,65	0,65	0,65	0,72	0,00																	
Terzibaba	0,68	0,54	0,70	0,67	0,65	0,65	0,74	0,53	0,00																
A-368	0,54	0,70	0,57	0,53	0,67	0,55	0,42	0,67	0,67	0,00															
Sehirali	0,58	0,75	0,50	0,54	0,66	0,54	0,50	0,63	0,72	0,59	0,00														
Kantar-05	0,67	0,60	0,65	0,68	0,68	0,63	0,69	0,57	0,54	0,63	0,69	0,00													
A-335	0,55	0,69	0,55	0,54	0,70	0,57	0,44	0,70	0,66	0,39	0,58	0,65	0,00												
Onceler	0,50	0,73	0,60	0,49	0,66	0,51	0,47	0,63	0,68	0,51	0,50	0,67	0,55	0,00											
Akmen	0,69	0,55	0,69	0,68	0,63	0,66	0,67	0,54	0,42	0,63	0,71	0,58	0,67	0,71	0,00										
E-5855	0,57	0,68	0,59	0,50	0,67	0,58	0,54	0,62	0,67	0,55	0,54	0,68	0,54	0,51	0,68	0,00									
Goynuk	0,54	0,68	0,42	0,47	0,62	0,41	0,51	0,69	0,65	0,50	0,46	0,66	0,51	0,51	0,63	0,53	0,00								
Elkoca-05	0,57	0,72	0,59	0,41	0,60	0,53	0,59	0,69	0,62	0,55	0,57	0,63	0,57	0,49	0,66	0,53	0,44	0,00							
A-328	0,54	0,70	0,49	0,50	0,65	0,53	0,46	0,71	0,69	0,41	0,54	0,63	0,42	0,57	0,68	0,55	0,37	0,50	0,00						
A-3	0,55	0,65	0,50	0,54	0,61	0,46	0,53	0,68	0,61	0,42	0,50	0,67	0,44	0,53	0,65	0,57	0,35	0,54	0,46	0,00					
Aras-98	0,59	0,72	0,57	0,47	0,58	0,50	0,54	0,62	0,69	0,60	0,54	0,61	0,57	0,54	0,68	0,55	0,50	0,53	0,47	0,59	0,00				
A-244	0,49	0,68	0,59	0,50	0,67	0,50	0,51	0,60	0,69	0,50	0,49	0,68	0,46	0,49	0,72	0,50	0,50	0,53	0,47	0,51	0,55	0,00			
A-69	0,65	0,62	0,62	0,59	0,61	0,61	0,65	0,59	0,68	0,66	0,60	0,65	0,65	0,60	0,71	0,51	0,61	0,63	0,59	0,62	0,57	0,57	0,00		
A-308	0,66	0,49	0,68	0,60	0,60	0,69	0,68	0,60	0,65	0,65	0,66	0,63	0,66	0,66	0,63	0,60	0,65	0,65	0,65	0,63	0,69	0,58	0,54	0,00	
A-303	0,68	0,61	0,72	0,67	0,69	0,73	0,72	0,58	0,60	0,65	0,70	0,66	0,63	0,70	0,59	0,62	0,69	0,69	0,65	0,66	0,67	0,67	0,70	0,62	0,00

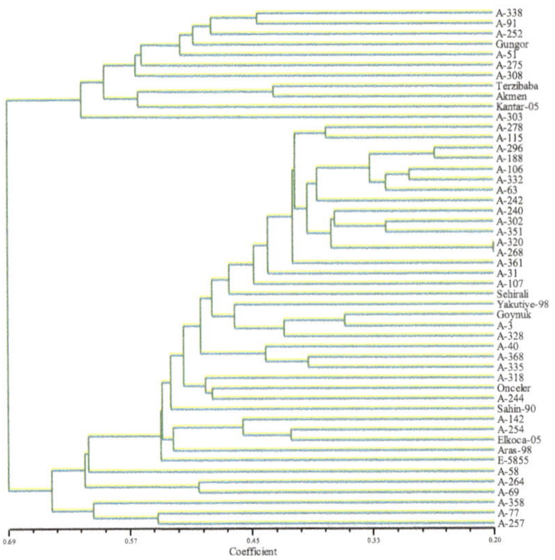

Figure 3. Phenogram constructed using the UPGMA method.

DISCUSSION

Information about the genetic variability of genotypes helps decision making for conservation activities, which range from collecting and managing through identifying genes to adding value to genetic resources and for breeding purposes, to predict the ability to combine or to rapidly verify the breeding material (Vicente et al., 2005). Therefore, assessment of genetic variability of germplasm is a first step, named as pre-breeding, for improvement and development of superior cultivars.

A range of molecular techniques can assess crop genetic diversity, however among the most ideal for distinguishing closely related germplasm are microsatellites, which are highly informative markers that detect length polymorphisms at loci with simple sequence repeats (Powell et al., 1996). Their advantages for diversity studies include uniform genome coverage, high levels of polymorphism, codominance, and an easy-to-implement, specific PCR-based assay (Pejic et al., 1998).

In this study, genotypes collected from Northeast Anatolia of Turkey were evaluated at molecular level via SSR marker system. Though common bean genotypes collected from the Black Sea region (Balkaya, 1999; Balkaya and Yanmaz, 2002) and Van province region (Sarikamis et al., 2009) were previously characterized, the common bean ecotypes from Northeast Anatolia of Turkey had not been genetically characterized before. Therefore, these are important first steps regarding better understanding and maintenance of common bean germplasm of the region.

Genomic SSRs selected on the basis of high polymorphism information content (Blair et al. 2003, 2006; Yu et al., 1999, 2000; Gaitan-Solis et al. 2002) successfully helped discriminate genotypes in the present study. Successful amplifications of expected sizes were obtained. The results of this study indicate that SSR analysis could be successfully used for the estimation of genetic diversity among common bean genotypes and can be potentially incorporated into future studies that examine the diversity within a larger collection of bean genotypes from diverse regions. We also hope that the results of this study will benefit current bean breeding efforts in Turkey as well as maintain the genetic integrity of the genetic resources.

LITERATURE CITED

Balkaya, A., 1999. A Research on Collection of Genetic Resources, Determination of Phenological and Morphological Characteristics and Selection of Suitable Types for Fresh Consumption by Pedigree Selection in Green Bean (*Phaseolus vulgaris* L.) in the Black Sea Region. PhD Thesis, Ondokuz Mayis University, Samsun, Turkey.

Balkaya, A., R. Yanmaz, 2002. Morphological properties of cultivar nominates selected Black Sea Region bean populations and identification by protein markers. Ankara Univ. J. Agric. Sci. 9: 182-188.

Blair M.W., F. Pedraza, H.F. Buendia, E. Gaitan-Solis, 2003. Development of a genome-wide anchored microsatellite map for common bean (*Phaseolus vulgaris* L.). Theor. Appl. Genet. 107: 1362-1374.

Blair, M.W., M.C. Giraldo, H.F. Buendia, E. Tovar, M.C. Duque, S.E. Beebe, 2006. Microsatellite marker diversity in common bean (*Phaseolus vulgaris* L.). Theor. Appl. Genet. 113: 100-109.

Buso, G.S.C., Z.P.S. Amaral, R.P.V. Brondani, M.E. Ferreira, 2006. Primer note: microsatellite markers for the common bean *Phaseolus vulgaris*. Molecular Ecology Notes 6: 252–254.

Dellaporta, S.L., J. Wood, J.B. Hicks, 1983. A plant DNA mini-preparation: version II. Plant Molecular Biology Reporter 1: 19-21.

Elkoca, E., K. Haliloglu, F. Kantar, C. Eken, F. Donmez, M. Aydin, 2010. Genetic diversity of common beans (*Phaseolus vulgaris* L.) grown in the northeast region of Turkey based on morphological traits. 5th International Food Legumes Research Conference (IFLRC V) & 7th European Conference on Grain Legumes (AEP VII), Legumes for Global Health, Legume Crops and Products for Food, Feed and Environmental Benefits, 26-30 April, 2010, Antalya, Turkey, pp: 36.

FAO, 2010. http://faostat.fao.org/

Gaitan-Solis E., M.C. Duque, K.J. Edwards, J. Tohme, 2002. Microsatellite repeats in common bean (*Phaseolus vulgaris*): Isolation, characterization, and cross-species amplification in *Phaseolus* ssp. Crop Sci. 42: 2128-2136.

Gomez, O.J., M.W. Blair, B.E. Frankow-Lindberg, U. Gullberg, 2005. Comparative study of common bean (*Phaseolus vulgaris* L.) landraces conserved *ex situ* in genebanks and *in situ* by farmers. Genetic Resources and Crop Evolution 52: 371-380.

Kandemir, N., G. Yılmaz, Y.B. Karan, D. Borazan, 2010. Isolation of different genotypes in 'başçiftlik beyazı' potato landrace using ssr markers. Turkish Journal of Field Crops 15(1): 84-88.

Lioi, L., A.R. Piergiovanni, D. Pignone, S. Puglisi, M. Santantonio, G. Sonnante, 2005. Genetic diversity of some surviving on-farm Italian common bean (*Phaseolus vulgaris* L.) landraces. Plant Breeding 124: 576-581.

Marotti, I., A. Bonetti, M. Minelli, P. Catizone, G. Dinelli, 2007. Characterization of some Italian common bean (*Phaseolus vulgaris* L.) landraces by RAPD, semirandom and ISSR molecular markers. Genetic Resources and Crop Evolution 54: 175-188.

Martins, S.R., F.J. Vences, L.E.S. Miera, M.R. Barroso, V. Carnide, 2006. RAPD analysis of genetic diversity among and within Portuguese landraces of common white bean (*Phaseolus vulgaris* L.). Scientia Horticulturae 108: 133-142.

Masi, P., P.L.S. Zeuli, P. Donini, 2003. Development and analysis of multiplex microsatellite markers set in common bean (*Phaseolus vulgaris* L.). Molecular Breeding 11: 303-313.

Nei, M., W.H. Li, 1979. Mathematical model for studying genetic variation in terms of restriction endonucleases. Proc. Natl. Acad. Sci. USA 76: 5269-5273.

Nodari, R.O., E.M.K. Koinange, J.D. Kelly, P. Gepts, 1992. Towards an integrated linkage map of common bean: 1. Development of genomic DNA probes and levels of restriction fragment length polymorphism. Theor. Appl. Genet. 84: 186-192.

Ocampo, C.H., J.P., Martin, M.D. Sanchez-Yelamo, J.M. Ortiz, O. Toro, 2005. Tracing the origin of spanish common bean cultivars using biochemical and molecular markers. Genetic Resources and Crop Evolution 52: 33-40.

Pejic I., P. Ajmone-Marsan, M. Morgante, V. Kozumplick, P. Castaglioni, G. Taramino, M. Motto, 1998. Comparative analysis of genetic similarity among maize inbred lines detected by RFLPs, RAPDs, SSRs and AFLPs. Theor. Appl. Genet. 97: 1248-1255.

Powell W., M. Morgante, C. Andre, M. Hanafey, J. Vogel, S. Tingey, A. Rafalski, 1996. The comparison of RFLP, RAPD, AFLP and SSR (microsatellite) markers for germplasm analysis. Molecular Breeding 2: 225-238.

Rohlf, F.J., 2000. NTSYS-pc: Numerical Taxonomy and Multivariate Analysis System. Version 2.1 Exeter Publications, New York, USA.

Saghai Maroof, M.A., R.M. Biyashev, G.P. Yang, Q. Zhang, R.W. Allard, 1994. Extraordinarily polymorphic microsatellite DNA in barley: species diversity, chromosomal locations, and population dynamics. Proc. Natl. Acad. Sci. USA 91: 5466-5470.

Santalla, M., A.P. Rodino, A.M. De Ron, 2002. Allozyme evidence supporting southwestern Europe as a secondary center of genetic diversity for the common bean. Theor. Appl. Genet. 104: 934-944.

Sarikamis, G., F. Yasar, M. Bakir, K. Kazan, A. Ergul, 2009. Genetic characterization of green bean (*Phaseolus vulgaris*) genotypes from eastern Turkey. Genetics and Molecular Research 8(3): 880-887.

Singh, S.P., R. Nodari, P. Gepts, 1991. Genetic diversity in cultivated common bean: I. allozymes. Crop Sci. 31: 19-23.

Svetleva, D., G. Pereira, J. Carlier, L. Cabrita, J. Leitao, D. Genchev, 2006. Molecular characterization of *Phaseolus vulgaris* L genotypes included in Bulgarian collection by ISSR and AFLP analyses. Scientia Horticulturae 109: 198-206.

Tautz, D., M. Renz, 1984. Simple sequences are ubiquitous repetitive components of eukaryotic genomes. Nucleic Acids Research 12: 4127-4138.

Vicente, M.C., F.A. Guzman, J. Engels, V. Ramanatha Rao, 2005. Genetic characterization and its use in decision making for the conservation of crop germplasm. The Role of Biotechnology, 5-7 March, 2005, Villa Gualino, Turin, Italy, pp: 122-128.

Yu K., S.J. Park, V. Poysa, 1999. Abundance and variation of microsatellite DNA sequences in beans (*Phaseolus* and *Vigna*). Genome 42: 27-34.

Yu K., S.J. Park, V. Poysa, P. Gepts, 2000. Integration of simple sequence repeat (SSR) markers into a molecular linkage map of common bean (*Phaseolus vulgaris* L.). Journal of Heredity 91: 429-434.

24

ESTIMATING GENOME SIZE AND CONFIRMING PLOIDY LEVELS OF WILD TETRAPLOID ALFALFA ACCESSIONS (*Medicago sativa* subsp. × *varia)* USING FLOW CYTOMETRY

Muhammet SAKIROGLU[*1], *Muge Mavioglu KAYA*[2],

[1]*Kafkas University, Department of Bioengineering, Kars, TURKEY*
[2]*Kafkas University, Department of Biology, Kars, TURKEY,*
** Corresponding author: msakiroglu@kafkas.edu.tr*

ABSTRACT

The taxonomic group *Medicago sativa-falcata* continuum includes the important cultivated forage legume, alfalfa, along with a number of other perennial, outcrossing, and morphologically differentiated subspecies at diploid and tetraploid levels. Prior information of morphology, ploidy, and genome size is vital for accurate classification of the taxa included in the complex and thus for effective usage of genetic resources in alfalfa breeding programs. The United States Department of Agriculture-National Plant Germplasm System (USDA-GRIN) has an extensive collection of the members of *Medicago sativa-falcata* continuum gathered from all the centres of diversity. However, accessions classified in the complex are occasionally misidentified. Furthermore, the accessions identified as *M. sativa* subsp. ×*varia* (T. Martyn) Arcang. in the USDA GRIN collections have not been evaluated based on morphological traits, ploidy level or genome size. In this study, we evaluated morphological traits and determined ploidy levels and genome size of plants from 25 wild accessions classified as *Medicago sativa* L. subsp. ×*varia* using flow cytometry. All of the accessions classified as subsp. ×*varia* were found to be tetraploid; however, deviations from flower colour, pod shape expectations were observed. This will be a major step toward effective utilization of germplasm resources classified as *M. sativa* subsp. ×*varia*.

Keywords: Alfalfa; flow cytometry; genome size; *M. sativa-falcata* complex*; M. sativa* subsp. ×*varia;* ploidy

INTRODUCTION

The nuclear DNA content of an unreplicated diploid cell (in G1 phase) is considered to have a 2C DNA amount (Swift, 1950) and this value is used to determine genome size of plant species. The knowledge of C-value is vital for many scientific disciplines, including molecular biology, systematic, and ecology (Dolezel and Bartos, 2005). Despite the wide range of usage, C-value of only less than 2% of all angiosperms are known (Hanson et al., 2003). Ploidy is another significant aspect of plant genome and plays an important role in evolution and diversity (Grant, 1971; Wang et al., 2006; Gaeta et al., 2007). Allopolyploidy arises from interspecific hybridity and stabilizes the hybrid genome whereas autopolyploidy occurs within species in which there are more than two sets of homologous chromosomes (Grant, 1971; De Laat et al., 1987). In addition to the evolutionary implications, polyploidy has a tremendous significance for agriculture. A number of economically important crops such as wheat, sugar cane, potato, coffee, alfalfa, and cotton are polyploids (Leitch and Bennett, 1997).

The *M. sativa* species complex (or sometimes the *M. sativa-falcata* continuum) includes the important cultivated forage legume, alfalfa, and a number of other perennial, outcrossing, and morphologically differentiated but often interfertile taxa at diploid and tetraploid levels. The taxonomic status of the taxa included in the complex has been unsettled until recently (Sakiroglu et al., 2010). Some authors have considered the taxa as species (Sinskaya, 1961; Lesins and Lesins, 1982; Ivanov, 1988) and others considered them as subspecies (Quiros and Bauchan, 1988). Majority of the taxa has been demoted to the subspecies level recently and the nomenclature has been widely adopted (Sakiroglu et al., 2010). Diploid (2n=2x =16) subspecies include *Medicago sativa* L. subsp. *falcata* (L.) Arcang. (yellow flowers, sickle shaped pods), *M. sativa* subsp. *caerulea* (purple flowers, coiled pods), and their natural hybrid, *Medicago sativa* L. subsp. ×*hemicycla* (Grossh.) C.R. Gunn. The analogous set of tetraploid (2n=4x=32) subspecies have similar distinguishing phenotypes to the diploids and include *M. sativa* subsp. *falcata*, *M. sativa* subsp. *sativa* L., and the tetraploid hybrid *Medicago sativa* L. subsp. ×*varia* (T. Martyn) Arcang. (Small, 1985; Quiros and Bauchan, 1988). The key characters distinguishing among the member of the complex are morphology (flower colour, pod shape, and pod pubescence) and ploidy.

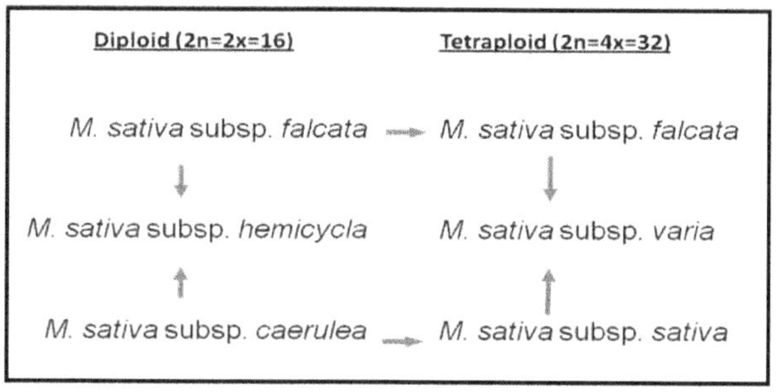

Figure 1. Taxonomic relationship among the members of *Medicago sativa* species complex

Chromosome counting of stained root tips under microscope was long regarded as the convention method for ploidy determination in plants (Sakiroglu and Brummer, 2011); however, the method has been noted as laborious and occasionally misleading (Brummer et al., 1999; Tuna et al., 2001). The two early approaches that were employed to determine the 2C DNA content of a given organism were the analysis of DNA extracted from a large number of cells and the measurements taken from the individual nuclei (Dolizel and Bartos, 2005). The estimate obtained from the first approach does not represent the 2C DNA amount, since source tissue could contain cells at different phases of the cell cycle and thus different DNA amounts. Although the estimation based on single nucleus approach yields higher precision, the technique is noted to be very demanding (Dolizel and Bartos, 2005). Flow cytometry serves as an accurate and fast alternative tool for plant scientists to determine ploidy level and to estimate the genome size of *M. sativa* accessions (Blondon et al., 1994; Brummer et al., 1999; Sakiroglu and Brummer, 2011). United States Department of Agriculture, Germplasm Resources Information Network (USDA-GRIN) maintains one of the largest alfalfa germplasm collections with more than 4000 accessions collected from the entire distribution range of all the members of the complex. However, the accessions are occasionally misidentified in the USDA-GRIN system (Sakiroglu and Brummer, 2011). The earlier attempts to determine ploidy level of *Medicago sativa* accessions either focused on subspecies *falcata* (Brummer et al., 1999) or diploid subspecies (Sakiroglu and Brummer, 2011). The evaluated accessions in the two above mentioned studies collectively correspond to less than 3% of the entire collection. Hence, majority of accessions still lack the information of ploidy level. As one of the subspecies considered in the complex, *Medicago sativa* L. subsp. ×*varia* has been incorporated into breeding material and has significant economic value. Accessions collected in the USDA-GRIN system and denoted as subsp. ×*varia* have not been evaluated based morphology and ploidy. The only genome size estimate of the subsp. ×*varia* is estimated from a single plant and extrapolation of that single plant estimate to the entire subspecies could be misleading. Therefore, there is need to evaluate subsp. ×*varia* accessions based on morphology, ploidy, and genome size.

In this study, we evaluate morphological traits and ploidy levels of 25 accessions of *Medicago sativa* L. subsp. ×*varia* gathered from one of the centre of diversity and estimated genome size of accessions using flow cytometry.

MATERIALS AND METHODS

We used 25 accessions from the USDA National Plant Germplasm System identified in GRIN that were originally collected from Turkey and denoted as *M. sativa,* subsp. ×*varia.* All the accessions are also listed as local landraces. A tomato plant (*Solanum lycopersicum* L.) from cultivar SC2121 (Arumuganathan and Earle, 1991), a diploid (CADL), and a tetraploid (Buldog 505) plant of known ploidy (Brummer et al., 1999, Sakiroglu and Brummer, 2011) were selected as diploid and tetraploid standards and genome sizes of all the samples were compared to the two standards.

Alfalfa seedlings were grown in Kafkas University greenhouse for about six weeks. Young leaves were harvested from each plant for flow cytometry analyses. Each individual sample consisted of 10 mg of leaf tissue of tomato plant and 15 mg of leaf tissue of individual genotypes from each accession. Each of the individual genotype from each accession was run independently.

The procedure explained in Galbraith et al. (1983) was used for the flow cytometry. The leaf material was chopped with a one-sided sharp razor blade in a Petri dish containing chopping buffer prepared according to Galbraith et al. (1983) to isolate nuclei. After chopping, the buffer, containing cell constituents and large tissue remnants, was passed sequentially through nylon filters of 50 μm and 20 μm mesh size to separate nuclei from the cell debris. The buffer with nuclei was then centrifuged at high speed (800 rpm for 5 minutes), the supernatant was discarded, and the pellet was resuspended in 100 μl propidium iodide (PI) staining solution of 100 μg/ml. All the sample preparation procedures were performed in darkness on ice until the analyses were completed.

The samples were analyzed on a Cytomics FC 500 (Beckman-Coulter, Fullerton, CA) flow cytometer at the Flow Cytometry Facility of Trakya University Hematology Lab with a wavelength of 488 nm. Mean DNA content was based on analysis of 10000 nuclei. Ploidy levels of accessions were determined based on diploid and tetraploid standards (CADL and Buldog 505) and the DNA content was determined based on tomato plant.

Since the morphological traits such as flower colour and the pod shape are employed to assign accessions into subspecies, we recorded flower colour and pod shape from each genotype under greenhouse conditions. The flower colour of each accession was visually classified following Barnes (1972). For tetraploid subspecies, yellow flowered accessions were considered as subsp. *falcata*, purple flowered accessions as subsp. *sativa,* and the variegated flowers as subsp. ×*varia.* The pod shape data was gathered following Sakiroglu et al. (2010). Briefly, pod shape among the members of *Medicago sativa-falcata* complex ranges from straight or falcate pods to highly coiled pod shape. We recorded the coil number for ten individual pods from each genotype and computed an average coiling value of the accession across all genotypes. Pods were scored in 1/4 coil intervals.

Table 1. Wild diploid alfalfa accessions along with geographic coordinates and the subspecies information.

No	Population	Taxonomic status given by USDA	Location of Collection	Number of plants assayed
1	PI 171719	Medicago sativa subsp. ×varia	Tokat/Turkey	3
2	PI 172429	Medicago sativa subsp. ×varia	Ardahan/Turkey	3
3	PI 206281	Medicago sativa subsp. ×varia	Sivas/Turkey	3
4	PI 206282	Medicago sativa subsp. ×varia	Sivas/Turkey	3
5	PI 238145	Medicago sativa subsp. ×varia	Tunceli/Turkey	3
6	PI 238150	Medicago sativa subsp. ×varia	Sivas/Turkey	3
7	PI 238151	Medicago sativa subsp. ×varia	Sivas/Turkey	3
8	PI 383693	Medicago sativa subsp. ×varia	Pertek/Turkey	3
9	PI 464800	Medicago sativa subsp. ×varia	Sivas/Turkey	3
10	PI 464801	Medicago sativa subsp. ×varia	Elazığ/Turkey	3
11	PI 464802	Medicago sativa subsp. ×varia	Bingöl/Turkey	3
12	PI 464803	Medicago sativa subsp. ×varia	Kars/Turkey	3
13	PI 464804	Medicago sativa subsp. ×varia	Kars/Turkey	3
14	PI 464805	Medicago sativa subsp. ×varia	Kars/Turkey	3
15	PI 464806	Medicago sativa subsp. ×varia	Kars/Turkey	3
16	PI 464807	Medicago sativa subsp. ×varia	Kars/Turkey	3
17	PI 464808	Medicago sativa subsp. ×varia	Kars/Turkey	3
18	PI 464809	Medicago sativa subsp. ×varia	Erzincan/Turkey	3
19	PI 464810	Medicago sativa subsp. ×varia	Sivas/Turkey	3
20	PI 464811	Medicago sativa subsp. ×varia	Sivas/Turkey	3
21	PI 464812	Medicago sativa subsp. ×varia	Yozgat/Turkey	3
22	PI 464813	Medicago sativa subsp. ×varia	Yozgat/Turkey	3
23	PI 464814	Medicago sativa subsp. ×varia	Ankara/Turkey	3
24	PI 577511	Medicago sativa subsp. ×varia	Erpeler/Turkey	3
25	PI 577512	Medicago sativa subsp. ×varia	Erzincan/Turkey	3
26	Buldog 505	Cultivar	Tetraploid Standard	2
27	CADL	Cultivar	Diploid Standard	1

RESULTS AND DISCUSSION

Ploidy level of the accessions:

In order to assess the ploidy level of the accession, we compared the fluorescence density of samples to the previously known diploid and tetraploid standards. Fluorescence densities of diploid and tetraploid standards were first detected. The peak from diploid standard was occurred at 183 nm whereas the peak of the tetraploid standard was occurred at 361 nm (Figure 2a and 2b). The peaks obtained from the samples were compared to the ones from standard plants (Figure 2a and 2b). All the accessions were found to have fluorescent density equal to the tetraploid standard confirming tetraploid status of the accessions. We did not observe any within accession ploidy level variations (Table 2).

Previously, extensive deviations at the ploidy level among the accession classified as diploid hybrid subspecies *Medicago sativa* L. subsp. ×*hemicycla* were detected (Sakiroglu and Brummer 2011). Majority of the accessions were proposed to be reclassified as *Medicago sativa* L. subsp. ×*varia.* A similar pattern was observed among other subspecies (Sakiroglu et al., 2010; Sakiroglu and Brummer, 2011; Havananda et al., 2010). Here, we did not observe any deviations from the expected ploidy level.

DNA content of the accessions

In order to estimate the DNA content of accessions, the samples including the tomato as the standard plant were run in flow cytometry. DNA content of the subsp. ×*varia* accessions were estimated based on relative florescent density of each sample compared to tomato plant. We initially targeted to estimate DNA content of three plants per accession; however, due to ambiguities of peaks from tomato plant, we could not estimate total nuclear DNA of some individual genotypes. Therefore, DNA content of seven accessions was estimated using two individual genotypes, and that of six accessions was estimated using a single genotype. Estimated DNA content was ranged from 2.85 pg to 4.99 pg with a mean of 3.25 pg (Table 3). These values show a large within subspecies genome size variation. The large intraspecific genome size variations were reported previously both among members of the complex (Blondon 1994) and in other species (Dolizel and Bartos, 2005). The only available genome size estimate for the subsp. ×*varia* is the early estimate obtained by Blondon et al. (1994) and it is based on one single plant. The 2C value of subsp. ×*varia* was 3.47pg. The estimate of genome in our study is 3.25 is based on average of 56 plants and is probably more robust. Given the sampling nature of two studies and extent of the variation within subsp. ×*varia,* the difference is not unexpected. Therefore, we conclude that

Table 2. Information on the ploidy level, flower colour, and pod coiling of the Medicago sativa accessions evaluated.

No	Accession	Subspecies	Ploidy level	Flower Colour	Average pod coiling
1	PI 171719	*Medicago sativa* subsp. ×*varia*	32	Purple	1.24
2	PI 172429	*Medicago sativa* subsp. ×*varia*	32	Purple	NA
3	PI 206281	*Medicago sativa* subsp. ×*varia*	32	Purple	1.46
4	PI 206282	*Medicago sativa* subsp. ×*varia*	32	Purple	1.85
5	PI 238145	*Medicago sativa* subsp. ×*varia*	32	Purple	0.98
6	PI 238150	*Medicago sativa* subsp. ×*varia*	32	Purple	1.03
7	PI 238151	*Medicago sativa* subsp. ×*varia*	32	Purple	1.63
8	PI 383693	*Medicago sativa* subsp. ×*varia*	32	Purple	1.15
9	PI 464800	*Medicago sativa* subsp. ×*varia*	32	Purple	1.28
10	PI 464801	*Medicago sativa* subsp. ×*varia*	32	Purple	1.56
11	PI 464802	*Medicago sativa* subsp. ×*varia*	32	Purple	1.40
12	PI 464803	*Medicago sativa* subsp. ×*varia*	32	Purple	1.32
13	PI 464804	*Medicago sativa* subsp. ×*varia*	32	Purple	1.58
14	PI 464805	*Medicago sativa* subsp. ×*varia*	32	Purple	1.13
15	PI 464806	*Medicago sativa* subsp. ×*varia*	32	Yellow/Purple	NA
16	PI 464807	*Medicago sativa* subsp. ×*varia*	32	Purple	1.65
17	PI 464808	*Medicago sativa* subsp. ×*varia*	32	Purple	1.18
18	PI 464809	*Medicago sativa* subsp. ×*varia*	32	Purple	1.28
19	PI 464810	*Medicago sativa* subsp. ×*varia*	32	Purple	NA
20	PI 464811	*Medicago sativa* subsp. ×*varia*	32	Purple	0.68
21	PI 464812	*Medicago sativa* subsp. ×*varia*	32	Purple	NA
22	PI 464813	*Medicago sativa* subsp. ×*varia*	32	Purple	0.87
23	PI 464814	*Medicago sativa* subsp. ×*varia*	32	Purple	NA
24	PI 577511	*Medicago sativa* subsp. ×*varia*	32	Purple	1.24
25	PI 577512	*Medicago sativa* subsp. ×*varia*	32	Purple	1.18
26	Buldog 505	Cultivar	32		
27	CADL	Cultivar	16		

NA: Not available

genome size estimations be conducted on large number of plants before the estimate is accurately established.

There is a significant ongoing discussion on whether the observed interspecies genome fluctuations are real or due to experimental artifacts and the discussions are comprehensively reviewed by Greilhuber (2005). Predominantly phenolic endogenous staining inhibitors in the plant material have been proposed to affect the genome size measurements (Greilhuber, 1986, 1988). Nonetheless, since we are reporting an average estimate of genome size for subsp. ×*varia*, the overall genome size should provide a robust estimate for the subspecies despite the evident variation.

Morphological traits

Flower colour of all the accessions evaluated was purple with the exception of PI464806. Within accession flower colour variation was detected from PI464806 one of the individual genotype found to have yellow flowers (Table 2). The within accession flower colour variations are not unusual given the outcrossing nature of the *M. sativa*, particularly when the location of collection is considered. The Eastern Anatolia is considered as one the centre of diversity for the members of *M. sativa* species complex (Sinskaya 1961; Lesins and Lesins, 1982; Quiros and Bauchan, 1988; Small et al., 1990; Sakiroglu and Brummer, 2012). All the subspecies are grown sympatricaly and the mixtures during the seed collection is very likely. The mixture at the seed collection sides during the seed increase could also be the cause of the within accession flower colour variations. The other important morphological trait that is used for classification of the accessions is pod shape and it is measured quantitatively. The average pod coiling of the putative subsp. ×*varia* accessions are measured from 20 accessions and ranged from 0.68 to 1,85 coils with an average of 1,28 (Table 3). Accessions of *M. sativa* subsp. ×*varia* are ought to have variegated flower colour which is distinct morphology of hybrid subspecies. The purple flowers however are distinctive characteristic of subsp. *sativa*. The classification of hybrid subspecies solely based on flower colour or pod coiling data could be misleading because these two morphological traits are governed by a small number of genes. In a previous study evaluating diploid subspecies using genome-wide marker data, it was

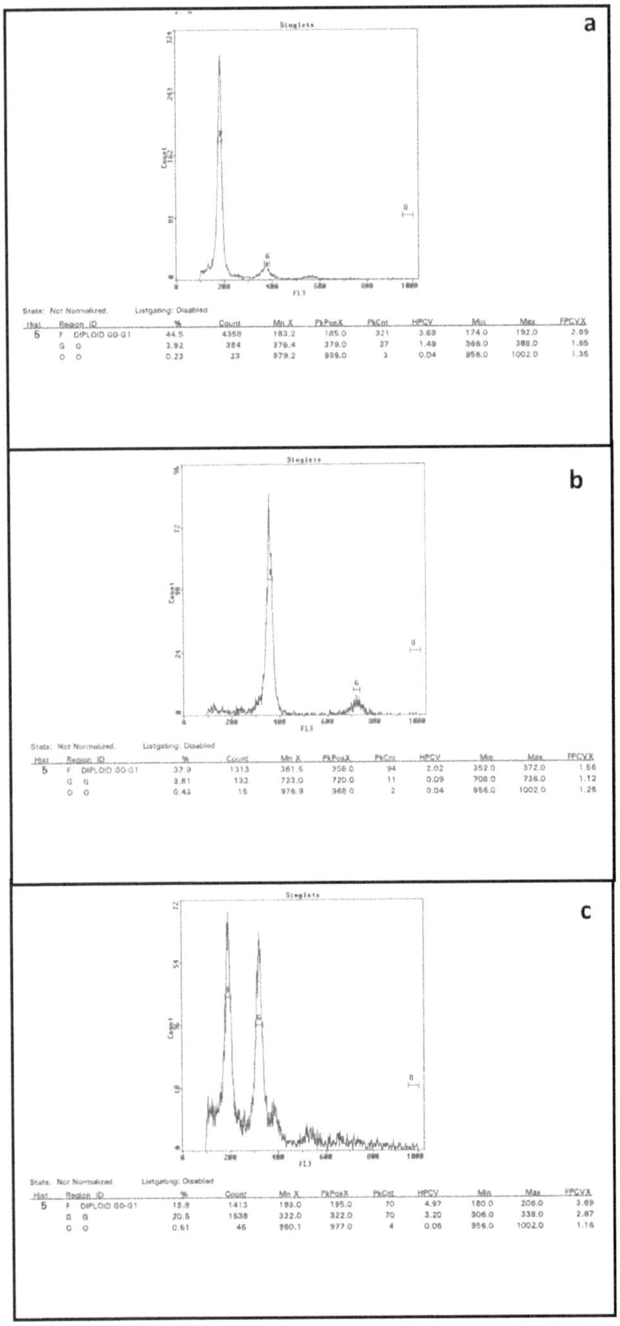

Figure 2. Flow cytometry results as a function of florescence density (nm) of (a) diploid standard CADL, (b) tetraploid standard Bulldog 505, and (c) comparison between tomato (*Solanum lycopersicum*) and *M. sativa* subsp. *varia.*

found that the flower colour could be a distinctive criterion among subspecies only when it is considered with pod coiling data (Sakiroglu et al., 2010). A comparison of genome composition with morphology, the study concluded that in addition to the variegated flower colour, accessions with either yellow or purple flower colour could also be hybrid. Yellow flowers and pods with more than one coil

Table 3. DNA content (pg) of 56 individual genotypes from 25 accessions and the percentage of total genetic variance attributable to each source of variation.

No	Accession	Florescence density of the standard	Florescence density of the sample	DNA content (pg)
1	PI 171719	212	370	3.31
		202	349	3.29
		234	380	3.08
2	PI 172429	219	359	3.12
		237	386	3.10
		228	383	3.20
3	PI 206281	183	315	3.27
		182	312	3.26
4	PI 206282	191	328	3.26
5	PI 238145	212	387	3.47
		220	361	3.12
		204	319	2.97
6	PI 238150	211	364	3.28
		208	360	3.29
7	PI 238151	140	210	2.85
		225	371	3.13
		211	358	3.22
8	PI 383693	203	356	3.33
		201	365	3.45
		190	334	3.35
9	PI 464800	199	351	3.34
		201	366	3.46
10	PI 464801	213	372	3.32
		197	343	3.31
		203	336	3.15
11	PI 464802	207	347	3.18
		178	324	3.46
		153	243	3.02
12	PI 464803	212	416	3.72
13	PI 464804	206	365	3.37
		210	372	3.37
14	PI 464805	188	305	3.08
15	PI 464806	198	339	3.26
16	PI 464807	167	271	3.08
		254	409	3.06
		196	351	3.40
17	PI 464808	226	385	3.23
18	PI 464809	212	358	3.20
		230	373	3.08
		228	366	3.05
19	PI 464810	225	356	3.01
20	PI 464811	191	303	3.01
		248	396	3.04
		235	376	3.05
21	PI 464812	221	366	3.15
		218	353	3.08
		193	297	2.92
22	PI 464813	184	341	3.52
		177	325	3.49
23	PI 464814	195	512	4.99
		192	323	3.20
24	PI 577511	202	351	3.30
		195	322	3.14
25	PI 577512	229	405	3.35
		202	331	3.12
		204	353	3.29
	Mean	212	370	3.25
	Std	21	43	0.29

lead the conclusion that an accession could be classified as hybrid subsp. ×*hemicycla*. The purple flowered accessions could be classified as hybrid subsp. ×*hemicycla* if the pods have fewer than 1.5 coils. Since tetraploid taxa are analogous to the diploid taxa, it is highly likely that a similar morphological relationship is present among tetraploids. Based on this analogy, five out of 20 accessions (PI206282, PI 238151, PI 464801, PI 464804, and PI 464807) are above 1.5 coils and could be reclassified as subsp. *sativa*. However, molecular confirmation of the analogy must be obtained before deriving a definite conclusion and such a confirmation effort with molecular tools could be an interesting research venue.

The current germplasm collections gathered and maintained by USDA provides an excellent source to explore *Medicago sativa* species complex via studying different genomic aspects of the complex. However, the evaluation of accessions based on morphology and genomics is the prerequisite for effective usage in plant breeding. Flow cytometry has been an effective tool employed in estimation of genome size and determination of the ploidy levels in plant populations.

ACKNOWLEDGEMENTS

We would like to thank Dr. Metin Tuna, Dr. Ilker Nizam, and Dr. Aysel Guven for their assistance and TUBITAK (The Grant Number 109O860 awarded to MS) and Kafkas University Scientific Research Funds (The grant awarded to MS) for funding this project. This publication is derived from Muge Mavioglu Kaya's MS thesis.

LITERATURE CITED

Arumuganathan, K., E.D. Earle, 1991. Nuclear DNA content of some important plant species. Plant Molecular Biology Reporter. 9 (3):208-218.

Barnes, D.K., 1972. A system for visually classifying alfalfa flower color. USDA-ARS Agric. Handbook No. 424. U.S. Gov. Print. Office, Washington, DC.

Blondon, F., D. Marie, S. Brown, A. Kondorosi, 1994. Genome size and base composition in *Medicago sativa* and *M. truncatula* species. Genome. 37(2): 264-270.

Brummer, E.C., P.M. Cazcarro, D. Luth, 1999. Ploidy determination of alfalfa germplasm accessions using flow cytometry. Crop Science. 39(4):1202.

De Laat, A.M.M., W. Göhde, M. Vogelzang. 1987. Determination of ploidy of single plants and plant populations by flow cytometry. Plant Breed. 99:303-307.

Dolezel, J., J. Bartos, 2005. Plant DNA flow cytometry and estimation of nuclear genome size. Annals of Botany. *95*(1):99-110.

Gaeta, R.T., J. C. Pires, F. Iniguez-Luy, E. Leon, T.C. Osborn, 2007. Genomic changes in resynthesized *Brassica napus* and their effect on gene expression and phenotype. Plant Cell. 19(11): 3403-3417.

Galbraith, D.W., K.R. Harkins, J.M. Maddox, N.M. Ayres, D.P. Sharma, E. Firoozabady, 1983. Rapid flow cytometric analysis of the cell cycle in intact plant tissues. Science. 220:1049-1051.

Grant, V., 1971. Plant Speciation. Columbia University Press. New York.

Greilhuber, J., 1986. Severely distorted Feulgen DNA amounts in Pinus (Coniferophytina) after nonadditive fixations as a result of meristematic self-tanning with vacuole contents. Canadian Journal of Genetics and Cytology. 28: 409–415.

Greilhuber, J., 1988. 'Self-tanning'—a new and important source of stoichiometric error in cytophotometric determination of nuclear DNA content in plants. Plant Systematics and Evolution. 158: 87–96.

Greilhuber, J., 2005. Intraspecific variation in genome size in angiosperms: identifying its existence. Annals of Botany. 95: 91–98.

Hanson, L., R.L. Brown, A. Boyd, M.A.T. Johnson, M.D. Bennett, 2003. First nuclear DNA C-values for 28 Angiosperm genera. Annals of Botany. 91: 31–38.

Havananda, T., E.C. Brummer, I.J. Maureira-Butler, J.J. Doyle, 2010. Relationships among diploid members of the *Medicago sativa* (Fabaceae) species complex based on chloroplast and mitochondrial DNA sequences. Systematic Botany. 35(1): 140-150.

Ivanov, A.I.,1988. Alfalfa. Amerind Publishing Co. New Delhi.

Leitch, I.J., M.D. Bennett, 1997. Polyploidy in angiosperms. Trends in Plant Science. 2(12):470-476.

Lesins, K.A., I. Lesins, 1982. Genus Medicago (Leguminosae), a taxogenetic study. Plant Ecology. 50(2): 92-92.

Quiros, C.F., G.R. Bauchan, 1988. The genus *Medicago* and the origin of the *Medicago sativa* complex. In: Hanson A.A., Barnes D.K., Hill R.R. (eds), Alfalfa and alfalfa improvement. ASA-CSSA-SSSA, Madison, WI.

Sakiroglu, M., J.J. Doyle, E.C. Brummer, 2010. Inferring population structure and genetic diversity of broad range of wild diploid alfalfa (Medicago sativa L.) accessions using SSR markers. Theoretical and Applied Genetics. 121(3): 403-415.

Sakiroglu, M., E.C. Brummer, 2011. Clarifying the ploidy of some accessions in the USDA alfalfa germplasm collection. Turkish Journal of Botany. 35:509-519.

Sakiroglu, M., E.C. Brummer, 2012. Presence of phylogeographic structure among wild diploid alfalfa accessions (Medicago sativa L. subsp. microcarpa Urb.) with evidence of the center of origin. Genetic Resources and Crop Evolution (in print).

Sinskaya, E.N., 1961. Flora of Cultivated Plants of the USSR: XIII. Perennial Leguminous Plants. (Part 1, translated 1961.) Israel Program of Scientific Translations, Jerusalem.

Small, E., 1985. Morphological differentiation in *Medicago sativa* s. l. in relation to ploidy. Canadian Journal of Botany. 63:1747–1752.

Small, E., M. Jurzysta, C. Nozzolillo, 1990. The evolution of hemolytic saponin content in wild and cultivated alfalfa (*Medicago sativa*, Fabaceae). Economic Botany. 44:226-235.

Swift, H., 1950. The constancy of desoxyribose nucleic acid in plant nuclei. Proc Natl Acad Sci USA. 36(11): 643–654.

Tuna, M., K.P. Vogel, K. Arumuganathan, K.S. Gill, 2001. DNA content and ploidy determination of bromegrass germplasm accessions by flow cytometry. Crop Science. 41(5):1629.

Wang, J., L. Tian, H. Lee, N.E. Wei, H. Jiang, B. Watson, A. Madlung, T.C. Osborn, R.W. Doerge, L. Comai, Z. J. Chen, 2006. Genomewide nonadditive gene regulation in *Arabidopsis* allotetraploids. Genetics. 172(1): 507-517.

EFFECTS OF VARIABLE NITROGEN SOURCE AND RATE ON LEAF AREA INDEX AND TOTAL DRY MATTER ACCUMULATION IN MAIZE (*Zea mays* L.) GENOTYPES UNDER CALCAREOUS SOILS

AMANULLAH[1], Shahen SHAH[1], Zahir SHAH[2], Shad Khan KHALAIL[1], Amanullah JAN[1], Muhammad Tariq JAN[1], Muhammad AFZAL[2], Habib AKBAR[1], Hamayoon KHAN[1], Hidayat-ur-RAHMAN[3], Khalid NAWAB[4], FARHATULLAH[3], Fida MUHAMMAD[3], Zahid HUSSAIN[1], Khair Muhammad KAKAR[1], Imran KHAN[1]*

[1]Department of Agronomy, [2] Department of Soil & Environmental Sciences, [3]Department of Plant Breeding & Genetics, [4]Department of Agricultural Extension Education & Communication
The University of Agriculture Peshawar, PAKISTAN

**Corresponding author: amanullah@aup.edu.pk*

ABSTRACT

Leaf area index (LAI) and total dry matter are important growth and yield-determining factors in maize (*Zea mays* L.). Field experiment was conducted to investigate the impact of different N-fertilizer sources [urea, calcium ammonium nitrate (CAN) and ammonium sulphate (AS)] and rates (50, 100, 150 and 200 kg ha^{-1}) on mean single leaf area (MSLA), number of leaves plant^{-1} (NLPP), LAI and total dry matter (TDM) of maize genotypes "Local cultivars (Azam and Jalal) vs. hybrid (Pioneer-3025)". The experiment was conducted at the Agriculture Research Farm of The University of Agriculture Peshawar during summer 2008-10. The results indicated that all the experimental plots treated with N produced higher MSLA, NLPP, LAI and TDM than control (N not applied). The MSLA, NLPP, LAI and TDM increased with higher N rates (150 and 200 kg N ha^{-1}). Application of CAN and AS produced higher MSLA, NLPP, LAI and TDM than urea in the 1st year, but no significant differences were observed in the 2nd year. The hybrid (P-3025) produced higher MSLA, NLPP, LAI and TDM than Azam and Jalal. The increase in MSLA and NLPP showed positive relationship with LAI. The TDM increased with increase in LAI. Application of N either as CAN or AS at the rate of 200 kg ha^{-1} was recommended for the high yielding hybrid maize but 150 kg N ha^{-1} was most suitable for the low yielding local cultivars in the study area.

Key words: maize, *Zea mays* L., genotypes, N levels, N source, leaf area index, total dry matter

INTRODUCTION

Leaf area index (LAI) is a measure of leafiness per unit ground area and denotes the extent of photosynthetic machinery. Leaf area index influences the interception and utilization of solar radiation, and consequently growth and yield (Amaullah et al., 2007 and 2008). According to Fageria et al. (2006), LAI is an important yield-determining factor for field crops because LAI is a major determinant of light interception and transpiration. Rapid leaf area expansion is a desirable trait in the early growth stages of cereal crops grown in low-rainfall areas, because it leads to rapid canopy closure, thereby reducing the evaporation from the soil surface, and thus increasing crop water-use efficiency (Richards et al., 2002). In more favorable conditions, fast canopy development will make the crop more competitive with weeds for light interception (Lemerle *et al.*, 2001). Moreover, Boogaard et al. (1996) showed, in a controlled-environment study, that a fast leaf area expansion rate in wheat was positively correlated with above-ground biomass and grain yield. Our previous research on maize (*Zea mays* L.) indicated that increase in nitrogen (N) rate and number of splits led to an increase LAI (Amanullah et al., 2007), light interception (Amanullah et al., 2008); grain yield (Amanullah et al., 2009a), leaf area and total biomass (Amanullah et al., 2009b), net returns (Amanullah et al., 2010a) and dry matter portioning (Amanullah and Shah, 2011). Breadth of the area per leaf profile decreased under high soil nitrogen level and high plant density but leaf area, LAI and grain yield increased with higher rate of N (Oscar and Tollenar, 2006).

Efficient use of N for maize production is important for increasing grain yield, maximizing economic returns and minimizing NO_3 leaching to ground water (Gehl et al., 2005). Site-specific N application to maize is a way of maximizing yield potential while minimizing fertilizer cost (Kahabka et al., 2004). Efficient fertilizer use can be defined as maximum returns per unit of fertilizer applied (Mortvedt et al., 2001). For the efficient management of N in the cropping systems, adequate rate, appropriate source and timing of application during crop growth cycle play an important role (Fageria et al., 2006). Urea, calcium ammonium nitrate (CAN) and ammonium sulfate (AS) urea are the three main N fertilizer sources for crop production in Pakistan. But there is no research work on the profitability of these N-fertilizers sources. However, our recent research work (Amanullah et al., 2012) reported the profitability of different phosphorus phosphatic fertilizers sources. Applications of either DAP (di-ammonium phosphate) or SSP (single super phosphate) was more profitable in terms net returns as compared to nitrophos (NP) and control plots (P not applied). The selection of fertilizers by the growers commonly depends upon price-the least costly fertilizer per kilogram of plant food is the one commonly selected (Plaster, 1992). Though the existing maize genotypes have a high yield potential, soil and climatic conditions of Pakistan are very ideal for its production, yet yield per hectare is very low particularly in the Northwest Pakistan (2000 kg ha^{-1}). The causes of yield gap include injudicious use of N fertilizer and growing of low yielding maize genotypes by the growers. In order to bridge this gap in maize productivity, the package of latest production technology involving the use of high yielding maize hybrids and low cost N fertilizers at appropriate level needs to be find out and used to increase maize productivity and profitability. There is lack of research on the interactive effects of maize genotypes, level and source of N fertilizers. For sustainable maize production, research on the interactive effect of genotypes into N-fertilizer source (NS x G), genotypes into N rates (NR x G) and N source into N rates (NS x NR) is indispensable. This experiment was therefore designed with an objective to investigate impact of different N fertilizer sources applied at various levels on the MSLA, NLPP, LAI and TDM of maize hybrid (Pioneer-3025) in comparison with the two local cultivars (Jalal and Azam) as checks.

MATERIALS AND METHODS

Site description

Field experiment was conducted at the Agriculture Research Farm of the University of Agriculture Peshawar during summer 2008 and 2010. The experimental farm is located at 34.01° N latitude, 71.35° E longitude at an altitude of 350 m above sea level in Peshawar valley. Peshawar is located about 1600 km north of the Indian Ocean and has continental type of climate. The research farm is irrigated by Warsak canal from river Kabul. Soil texture is clay loam, low in organic matter (0.87 %), extractable phosphorus (6.57 mg kg^{-1}), exchangeable potassium (121 mg kg^{-1}), and alkaline (pH 8.2) and is calcareous in nature (Amanullah et al., 2009a). Weather data for the maize growing periods in the two years are given in Table 1.

Table 1. Weather data of maize growing periods in 2008 and 2010 at Peshawar- Pakistan.

Weather Data	Growing season 1st year				Growing season 2nd year			
	July	August	Sept.	Oct.	July	August	Sept.	Oct.
Mean Temperature ^{0}C	31	30	28	25	31	29	24	24
Max Temperature ^{0}C	36	35	34	32	34	33	34	32
Min Temperature ^{0}C	26	25	22	19	26	26	21	19
Precipitation(mm)	37	274	38	1	409	125	4	0
Mean Humidity (%)	66	71	63	60	75	80	63	65

Experimentation

A field experiment was conducted in a randomized complete block (RCB) design with split-plot arrangement using three replications. Factorial experimental treatments were four N (nitrogen) rates [N_1 = 50 kg ha^{-1}, N_2 = 100 kg ha^{-1}, N_3 = 150 kg ha^{-1} and N_4 = 200 kg ha^{-1}] and three N-fertilizer sources [S_1 = Urea {$CO(NH_2)_2$ having 46% N}, S_2 = Calcium Ammonium Nitrate {($CaCO_3.NH_4NO_3$) having 26% N and 10% Ca)} and S_3 = Ammonium Sulphate {(NH_4SO_4) having 21% N and 24% S}] applied to main plots, while three maize genotypes [G_1 = Jalal, G_2 = Azam and G_3 = Pioneer-3025] were kept in sub plots. One control plot (N not applied) was also used in each replication as check. A sub-plot size of 4.2 m by 5 m, having 6 rows, 5 m long and 70 cm apart was used. A uniform basal dose of 60 kg P ha^{-1} as single super phosphate (18% P_2O_5), and 60 kg K ha^{-1} as sulphate of potash (50% K_2O) was applied and mixed with the soil during seedbed preparation. Nitrogen in the form of urea, calcium ammonium nitrate (CAN) and ammonium sulphate (AS) were applied in two equal splits i.e. 50 % at sowing and 50 % at 2nd irrigation (30 days after emergence). Numbers of leaves of five plants in each treatment were counted and then average number of leaves per plant (NLPP) was worked. Then leaf lengths and widths of the three middle leaves of each five plants in each treatment were measured and average length and width was calculated. Data on mean single leaf area (MSLA), and leaf area index (LAI) were calculated according to the following formulae:

Mean single leaf area (MSLA) = *Leaf length x leaf width x 0.75* (cm^2)

Leaf area plant^{-1} (LAPP)= *MSLA x leaves plant^{-1}*(cm^2)

Leaf area index (LAI) = *LAPP x plants m^{-2}* (no unit)

At harvest maturity, the four central rows were harvested; the material was dried up to constant weight and weighed, and then converted into kg ha^{-1} using the following formula:

Total dry matter (TDM) = *Dry matter* m^{-2} *x 10,000 (kg* ha^{-1}*)*

Statistical analysis

Data were subjected to analysis of variance (ANOVA) according to the methods described by Steel et al. (1997), and means between treatments was compared by least significant difference ($P \leq 0.05$). The analysis of variance for the two years data is given in Table 2. The mean values are given in Table 3, while different significant interactions are reported in figures.

Table 2 . Analysis of variance table for statistical analysis of the data during the two years at Peshawar-Pakistan.

Source of variance	Degree of freedom	Probability ($P \leq 0.05$) 1st year				2nd year			
		MSLA cm²	NLPP -	LAI -	TDM kg ha⁻¹	MSLA cm²	NLPP -	LAI -	TDM kg ha⁻¹
Replications	2	-	-	-	-	-	-	-	-
Treatments	12	0.000	0.000	0.000	0.000	0.000	0.000	0.000	0.000
N sources (NS)	2	0.004	0.000	0.000	0.000	0.443	0.551	0.333	0.161
N rates (NR)	3	0.000	0.000	0.000	0.000	0.000	0.008	0.000	0.000
NS x NR	6	0.002	0.073	0.005	0.004	0.000	0.117	0.001	0.152
Control vs. rest	1	0.000	0.000	0.000	0.000	0.000	0.000	0.000	0.000
Error I	24								
Genotypes (G)	2	0.000	0.000	0.000	0.000	0.000	0.000	0.000	0.000
NS x G	4	0.422	0.049	0.167	0.234	0.008	0.843	0.096	0.921
NR x G	6	0.050	0.000	0.000	0.005	0.000	0.064	0.000	0.071
NS x NR x G	12	0.714	0.344	0.695	0.112	0.000	0.209	0.000	0.494
Error II	52								
Total	116								

Where:
MSLA=mean single leaf area in cm^2
NLPP=number of leaves $plant^{-1}$
LAI=leaf area index
TDM=total dry matter in kg ha^{-1}
ns=non significant

Table 3. Control (N not applied) vs. rest (all N applied plots), N rate, N source and genotypes influence on MSLA, NLPP, LAI and TDM in maize grown on calcareous soil (mean effects).

Treatments	1st Year MSLA cm²	NLPP -	LAI -	TDM kg ha⁻¹	2nd Year MSLA cm²	NLPP -	LAI -	TDM kg ha⁻¹
Control	298.89	10.11	1.97	6429	320.33	11.11	2.35	8814
Rest	380.06	13.40	3.38	9962	417.06	13.70	3.79	11379
N rate (kg ha⁻¹)								
50	353.15	11.78	2.72	8575	367.62	13.22	3.18	9729
100	335.11	13.00	2.86	9883	398.00	13.63	3.60	10929
150	418.37	14.48	3.97	10543	450.68	14.07	4.22	12187
200	413.59	14.33	3.96	10846	451.96	13.89	4.16	12669
LSD ($P \leq 0.05$)	**23.48**	**0.43**	**0.22**	**456**	**21.17**	**0.48**	**0.25**	**985**
N Fertilizer Source								
Urea	375.33	13.11	3.26	9472	414.81	13.72	3.78	11050
CAN	400.56	13.89	3.68	10352	423.62	13.81	3.87	11839
AS	364.28	13.19	3.20	10062	412.77	13.58	3.71	11247
LSD ($P \leq 0.05$)	**20.33**	**0.37**	**0.18**	**395**	**ns**	**ns**	**ns**	**ns**
Maize Genotypes								
Azam	349.03	12.13	2.80	8980	361.26	12.28	2.88	10123
Jalal	352.42	12.36	2.85	9651	401.47	12.58	3.30	11173
Pioneer-3025	438.72	15.53	4.48	11254	488.46	16.25	5.19	12841
LSD ($P \leq 0.05$)	**24.18**	**0.38**	**0.22**	**237**	**13.47**	**0.38**	**0.15**	**639**

Where:
MSLA = mean single leaf area in cm^2
NLPP = number of leaves $plant^{-1}$
LAI = leaf area index
TDM = total dry matter in kg ha^{-1}
ns = non significant

RESULTS AND DISCUSSIONS

Year effects

Mean single leaf area (MSLA), number of leaves per plant (NLPP), leaf area index (LAI), and total dry matter (TDM) were higher in 2nd than the 1st year (Table 3). The lower MSLA, NLPP, LAI and TDM of maize in first year could be due to less precipitation (350 mm in 2008) during the time of emergence and early growth stage of maize (Table 1). Moreover, irrigation water was not available in the first year of the experiment during the early growth of maize, and only less irrigation water was applied from the tube well. On the other hand, in the second year, the higher amount of rainfall during the early growth stage of maize (409 mm in July, 2010) had positive impact on the maize that might have been resulted in the higher MSLA, NLPP, LAI and TDM of maize in the second year. Aliu et al. (2009) in their field experiments found the highest value of LAI for maize inbred L7 (3.9) in 1st year and 4.0 in 2nd year, whereas the lowest LAI was noted for maize inbred L10 (2.52) in 1st year, while in 2nd year the lowest LAI (2.6) was obtained for inbred L8. Amanullah et al. (2007) found significant increase in LAI of maize "Azam" in 2002 (4.22) as compared to LAI in 2003 (3.79) because of the fluctuation in the precipitation of the two years. Lloyd et al. (1997) noted that increase in yield with N application at all sites. But at three sites (B6, B8 and C7), yield was greater with ammonium nitrate, while at two sites (B9 and C13) yield was greater with application of urea. Wang et al. (2007) earlier reported that understanding concepts of ideal soil fertility level and response to nutrient management provide practical guidelines for improving nutrient management under the variable rainfall conditions. While, Okalebo et al. (2006) suggested that site specific recommendations are needed for maize because of its differential response to nutrient inputs which varied widely within and across agro-ecological zones. According to Harold et al. (2006) maize N availability varied greatly from year to year based on weather conditions.

Nitrogen rates effects

Nitrogen rates had significant effects on all parameters under study in both years (Table 2). All the experimental plots treated with N had higher LAI than the control plots (N not applied) in both years (Table 3). Nitrogen application on N deficient soil increased both MSLA and NLPP of maize that resulted in the higher LAI as well as TDM. In both years, the MSLA, NLPP, and LAI increased significantly ($P \leq 0.05$) with increase in N rate up to 150 kg N ha^{-1} and further increase in N up to the highest 200 kg N ha^{-1} did not increased the parameters under study (Table 3). However, the highest TDM was obtained with the highest rate of 200 kg N ha^{-1} in both years. Although, the differences in TDM at 150 and 200 kg N ha^{-1} in both years were significantly not different from each other. Similarly, the differences in MSLA at 150 and 200 kg N ha^{-1} in 2nd year were significantly not different from each other. Our previous published research (Amanullah et al., 2007; Amanullah et al., 2009b) indicated that on N deficient soils in Northwestern Pakistan, the MSLA, NLPP, LAI and TDM in maize increased with application of higher N rates (120 and 180 kg ha^{-1}) as compared to the lower N rate (60 kg ha^{-1}). The increase in N level increased LAI (Amanullah et al., 2007) and showed positive relationship with increase crop growth rate and light interception (Amanullah et al., 2008). The increase in NLPP and leaf area $plant^{-1}$ increased TDM in maize (Amanullah et al., 2009b).

Nitrogen source effects

Nitrogen sources had significant effects on all parameters in 1st year but had no significant effects in 2nd year (Table 2). Application of CAN (calcium ammonium nitrate) produced significantly higher MSLA, NLPP, LAI and TDM than urea and AS (ammonium sulphate) in the 1st year (Table 3). The higher MSLA, NLPP, LAI and TDM produced with CAN than other two fertilizers in 1st year probably may be due to the less rainfall (345 mm) because CAN was better N-fertilizer source under moisture stress condition than other two N-fertilizers. In the 2nd year, the higher rainfall (538 mm) probably had the same positive impact on the performance of all three N-fertilizers sources and so the differences in all the parameters under study were statistically not different from each other. According to Chien et al. (2011), AS is the best N-fertilizer source which contains frees sulfur and had many potential agronomic and environmental benefits over urea and ammonium nitrate. In Northwestern Pakistan where most of the soils are calcareous in nature, AS because of its free sulfur content, could be the most beneficial N-fertilizer in terms of reducing soil pH and increasing seed oil content (data not shown). However, because of its (AS) highest N cost (1.43 and 1.91 USD kg^{-1} N in 1st and 2nd year, respectively; one USD = 100 PKR) as compared to other sources of N i.e. CAN (0.77 and 0.97 USD kg^{-1} N in 1st and 2nd year, respectively) and urea (0.44 and 0.56 USD kg^{-1} N in 1st and 2nd year, respectively) the poor maize growers in the region can't afford to use AS. Moreover, the transportation charges of AS was more than urea and CAN. Lloyd et al. (1997) reported that urea (200 USD per ton) is a less expensive form of N fertilizer than ammonium nitrate (260 USD per ton). However, urea has been considered to be less effective than other N fertilizers, due to N loss by ammonia volatilization, especially when used on soils of high pH or low CEC (Terman, 1979). Our previous research work (Amanullah et al., 2010b) on maize regarding the MSLA, NLPP and LAI response to phosphorus (P) fertilizers sources indicated that plots applied either with DAP (Di-ammonium phosphate) or SSP (single super phosphate) were better in terms of LAI and TDM accumulation as compared to the application of nitrophos (NP) and control plots (P not a). The increase in the LAI and TDM in maize was attributed to the increase in the leaf area with application of SSP or DAP as compared to smaller leaf area with application of NP (nitrophos) and in the zero-P control plots. Lloyd et al. (1997) noted that increase in yield with N application at all sites. But at three sites (B6, B8 and C7), yield was greater with ammonium nitrate, while at two sites (B9 and C13) yield was greater with application of urea. In some area with higher soil pH, AS because of its residual acidity, has been discontinued, and therefore not recommended (Kurtz, 2004). The advantages of AS include its low hygroscopicity and source of both N (21%) and S (24%), and strongly acid forming reaction in soil can be advantageous in high pH soils and can be more economical N fertilizer source where S is required (Halvin et al., 2009). In some other areas, the most widely used N source was CAN, due to its very low residual acidity, and calcium content (10 %) which particularly, in the savanna areas helps to neutralize soil acidity (Sas, 2006). Amanullah et al. (2013) reported that all foliar application of various N-sources (urea, CAN, and AS) had significantly produced higher biomass yield (TDM) than control (water spray only). But among the N source, urea (9283 kg ha^{-1}) and AS (9254 kg ha^{-1}) produced higher biomass yield than CAN (8703 kg ha^{-1}) and control (7973 kg ha^{-1}).

Genotypes effects

The maize hybrid (P-3025) had significantly ($P \leq 0.05$) higher MSLA, LAPP, LAI and TDM than the two local cultivars (Azam and Jalal) in both years (Table 3). Between the two local cultivars, Jalal had relatively higher MSLA, NLPP, LAI and TDM than Azam, but the differences were statistically not different. The possible reason of higher LAI and TDM in case of P-3025 was attributed to its more MSLA and NLPP. Moreover, the P-3025 had delayed maturity, taller plants and higher leaf area $plant^{-1}$ than the two local cultivars (unpublished data). This two years study confirmed that maize hybrid was more efficient in terms of LAI and TDM accumulation because its more NLPP, higher MSLA, leaf area $plant^{-1}$ and its more DM partitioning than the local cultivars. Difference in the LAI of maize genotypes were earlier reported by Aliu et al. (2009) and Subedi and Ma (2005). Fageria et al. (2008) suggested that in the 21st century, nutrient efficient plants will play a major role in increasing crop yields compared to the 20th century, mainly due to limited land and water resources available for crop production, higher cost of inorganic fertilizers, declining trends in crop yields globally, and increasing environmental concerns. Baligar et al. (2001) reported that efficiency of acquisition, transport and utilization of nutrients varies with crop species and genotype/cultivar within species, and their interactions with the environment. Differential response of maize genotypes regarding leaf area index has also been reported earlier by Azadgoleh and Kazmi (2007) and Luque *et al.* (2006). Differences in N use efficiency, harvest index, economic yield and net returns in different maize genotypes was also reported by Mkhabela et al. (2001) and (Hokmalipour et al., 2010). The cultivar Korduna which had the highest harvest index, kernels number/row, kernels number/ear and weight of 1000 kernels gained the highest NUE. In Brazil, Carvalho et al. (2012) concluded from their study on 21 maize genotypes, the three promising genotypes viz. GEN 03, GEN 10 and GEN16 were considered the most efficient genotypes in terms of agronomic N use efficiency and yield.

Nitrogen source x rates

Interaction of NS x NR (N source into rates) had significant effects ($P \leq 0.05$) on MSLA, in both years (Table 2). The MSLA increased to maximum with CAN at N1, and with urea at N3 in both years (Fig 1a). At N4, the MSLA increased with AS in both years, while CAN in the 1st year (Fig 1a). The NLPP were not affected by NS x NR in both years (Table 2) which is clear from Fig 1b. Interaction of NS x NR had significant effects ($P \leq 0.05$) on LAI in both years (Table 2). Likewise MSLA, the LAI increased with CAN at N1 in both years (Fig 1c). Urea and CAN increased LAI than AS when applied at N4; on the other hand, urea and CAN were best in terms of LAI at N3 (Fig 1c). The TDM varied significantly only in the first year but not in the second year (Table 2). In year one, CAN and AS had higher TDM than urea at N1 and N2, and CAN at N3 and N4 (Fig 1d). The increase in TDM with application of CAN and AS may be due to the increase in MSLA and LAI. Abayomi et al. (2006) reported that N application had greater improve leaf growth and hence yield of maize. Fageria et al. (2011) found that the higher and lower N rate of AS produced higher LAI and most of the plant growth and yield components, while the intermediate N rates (125 to 275 mg N kg^{-1}) of urea was slightly better compared to AS for grain production. The discrepancies in our results and that of Fageria et al. (2011) was due to the differences in the crop species used, amount of N sources applied, and the climatic conditions.

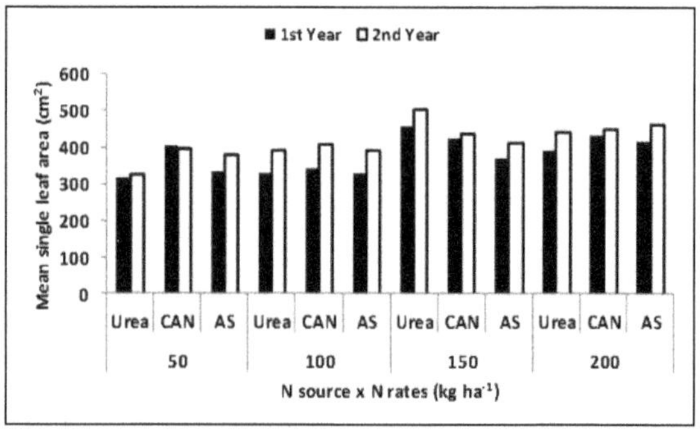

Figure 1a. Interactive effects of N source x N rates on the mean single leaf area (cm^2) of maize grown on calcareous soil

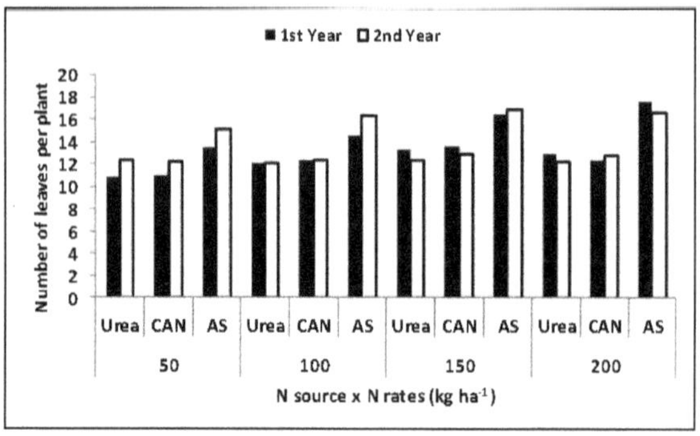

Figure 1b. Interactive effects of N source x N rates on the number of leaves $plant^{-1}$ of maize grown on calcareous soil

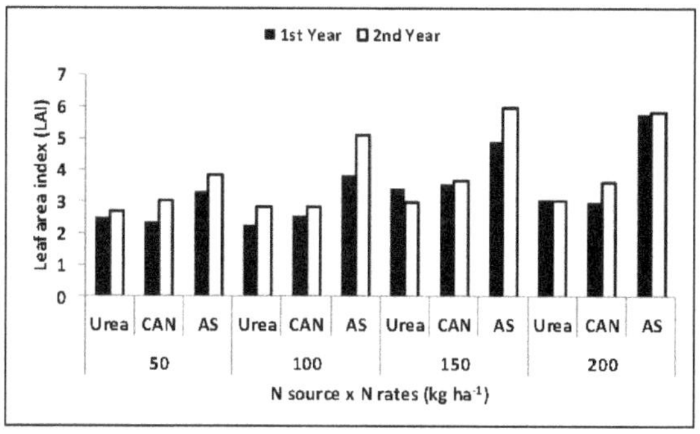

Figure 1c. Interactive effects of N source x N rates on the leaf area index of maize grown on calcareous soil

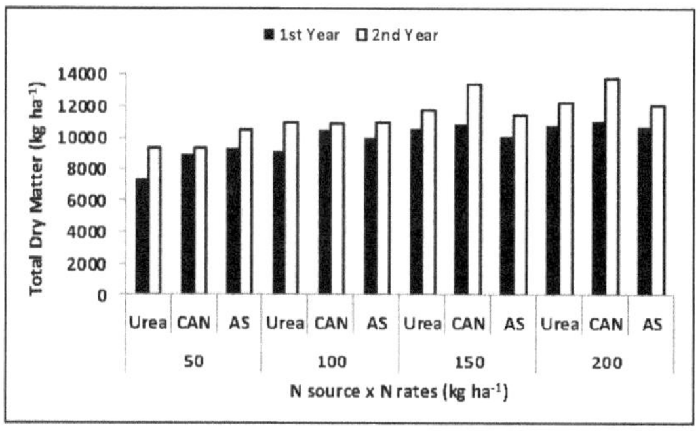

Figure 1d. Interactive effects of N source x N rates on the total dry matter (kg ha^{-1}) of maize grown on calcareous soil

Nitrogen Rates x Genotypes

Interaction of NR x G (N rates into genotypes) had significant effects ($P \leq 0.05$) on MSLA in both years (Table 2). The MSLA increased to maximum in genotype (P3025) at all four levels of N as compared with the two local cultivars (Fig 2a). The MSLA in the local cultivars increased up to N3 and further increase in N decreased it. On the other hand, the MSLA increased with N4 than N3 in year one, but it was statistically at par with each other in year two when applied with N3 or N4 (Fig 2a). The NLPP were significantly affected by NR x G in year one but not in year two (Table 2). In year one, the NLPP in the local cultivars increased up to N3 and further increase in N decreased it. On the other hand, the NLPP increased significantly with N4 than N3 in the 1st year, but it was statistically the same with each other in the 2nd year when applied with N3 or N4 (Fig 2b). Interaction of NS x NR had significant effects ($P \leq 0.05$) on LAI in both years (Table 2).

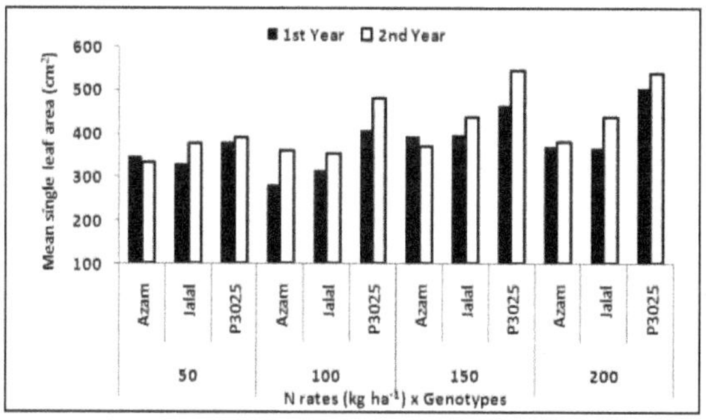

Figure 2a. Interactive effects of N rate x genotypes on the mean single leaf area (cm^2) of maize grown on calcareous soil

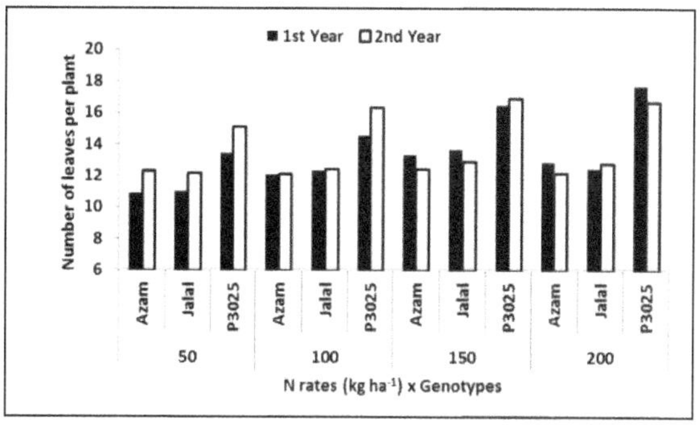

Figure 2b. Interactive effects of N rate x genotypes on the number of leaves plant^{-1}of maize grown on calcareous soil

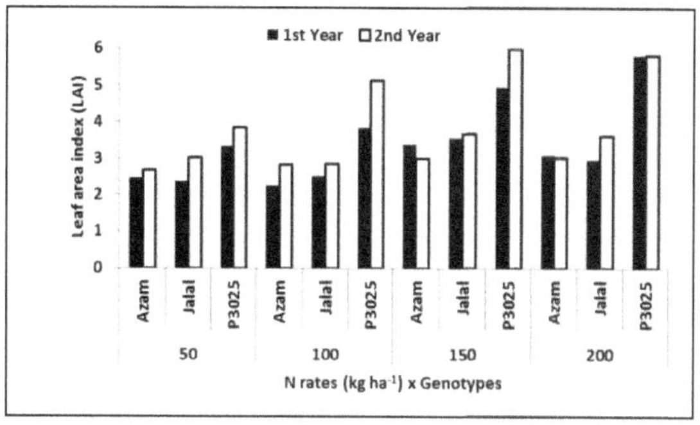

Figure 2c. Interactive effects of N rate x genotypes on the leaf area index of maize grown on calcareous soil

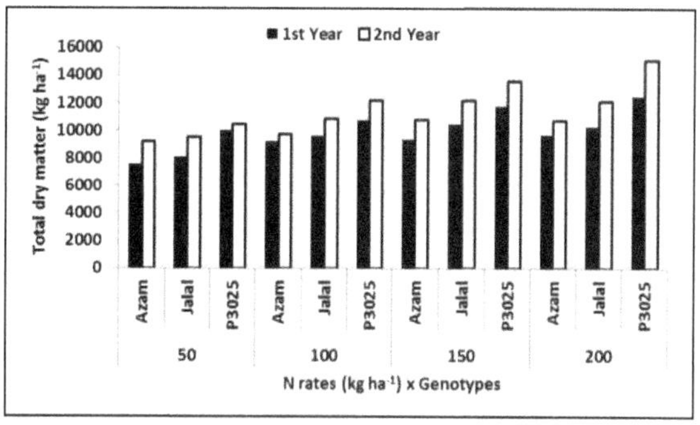

Figure 2d. Interactive effects of N rate x genotypes on total dry matter (kg ha^{-1}) of maize grown on calcareous soil

Likewise MSLA and NLPP, the LAI increased to maximum in genotype (P3025) at all four levels of N as compared with the two local cultivars (Fig 2c). The LAI in the local cultivars was highest at N3 and further increase or decrease in N decreased LAI. In the 1st year, the LAI in P3025 increased significantly wit increase in N rate up to N4, but in the second year the LAI in P3025 increased up to N3 and then decreased with further increase in N up to N4 (Fig 2c). Interaction of NS x NR had significant effects ($P \leq 0.05$) on TDM in year one and not in year two (Table 2). The TDM was generally higher in the 2nd year than 1st year and showed positive relationship with MSLA, NLPP and LAI. In the 1st year, the P3025 had higher TDM than the two local cultivars at all four levels of N. At the two lower N rates (N1 and N2) the TDM of the two local cultivars was almost the same, but at the two higher rates (N3 and N4) the TDM of Jalal was greater than Azam (Fig 2d). Frank et al. (2004) found a significant cultivar × N rate interaction at all sites for turfgrass visual quality and color, and that these ratings decreased with decrease in N rates. Amanullah et al. (2012) reported that the higher average yield in the Punjab Province than Khyber Pakhtunkhwa Province in Pakistan is due to the use of hybrid maize and efficient fertilizer use by the farmers in Punjab than Khyber Pakhtunkhwa. Wu et al. (2010) reported that the higher rates of N were not necessary for the tall fescue [*Schedonorus phoenix* (Scop.) Holub] that resulted in increased nitrate leaching.). In Indonesia, among the four genotypes tested showed significant interaction of genotypes into N rates on N uptake efficiency was observed and the highest N uptake efficiency was attained by Bisma, followed by Pioneer and Arjuna, while Madura local had the least N uptake efficiency in all rates of N fertilizer (Rahman and Koentjoro, 2011). Thus, genotypes with differences in yield potential may have differences in N accumulation and NUE (Sinclair and de Wit, 1975). Abayomi et al. (2006) found that N rates had greater impact on the yield of maize than genotypes effects. They reported that yield increased while increasing N rates among all maize genotypes, and the increase was more due to the increase in leaf growth with increasing N rates.

Nitrogen source x genotypes

Interaction of NS x G (N source into genotypes) had significant effects ($P \leq 0.05$) on MSLA in the 2nd year and not in the 1st year (Table 2). The MSLA increased to maximum in genotype (P3025) at all three sources of N as compared with the two local cultivars (Fig 3a). In the 2nd year of the experiment, the MSLA in the P3025 was statistically the same with each other at all three sources of N. The local cultivar "Jalal" had higher MSLA than "Azam" when applied with CAN. There were no significant differences in the MSLA of Azam and Jalal with application of urea and AS (Fig 3a). Amanullah et al. (2013) reported that foliar application of various N-sources (urea, CAN, and AS) had no significant effect on the MSLA of maize. In contrast to MSLA, the NLPP were significantly affected by NR x G in the 1st year but not in the 2nd year (Table 2). The NLPP increased to maximum in genotype (P3025) at all three sources of N as compared with the two local cultivars (Fig b. The two local cultivars "Jalal and Azam" produced similar NLPP at all three sources of N (Fig 3b). Interaction of NS x G (N source into genotypes) had no significant effects ($P \leq 0.05$) on LAI in both years (Table 2). The LAI increased to the highest level in genotype (P3025) at all three sources of N as compared with the two local cultivars (Fig 3c).

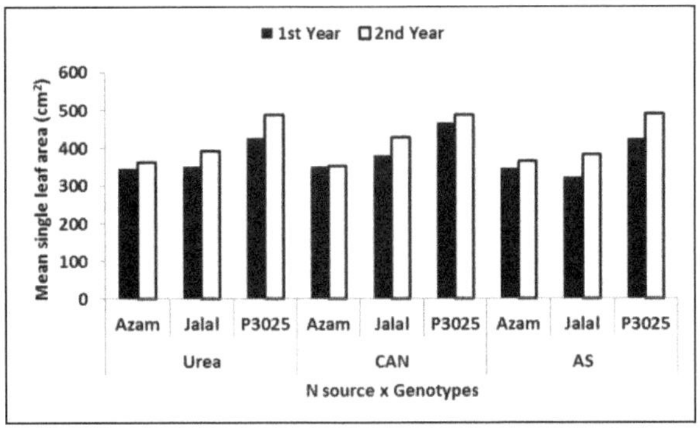

Figure 3a. Interactive effects of N source x genotypes on the mean single leaf area (cm^2) of maize grown on calcareous soil

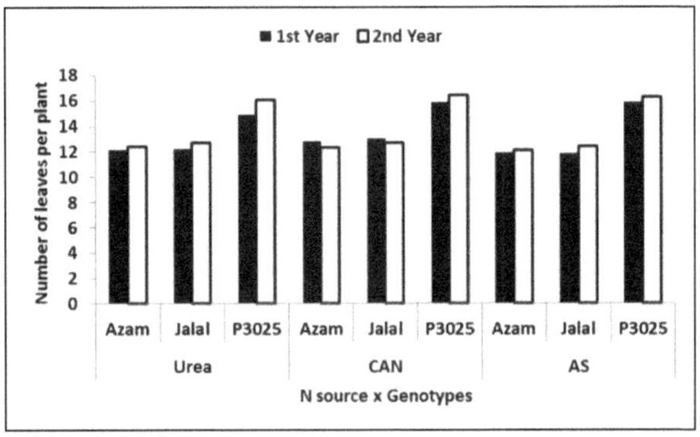

Figure 3b. Interactive effects of N source x genotypes on the number of leaves $plant^{-1}$ of maize grown on calcareous soil

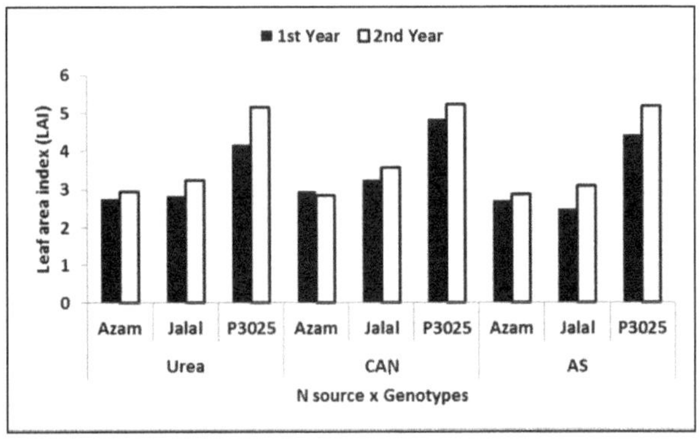

Figure 3c. Interactive effects of N source x genotypes on the leaf area index of maize grown on calcareous soil

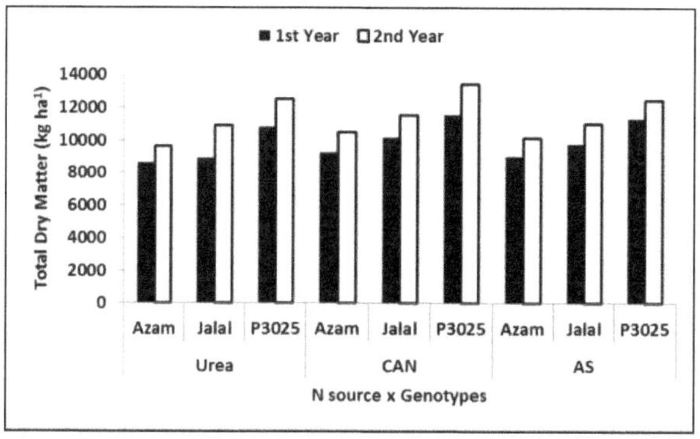

Figure 3d. Interactive effects of N source x genotypes on the total dry matter (kg ha^{-1}) of maize grown on calcareous soil

In the 2nd year of the experiment, all the genotypes produced higher LAI with different N sources than in 1st year (Fig 3c). Interaction of NS x G (N source into genotypes) had no significant effects ($P \leq 0.05$) on TDM in the two years (Table 2). However, the TDM increased to the highest level in genotype (P3025) at all three sources of N as compared with the two local cultivars (Fig 3d). The TDM was generally higher in the 2nd year than in 1st year (Fig 3d). Frank et al. (2004) found a significant cultivar × N rate interaction at all sites for turf grass visual quality and color, and that these ratings decreased with decrease in N rates. Wu et al. (2010) reported that among the N sources, slow-release N fertilizers were better for the tall fescue because of less nitrate leaching than the fast-release N sources. The higher TDM in the hybrid maize in this study was attributed to its higher LAI than the two local varieties. Amanullah et al. (2008) reported that the higher LAI in maize intercepted more light and increased the biomass in maize.

CONCLUSIONS

The results of this study confirmed the existence of significant variability in MSLA, NLPP, LAI and TDM accumulation among different maize genotypes depending on the base of their inheritance, differences in N-fertilizer source and rates as well as differences in the weather conditions in the two years. The decline in price of AS could make it most profitable and beneficial N-fertilizer source on calcareous soils with high pH because its free available sulfur (24 % S) is important for improving crop growth, increasing LAI and TDM. The decline in CAN price could also make it more beneficial N fertilizer source for improving growth and yield on acidic soils and/or soils having Ca deficiency. However, the favorable manufacturing, handling, storage, low transportation charges, marketing, easy availability and low N cost make urea the most competitive and beneficial N fertilizer source in terms highest net returns (unpublished data). Growing maize hybrid could increase maize productivity, grower's income, and N use efficiency (unpublished data) due to its higher MSLA, NLPP, LAI and TDM. The higher LAI of maize hybrid in this study was attributed to increased MSLA and NLPP indicating that leaf characteristics are the most important criteria for selection maize genotypes for higher LAI and TDM accumulation. The variation LAI and TDM in the two years of studies was attributed to the fluctuation in weather conditions. This problem poses a challenge for the development of technical recommendations targeted for diverse environments. In the light of the research results, ammonium sulphate was considered the most suitable N-fertilizer source for improving growth and total dry matter of maize under calcareous soils. Application of N at the rate of 200 kg ha^{-1} was recommended for the high yielding hybrid maize but 150 kg N ha^{-1} was most suitable for the low yielding local cultivars in the study area.

ACKNOWLEDGEMENTS

Financial support from The University of Agriculture Peshawar is greatly acknowledged for sponsoring this project. I am also highly thankful to Prof. Dr. Paigham Shah (retired) for the statistical analysis of the data.

LITRATURE CITED

Abayomi, Y., A. Arijenja, and I.A. Kolawole. 2006. Comparative leaf growth and grain yield reponses of hybrid and open pollinated maize genotypes to nitrogen fertilizers application. Agrosearch. 8(1): 13-25.

Aliu, S., S. Fetahu, A. Salillari, and S. Kaçiu. 2009. Genetic Variability of some Maize Inbred Lines (*Zea mays* L.) in Agroecological Conditions of Kosovo. Not. Bot. Hort. Agrobot. Cluj 37: 209-214.

Amanullah, A.Z. Khan, A. Jan, and Z. Shah. 2013. Foliar application of nitrogen at different growth stages influences the phenology, growth and yield of maize (*Zea mays* L.). Soil & Environment. 32(2): 135-140.

Amanullah, M. Asif, and L.K. Almas. 2012. Agronomic efficiency and profitability of P-fertilizers applied at different planting densities of maize in northwest Pakistan. J. Plant Nutr. 35: 331-341.

Amanullah and P. Shah. 2011. Nitrogen rates and its time of application influence dry matter partitioning and yield in maize planted at low and high densities. J. Plant Nutr. 34: 224-242.

Amanullah, L.K. Almas and P. Shah. 2010a. Timing and rate of nitrogen application influence profitability of maize planted at low and high densities in Northwest Pakistan. Agron. J. 102: 575-579.

Amanullah, M. Asif, K. Nawab and Z. Shah. 2010b. Impact of planting density and p-fertilizer source on the growth analysis of maize. Pak. J. Bot. 42: 2349-2357.

Amanullah, R.A. Khattak and S.K. Khalil. 2009a. Effects of plant density and N on phenology and yield of maize. J. Plant Nutr.. 32: 245-259.

Amanullah, K.B. Marwat and P. Shah. 2009b. Nitrogen levels and its time of application influence leaf area, height and biomass of maize planted at low and high density. Pak. J. Bot. 41: 761-768.

Amanullah, H. Rahman, Z. Shah and P. Shah. 2008. Effects of plant density and N on growth dynamics, light interception and yield of maize. Arch. Agron. Soil Sci. 54: 401-411.

Amanullah, K. Nawab and A. Ali. 2007. Response of specific leaf area (SLA), leaf area index (LAI) and leaf area ratio (LAR) of maize (*Zea mays* L.) to plant density, rate and timing of nitrogen application. World Appl. Sci. J. 2: 235-243.

Azadgoleh, M.A.E., and Z. Kazmi. 2007. A study of the planting pattern and density effects on yield and physiological growth parameters in two corn cultivars. Ecology, Envi. Conser. 13: 467-472.

Baligar, V. C., N. K. Fageria, and Z. L. He. 2001. Nutrient use efficiency in plants. Comm. Soil Sci. Plant Anal. 32: 921–950.

Boogaard, V.R., E.J. Veneklaas, J. Peacock, and H. Lambers. 1996. Yield and water use of wheat (*Triticum aestivum* L.) cultivars in a Mediterranean environment: effects of water availability and sowing density. Plant Soil. 181: 251–262.

Carvalho, E.V., F.S. Afférri, J.M. Peluzio, M.A. Dotto, and L.L. Cancellier, 2012. Nitrogen use efficiency in corn (*Zea mays* L) genotypes under different conditions of nitrogen and seeding date. Maydica. 57: 43-48.

Chien, S.H., M.M. Gearhart and S. Villagarcı. 2011. Comparison of ammonium sulfate with other nitrogen and sulfur fertilizers in increasing crop production and minimizing environmental impact: a review. Soil Sci. 176: 327-335.

Fageria, N.K., A. Moreira, and A.M. Coelho. 2011. Yield and yield components of upland rice as influenced by nitrogen sources. J. Plant Nutr. 34: 361-370.

Fageria, N.K., V.C. Baligar and Y. C. Li. 2008. The role of nutrient efficient plants in improving crop yields in the twenty first century. J. Plant Nutr. 31: 1121-1157.

Fageria, N. K., V. C. Baligar, and R. B. Clark. 2006. Physiology of Crop Production. New York: The Haworth Press.

Frank, K.W., R.E. Gaussoin, T.P. Riordan, R.C. Shearman, J.D. Fry, E.D. Miltner, and P.G. Johnson. 2004. Nitrogen rate and mowing height effects on turf-type buffalo grass. Crop Sci. 44:1615-1621.

Gehl, R. J., J. P. Schmidt, L. D. Maddux, and W. B. Gordon. 2005. Corn yield response to nitrogen rate and timing in sandy irrigated soils. Agron. J. 97: 1230-1238.

Havlin, J.H., J.D. Beaton, S.L. Tisdale, and W.L. Nelson, 2009. *Soil Fertility and Fertilizers*. 7th Edition, PHI Learning, Private Limited, New Delhi, India.

Harold, M.V.E., B.K.J. Sogbedji, J. Melkonian, R.S. Dharmakeerthi, H. Dadfar, and I. Tan. 2006. Nitrogen Management under Maize in Humid Regions: The 18th World Congress of Soil Science held on July 9-15, 2006) at Philadelphia, Pennsylvania, USA.

Hokmalipour, S., M.S. Janagard, M.H. Darbandi, F.P. Ashenaee, M. Hasanzadeh, M.S. Seiedi, and R. Shabani, 2010. Comparison of agronomical nitrogen use efficiency in three cultivars of corn as affected by nitrogen fertilizer levels. World Appl. Sci. J. 8: 1168-1174.

Kahabka, J. E., H. M. Van, E. J. McClenahan, and W. J. Cox. 2004. Spatial Analysis of Maize Response to Nitrogen Fertilizer in Central New York. Prec. Agric. 5: 463-476.

Kurtz, C.O., 2004. The influence of organic manure on growth and yield components of maize in South Asia. Crop Sci. Res. 20: 421 – 426.

Lemerle, D., G.S., Gill, C.E. Murphy, S.R. Walker, R.D. Cousens, S. Mokhtari, S.J. Peltzer, R. Coleman, and D.J. Luckett. 2001. Genetic improvement and agronomy for enhanced wheat competitiveness with weeds. Australian J. Agric. Res. 52: 527–548.

Luque, S.F., A.G. Cirilo and M.E. Otegui. 2006. Genetic gains in grain yield and related physiological attributes in Argentine maize hybrids. Field Crops Res. 95: 383-397.

Lloyd, A., J. Webb, J. R. Archer and R S. Bradley. 1997. Urea as a nitrogen fertilizer for cereals . The J. Agric. Sci. 128: 263-271.

Mkhabela, M.S., M.S. Makhabela and P. Shikhulu. 2001. Response of maize cultivars to different levels of N application in Swaziland. Proceedings of 7th Eastern and Southern Africa Regional Maize Conference. 11th to 15th February, 2001. pp: 377-381.

Mortvedt, J. J., L. S. Murphy, and R. H. Follett. 2001. Fertilizer technology and application. Willoughby,OH: Meister Publishing Co.

Okalebo, J.R., C.O. Othieno P.L. Woomer, N.K. Karanja, J.R.M. Semoka, .A. Bekunda, D.N. Mugendi, R.M. Muasya, A. Bationo and E.J. Mukhwana. 2006. Available technologies to replenish soil fertility in East Africa Nutr. Cycl. Agro. 76: 153-170.

Oscar, R.V., and M. Tollennar, 2006. Effect of genotype, nitrogen, plant density and row spacing on the area-per-leaf profile in maize. Agron. J. 98: 94-99.

Plaster, E. J. 1992. Soil Science and Management, New York: Delmar Publishers Inc.

Rahman, M. and Y. Koentjoro, 2011. Genetic variation and agronomic traits associated with nitrogen use efficiency in maize. J. Nature Stud. 9: 47-52.

Richards, R.A., G.J. Rebetzke, A.G. Condon, and A.F. van Herwaarden. 2002. Breeding opportunities for increasing the efficiency of water use and crop yield in temperate cereals. Crop Sci. 42: 111–121.

Sas, K.M. 2006. Effects of different N sources on growth and yield of sweet potato in the tropics. *Fertilizer and Soil Fertility Research*.6: 321–326.

Sinclair, T.R. and C.T. de Wit, 1975. Photosynthate and nitrogen requirements for seed production by various crops. Sci. 189: 565-567.

Steel, R.G.D., J.H. Torrie and D.A. Deekey, 1997. *Principles and Procedures of Statistics: A Biometrical Approach*, 3rd edition. McGraw Hill Book Co. Inc. New York.

Subedi, K.D., and B.L.Ma. 2005. Ear position, leaf area and contribution of individual leaf to grain yield in conventional and leafy maize hybrids. Crop Sci. 45: 2246-2257.

Terman, G. L. 1979. Volatilization losses of nitrogen as ammonia from surface-applied fertilizers, organic amendments, and crop residues. Adv. Agron. 31: 189-223.

Wang, X., W.B. Hoogmoed, D. Cai, U.D. Perdok and O. Oenema. 2007. Crop residue, manure and fertilizer in dryland maize under reduced tillage in northern China: II nutrient balances and soil fertility. Nutr. Cycl. Agro. 79: 17-34.

Wu, L., R. Green, G. Klein, J.S. Hartin and D.W. Burger. 2010. Nitrogen source and rate influence on tall fescue quality and nitrate leaching in a Southern California lawn. Agron. J. 102: 31-38.

NONPARAMETRIC STABILITY ANALYSIS IN MULTI-ENVIRONMENT TRIAL OF CANOLA

S.M.Mahdi MORTAZAVIAN, Shiva AZIZI-NIA*

Department of Agronomy Sciences and Plant Breeding, College of Aburaihan, University of Tehran, Pakdasht, IRAN
** Corresponding author: mortazavian@ut.ac.ir*

ABSTRACT

Rapeseed is the world's second most important source of vegetable oils. Development of genotypes having high seed yield with stable performance is of paramount importance. In the present investigation seventeen genotypes were grown at seven locations during two growing seasons in semi-cold regions of Iran. Data recorded on seed yield were subjected to different nonparametric measures which do not require distributional assumptions. Results of nonparametric tests of G, E and GE interaction and a combined ANOVA across environments showed there were both cross over and non-cross over interactions for G and E and only non-cross over type for GE interaction. In this study, high values of Top (proportion of environments in which a genotype ranked in the top third) and mean of rank were associated with high mean yield. However Rank-sum measure was successful to detect genotypes showing simultaneous high yield and yield stability. Cluster analysis and principal component (PC) analysis help to group genotypes and measures and they revealed association among different statistics. Finally, among nonparametric measures, Top, $S_i^{(1)}$ and Rank-sum statistics of nonparametric procedures were found to be useful in detecting the stability of the genotypes. According to these parameters Geronimo was found as stable and high yield canola genotype.

Keywords: ***Brassica napus*, Multi-environment trial, nonparametric measures.**

INTRODUCTION

Seeds from rapeseed (*Brassica napus* L.) are used for oil extraction, which makes rapeseed the world's second most important source of vegetable oils after soybean (FAO,2012).During these years, the importance of rapeseed has significantly increased in Iran, mainly due to the demand of oil and some transport sanctions. Worldwide canola production is approximately 50 million tons and covers a total area of 27 million hectares (FAOSTAT, 2010). The total area of canola in Iran in the 2010-2011 season was 93000 ha which produced 190000 tons rapeseed (Agricultural Ministry Statistics, 2011). Crops are largely influenced by climate conditions during growing season; hence, even minor climate changes may cause highly variation on crop yield. The environment is the all surrounding conditions including location and year. Any difference in location properties such as soil texture, pH, depth, organic matter, and so on affect cultivar growth and development. The genotype × environment interaction (GEI) is the response of each cultivar to variations in the environment Crossa et al. (1991). The GE interaction has been one of the principal subjects of study in breeding, allowing the generation of different methodologies for genetic improvement. It is perquisite to evaluate a cultivar behavior during different years and locations to find out cultivars with general or specific adaptation before release or any recommendation. There are two major parametric and non-parametric approaches to study GE interaction and determining adaptation of genotypes (Huehn, 1996). A parametric approach relies on large number of assumptions about genotypic, environmental, and GE interaction like normal distribution of errors and interaction effects. Any change in these assumptions may give inaccurate results (Huehn, 1990a). Nonparametric Stability Analysis (NPSA) based on ranks are far from any specific assumptions and provide an alternative useful approach (Huehn, 1990a). These methods provide a lot of flexibility for plant breeders for the simultaneous selection for both mean yield and stability (Karimizadeh et al., 2012). The nonparametric statistics are easy to use and appears to be able to extract a large portion of the GE interaction. Meanwhile, interpretation of GE interaction and yield stability is feasible in different crops using nonparametric strategy. Several nonparametric measures and later modifications proposed till now (Hühn, 1979; Hühn and Nassar, 1989; Kang and Pham, 1991; Lu, 1995; Nassar and Huehn, 1987; Thennarasu, 1995). These parameters are mainly based on the ranks, r_{ij}, or adjusted ranks, r^*_{ij}, of genotypes in each environment. The major criterion to classify a genotype as stable is similar rankings across environments. Some measures consider only stability of genotypes like $S_i^{(3)}$ and $S_i^{(4)}$ while others like $S_i^{(1)}$, $S_i^{(2)}$ and Kang's (1988) rank-sum combine both yield and stability to propose an ideal genotype (Kang and Pham, 1991). In practice, the ideal cultivar would be the one with high seed yield and high stability when evaluated across different environments (Yan et al., 2007). Thennarasu (1995) proposed the four non-parametric stability measures, $NP_i^{(1)}$, $NP_i^{(2)}$, $NP_i^{(3)}$ and $NP_i^{(4)}$ based on adjusted rank calculated from adjusted mean yield ($x_{ij}^* = x_{ij} - \bar{x}_{i.}$). In this case, the rank removes the effect of the environment (E) and genotypes (G) and is only depend on GE and error effects. There are two different types of genotype × environment interaction, conceptually (Baker, 1988). Cross over (or quantitative) type of interaction referred to change in rank orders of genotypes across environments while non-cross over (or qualitative) interaction referred to difference in rank amounts of genotypes. The breeders are mainly interested in the former type. Some procedures have been developed to reveal the type of GE interaction (Azzalini and Cox, 1984; Bredenkamp, 1974; Hildebrand, 1980; Kroon and Laan, 1981; Kubinger, 1986). In the methods of Bredenkamp (1974), Hildebrand (1980) and Kubinger (1986), interactions are defined as deviations from the additivity of main effects for genotypes and environments. In the other hand, using these methods non-cross over interactions are defined. The methods of Kroon and Laan (1981) and Azzalini and Cox (1984) define interactions according to the cross over interactions model. So, using these methods cross over GEI is evaluated. Since nonparametric test is robust to ANOVA assumptions results obtained from these methods can be more reliable (Truberg and Hühn, 2002). Association among stability measures helps breeder to choose the best and most informative method obtaining reliable prediction of cultivar behaviors and also to fit the statistic and dynamic concepts of stability (Sabaghnia et al., 2006).

The objectives of this study were to (i) identify canola genotypes that have both high yield and stability across different test environments of Iran's cold and semi cold areas, (ii) apply nonparametric tests to investigate the cross over and non-cross over GE interaction in multi-environment trials, and (iii) study the relationships among different nonparametric stability statistics.

MATERIALS and METHODS

Plant material

17 winter rapeseed cultivars were used as plant material. Cultivars were grown in 14 environments (seven locations during two growing seasons). The experimental layout was randomized complete block design with four replications in each environment. The seed was sown in four rows, 5 m long, with a between-row spacing of 30 cm. All agricultural practices were applied based on recommended protocols for all locations. Grain yield of each cultivar was recorded on a plot basis.

Nonparametric Stability Analysis (NPSA)

A parametric combined analysis of variance (ANOVA) (F-test) and three non-parametric statistical procedures including Kroon and Laan (1981), Bredenkamp (1974) and Hildebrand (1980) were used to test the significance of GEI, G and E (Hühn and Léon, 1995). The test statistics for rank-change interactions, genotypes and environments are approximately x^2 distributed with $(l-1) \times (m-1)$, $(l-1)$ and $(m-1)$ degrees of freedom, respectively, where l is the number of genotypes, and m the number of environments (Hühn and Léon, 1995). For a two-way dataset with k genotypes and n environments, it was denoted the phenotypic value of ith genotype in jth environment as x_{ij}, where $i=1,2,\ldots,k;$ $j=1,2,\ldots,n;$ r_{ij} as the rank of the ith genotype in the jth environment, and $\bar{r}_{ij}$ as the mean rank across all environments for the ith genotype. Huehn (1990b) proposed a corrected mean yield on the nonparametric measures consists on $S_i^{(1)}$ and $S_i^{(2)}$ to remove the effect of genotype from phenotypic value. $S_i^{(3)}$ and $S_i^{(6)}$ are the sum of the absolute deviations and sum of squares of ranks, respectively. The genotype with the highest adjusted yield was given a rank of 17 and a genotype with the lowest adjusted yield was assigned a rank of 1. A genotype is stable over environments if its ranks are similar over environments; i.e. maximum stability occurs with equal ranks over environments.

The formula for calculating the statistics based on yield ranks of genotypes in each environment are expressed as follows (Hühn, 1979):

$$S_i^{(1)} = \sum_{j=1}^{n-1} \sum_{j'=j+1}^{n} |r_{ij} - r_{ij'}| / [n(n-1)] \qquad S_i^{(2)} = \frac{\sum_{j=1}^{n} (r_{ij} - \bar{r}_{i.})^2}{n-1}$$

$$S_i^{(3)} = \frac{\sum_{j=1}^{n} (r_{ij}^* - \bar{r}_{i.}^*)^2}{\bar{r}_{i.}^*} \qquad S_i^{(4)} = \sqrt{\frac{\sum_{j=1}^{n} (r_{ij} - \bar{r}_{i.})^2}{n}}$$

$$S_i^{(5)} = \frac{\sum_{j=1}^{n} |r_{ij} - \bar{r}_{i.}|}{n}$$

Where r_{ij}^* are ranks of genotypes in each environment based on corrected mean and r_{ij} are ranks of genotypes in each environment based on the mean yields, X_{ij}.

Fox et al. (1990) suggested nonparametric superiority index (SI), Top, Mid and Low. They expressed widely adapted genotypes fall into top third section of ranks after stratified ranking of the genotypes at each environment. A genotype with high value Top is considered as a widely adapted genotype. Top, Mid and Low are the proportion of ranks the genotype occurs in the top third, middle third and bottom third of the ranks.

$$S_i^{(6)} = \frac{\sum_{j=1}^{n} |r_{ij}^* - \bar{r}_{i.}^*|}{\bar{r}_{i.}^*}$$

Kang's (1988) rank-sum is another nonparametric stability statistics where both the genotype mean rank and Shukla's (1972) stability variance are used as selection criteria. This statistics assigns a weight of one to both mean yield and stability and enables the identification of high yielding and stable genotype, simultaneously.

Thennarasu (1995) proposed the use of four nonparametric measures based on the corrected ranks of genotypes in each environment. The ranks of genotypes in each environment were adjusted as ($x_{ij}^* = x_{ij} - \bar{x}_{i.}$).

Thennarasu's (1995) nonparametric stability statistics are:

$$NP_i^{(1)} = \frac{\sum_{j=1}^{n} |r_{ij}^* - Md_i^*|}{n}$$

$$NP_i^{(2)} = \frac{\sum_{j=1}^{n} |r_{ij}^* - Md_i^*| / Md_i}{n}$$

$$NP_i^{(3)} = \frac{\sqrt{\sum_{j=1}^{n} (r_{ij}^* - \bar{r}_{i.}^*)^2 / n}}{\bar{r}_{i.}^*}$$

$$NP_i^{(4)} = \frac{2}{n(n-1)} \left[\sum_{j=1}^{n-1} \sum_{j'=j+1}^{n} |r_{ij}^* - r_{ij'}^*| / \bar{r}_{i.} \right]$$

In the above formulas, r_{ij}^* is the rank of X^* and M_{di}^* are the mean and median ranks for adjusted values. Spearman's coefficient of rank correlation was employed (Steel and Torrie, 1980) to statistically compare the stability indices used in this study. To understand relationships among stability methods, principal component analysis (PCA) based on rank correlation matrices and to group different parameters into clusters, hierarchical cluster analysis with squared Euclidean distance and Ward's method were performed using IBM SPSS Statistics 21. All nonparametric stability statistics calculated using SAS-based program, and Microsoft Excel (Hussein et al., 2000; Lu, 1995).

RESULTS AND DISCUSSION

Analysis of variance and nature of GE interaction

The parametric combined analysis of variance indicated that the main effects of genotypes, environments, locations and years were highly significant (Table 1). Also, all the second order interactions including the GE interaction effect was significant ($P < 0.01$). The significant genotype × location interaction implied that the responses of the genotypes changed depending on the environmental conditions. These finding are in agreement with the results reported by Balalić and Zorić (2013). There was no significant genotype × year interaction that means genotype behavior in a location between years is not significant. The large seed yield variation due to environment is the main source of variation in most of the multi-environment trials (Gauch and Zobel, 1997). In present study, the canola seed yield was affected by environment, which accounted for 91 % of sum of squares (E+G+GE), whereas G and GE captured 2% and 7% of sum of squares (E+G+GE), respectively. Normally, environment explains most (80% or higher) of the total yield variation, while genotype and GE interaction are usually small (Yan and Kang, 2003). The results in this investigation were in agreement with other reports of multi-environment yield trials. However, effective interpretation and utilization of a multi-environment trial dataset in making selection decisions remains a major challenge to researchers.

Table 1. Combined analysis of variance of 17 canola genotypes across 14 environments

Source of variation	d.f.	Sum of squares	Mean of squares	(%) Explained
Environments (E)	13	1371.82	105.52 **	90.84
Place (P)	7	1126.76	160.96 **	
Year (Y)	1	14.57	14.57 **	
P×Y	5	242.95	48.59 **	
Rep (P×Y)	41	34.75	0.84 **	
Genotypes (G)	16	30.88	1.93 **	2.04
G×E	208	107.34	0.51 **	7.10
G×P	112	57.80	0.51 **	
G×Y	16	6.52	0.40 ns	
G×P×Y	80	42.37	0.52 *	
Error	656	246.79	0.37	
Total	934	1796.63		

** Significant at the 0.01 probability level; ns Non-Significant at the 0.05 probability level

The test statistics of the different nonparametric statistical procedures including Bredenkamp (1974) and Hildebrand (1980) for non-cross over GE interaction indicated the presence of this interaction type in canola dataset. Kroon and Laan (1981) procedure showed that there was only non-cross over or non-additive GE interaction type in this investigation (Table 2). Lack of cross over interaction for quantitative traits even in the presence of significant GE interaction has been reported (Kang, 2002). Although these findings are in agreement with the conventional ANOVA, but provide more specific information about the nature of GE interactions. This advantage is also noted in studies of Karimzadeh et al. (2013) and Sabaghnia et al., (2013). Cooper and Byth (1996) described that the larger the degree of GEI, the more dissimilar the genetic systems controlling the physiological processes conferring adaptation to different environments. The complexity of genotype behavior in different environments make difficulties to have selection in any specific environment (Karimzadeh et al., 2013; Sabaghnia et al., 2008; Yau, 1995). The value ratio obtained by the application of non-parametric methods is not in agreement with the ratio presented by Huehn (1996), considering that the x^2-values of Bredenkamp had the lowest numerical values in all tested traits. Delic et al., (2009) was also found disagreement in Kroon and Laan method with Huhn results. Non-cross over interactions may mean that genotypes are genetically heterogeneous but test environments are more or less homogeneous or that genotypes are genetically homogeneous but environments are heterogeneous (Kang, 2002).

Table 2. Test statistics for a test of genotype × environment interaction of 17 Canola genotypes over 14 environments.

Nonparametric test	d.f.	Statistic □^{2}
van der Laan-de Kroon	208	214.33 ns
Bredenkamp	208	5626.86 **
Hildebrand	208	255.44 *

*, ** Significant at the 0.05 and 0.01 probability levels, respectively; ns Non-Significant at the 0.05 probability level

The relatively large magnitude of GE interaction for grain yield of 17 canola genotypes tested across seven locations were larger than that of genotypic main effect (three times), but smaller than that of environment main effect (Table 1). The studied genotypes showed cross over types of GE interaction while environments showed both cross over and non-cross over types of GE interaction (Table 3). The relative contributions of G and GE interaction effects to the total variation for grain yield found in this study are in agreement with those found in other crop adaptation studies such as cereals or food legumes in rain-fed environments (Alagarswamy and Chandra, 1998; Balalić et al., 2011; Ceccarelli and Grando, 1991; Mohammadi et al., 2007; Sabaghnia et al., 2008; Živanović et al., 2012). Therefore, it would be very difficult to achieve an indirect response to selection over all of the canola target population of environments from selection in a few environments, ignoring the observed GE interactions.

Table 3. Test statistics for a test of genotype and environments of 17 Canola genotypes over 14 environments

Nonparametric test (G)	d.f.	Statistic □^{2}
van der Laan-de Kroon	16	35.46 **
Bredenkamp	16	20.63 ns
Hildebrand	16	0.56 ns
Nonparametric test (E)	d.f.	Statistic □^{2}
van der Laan-de Kroon	13	1289.48 **
Bredenkamp	13	644.49 **
Hildebrand	13	2954.04 **

** Significant at the 0.01 probability level; ns Non-Significant at the 0.05 probability level

Nonparametric Stability Analysis (NPSA)

According to mean yield, genotype Ebonite (3410 kg ha^{-1}), Elite (3190 kg ha^{-1}) and ARC-5 (3316 kg ha^{-1}) were the high yielding genotypes (Table 4). It has been suggested that, in unfavorable environments, plant breeders should look at GE interaction in a different way (Stroup et al., 1993). Yield stability is more important than mean yield in these environments. According to Ceccarelli (1996), in an unfavorable environment, the lowest mean yield of landraces was always higher than the lowest mean yield of non-landraces. This particular property of landraces is identifiable as stability by farmers but it would have been missed in a selection process within a breeding program to select landraces for unfavorable environments (Ceccarelli, 2000). According to the mean of ranks, ARC-5 and Ebonite showed highest yield over environments while regarding to corrected mean of ranks, Licord and Opera were the top rank. The effect of correction and removing the genotype effect from phenotypic data is in agreement with Karimizadeh et al. (2012). Ketata *et al.* (1989) proposed two ranking methods according to mean and standard deviation of ranks and Cravero *et al.* (2010) reported advantages of these nonparametric procedures in phenotypic stability studies. According to corrected and uncorrected CV nonparametric measures contain mean and standard deviation of ranks, genotype Geronimo identified as the high stable genotype (Table 3). Genotypes Geronimo and Sunday with minimum first two nonparametric stability statistics of Huehn (1990a) which are known as $S_i^{(1)}$ and $S_i^{(2)}$ were the most stable genotypes (Table 4). The $S_i^{(1)}$ estimates are based on all possible pair-wise rank differences, while $S_i^{(2)}$ is based on variances of ranks for each genotype across environments (Hühn and Nassar, 1989; Nassar and Huehn, 1987). These statistics ranked genotypes similarly for stability. The similarity result of these two parameters was also found in Balalić and Zorić (2013). Genotypes Licord, Okapi, Elite and Ebonite were most unstable. The significance tests for $S_i^{(1)}$ and $S_i^{(2)}$, developed by Nassar and Huehn (1987). For each canola genotype, $Z_i^{(1)}$ and $Z_i^{(2)}$ values were calculated according to the ranks of adjusted data and summed over genotypes to obtain Z values. Since $Z_i^{(1)}$ sum = 30.72 and $Z_i^{(2)}$ sum = 35.81 were more than the critical value of x^2 sum = 27.58, there were significant differences among 17 canola genotypes across 14 environments (Table 2). Inspecting the individual Z values, it was found that some genotypes were significantly unstable relative to others, because they revealed large Z values, in comparison with the critical value $X^2_{0.05, df=1}$ = 3.84. For practical applications Huehn (1990b), preferred the use of $S_i^{(1)}$ against $S_i^{(2)}$.

Table 4 . Estimation and test of nonparametric stability measures for 17 Canola genotypes across environments

Genotype	Mean Yield (Kg ha^{-1})	MR	CMR	SD	CSD	CV	CCV	$S_i^{(1)}$	$Z_i^{(1)}$	$S_i^{(2)}$	$Z_i^{(2)}$
Ebonite	3410.00	13.64	8.14	3.89	6.00	0.29	0.74	7.033	3.227	35.978	3.688
Elite	3190.50	10.93	8.00	5.17	6.05	0.47	0.76	6.956	2.879	37.456	4.654
Sinatra	2908.07	7.79	9.36	3.70	4.13	0.48	0.44	4.868	1.019	17.017	1.254
Sahara	2950.36	7.00	8.64	4.64	4.92	0.66	0.57	5.725	0.010	24.247	0.002
Celsius	2941.00	7.14	8.86	4.75	4.42	0.67	0.50	5.143	0.427	19.517	0.517
Sunday	3003.14	8.79	8.93	3.87	3.89	0.44	0.44	4.637	1.713	15.824	1.718
Modena	3040.86	9.43	9.29	4.72	4.75	0.50	0.51	5.604	0.003	22.528	0.056
Geronimo	3059.64	10.07	8.43	3.10	3.32	0.31	0.39	3.692	6.419	11.033	4.322
Opera	2666.64	4.86	10.36	3.30	5.75	0.68	0.55	6.670	1.759	33.017	2.090
ARC-5	3316.43	12.57	8.07	5.00	6.01	0.40	0.74	6.967	2.927	35.209	3.230
ARC-2	2969.21	8.29	8.79	4.53	4.35	0.55	0.50	5.044	0.611	18.951	0.655
Milena	3092.50	11.07	9.50	4.57	4.59	0.41	0.48	5.374	0.126	21.039	0.225
Dexter	2870.93	7.21	9.86	4.61	4.74	0.64	0.48	5.538	0.020	22.440	0.063
SLM046	3036.71	9.79	8.64	3.70	4.29	0.38	0.50	4.967	0.777	17.758	1.002
Zarfam	3111.29	11.14	8.29	4.00	4.46	0.36	0.54	5.297	0.206	19.912	0.430
Okapi	2810.36	6.29	9.64	5.88	6.07	0.93	0.63	7.110	3.595	36.863	4.253
Licord	2772.57	6.86	10.21	5.72	6.42	0.83	0.63	7.374	5.008	41.258	7.656
								Test Statistics			
	Grand Mean			$E(S_i^{(1)})$	$E(S_i^{(2)})$	$Var(S_i^{(1)})$	$Var(S_i^{(2)})$	x^2 Z_1,Z_2		x^2 sum	
	3008.84			5.647	24	0.595	38.901	8.844		27.587	

Table 4. Continue

Genotype	$S_i^{(3)}$	$S_i^{(6)}$	$NP_i^{(1)}$	$NP_i^{(2)}$	$NP_i^{(3)}$	$NP_i^{(4)}$	PI	ProbP	TOP	MID	LOW	RS
Ebonite	55.00	10.13	5.29	0.34	1.226	0.135	0.02	0.96	85.71	14.29	0.00	9
Elite	49.06	8.93	5.29	0.42	1.237	0.172	0.08	0.43	42.86	42.86	14.29	17
Sinatra	17.46	4.17	3.50	0.58	1.086	0.086	0.27	0.00	14.29	14.29	71.43	20
Sahara	25.45	4.73	3.79	0.54	1.141	0.102	0.27	0.00	14.29	14.29	71.43	26
Celsius	26.55	5.10	3.43	0.53	1.110	0.093	0.25	0.00	28.57	42.86	28.57	17
Sunday	20.31	4.66	2.93	0.31	1.085	0.064	0.17	0.04	0.00	71.43	28.57	12
Modena	33.77	6.43	3.71	0.37	1.115	0.090	0.19	0.02	42.86	28.57	28.57	16
Geronimo	15.76	4.41	2.43	0.24	1.070	0.057	0.15	0.07	28.57	57.14	14.29	8
Opera	11.54	2.77	4.93	1.10	1.134	0.106	0.48	0.00	0.00	14.29	85.71	30
ARC-5	59.95	10.63	5.21	0.35	1.231	0.143	0.07	0.52	85.71	0.00	14.29	18
ARC-2	27.47	5.50	3.36	0.35	1.108	0.091	0.22	0.01	14.29	42.86	42.86	11
Milena	39.10	7.03	3.79	0.32	1.103	0.096	0.16	0.05	42.86	42.86	14.29	15
Dexter	25.62	5.10	3.86	0.55	1.102	0.089	0.34	0.00	14.29	42.86	42.86	25
SLM046	21.50	5.44	3.64	0.40	1.108	0.097	0.17	0.04	28.57	42.86	28.57	12
Zarfam	30.29	6.08	3.71	0.31	1.127	0.106	0.13	0.11	28.57	71.43	0.00	10
Okapi	38.52	5.98	5.21	1.30	1.170	0.163	0.46	0.00	28.57	0.00	71.43	32
Licord	38.21	6.10	5.50	0.85	1.169	0.096	0.41	0.00	0.00	57.14	42.86	28

MR = Mean of Rank; CMR = Corrected Mean of Rank; SD=Standard Deviation; CSD= Corrected Standard Deviation; CV= Coefficient of Variation; CCV= Corrected Coefficient of Variation; PI = Superiority measure of Lin and Binns (1988); ProbPI= The significance level of the test on PI; TOP, MID, and LOW are the parameters of Fox et al. (1990); RS = the Rank-Sum of Kang (1988)

This stability parameter is very easy to compute and allows a clear and relevant interpretation (mean absolute difference between environments). Furthermore, an efficient test of significance is also available. Nonparametric statistics of $S_i^{(3)}$ and $S_i^{(6)}$ combine yield and stability based on yield ranks of genotypes in each environment (Hühn, 1979). Genotypes Opera, Geronimo and Sinatra based on $Si^{(3)}$ and $Si^{(6)}$ statistics were identified as the most stable genotypes (Table 4). The numerical value of $Si^{(3)}$ is determined by both yield and stability. Among these the most stable genotypes, genotype Geronimo had relatively high mean yield (Table 4). According to Nassar and Huehn (1987), Flores et al. (1998), Scapim et al. (2000) and Segherloo Ebadi et al. (2008), $S_i^{(1)}$ and $S_i^{(2)}$ nonparametric stability statistics considered in the sense of homeostasis and show static or biological concept of stability. Sabaghnia et al. (2006), reported that $S_i^{(3)}$ and $S_i^{(6)}$ nonparametric stability statistics are associated with the static or biological concept of stability. Based on Top measure, genotypes Ebonite and ARC-5 were the most favorable genotypes while according to Mid measure, genotypes Sunday and Zarfam were detected the most favorable genotypes (Table 4). Considering all three Top, Mid and Low values as the nonparametric superiority index (PI) of Fox et al. (1990), genotypes Ebonite, Elite and ARC-5 were the most favorable genotypes from both stability and mean yield aspects. Significance level (p-value) for superiority measure of each genotype (PI) is also shown as ProbP in table 4. The lowest value for rank-sum nonparametric stability method (Kang, 1988) indicates maximum stability for a certain genotype. According to this statistics, genotypes Geronimo, Ebonite and Zarfam were detected as the most stable and high yield genotypes (Table 4). Kang and Pham (1991) studied several stability methods for simultaneous selection for yield and stability. They found that the rank–sum method of Kang (1988) would be useful tool for selecting simultaneously for both mean yield and yield stability that is in agreement with our result. Kaya and Taner (2003) have reported the dynamic concept of this method like to nonparametric stability measure of Fox et al. (1990). According to all nonparametric stability statistics of Thennarasu (1995), genotypes Sunday and Geronimo had the lowest of NP_i and so the most stable genotypes (Table 4). These most stable genotypes, were among high mean yield genotypes. Although based on Sabaghnia et al. (2006), all nonparametric stability statistics of Thennarasu (1995) had static concept stability, but some of the most stable genotypes had high mean yield an dso reflected dynamic concept of stability. Segherloo Ebadi et al. (2008) reported that NP1 and NP4 are associated with the static or biological concept of stability. Among different nonparametric stability statistics, only Top measure and rank-sum procedure were related to the agronomic concept of yield stability (Flores et al., 1998; Sabaghnia et al., 2012). Our results are in a good agreement with the finding of the other researchers who used various nonparametric stability statistics in different crops. Traditionally, stability has been divided into distinct concepts including dynamic and static concepts (Becker, 1981). For many decades, most plant breeders used the static or biological stability concept to explain a genotype which indicates a relatively constant mean yield, independent of various environmental conditions. However, this stability concept is not acceptable to most plant researchers, who would prefer a dynamic concept of stability. In this concept of stability, it is not needed that the genotypic response to environmental conditions should be equal for all genotypes (Becker and Leon, 1988). In recent decades, most plant breeders prefer to use dynamic or agronomic concept of yield stability for GE interaction investigation and identification of the most stable genotype(s). Considering most of the nonparametric stability statistics, genotype Geronimo was the most stable genotypes. Regarding to importance of $S_i^{(1)}$ nonparametric measure some researcher use plot of mean yield of genotypes against non-parametric measure $S_i^{(1)}$ to enhance visual efficiency of genotype recommendation (Kaya and Taner, 2003). A stable genotype is one that provides high yield and consistent performance across locations (Ebadi Segherloo et al. 2008, Balalic et al. 2011 and 2013). According to this definition, only canola genotypes in section 4 can be considered as stable and with high seed yield (Figure 1).

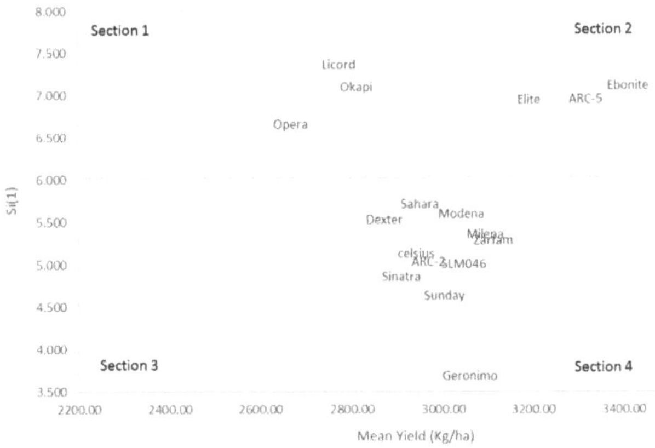

Fig 1. Plot of Si (1) vs. mean yield (kg ha^{-1}) for 18 canola genotypes over environments

Cluster Analysis on NPSA

To better reveal associations among evaluated genotypes, the two-way data of genotypes' mean yield and ranks based on different nonparametric stability measures, was performed further using a clustering procedure. A hierarchical cluster analysis with squared Euclidean distance as dissimilarity measure and Ward's method revealed three major distinct groups among 17 genotypes (Figure 2). The square root of the SS difference between values for canola genotypes was used as Euclidean distance. In Ward's procedure, the dissimilarity between two clusters is shown by the "loss of information" from joining the two clusters with this loss of information measured by the increase in error sum of squares. First cluster consisted of genotypes Ebonite, Elite and ARC-5 which were high mean yield and unstable or semi-stable genotypes. Second cluster consisted of genotypes Sahara, Opera, Okapi and Licord which were low mean yield and stable or semi-stable genotypes. Third cluster consisted of all other genotypes which were moderate mean yield and moderate or stable genotypes. Although the most favorable genotype Geronimo was belong to third cluster, but it seems that this genotype as one of the most stable genotypes which had acceptable mean yield and so could be regarded for commercial release.

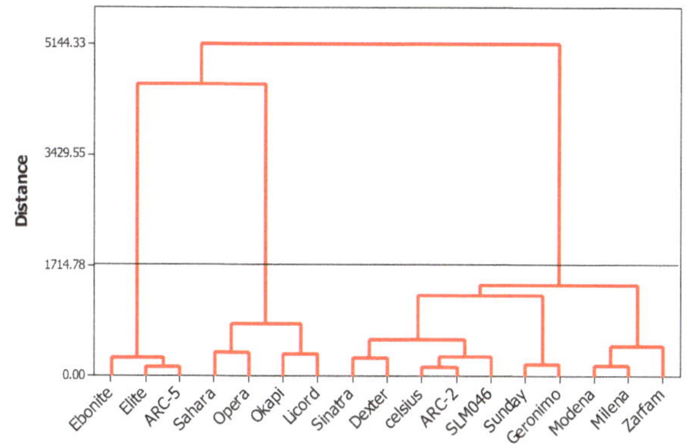

Figure 2. Hierarchical cluster analysis of the 17 canola genotypes based on Ward's method

Principal component analysis

The level of association among stability estimates of different models is indicative of whether one or more estimates should be obtained for reliable predictions of cultivar behavior, and also could help to choose the best adjusted and most informative stability parameter(s) to fit concept of stability (Duarte and Zimmermann, 1995). Principal component analysis based on rank correlation matrices was performed to understand the relationship among the nonparametric statistics. For better visualization, the first two principal components (PC1 and PC2) were plotted against each other (Figure 3). These first two principal components explained 83% (43% and 40% by PC1 and PC2, respectively) of the variances in nonparametric stability statistics. Biplot classified the stability measures in 4 groups. The lines that connect the stability estimates to the biplot origin are called stability vectors. An acute angle between the vectors of two stability indices approximates the positive correlation between them while an obtuse angle reveals independence or very weak correlation between these stability measures. Mean yield shows a highly significant negative rank correlation with corrected mean yield. This result was also shown by Karimzadeh and his colleagues (2012). In this plot, the Low measure (LOW) and the nonparametric superiority index (PI) of Fox et al. (1990) were correlated with CV, $NP_i^{(2)}$, Rank-sum and corrected mean of rank (CMR). Sabaghnia et al. (2006) reported that the Low measure of Fox et al. (1990) and NP2 are similar in concept to GE interaction measures as it defines stability in the sense of biological concept. Top and mean rank (MR) were correlated with each other and with total mean yield. Also there is significant negative correlation between mean of yield (MY) and corrected mean of rank, Low, PI, CV, $NP_i^{(2)}$, and RS measures. However, Mahtabi et al. (2013) found significantly and negatively correlated between mean yield and $S_i^{(3)}$, $S_i^{(6)}$, $NP_i^{(2)}$ and $NP_i^{(4)}$ statistics that is in contradiction with our result. They referred the high correlation between mean yield and stability statistics to higher values of these statistics for high yielding genotypes. The non-significant and negative significant correlations between mean yield and stability parameters suggest that stability statistics provide information that cannot be gleaned from average yield (Mekbib, 2003). Kang and Pham (1991) reported that $S_i^{(3)}$ and $S_i^{(6)}$ have significant correlation with yield and stability statistics of ecovalance and deviation from regression. Meanwhile, they showed $S_i^{(6)}$ has more strongly correlated with mean yield. Positive correlation between the stability statistics $S_i^{(1)}$ and $S_i^{(2)}$ (Mohammadi et al., 2007; Scapim et al., 2000; Segherloo Ebadi et al., 2008; Zali et al., 2011) and between $S_i^{(3)}$ and $S_i^{(6)}$ (Kang and Pham, 1991; Mohammadi et al., 2007; Segherloo Ebadi et al., 2008) were reported. Our data confirm these reports in canola. Kang and Pham (1991) reported that the parameters $S_i^{(1)}$ and $S_i^{(2)}$ were nearly perfectly associated for grain yield in winter wheat. The same results, high correlations between stability parameters, are reported by Yue et al. (1997) in soybean, Scapim et al. (2000) and Delic et al. (2009) in maize, Sabaghnia et al. (2006) in lentil, Mohammadi et al. (2007) and Akcura et al. (2009) in wheat, and Segherloo Ebadi et al. (2008) in chickpea. We showed that this high correlation can be found in canola. Kaya and Taner (2003) pointed out that the method of Fox et al. (1990) is associated with the dynamic concept of stability. Also, Sabaghnia et al. (2006) and Dehghani (2008) noted that the PI nonparametric measure of stability is similar in concept to GE interaction measures as it defines stability in the sense of agronomical concept of stability.

As conclusions, producing constant yield that means yield stability and economic profitability is an important and complicated issues for plant breeders and also agronomists. The main factor behind the stability and the important aspect of any plant breeding program is GE interaction. Hence, awareness of the nature of GE interaction and yield stability is of paramount importance for canola and the other crops breeders and farmers. Successful genotypes need to be adapted to a broad range of environmental conditions in order to ensure their economic profitability everywhere. The nonparametric methods provide the opportunity of detect the nature of GE interaction. Moreover, the nonparametric method has some advantages over parametric methods. They are not limited to the assumption of normality or any distribution. They are easy to understand and the computations and interpretations are easier than those for the parametric methods. Results of different nonparametric tests for GE interactions were equivalent with conventional parametric method (combined ANOVA) in this research.

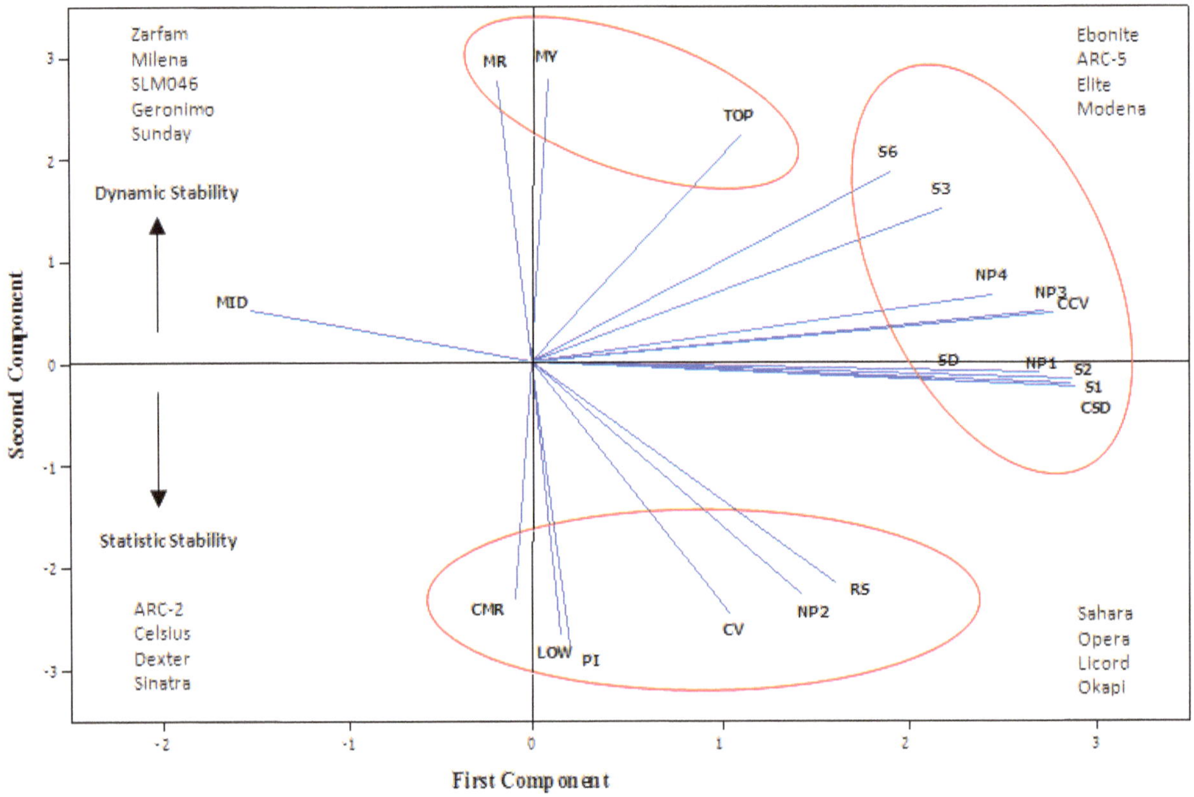

Figure 3. Biplot analysis of non-parametric indicators of phenotypic stability in Canola genotypes

These tests revealed existence of non-cross over GE interaction in canola genotypes multi-environment trials. The presence of significant GE interaction showed that differences among genotypes were not stable from one year to another, and these interactions are unavoidable in agricultural studies specially in multi-environment trials (Sabaghnia et al., 2010; Yan and Kang, 2003). Huehn and Leon (1995) found the same result and recommended Hildebrand and Kuninger nonparametric tests for interactions. In addition, Truberg and Huehn (2000) proposed Kroon and Laan test for cross over interaction. The genotypes used in this study did not exhibit a uniform stability and response pattern to different environments. On the basis of this assumption, genotypes with a minimal variance for yield across different environments are considered stable. Simultaneous selection for both mean yield and stability is an important consideration in breeding programs (Yan and Kang, 2003). Most plant breeders prefer simultaneous selection for mean yield and stability because the selected genotypes must have high mean values coupled with stable performance. This study showed that using corrected data on nonparametric measures of phenotypic stability is different from using original data. Similar was result found byKarimizadeh et al., 2012. They pointed out that using corrected data change Hahn's nonparametric measures nature to a static concept of stability. There is good potential in nonparametric stability methods to identify favorable genotypes in plant breeding programs. Therefore, using the agronomic or dynamic concept of stability is the better use of original dataset because the statistics of mean of ranks (MR), and Top were highly correlated with mean yield and indicated the dynamic concept of stability. The PCA biplot distinguish between measures based on static (biological) concept and dynamic (agronomic) concept of stability (Sabaghnia *et al.* 2006). The statistic Fox Top measure is related to dynamic stability and other remaining measures are associated with static stability. Researchers believe nonparametric statistics related to high seed yield performance define stability with dynamic concept (Kang and Pham, 1991; Mohammadi *et al.*, 2007 and Sabaghnia *et al.*, 2006). This research revealed that four non-parametric statistics of Huehn ($S_i^{(1)}$, $S_i^{(2)}$, $S_i^{(3)}$ and $S_i^{(6)}$) and three Thennarasu's statistics ($NP_i^{(1)}$, $NP_i^{(3)}$ and $NP_i^{(4)}$) clustered together and they classify genotypes as stable or unstable in a similar manner. Therefore, only one of these statistics can be applied for selecting stable genotypes in a breeding program. Among them $S_i^{(1)}$ is offered because of easy to use and simple interpretation (Huehn, 1990, 1996). According to these statistics, genotypes Geronimo and Opera had the smallest ranks and regarded as the most stable genotypes. Considering nonparametric measures including PI and King's' Rank-sum measures, genotypes Ebonite, ARC-5 and Geronimo had the smallest ranks and regarded as the most stable genotypes with high seed yield. According to statistics of Mean yield, Mean of rank and Top measures, genotypes Licord, Opera and Okapi had the smallest ranks and regarded as the most stable genotypes. Top and Rank-sum measures were positively correlated with high seed yield ($p<0.01$). Therefore, these could be recommended as useful measures for cultivar selection. Positive correlation between Rank-sum and seed yield was also reported by Zali et al, in 2011. According to the most of the nonparametric stability statistics genotype Geronimo was the most stable genotype which had relatively high mean yield. In conclusion, this genotype is ideal candidate in this regard as they had the desirable characteristics of high stability with high yield for all test environments or similar regions. Study on the association among nonparametric stability statistics is essential to make any recommendations (Dehghani, 2008; Karimzadeh et al., 2013). Among nonparametric measures, Top, $S_i^{(1)}$ and Rank-sum statistics of nonparametric procedures were found to be useful in detecting the stability of the genotypes.

ACKNOWLEDGMENTS

Thanks to Prof. Dr. A. Bjornstad and Dr. M.A. Hussein for their helpful comments and providing SAS programs used for this research. Sincere gratitude goes to Iran's Agricultural Research Center and its Agricultural Research Stations for providing plant materials, experimental sites, and technical assistance.

LITERATURE CITED

Akcura, M., Y. Kaya and S. Taner. 2009. Evaluation of durum wheat genotypes using parametric and nonparametric stability statistics. Turk J Field Crops 14: 111-122.

Alagarswamy, G. and S. Chandra. 1998. Pattern analysis of international sorghum multi-environment trials for grain-yield adaptation. Theoretical and Applied Genetics 96: 397-405.

Anonymus. 2012. FAO - Trade and Markets Division. Food Outlook, Oilseeds market summary.

Anonymus. 2011. Agricultural Ministry Statistics of Iran, Agricultural products outlook, Oilseeds summary.

Azzalini, A. and D.R. Cox. 1984. Two new tests associated with analysis of variance. Journal of the Royal Statistical Society. Series B (Methodological): 335-343.

Baker, R.J. 1988. Tests for crossover genotype-environmental interactions. Canadian Journal of Plant Science 68: 405-410.

Balalić, I. and M. Zorić. 2013. Non-Parametric Stability Analysis of Sunflower Multi-Location Trials. 314-330.

Balalić, I., M. Zorić, V. Miklič, N. Dušanić, S. Terzić and V. Radić. 2011. Non-parametric stability analysis of sunflower oil yield trials. Helia 34: 67-77.

Becker, H.C. 1981. Correlations among some statistical measures of phenotypic stability. Euphytica 30: 835-840.

Becker, H.C. and J. Leon. 1988. Stability analysis in plant breeding. Plant Breeding 101: 1-23.

Bredenkamp, J. 1974. The nonparametric examination of interactions. Psychologische Beitrage 16: 398-416.

Ceccarelli, S. 2000. Decentralized-participatory plant breeding: adapting crops to environments and clients. In Decentralized-participatory plant breeding: adapting crops to environments and clients, Proceedings of the 8th Int. Barley Genetics Symposium, Adelaide, 159-166.

Ceccarelli, S. and S. Grando. 1991. Selection environment and environmental sensitivity in barley. Euphytica 57: 157-167.

Crossa, J., P.N. Fox, W.H. Pfeiffer, S. Rajaram and H.G. Gauch. 1991. AMMI adjustment for statistical analysis of an international wheat yield trial. Theoretical and Applied Genetics 81: 27-37.

Dehghani, H. . 2008. Estimating yield stability by nonparametric stability analysis in maize (Zea mays L.). Plant Breeding and Seed Science 58: 61-77.

Delic, N., G. Stankovic and K. Konstantinov. 2009. Use of non parametric statistics in estimation of genotypes stability. Maydica 54: 155.

Duarte, J. B. and M. Zimmermann. 1995. Correlation among yield stability parameters in common bean. Crop Science 35: 905-912.

FAOSTAT. 2010. (http://www.fao.org/corp/statistics/en/). Food and Agriculture Organisation of the United Nations.

Flores, F., M.T. Moreno and J.I. Cubero. 1998. A comparison of univariate and multivariate methods to analyze G× E interaction. Field Crops Research 56: 271-286.

Fox, PN, B Skovmand, BK Thompson, HJ Braun and R Cormier. 1990. Yield and adaptation of hexaploid spring triticale. Euphytica 47: 57-64.

Gauch, HG and R.W. Zobel. 1997. Identifying mega-environments and targeting genotypes. Crop Science 37: 311-326.

Hildebrand, H. 1980. Asymptotisch verteilungsfreie Rangtests in linearen Modellen. Medizinische Informatik und Statistik 17: 344-349.

Huehn, M. 1996. Nonparametric analysis of genotype x environment interactions by ranks. Genotype by environment interaction. CRC Press, Boca Raton, FL: 213-228.

Huehn, M. 1990a. Nonparametric measures of phenotypic stability. Part 1: Theory. Euphytica 47: 189-194.

Huehn, M. 1990b. Nonparametric measures of phenotypic stability. Part 2: Applications. Euphytica 47: 195-201.

Hühn, M. 1979. Beitrage zur erfassung der phanotypischen stabilitat. EDV Med Biol 10: 112-117.

Hühn, M. and J. Léon. 1995. Nonparametric analysis of cultivar performance trials: experimental results and comparison of different procedures based on ranks. Agronomy journal 87: 627-632.

Hühn, M. and R. Nassar. 1989. On tests of significance for nonparametric measures of phenotypic stability. Biometrics: 997-1000.

Hussein, M. A., A. Bjornstad and A.H. Aastveit. 2000. SASG× ESTAB: a SAS program for computing genotype× environment stability statistics. Agronomy journal 92: 454-459.

Kang, M.S. 2002. 15 Genotype–Environment Interaction: Progress and Prospects. In: Quantitative genetics, genomics, and plant breeding, ed. Kang, M.S., 219, CABI Publication, New York.

Kang, M.S. 1988. A rank-sum method for selecting high-yielding, stable corn genotypes. Cereal Res. Commun 16: 113-115.

Kang, M.S. and H.N. Pham. 1991. Simultaneous selection for high yielding and stable crop genotypes. Agronomy journal 83: 161-165.

Karimizadeh, R., M. Mohammadi, N. Sabaghnia and M.K. Shefazadeh. 2012. Using different aspects of stability concepts for interpreting genotype by environment interaction of some lentil genotypes. Aust. J. Crop Sci 6: 1017-1023.

Karimzadeh, R., M. Mohammadi, N. Sabaghnia, M.K. Shefazadeh, T. Hosseinpour and M. Armion. 2013. Exploring of genotype by environment interaction by nonparametric stability procedures. Natura montenegrina 12: 181-204.

Kaya, Y. and S. Taner. 2003. Estimating genotypic ranks by nonparametric stability analysis in bread wheat (Triticum aestivum L.). Journal of Central European Agriculture 4: 47-54.

Kroon, J. and P. Laan. 1981. Distribution-free test procedures in two-way layouts; a concept of rank-interaction. Statistica Neerlandica 35: 189-213.

Kubinger, K.D. 1986. A Note on Non-Parametric Tests for the Interaction in Two-Way Layouts. Biometrical Journal 28: 67-72.

Lu, H.Y. 1995. PC-SAS program for estimating Huehn's nonparametric stability statistics. Agronomy journal 87: 888-891.

Mahtabi, E., E. Farshadfar and M.M. Jowkar. 2013. Non parametric estimation of phenotypic stability in Chickpea (Cicer arietinum L.). International Journal of Agriculture and Crop Sciences 5: 888-895.

Mekbib, F. 2003. Yield stability in common bean (Phaseolus vulgaris L.) genotypes. Euphytica 130: 147-153.

Mohammadi, R., A. Abdulahi, R. Haghparast, M. Aghaee and M. Rostaee. 2007. Nonparametric methods for evaluating of winter wheat genotypes in multi-environment trials. World Journal of Agricultural Sciences 3: 137-242.

Nassar, R. and M. Huehn. 1987. Studies on estimation of phenotypic stability: Tests of significance for nonparametric measures of phenotypic stability. Biometrics: 45-53.

Sabaghnia, N., H. Dehghani, B. Alizadeh and M. Mohghaddam. 2010. Interrelationships between seed yield and 20 related traits of 49 canola (Brassica napus L.) genotypes in non-stressed and water-stressed environments. Spanish Journal of Agricultural Research 8: 356-370.

Sabaghnia, N., H. Dehghani and S.H. Sabaghpour. 2006. Nonparametric methods for interpreting genotype× environment interaction of lentil genotypes. Crop Science 46: 1100-1106.

Sabaghnia, N., R. Karimizadeh and M. Mohammadi. 2012. The use of corrected and uncorrected nonparametric stability measurements in durum wheat multi-environmental trials. Spanish Journal of Agricultural Research 10: 722-730.

Sabaghnia, N., S.H. Sabaghpour and H. Dehghani. 2008. The use of an AMMI model and its parameters to analyse yield stability in multi-environment trials. The Journal of Agricultural Science 146: 571-581.

Scapim, C.A., V.R. Oliveira, C.D. Cruz, C.A.B. Andrade and M.C.G. Vidigal. 2000. Yield stability in maize (Zea mays L.) and correlations among the parameters of the Eberhart and Russell, Lin and Binns and Huehn models. Genetics and Molecular Biology 23: 387-393.

Segherloo Ebadi, A. , S.H. Sabaghpour, H. Dehghani and M. Kamrani. 2008. Non-parametric measures of phenotypic stability in chickpea genotypes (Cicer arietinum L.). Euphytica 162: 221-229.

Shukla, G.K. 1972. Some statistical aspects of partitioning genotype-environmental components of variability. Heredity 28: 237-245.

Steel, R.G.D. and J.H. Torrie 1980. Principles and procedures of statistics, a biometrical approach: McGraw-Hill Kogakusha, Ltd.

Stroup, W.W., P.E. Hildebrand, C.A. Francis, J. Ragland and R. Lal 1993. Farmer participation for more effective research in sustainable agriculture. In Farmer participation for more effective research in sustainable agriculture, Technologies for sustainable agriculture in the tropics: Proceedings of two international symposia held in San Antonia, TX and Denver, CO, 1990 and 1991 respectively., 153-186: American Society of Agronomy (ASA).

Thennarasu, K. 1995. On Certain Non-Parametric Procedures For Studying Genotype-Environmentinteractions. And Yield Stability, IARI, Division of Agricultural Statistics: New Delhi.

Truberg, B. and M. Hühn. 2002. Contributions to the analysis of genotype× environment interactions: Experimental results of the application and comparison of clustering techniques for the stratification of field test sites. Journal of Agronomy and Crop Science 188: 113-122.

Yan, W and MS Kang 2003. GGE biplot analysis: A graphical tool for breeders, geneticists, and agronomists CRC Press.

Yan, W., M.S. Kang, B. Ma, S. Woods and P.L. Cornelius. 2007. GGE biplot vs. AMMI analysis of genotype-by-environment data. Crop Sci 47: 643-655.

Yau, S.K. 1995. Regression and AMMI analyses of genotype× environment interactions: An empirical comparison. Agronomy journal 87: 121-126.

Yue, G.L., K.L. Roozeboom, W.T. Schapaugh and G.H. Liang. 1997. Evaluation of soybean cultivars using parametric and nonparametric stability estimates. Plant Breeding 116: 271-275.

Zali, H., E. Farshadfar and H. Sabaghpour. 2011. Non-parametric analysis of phenotypic stability in chickpea (Cicer arietinum L.) genotypes in Iran. Crop Breeding Journal 1: 85-96.

Živanović, T., G. Branković, M. Zorić, G.Š. Momirović, S. Janković, S. Vasiljević and J. Pavlov. 2012. Effect of recombination in the maize breeding population with exotic germplasm on the yield stability. Euphytica 185: 407-417.

SOME AGRICULTURAL AND QUALITY PROPERTIES OF ULUBAG ECOTYPE LINES OF WILD ORCHARDGRASS (*Dactylis glomerata* L.)

*Esin SAHIN** , *Metin TOSUN* ,*Kamil HALILOGLU*

Department of Field Crops, Faculty of Agriculture, Ataturk University. Erzurum, TURKEY
**Corresponding author: esinsahin@atauni.edu.tr*

ABSTRACT

Research was conducted in Erzurum ecologic conditions. Eight lines of Ulubag ecotype that were selected in the previous selection process and 1 control cultivar were used in this research. Lines used in the research were compared to each other and the control variety. Fresh hay yield, dried hay yield, crude protein content and yield, plant nutritional elements, ADF and NDF rates, and seed yield traits were investigated in this research. In Ulubag ecotypes, 6 lines in fresh hay yield and dried hay yield, 1 line in crude protein content, 4 lines in crude protein yield, 1 line in P content, 1 line in Mg content, 7 lines in K content, 1 line in Ca content and 4 lines in seed yield gave better result than control. On the other hand, ADF rates in 4 lines were lower than control variety. However, NDF rates in all lines were better than the control variety. It is concluded that lines 3, 9, 14, and 18 in Ulubag ecotype can be used for new variety development, in terms of hay yield and quality.

Key words: Agronomy, crude protein, selection.

INTRODUCTION

Orchardgrass (*Dactylis glomerata* L.) is a bunch-type, tall-growing, cool-season perennial grass. It is one of the most productive cool-season grasses, tolerant to shade, fairly drought resistant with moderate winter hardiness (Serin and Tan, 1998). It is a very important species in pasture lands in the regions where temperate and humid climatic characteristics are prevalent. This species does not form a dense pasture structure and develops in the form of a bunch; therefore, it has a branching, deeper root formation system. This species can compete with weeds and other grassy species in the regions it is adapted to. Since it is an early rising species in the spring, it is suitable to be used as early spring meadow pasture (Serin and Tan, 1998).

It was reported in the study carried out by Tosun et al., (1996) with the aim of determining some chemical features of plant materials, crop and seed yield of wild orchard grass under irrigated conditions of Erzurum between 1992 and 1995 that meant fresh hay yield was 413.85g per, hay yield was 120.78g per plant, crude protein rate was 13.38%, crude protein yield was 16.35g plant and seed yield was 24.45g plant.

ADF (Acid Detergent Fibre) is composed of cell membranes including the component of alkali soluble lignin, alkali insoluble lignin, fibre bounded nitrogen, minerals insoluble in cellulose and solvents. ADF is the best way of measuring fibre in plant material (Van Soest et al., 1991). NDF (Neutral Detergent Fibre) is composed of cell membrane elements including hemicelluloses, alkali soluble lignin, alkali insoluble lignin, fibre bound nitrogen and minerals insoluble in cellulose and solvents (Van Soest et al., 1991). Both ADF and NDF are the characteristics used for determination of quality in forages. It was stated in a study aimed to determine the quality and mineral content of forages growing in the Mediterranean region that average rates of crude protein, NDF, nitrogen, potassium, magnesium, sodium, calcium and phosphorus were 13.60%, 48.63%, 2.18%, 3.51%, 0.41%, 1.89%, 0.42% and 0.31% in orchardgrass (Convertini et al., 1999).

MATERIALS AND METHODS

This study was carried out under the ecological conditions of Erzurum. Lines of Ulubag populations determined to be promising and one variety (USA breed as standard) were used to make comparisons in previous studies carried out over the plant materials of orchardgrass taken from eight different locations in Erzurum (Tosun and Sagsoz 1994; Sagsoz et al. 1996; Tosun et al. 1996; Tosun et al., 1997).

The amounts of annual rainfall were close to each other in 2005 (414.1mm) and 2006 (410.4mm) while in 2007 it was higher (433.7mm) than previous two years. Average annual temperature in 2005, 2006 and 2007 was 5.1, 6.4 and 5.5°C, respectively while long term average temperature was 5.7°C. Soil pH in experimental area is 7.45 and reflects light alkali reaction. Structure of soil is sandy, clayish and loamy and poor in organic material (1.37%) (Sagsoz et al., 1996).

Experiment was designed in the complete randomized block design with three replications. Plots were arranged in the length of 6 m and 6 plant lines (with the interval distance of 50 cm). Spring fertilisation was 15kg N da^{-1} and autumn fertilisation was 10kg P_2O_5 da^{-1}. Plots were irrigated and weeds were controlled when needed. Crops from 50 cm inside the upper and lower side of the plot and in one line along the longer plot side were removed from the area as side effect while harvesting the plot. Half of the plot was harvested for hay yield and the other half for seed yield.

Parameters studied in the experiment

Fresh hay yield and hay yield

Green plant material cut from the area in the plot at the beginning of flowering period after removing side effect was weighed without allowing for water loss. Data was converted into kg da^{-1}. 500g samples were taken from the harvested plants in order to estimate the hay yield, and then dried in an oven at 78°C for 24 hours and weighed using a sensitive scale. Dry matter yield was estimated using data obtained by considering green plant matter yield in kg/da.

Crude protein rate and yield

Samples used for the determination of hay yield were weighed by dividing 0.5g parts after grinding and the nitrogen total was estimated according to Kjeldahl method. Crude protein rates of plant material were estimated by multiplying nitrogen rates calculated with coefficient 6.25 (Yeldan, 1984). Crude protein yield was calculated in kg/da by multiplying the hay yield with crude protein rate.

Mineral content analysis of plant samples

Dried plant samples were subjected to fresh decomposition by treating 0.5g of material with nitric perchloride acid. Ratios of Na, K, P, Ca and Mg were determined in the solutions (Kacar, 1972).

Determination of ADF (acid detergent fibre) and NDF (neutral detergent fibre) rates

An analysis method was applied to the 0.8g and 0.9g of grinded plant samples in ANKOM fibre analysing device. Samples were washed off in acetone; stored and dried at 105°C overnight; cooled in desiccators; weighed and ADF and NDF rates were determined (Anonymous, 1995).

Seed yield

Seeds of plants in the other half of the plot area remaining after the harvesting for the determination of plant material were obtained by cutting plants after taking out the side effects. Seeds were weighed and seed yield was estimated in kg da^{-1}.

Statistical analysis of data

Data was analysed using the GLM procedure in the SAS (SAS 1999) program.

RESULTS AND DISCUSSION

Fresh hay yield and hay yield

Differences in genotypes were statistically insignificant for fresh plant hay yield between the first (2006) and second (2007) experimental years and an average was taken. Average fresh hay yield of genotypes was 1094.04 kg da^{-1} and the highest yield in the first year (1260.53 kg da^{-1}) was obtained from line 3. Yield of fresh hay in the second year was higher than the first year in all genotypes and difference between the two years was statistically significant *($P<0.01$)*.

Second year average fresh hay yield was determined to be 2062.68 kg da^{-1}. Line 9 was the highest in two – year average values with 1800.87 kg da^{-1} while the lowest yield was obtained from line 1 with 1301.07 kg da^{-1}. Two – year average yield was 1578.36 kg da^{-1} (Table 1).

Table 1. Fresh and hay yields of orchardgrass genotypes

Genotype	Fresh hay yield (kg da^{-1})			Hay yield (kg da^{-1})		
	2006 year	2007 year	Average	2006 year	2007 year	Average
1	1146.67	1455.47	1301.07	467.73	385.17	426.45
3	1260.53	2176.80	1718.67	515.77	541.80	528.78
6	1100.00	1847.20	1473.60	417.40	498.60	458.00
9	1138.13	2463.60	1800.87	431.10	575.03	503.07
11	1116.80	1924.53	1520.67	416.03	521.03	468.53
14	1077.07	2148.27	1612.67	444.07	571.40	507.73
18	978.13	2505.07	1741.60	350.67	599.10	474.88
19	1064.00	2043.73	1553.87	421.87	524.67	473.27
Control	965.07	1999.47	1482.27	428.57	498.23	463.40
Average	1094.04	2062.68	1578.36	432.58	523.89	478.24

As in fresh hay yield, differences between hay yield in the first and second years and two – year average of genotypes were statistically insignificant while those between the years were very significant *($P<0.01$)*. In the first year, average hay yield of genotypes was 432.58 kg da^{-1} and the highest hay yield was obtained from line 3 (515.77 kg da^{-1}). In the second year, average hay yield was 523.89 kg da^{-1} and line 18 was in the first row (599.10 kg da^{-1}). The two year average hay yield was 478.24 kg da^{-1}. Dry plant material yield of control was below the average with 498.23 kg da^{-1} being in the 8th row (Table 1).

No recorded differences between the ecotypes used in the experiment for plant material yield may indicate that selection studies carried out in future years will be successful. The final (third) stage of selection started previously was evaluated in the scope of the present experimental study. Until the final stage of mentioned selection study, better lines were determined after each selection cycle and pollination was performed among them. Variability existing between the lines was narrowed. The highest two – year average hay yield was determined to be 622.5 kg da^{-1} in a study (Serin, 1991) using orchardgrass as control (USA Breed) under ecological conditions of Erzurum. Results of the present study were slightly lower than the results of the outlined study, which might have resulted from the different years and locations of the studies. As already known, yield is a quantitative character affected largely by environment. Therefore, different amount of rainfall might have affected this parameter.

Crude protein rate and yield

Differences between genotypes for crude protein rate in the first and second years and two – year average were statistically insignificant. However, differences between the years were very significant and *($P<0.01$)* and interaction between year x genotype was significant *($P<0.05$)*. In the first year of study, average crude protein rate of genotypes was 10.58%, line 9 was in the first row with 11.92%. In the second year, average crude protein rate was 13.37% and higher than the first year. Line 3 exhibited the highest rate (15.03%) (Table 2). When considered two year average values, average crude protein rate was 11.98%.

Table 2. Crude protein content and crude protein yield of orchardgrass genotypes.

Genotype	Crude protein content (%)			Crude protein yield (kg da^{-1})		
	2006 year	2007 year	Average	2006 year	2007 year	Average
1	9.70	14.28	11.99	45.21 **a-c***	54.22	49.72
3	9.87	15.03	12.45	50.81 **a**	81.46	66.14
6	10.39	11.60	10.99	43.07 **a-c**	57.53	50.30
9	11.92	12.89	12.41	51.15 **a**	73.88	62.52
11	10.08	12.39	11.23	41.23 **bc**	63.07	52.15
14	10.76	13.56	12.16	47.55 **ab**	77.98	62.77
18	10.57	14.69	12.63	36.44 **c**	87.16	61.80
19	10.22	12.66	11.44	42.58 **a-c**	65.89	54.23
K	11.72	13.26	12.49	49.33 **ab**	66.44	57.88
Average	10.58	13.37	11.98	45.26	69.74	57.50
LSD				**9.08**		

*Differences were significant in the same column with a different letter ($p<0.05$).

Even though the differences between genotypes for crude protein yield were significant in the first year *($P<0.05$)*, they were insignificant in the second year and the two year average. Average crude protein yield in the first year was 45.26 kg da^{-1} and the highest value was 51.15 kg da^{-1} in line 9. Crude protein yield in the second year was higher than the first year and differences between the years were significant *($P<0.01$)*. In the second year, the average crude protein yield was 69.74 kg da^{-1}. Line 18 exhibited the highest yield with 87.16 kg da^{-1}. When considered the two year average crude protein yield average of all genotypes was calculated to be 57.50 kg da^{-1}, line 3 exhibited the highest crude protein yield (66.14 kg da^{-1}) and control was in the fifth row with 57.88 kg da^{-1} (Table 2).

The crude protein rate was 13.38% in a study by Tosun et al. (1996) on orchard grass under irrigated conditions of Erzurum while ranging from 10.10 to 13.60% in other studies (Gordon et al., 1962; Reid et al., 1967; Aydin, 1994; Schoner and Pfeffer, 1996).

Mineral content of plant samples

The difference between genotypes for P (phosphorous) content in the first year was significant *($P<0.05$)* and average P content was 0.229%. The highest P rate was in line 19 (0.263%). The rate of P in the second year (0.327%) was higher than the first year and the annual difference was significant *($P<0.01$)*. The difference between genotypes in the second year was insignificant and the highest P rate was in line 19 with 0.367%. The second year average P rate was 0.327%.

When considering, the two year average, differences between genotypes for P rate were significant *($P<0.05$)*. Average P rate was 0.278% and line 19 was in the first row (0.315%) (Table 3).

Table 3. Phosphor and magnesium percentages of orchardgrass genotypes.

Genotype	P (%)			Mg (%)		
	2006 year	2007 year	Average	2006 year	2007 year	Average
1	0.260 **ab***	0.323	0.292 **ab***	0.117	0.107	0.112
3	0.193 **bc**	0.353	0.273 **ab**	0.133	0.110	0.122
6	0.133 **c**	0.310	0.222 **c**	0.127	0.113	0.120
9	0.233 **ab**	0.310	0.272 **ab**	0.143	0.110	0.127
11	0.250 **ab**	0.273	0.262 **bc**	0.123	0.110	0.117
14	0.220 **ab**	0.347	0.283 **ab**	0.127	0.103	0.115
18	0.243 **ab**	0.337	0.290 **ab**	0.120	0.107	0.113
19	0.263 **a**	0.367	0.315 **a**	0.120	0.107	0.113
Control	0.260 **ab**	0.323	0.292 **ab**	0.143	0.107	0.125
Average	0.229	0.327	0.278	0.128	0.108	0.118
LSD	**0.067**		**0.048**			

*Differences were significant in the same column with a different letter ($p<0.05$).

Differences between genotypes for Mg (magnesium) rate in the first and second years and the two years average was statistically insignificant, however; those between the years are significant *($P<0.01$)*. Average Mg rate was 0.108% in the second year. From the two year average, the average Mg rate of genotypes in the study was determined to be 0.118%, Line 9 exhibited the highest rate (0.127%) (Table 3).

The difference between K (potassium) rates of genotypes in the first year were significant *($P<0.05$)*. Average potassium rates in the first year were 2.109%. The second year average K rate (0.337%) was lower than the first year and the difference between the years was significant *($P<0.01$)*. When considering the average values, the difference between the K rates of genotypes was significant *($P<0.05$)*. In addition, interaction between year x genotype was also significant *($P<0.05$)*. Significant differences between genotypes and extremely significant differences the between years caused the interaction between the years x genotypes to be significant (Table 4).

Table 4. Potassium and calcium percentages of orchardgrass genotypes.

Genotype	K (%)			Ca (%)		
	2006 year	2007 year	Average	2006 year	2007 year	Average
1	2.217 **b***	0.327	1.272 **b***	0.193	0.107 **c****	0.150
3	2.043 **b**	0.357	1.200 **b**	0.137	0.107 **c**	0.122
6	2.027 **b**	0.350	1.188 **b**	0.213	0.217 **ab**	0.215
9	2.177 **b**	0.333	1.255 **b**	0.317	0.077 **c**	0.197
11	1.820 **b**	0.340	1.080 **b**	0.093	0.173 **b**	0.133
14	2.650 **a**	0.313	1.482 **a**	0.143	0.097 **c**	0.120
18	2.050 **b**	0.350	1.200 **b**	0.097	0.097 **c**	0.097
19	2.073 **b**	0.343	1.208 **b**	0.123	0.183 **ab**	0.153
K	1.920 **b**	0.323	1.122 **b**	0.160	0.240 **a**	0.200
Average	2.109	0.337	1.223	0.164	0.144	0.154
LSD	**0.412**		**0.209**		**0.060**	

*Differences were significant in the same column with a different letter ($p<0.05$).
**Differences were very significant in the same column with a different letter ($p<0.01$).

Differences between the genotypes for Ca (calcium) rates were insignificant in the first year and the two year average, but highly significant in the second year *($P<0.01$)*. The average Ca rate in the first year was 0.164% and line 9 exhibited the highest rate 0.317%. The second year average Ca rate was 0.144%. Control was in the first row for Ca rate (0.240%) while for the two year average line 6 was in the first row (0.215%) (Table 4).

The rates of K, Ca and Mg were found as 2.59 to 2.84%, 0.36 to 0.64% and 0.17 to 0.26% in some studies on orchardgrass (Copponet, 1964; Whitehead, 1966). In another study, Convertini et al., (1999) reported that orchardgrass had the contents of K, Mg, Ca and P in the rates of 3.51%, 0.41%, 0.42% and 0.31% respectively. Results in the present study were slightly lower than the values mentioned above. This situation is believed to be caused by both genotypes and different environmental conditions of the plants. In this respect, Cherney and Cherney (2005) reported that K fertilisation increases K rate in forage plants while reducing the content of P, Ca, Mg and Na.

ADF and NDF rates of plant samples

The first year average of ADF was 36.02% and the control had the highest rate with 37.11% while line 9 showed the lowest rate (34.68%). In the second year, the average ADF rate was 35.53% and lower than the first year, but their difference was not significant. In the second year, line 14 had the highest ADF rate (37.75%). The two year average ADF rate was 35.78% and the control was in the fifth row for the average ADF (35.76%) (Table 5).

Table 5. ADF and NDF percentages of orchardgrass genotypes.

Genotype	ADF (%)			NDF (%)		
	2006 year	**2007 year**	**Average**	**2006 year**	**2007 year**	**Average**
1	36.96	35.22	36.09	63.96	58.74	61.35
3	35.89	35.50	35.70	62.32	57.25	59.79
6	36.04	36.08	36.06	63.20	60.72	61.96
9	34.68	35.75	35.22	61.80	60.95	61.38
11	34.90	34.53	34.72	62.91	59.34	61.12
14	36.15	37.75	36.95	62.42	61.28	61.85
18	35.90	33.76	34.83	61.89	58.59	60.24
19	36.59	36.78	36.69	61.64	60.53	61.08
K	37.11	34.42	35.76	62.58	56.55	59.56
Average	36.02	35.53	35.78	62.53	59.33	60.93

The differences between the first and second years and the average NDF rates of genotypes were statistically insignificant, while those between years were extremely significant *($P<0.01$)*. Average NDF rate of genotypes in the first year was 62.53% and the highest NDF rate were obtained from line 1 (63.96%). Line 14 were in the first row (61.28%), in the second year for NDF, while the second year's average NDF was 59.33%. When considering the two year average data, line 6 exhibited the highest NDF rate (61.96%) and the lowest rate was obtained from the control (59.56%). Average NDF of the two years was 60.93 % (Table 5).

In the study of Holden et al. (2000) on orchardgrass, ADF and NDF rates were 28.3 and 48.3% respectively in the first year while in the second year they were 30.2% and 52.8%. Cherney and Cherney (2005) reported that NDF rates were between 61.0 and 65.0% in orchardgrass. Results of previous studies are parallel with those found in the present study. It is desirable for forage crops to have low ADF and NDF rates since these substances cause digestion difficulties and thus lowering food quality. From this point of view, it is an expected and desired situation in the present study that ADF and NDF rates of genotypes used are not different significantly from control, and content of ADF in some genotypes is lower than control.

Seed yield

Seed yield differences between genotypes were significant in the first year *($P<0.05$)* while insignificant in the second year and for the averages. However, differences between the years were significant *($P<0.01$)* and interaction between year and genotype were also significant *($P<0.05$)*. The average seed yield in the first year was 43.19 kg da^{-1} and the highest yield was obtained from line 14 with 55.40 kg da^{-1}. The average seed yield in the second year was lower than the first year (30.71 kg da^{-1}). For the average seed yield, line 18 was in the first row (39.14 kg da^{-1}). When considering the two year average seed yield, it was seen that the average of all genotypes was 36.95 kg da-1 and line 14 revealed the highest yield with 41.37 kg da-1 (Table 6).

Table 6. Seed yield of orchardgrass genotypes.

Genotype		Seed yield (kg da^{-1})		
		2006 year	**2007 year**	**Average**
1		47.37 **a-c***	23.03	35.20
3		33.30 **d**	32.47	32.88
6		37.57 **cd**	30.64	34.10
9		37.57 **cd**	38.30	37.93
11		40.53 **b-d**	29.63	35.08
14		55.40 **a**	27.34	41.37
18		41.13 **b-d**	39.14	40.14
19		50.37 **ab**	25.73	38.05
K		45.43 **a-c**	30.10	37.77
Average		43.19	30.71	36.95
LSD	**11.1**			

*Differences were significant in the same column with a different letter ($p<0.05$).

Four of the genotypes were superior to control variety for seed yield. Seed yield of year was considerably lower than the first second year. The average monthly temperature of March (9.1°C) was higher in 2007 (second year), than 2006 (1.2°C) and long term (-2.8°C). Metabolic activity in plants may start depending on this situation. The sudden decrease in average temperature in April to 1.4 °C (which was 7.2°C in 2006 and 5.3°C long term average) might have caused damage at the points where metabolic activity started. Therefore, it is thought that the number of plant stems decreased. It was observed that the number of bunched stems decreased. This situation may be the reason for the decrease in seed yield in the second year.

The highest seed yield was obtained to be 24.6 kg da^{-1} in the study of Serin et al. (1994) conducted under the ecological conditions of Erzurum. It was stated in another study by Gokkus et al. (1994) that average seed yield was 27.0 kg da^{-1}. Seed yield obtained in the present study is higher than that found in other studies under the ecological conditions of Erzurum. It is a desired situation in plant production to have high seed yields as well as high hay yields. From this point of view, it can be said that lines taken in selection experiment are promising.

In conclusion lines obtained from the last stage of selection program were evaluated in the present study that was aimed to develop a new cultivar using selection in Ulubag ecotype of orchardgrass native in Erzurum province. The main aim of the selection programme is to develop variety from the lines with higher quality plant material and yield (compared to control variety). It was determined that lines 3, 9, 14 and 18 belonging to Ulubag ecotype and used in experiment can be used to develop new varieties.

LITERATURE CITED

Acıkgoz, E., 1995. Forage Book. Department of Field Crops. Faculty of Agriculture. Uludag University. Bursa.

Andic, C., 1985. Plant species in rangelands and uplands of Erzurum region and their life forms and flowering periods. Journal of Faculty of Agriculture, Ataturk University. 16: 95-104.

Aydin, I., Z. Acar, F. Tosun, 1994. Adaptation studies of some perennial grass forage crops in Samsun conditions. Field Crop Symposium, 25-29 April, İzmir, pp. 61-64.

Anonymus, 1995. Acid detergent and neutral detergent fiber using ANKOM's fiber analyzer F200. Ankom Technology Corporation, Fairport, NY.

Ball, D.M., M. Collins, G.D. Lacefield, N.P. Martin, D.A. Mertens, K.E. Olson, D.H. Putnam, D.J. Undersander, M.W. Wolf, 2001. Understanding forage quality. American Farm Bureau Federation Publication 1-01, Park Ridge, IL.

Berg, C.C., R.R. Hill, 1989. Maturity effect on yield and quality of spring harvested orchardgrass forage. Crop Science, 29: 944-948.

Cherney, J.H., D.J.R. Cherney, 2005. Agronomic response of cool-season grasses to low-intensity harvest management and low potassium fertility. Agronomy Journal, 97:1216-1221.

Colburn, M.W., J.L. Evans, 1967. Chemical composition of the cell-wall constituent and acid detergent fiber fractions of forages. Journal of Dairy Science Vol.50, No.7. pp.1130-1135.

Convertini, G., D. Ferri, M. Maiorana, 1999. Quality and mineral composition of forages in Mediterranean environment, CIHEAM- Options Mediterraneennes, pp. 51-54.

Copponet, M., 1964. Changes in the mineral composition of pasture grasses under a grazing regime. Comptes Rendus Seances Academic Agriculture (France), 50:330-342.

Davies, I., 1976. Developmental characteristics of grass varieties in relation to their herbage production. I. Analysis of high-digesbility varieties of *Dactylis glomerata* at three stages of development. Journal Agriculture Science, 87:25-32.

Edward, B., Ph. D. Rayburn, 1998. Using a forage test to identify improvements in forage management. West Virginia University Extension Service, PO Box 6108, Morgantown.

Fisher, D.S., J.C. Burns, J.E. Moore, 1995. The nutritive evaluation of forage. In forages. Vol. II: The science of Grassland Agriculture (Ed.: R.F. Barnes, D.A. Miller and C.J. Nelson), Iowa State Univ. Press, Inc., 105-115.

Gordon, C.H., A.M. Decker, H.G. Wiseman, 1962. Some effect of nitrogen fertilizer, maturity and light on the composition of orchardgrass. Agronomy Journal, 54: 376-378.

Gokkus, A., Y. Serin, A. Koc, 1994. Effect of N given different time and doses on seed yield of orchardgrass. Field Crop Symposium, 25-29 April, İzmir. 3:128-131.

Holden, L. A., G.A. Varga, G.A. Jung, J.A. Shaffer, 2000. Comparison of 'grasslands puna' chicory and orchardgrass for multiple harvests at different management levels. Agronomy Journal, 92:191-194.

Kacar, B., 1972. Chemical Analyses of Plant and Soil. (Volume II) Analyses of Plant. Faculty of Agriculture Press. Ankara University, Ankara. No:453. p:50-72.

Kacar, B., 1984. Application of Plant Breeding Handbook. Faculty of Agriculture Press. Ankara University, No: 900. Application Handbook, No: 214 Ankara. 140p.

Kunelius, H.T., L.B. Mecleod, F.W. Calder, 1974. Effect of cutting management on yields digestibility, crude protein and persistence of timothy, bromegrass and orchardgrass. Canadian Journal Plant Science, 54:55-64.

Reid, R.L., G.A. Jung, C.M. Kinsey, 1967. Nutritive value of nitrogen-fertilized orchardgrass pasture at different periods of the year. Agronomy Journal, 59: 519-525.

Sagsoz, S., M. Tosun, I. Akgun, 1996. Determination of phenological, morphological and biological parameters of orchardgrass collected from different locations. 3th Turkish Ranglands and Forage Crops Symposium, Erzurum. pp. 527-534.

Schoner, F.J., E. Pfeffer, 1986. Influence of mowing season amount fed and physical structure on digestibility of organic matter in two grasses. Arch. Anim. Nutr. Berlin, 36:851-855.

Serin, Y., M. Tan, 1998. Forage Book. Faculty of Agriculture. Ataturk University. Erzurum. No:859, pp. 136-150.

Stratton, S.D., H.W. Ohm, 1989. Relationship between orchardgrass seed production Indiana and Oregon Crop Science, 29: 908-913.

Tosun, M., S. Sagsoz, 1994. Determination of morphological and biological parameters of naturally growing orchardgrass (*Dactylis glomerata* ssp hispanica (Roth) Nyman) in Erzurum region collected from different locations. Field Crop Symposium, Izmir, pp. 39-43.

Tosun, M., I. Akgun, S. Sengul, S. Sagsoz, 1997. Breeding of orcahrgrass *(Dactylis glomerata* L.) ecotypes collected from Erzurum region. 2th Field Crop Symposium, Samsun, pp. 461-466.

Tosun, M., S. Sagsoz, I. Akgun, 1996. Determination of seed and hay yield in wild orchadrgrass (*Dactylis glomerata* L.)

plants. 3th Turkish Ranglands and Forage Crops, Symposium. Erzurum, pp. 402-407.

Van Soest, P.J., J.B. Robertson, R.A. Lewis, 1991. Methods for dietary fiber, neutral detergent fiber, and nonstarch polysaccharides in relation to animal nutrition. Journal Dairy Science, 74: 3583-3597.

Whitehead, D.C., 1966. Nutrient minerals in grassland herbage. Review series 1/1966. Farnham Royal, Commonwealth Bureau of Pastures and Field Crops.

Yeldan, M., 1984. Forage and Breeding of Animal. Faculty of Agriculture Press (in Turkish). No: 293. Ankara University, Ankara, pp. 55-65.

EFFECTS OF NANO FERTILIZER APPLICATION AND MATERNAL CORM WEIGHT ON FLOWERING OF SOME SAFFRON (*Crocus sativus* L.) ECOTYPES

Reza AMIRNIA[1], *Mahdi BAYAT*[1*], *Mahdi TAJBAKHSH*[1]

Urmia University, Faculty of Agriculture, Department of Agronomy, Urmia, IRAN
[*] *Corresponding author: mahdibayat971@gmail.com*

ABSTRACT

To develop saffron planting in Urmia, West Azerbaijan, Iran, a split-split plot experiment based on CRBD was carried out in the Urmia University's research farm for two years (2013-14). Nanofertilizers (Fe, P, K and no-fertilizer (control)) as main plots, saffron ecotypes (Mashhad, Torbat-Heydarieh, Torbat-jam, Gonabad, Ghaen and Birjand) as subplots and maternal corm weight (6, 8, 10 and 12 g) as sub-sub plots were considered. Throughout the two years of the study, results showed significant differences between nanofertilizers levels, saffron ecotypes, maternal corm weight and their interactions in terms of all flowering traits. Results highlighted the importance of the nanofertilizers on improving saffron yield. In addition, it was also clear that Fe, P and K nanofertilizers all had positive effects on the saffron flowering. The results also illustrated that the high yield ecotypes Torbat-Heydarieh and Mashhad in Urmia region were due to similar climatic conditions of these regions. These results emphasized the importance of the mother corm weight on increasing saffron flowering was much better than both the nanofertilizer and saffron ecotype. As the mother corm weight increases from 6 to 12 grams, all the studied traits, including dry saffron yield and flower number also increased, by 5.17 times and 4.4 times, respectively. The results of stepwise regression and correlation coefficients noted that flower number, dry and fresh flower weight were the most effective traits on dry saffron yield. In total, it was concluded that saffron flowering traits are strongly influenced by environmental conditions and farm management. Therefore, the correct choice of saffron ecotypes, nanofertilizer and maternal corm weight are vital factors in farm management and economic saffron production.

Keywords: Correlation Coefficients, Saffron Ecotypes, Stepwise Regression, saffron yield.

[*]*The part of PhD thesis results of the corresponding author, MAHDI BAYAT*

INTRODUCTION

Saffron is a spice derived from the dried stigma of the *Crocus sativus* flower, which is a sterile triploid plant that propagated by means of corms (Botella, et al., 2002; Gómez-Gómez, 2012). From the ancient times, the red scarlet stigmas of saffron have been used mainly in drug applications, textile dye, incense, cosmetics and food purposes (Zdemür, et al., 2006; Juana, et al., 2009; Rubio-Moraga, et al., 2009, Anastasaki, et al., 2010). Recently, there has been increasing interest in the biological effects of the components of saffron. In particular, their potential medical applications, especially those based on their cytotoxic, mood elevating, liver detoxifying, analgesic and antitumor properties (Ma, et al., 2001; Abdullaev, 2003; Schmidt, et al., 2007; Siracusa, et al., 2010). For these reasons, saffron is the world's most expensive spice (Kafi, 2006). Saffron is currently being cultivated in Morocco, Spain, Iran, India, Pakistan, Turkey, Italy, Switzerland, Greece and Central Asia. Additionally, new cultivations have been created in Australia, Mexico, Argentine and New Zealand. Among the saffron producing countries, Iran is one of the main producers of saffron in the world (Jalali-Heravi, et al., 2010).

Much information is available on the use of saffron as a dye, aroma and for medicinal purposes, but there is a lack of information on its climatic requirements, agronomic management techniques and their effects on its quantitative and qualitative traits. From an agronomic point of view, saffron is actually very well adapted to different environmental conditions ranging from dry sub-tropical to continental climates (Azizbekova and Milyaeva, 1999; Sampathu, et al., 1984; Gresta, et al., 2009). Mollafilabi (2004) and Fernandez (2004) suggested that the optimal climatic conditions for this species are rainy autumns, mild winters and warm summers. The effect of the corm weight at planting the saffron quantitative yield has been studied by Salinger (1991); Rees (1992); Negbi (1999); De Juan, et al., (2003);

Molina, et al., (2004) and Renau-Morata, et al., (2012). These researchers reported that the weight of the mother corm has a significant effect on the vegetative development and the production of daughter corms. The effect of the different saffron corms proveniences on stigmas yield and qualitative traits of saffron has been studied by several researchers such as Molina, et al., (2005); Castillo, et al., (2005); Rubio-Moraga, et al., (2009); Siracusa et al., (2010); Anastasaki, et al., (2010) and Maggi, et al., 2011). These authors pointed out that the environment and climatic conditions, for example: temperature, soil and water content, have severe effect on both quantitative and qualitative traits of saffron.

Because of the limitation in arable lands and water resources, the development of agriculture sector is only possible by increasing of resources use efficiency with the minimum damage to production bed through effective use of modern technologies. Among these, nanotechnology has the potential to revolutionize the agricultural systems (Baruah and Dutta, 2009). Recent research on nanoparticles in a number of crops has evidenced for enhanced germination and seedling growth, physiological activities, gene expression and protein level indicating their potential use in crop improvement (Kole, et al., 2013). In this way, Azarpour, et al., (2013) reported nano iron fertilizers foliar spraying had significant effects at 1% probability level on fresh flower cover yield of saffron. In addition, Studies showed that the use of nanofertilizers causes an increase in nutrients use efficiency, reduces soil toxicity, minimizes the potential negative effects associated with over dosage and reduces the frequency of the application. Hence, nanotechnology has a high potential for achieving sustainable agriculture, especially in developing countries (Naderi and Danesh-Shahraki, 2013).

Besides the expanding employment opportunities, Saffron can contribute to the economy of Iran by earning and saving valuable foreign exchange. Therefore, the aim of this study is to investigate the agricultural practices with the hope of improving saffron yield and flowering traits, in order to determine the optimum levels of maternal corm weight, nanofertilizer and saffron ecotypes. It is our hope that this work will eventually establish a new standard of agricultural practices which will optimize Saffron yields in the Urmia region.

MATERIALS AND METHODS

To develop saffron (*Crocus sativus* L.) planting in Urmia, West Azerbaijan, Iran, a split-split plot experiment based on CRBD was carried out in the research farm of Urmia University for the years of 2013-2014. Nano fertilizers (Fe, P, K and no-fertilizer (control)) as main plots, saffron ecotypes (Mashhad, Torbat-Heydarieh, Torbat-jam, Gonabad, Ghaen and Birjand) (Table 1) as subplots and maternal corm weight (6, 8, 10 and 12 g) as sub-sub plots were considered. Six saffron samples from different regions of Iran's traditional saffron production areas were studied in this work. The geographic and climatic characteristics of each region are shown in Tables 1 and 2.

Table 1. Geographic characteristics of regions.

No	Region	Province	Country	Elevation (M)	Longitude (E)	Latitude (N)
1	Mashhad	Razavi Khorasan	Iran	999.2	38° 59′	16° 36′
2	Torbat Jam	Razavi Khorasan	Iran	1056	58° 41′	34° 21′
3	Gonabad	Razavi Khorasan	Iran	1450.8	13° 59′	16° 35′
4	Torbat Heidarish	Razavi Khorasan	Iran	950.4	35° 60′	15° 35′
5	Birjand	South Khorasan	Iran	1491	59° 12′	32° 52′
6	Gaean	South Khorasan	Iran	1432	59° 10′	33° 43′
7	Urmia	West Azerbaijan	Iran	1315.9	45° 05′	37° 32′

Table 2. Climatic characteristics of regions based on the 30 years average.

Station	Maximum Temperature (C°)	Minimum Temperature (C°)	Mean Daily Temperature (C°)	Relative Humidity (%)	Monthly Total Of Precipitation (Mm)	Days With Minimum Temperature Equal Zero And Below
Mashhad	21.1	7.1	14.1	55	255.2	89.3
Gonabad	23.8	10.7	17.3	37	143.6	49.2
Torbate Heydarieh	21.3	7.3	14.3	46	274.8	95.7
Torbate Jam	22.4	8.8	15.6	45	175.6	73.6
Birjand	24.5	8.4	16.5	36	170.8	76.2
Ghaen	22.3	6.3	14.3	37	175.8	93.8
Mean	22.6	8.1	15.4	42.7	199.3	22.6
Urmia	17.6	5.4	11.5	60	341	110.6

In terms of the soil characteristics at the research farm, the soil texture is a silty-loam structure containing 1.5% of organic carbon and 0.03% of nitrogen, in addition to 25 p.p.m of phosphorus and 534 p.p.m of potassium. The soil acidity was indicated at a pH of 7.7 at a depth of 0-30 cm. After preparing the field on April 15, 2013; the cultivation practices applied were those commonly used for this crop. Every plot contained 8 culture lines, each of which were 3 meters in length and placed 25 cm apart from each other. The total plot area was $6m^2$. Corm distances on the lines were 8 cm and placed at a depth of 15 cm. The density of the corm placement was 50 corms per m^2. To avoid marginal effects and to minimize error, plots were situated beside each other no closer than 50 cm. To enhance accuracy, margins were placed at the beginning and the end of plots as well as the 50 cm border between each plot. Nano fertilizers were sprayed twice in the months of January and February. Data were recorded for Dry Saffron Yield (kg ha^{-1}) (DSY), Flower Number (m^2) (FN), Fresh Stigma Weight (mg) (FSW), Dry Stigma Weight (mg) (DSW), Stigma Length (cm) (SL), Fresh Flower Weight (mg) (FWF) and Dry Flower Weight (mg) (DFW). The relationships between these traits using simple correlation coefficients were studied. Step-wise regression analysis was done to fit the best model for existent variation in dry saffron yield as the dependent variable. The cluster analysis based on Ward's method was also used to classify saffron ecotypes. Data analysis was done using SAS ver. 9.2 and SPSS ver. 21 programs.

RESULTS AND DISCUSSION

Analysis of Variance

The results of the variance analysis (Table 3) indicated significant differences between years, nano-fertilizers, saffron ecotypes, maternal corm weight and their interactions in terms of all flowering traits. This result implicated that flowering traits and dry saffron yield (DSY) were strongly influenced by environmental conditions and farm management. Therefore, the correct choice of saffron ecotypes, nanofertilizer and maternal corm weight can be considered the crucial factors in farm management and economic saffron production

Table 3. Variance analysis of flowering characteristics in saffron.

		Mean Square						
S.O.V	**Df**	**DSY (kg ha^{-1})**	**FN (m^2)**	**FSW (mg)**	**DSW (mg)**	**SL (cm)**	**FFW (mg)**	**DFW (mg)**
Y♣	1	252.02 **	69548.90 **	109.05 ns	29.88 *	8.51 ns	1918.18 ns	146.10 ns
Error $_a$	2	0.12	14.41	143.69	2.86	3.29	3505.13	86.80
F	3	3.98 **	278.18 **	123.51 **	7.80 **	8.90 **	5908.26 **	250.73 **
F*Y	3	2.24 **	263.94 **	113.46 **	7.23 **	1.29 **	880.03 *	50.02 *
Error $_b$	6	0.03	8.86	8.07	0.43	0.19	197.83	11.84
E	5	3.98 **	234.28 **	142.51 **	15.51 **	2.93 **	3320.89 **	47.53 **
E*F	15	0.18 **	14.92 **	23.89 **	1.07 **	0.44 *	581.16 **	32.95 **
E*Y	5	0.10 *	16.15 *	14.37 ns	0.33 ns	2.59 **	1139.30 **	99.78 **
E*F*Y	15	0.11 **	10.47 *	18.59 **	0.72 *	1.60 **	328.28 *	27.32 **
Error $_c$	40	0.04	5.83	7.71	0.34	0.22	162.33	6.30
S	3	99.15 **	27628.20 **	429.47 **	13.26 **	11.36 **	90321.92 **	599.54 **
S*F	9	0.44 **	56.34 **	1.18 ns	0.61 *	0.33 *	74.17 ns	2.40 ns
S*E	15	0.33 **	6.74 *	11.25 *	0.59 **	0.29 *	609.94 **	16.06 **
S*F*E	45	0.09 **	8.23 **	2.60 ns	0.24 ns	0.12 ns	80.14 ns	4.40 ns
S*Y	3	49.37 **	13738.30 **	12.63 *	0.25 ns	2.63 **	219.29 ns	4.06 ns
S*F*Y	9	0.35 **	50.99 **	1.12 ns	0.27 ns	0.08 ns	36.48 ns	1.86 ns
S*E*Y	15	0.21 **	6.34 ns	10.14 *	0.56 **	0.44 **	614.61 **	15.26 **
S*F*E*Y	45	0.09 **	7.84 **	2.99 ns	0.29 ns	0.15 ns	72.07 ns	4.43 ns
Error $_t$	336	0.04	3.81	5.43	0.26	0.15	108.80	5.95
CV (%)	---	14.0	7.8	8.4	9.8	10.9	13.1	5.7

♣ Abbreviations described in materials and methods
**: p> 0.01, *: p>0.05, ns: non-significant

Mean Comparison of the Main Effects

Year

The mean comparisons of the results across the two years are shown in Table 4. The results showed that all saffron flowering traits especially dry saffron yield (DSY), flower number (FN) and dry stigma weight (DSW) increased in 2014 compared to 2013. However, the improvement of the flowering traits in the second year was predictable, because the saffron corms were carried from the easternmost province of Iran (Khorasan Razavi and Khorasan South) to the westernmost province of Iran (West Azarbaijan, Urmia). As shown in tables 1 and 2, climate characteristics of the easternmost province of Iran (average maximum temperature 22.6 °C, average minimum temperature 8.1 °C, average daily temperature 15.4 °C, average relative humidity 42.7%, average monthly total precipitation 199.3 mm and average number of days with minimum temperature equal zero and below 22.6 days) were different from climate characteristics of the westernmost province of Iran (average maximum temperature 17.6 °C, average minimum temperature 5.4

°C, average daily temperature 11.5 °C, average relative humidity of 60.0%, average monthly total precipitation 341.0 mm and average number of days with minimum temperature equal zero and below 110.6 days). As a result, the saffron corms were highly stressed in the first year; consequently, the flowering traits and saffron yield were extremely reduced. Nonetheless, in the second year, saffron corms not only adapted to Urmia environmental conditions but also produced more daughter corms, so that saffron flowering traits, especially DSY and FN increased dramatically.

Table 4. Mean compression of flowering traits in saffron.

Treatment	DSY (kg ha^{-1})	FN (m^2)	FSW (mg)	DSW (mg)	SL (cm)	FFW (mg)	DFW (mg)
Year							
2013	0.71 b♣	13.91 b	27.25 a	5.01 b	3.44 a	338.52 a	42.02 a
2014	2.03 a	35.89 a	28.12 a	5.47 a	3.69 a	342.17 a	43.03 a
Nano Fertilizer							
Control	1.16 c	23.04 c	27.44 b-c	5.07 b	3.23 c	336.80 b	41.19 b
Fe	1.52 a	26.13 a	28.87 a	5.52 a	3.81 a	349.93 a	44.35 a
K	1.30 b	24.66 b	26.64 c	5.02 b	3.52 b	336.89 b	42.22 b
P	1.48 a	25.77 a	27.80 a	5.33 a	3.69 a	337.76 b	42.36 b
Mean of Nano Fertilizer	1.43	25.52	27.77	5.29	3.67	341.53	42.98
Mean - Control	0.27	2.48	0.33	0.22	0.44	4.73	1.79
Saffron Ecotype							
Birjand	1.28 d	25.15 b	27.24 b-c	4.87 d	3.44 b-c	336.53 b	42.42 a-b
Ghaen	1.15 e	22.83 c	26.25 c	4.82 d	3.37 c	336.29 b	41.42 b
Gonabad	1.18 e	23.26 c	26.46 c	5.06 c-d	3.73 a	333.04 b	42.35 a-b
Mashhad	1.52 b	26.25 a	28.37 a-b	5.54 b	3.43 b-c	347.76 a	42.39 a-b
Torbat-Heidarieh	1.67 a	26.70 a	29.27 a	5.85 a	3.79 a	345.57 a	43.26 a
Torbat-Jam	1.42 c	25.20 b	28.52 a	5.29 c	3.62 a-b	342.86 a	43.34 a
Cluster I	1.53	26.05	28.72	5.56	3.61	345.42	42.98
Cluster II	1.20	23.75	26.65	4.92	3.51	335.29	42.05
Maternal Corm Weight (gr)							
6	0.47 d	9.64 d	25.53 d	4.84 d	3.26 c	307.56 c	40.00 c
8	1.04 c	19.71 c	27.25 c	5.19 c	3.39 b	334.01 b	41.76 b
10	1.53 b	27.83 b	28.35 b	5.38 b	3.83 a	358.84 a	43.97 a
12	2.43 a	42.42 a	29.60 a	5.54 a	3.77 a	360.96 a	44.38 a
Ratio 12 on 6	5.17	4.40	1.16	1.14	1.16	1.17	1.11

♣ - Treatments with the same letter(s) don't have significant difference

Nano-fertilizer

The results of the mean comparison (Table 4) showed that nanofertilizers had a positive effect on all saffron flowering traits. In comparison to the control (no-fertilizer), the nanofertilizers significantly increased dry saffron yield (DSY), flower number (FN), fresh stigma weight (FSW), dry stigma weight (DSW), stigma length (SL), fresh flower weight (FFW) and dry flower weight (DFW) up to 270 g ha^{-1}, 2.48, 0.33 mg, 0.22 mg, 0.44 cm, 4.73 mg and 1.79 mg, respectively. Overall, these results emphasized the positive effects of all three nanofertilizers, Fe, P and K, in the improvement saffron flowering traits. Behzad et al., (1992) stated that organic matter and cow manure applications elevated the fertility of soils in saffron cultivation. Some of the earlier studies on saffron argued that chemical fertilization alone did not much improve flower yield, unless applied together with organic material such as cow manure and sawdust (McGimpsey, et al., 1997; Behnia, et al., 1999; Unal and Cavusoglu, 2005). Rico et al., (2011) stated that nanotechnology have positive effects on plants included enhanced germination percentage and rate; length of root and shoot, and their ratio; and vegetative biomass of seedlings in many crop plants. Also Liu et al., (2010) found that nanocomposites were safe for wheat seed germination, emergence and growth of seedlings.

Saffron ecotype

The results of mean comparison (Table 4) showed the ecotypes Torbat-Heydarieh (DSY = 1.67 kg ha^{-1}, FN = 26.7) and Mashhad (DSY = 1.52 kg ha^{-1}, FN = 26.25) were as the best saffron ecotypes in the weather conditions of Urmia. While ecotypes Ghaen (DSY = 1.15 kg ha^{-1}, FN = 22.83) and Gonabad (DSY = 1.18 kg ha^{-1}, FN = 23.26) did not thrive in the conditions and were considered the most inappropriate ecotypes in Urmia region. In this study, saffron ecotypes were collected from regions with different climatic characteristics, so the observed differences between the ecotypes were related to the different climatic characteristics. With the evaluation of the meteorological data (Tables 1 and 2) and performing the cluster analysis in terms of all climatic characteristics, it was found that the studied regions can be classified in two clusters (Figure 1). Cluster I contained the regions of Torbat-Jam, Birjand, Ghaen and Gonabad and Cluster II contained regions Mashhad, Torbat-Heydarieh and Urmia. These results confirmed that the climatic characteristics of the Urmia region had the most similarity to the climate characteristics of Mashhad and Torbat-Heydarih regions and had the lowest similarity to the climate characteristics of Torbat-Jam, Birjand, Ghaen and Gonabad regions. In total, it can conclude that for establishing a new saffron farm, considering the climate characteristics of the region is very important and plays an effective role in the improvement of the saffron yield. Rubio-Moraga, et al., (2009) reported that all accessions (43 ecotypes from 11 countries) appear as identical clones, not only because of morphological characteristics, but also at the molecular level. The observed differences in saffron quality are mainly due to the methodology followed in the processing of stigmas, independent of the species origin (Ordoudi and Tsimidou, 2004). On the other hand, Molina et al., (2004; 2005) reported that the number of flowers formed per corm and the time of flower emergence depended on the environment conditions such as temperature and time-lifted corm.

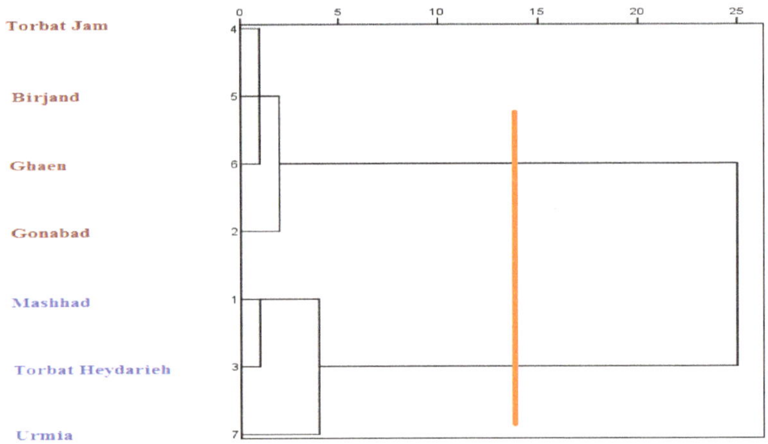

Figure 1. Dendrogram cities based on climatic characteristics using Ward's method.

Conversely, in order to classify the saffron ecotypes with respect to all studied flowering traits, the cluster analysis was used. The results showed six saffron ecotypes grouped in two clusters (Figure 2). Cluster I consisted ecotypes of Mashhad, Torbat-Heydarieh and Torbat-Jam and cluster II consisted ecotypes Birjand, Ghaen and Gonabad. When calculating the trait mean within each cluster (Table 4) it was revealed that all flowering traits of cluster I were superior to that of Cluster II. Therefore, it can be concluded that the ecotypes of cluster I were the most compatible with the Urmia climatic conditions, so these ecotypes can be introduced as the best saffron ecotypes in Urmia. It has previously been demonstrated that the geographical signals and regional agricultural practices, especially the use of fertilizers, in organic and conventional agriculture (Bateman and Kelly, 2007) effects the plant's production ability. Lage and Cantrell (2009) have grown saffron in eleven different experimental zones with differences in the altitudes, soils texture and climate. These results showed that environmental conditions have a great effect on saffron quality (Turhan, et al., 2007).

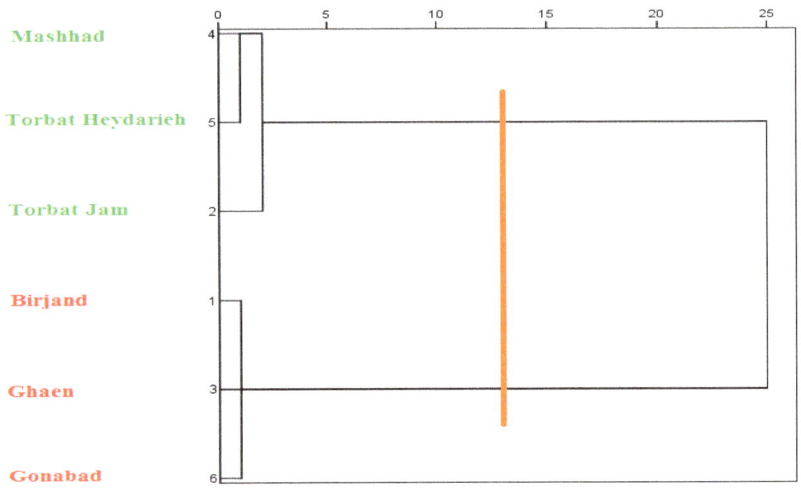

Figure 2. Dendrogram saffron ecotypes based on flowering characteristics using Ward's method.

Maternal corm weight

The results of the mean comparison of maternal corm weight are shown in Table 4. The results emphasized the importance of maternal corm weight on improving saffron flowering and yield, which was much more significant than nanofertilizer and saffron ecotype effects. As mentioned above, increasing the maternal corm weight from 6 to 12 grams produced a corresponding significant increase all studied traits. DSY, FN, FSW, DSW, SL, FFW and DFW in corms of 12 grams, compared with corms of only 6 grams increased 5.17, 4.4, 1.16, 1.14, 1.16, 1.17 and 1.11 times, respectively (Table 4). These results showed that corms of 12 grams not only had more flower buds, but that also this corm size increased the saffron yield in the first year directly. Additionally, because of their ability to produce more daughter corms with higher weight, the saffron yield was able to increase dramatically in the coming years. In summary, it can be advised that saffron corms 10 grams and above are most suitable for establishing a new farm, while corms 6 grams and smaller are not recommended for use. Mollafilabi (2004) pointed out that planting corms above 3 cm diameter with an approximate weight of 10 g produced the highest yields. On the other hand, Negbi (1999); De Mastro and Ruta, (1993); Molina, et al., (2004) stated that there is an increase in the number of flowers per corm when larger corms are used at planting. Renau-Morata, et al., (2012) conclude that maternal corm weight determines if the corm will flower or not, given that if the corm does not reach a certain weight, it only produces leaves.

Mean comparison of triple interaction effect (maternal corm weight × saffron ecotype × nano-fertilizer)

The results of the mean comparison (Table 5) showed that the interaction effects 12 g × Mashhad × Fe, 12 g × Torbat-Heydarieh × Fe and 12 g × Torbat-Jam × Fe gained rank 1, 2 and 3, respectively, in terms of all studied flowering traits. Therefore, these interaction effects were identified as the most suitable effects in the Urmia climatic condition. On the other hand, interactions effects 6 g × Birjand × control, 6 g × Ghaen × control and 6 g × Ghaen × K received the highest rank (94, 95 and 96 respectively) were found as the weakest and had the most unsuitable interaction effects. In total, it was highlighted that maternal corm weight, nanofertilizer and saffron ecotype had the important roles when determining the potential of saffron yield, respectively. Therefore we recommended for establishing a new saffron farm, attention to the following notes: (i) The selection of better quality and larger saffron corms and performing the proper planting density. (ii) Fertilizers to be used in optimal times and doses. (iii) Saffron ecotypes to be supplied from regions where there is the maximum climate similarity between regions. With observing the above notes, saffron yield not only increases in the first year directly, but also increases in the coming years due to improved yield components. Some authors have proposed that an increase in the weight of the produced corms could improve saffron production (Juan, et al., 2009). On the other hand, corm formation of saffron is one of the important characteristics because the corm is the only source for propagation (Turhan, et al., 2007).

Correlation Coefficients

The results of correlation coefficients (Table 6) showed that there were positive and significant correlations between all studied flowering traits. Therefore any improvement in flowering traits will increase dry saffron yield. The results clearly demonstrate that dry saffron yield (DSY) had the highest correlation with flower number (FN) ($r = 0.98^{**}$). As explained earlier, larger saffron corms can produce more DSY in the first year due to have more flower bulbs. Moreover, larger saffron corms have higher quality, so that these corms can produce more daughter corms and subsequently will significantly increase DSY in the coming years (Mollafilabi, 2004; Renau-Morata et al., 2012). Within the course of this study, it also became clear that the nanofertilizers, especially Fe and P improved yield components especially FN and consequently increased DSY. In general, it can be concluded that increasing maternal corm weight and the use nanofertilizers will increase the yield components as well as the dry saffron yield, both directly and indirectly.

Stepwise Regression

In order to determine the most influential traits on dry saffron yield, stepwise regression was used. In this analysis, DSY as the dependent variable and the other traits as independent variables were considered. On the other hand, to achieve a more realistic understanding of traits effects on DSY, we used step by step regression for each year individually, and for the two years overall. The results of the stepwise regression in cropping year 2013 (Table 7) showed that the flower number (FN), dry stigma weight (DSW), stigma length (SL), fresh stigma weight (FSW) and fresh flower weight (FFW) entered into the regression model, was able to justify more than 91% of the variation of DSY. While, the results of stepwise regression in cropping year 2014 (Table 7) pointed out that FN, DSW, DFW and FFW justified more than 99% of the variation of DSY. The regression equation of DSY in 2013 and 2014 were as follows:

$$DSY_{2013} = 0.494(FN) + 0.424(DSW) + 0.179(SL) - 0.186(FSW) + 0.215(FFW)$$

$$DSY_{2014} = 0.934(FN) + 0.214(DSW) - 0.044(DFW) - 0.041(FFW)$$

At last, the results of stepwise regression over the two cropping years (Table 7) highlighted that the traits FN, DSW and DFW were the most effective traits on DSY and justified more than 99% of the variation of DSY. The regression equation of DSY in total two cropping years was as follows:

$$DSY_{total} = 0.911(FN) + 0.181(DSW) - 0.046(DFW)$$

Table 5. Mean comparison of triple interaction effect (maternal corm weight × saffron ecotype × nanofertilizer) of flowering traits in saffron.

Weight	Ecotype	Fertilizer	DSY	FN	FSW	DSW	SL	FFW	DFW	Rank
6	Birjand	Fe	0.44	9.93	25.01	4.46	3.48	308.65	40.62	87
6	Birjand	P	0.44	10.06	25.67	4.28	3.16	301.00	37.23	91
6	Birjand	K	0.46	9.85	25.42	4.59	3.17	298.70	41.75	88
6	Birjand	Control	0.34	8.94	22.61	4.03	2.59	300.64	37.60	94
6	Torbat-Jam	Fe	0.52	10.31	25.92	5.05	3.33	323.62	42.54	73
6	Torbat-Jam	P	0.54	9.66	24.23	5.01	3.40	317.06	41.18	82
6	Torbat-Jam	K	0.44	9.56	25.23	4.63	3.44	311.50	40.41	86
6	Torbat-Jam	Control	0.50	10.59	25.52	4.86	2.79	311.37	40.54	90
6	Ghaen	Fe	0.38	7.95	24.93	4.94	3.20	308.65	41.27	89
6	Ghaen	P	0.35	8.24	24.71	4.44	3.35	301.00	38.18	92
6	Ghaen	K	0.31	7.88	22.93	3.95	2.82	298.70	34.45	96
6	Ghaen	Control	0.29	7.37	24.19	4.03	2.69	300.64	35.32	95
6	Mashhad	Fe	0.66	11.17	29.14	5.80	3.40	323.62	42.57	55
6	Mashhad	P	0.54	10.80	25.42	4.89	3.05	309.68	43.00	78
6	Mashhad	K	0.57	10.68	25.07	5.04	3.03	313.09	41.67	83
6	Mashhad	Control	0.55	10.75	26.98	4.97	3.14	317.15	38.51	77
6	Torbat-Heidarieh	Fe	0.73	12.03	28.15	5.47	3.69	327.98	42.54	52
6	Torbat-Heidarieh	P	0.66	11.37	27.22	5.49	3.48	314.79	41.80	65
6	Torbat-Heidarieh	K	0.55	10.76	24.97	5.06	3.73	298.10	39.13	75
6	Torbat-Heidarieh	Control	0.64	11.08	28.24	5.57	3.21	314.89	41.20	70
6	Gonabad	Fe	0.43	8.17	26.22	5.03	3.69	311.41	40.62	74
6	Gonabad	P	0.41	7.99	26.27	5.12	3.63	296.94	37.23	81
6	Gonabad	K	0.37	8.40	25.04	4.76	3.73	297.00	41.75	80
6	Gonabad	Control	0.28	7.77	23.72	4.64	3.06	275.39	37.60	93
8	Birjand	Fe	1.11	21.43	26.62	5.17	3.28	335.60	43.83	57
8	Birjand	P	1.00	20.13	26.97	5.00	3.52	326.39	40.74	69
8	Birjand	K	0.90	19.73	26.43	4.67	3.40	325.50	42.15	71
8	Birjand	Control	0.84	19.61	24.87	4.52	2.72	327.07	41.89	76
8	Torbat-Jam	Fe	1.17	20.70	30.59	5.48	3.47	341.00	43.83	40
8	Torbat-Jam	P	1.19	20.38	27.58	5.41	3.54	323.83	40.74	59
8	Torbat-Jam	K	1.02	19.04	28.33	5.21	3.54	334.83	42.15	53
8	Torbat-Jam	Control	0.97	19.04	29.39	5.39	2.90	321.29	41.89	63
8	Ghaen	Fe	0.96	18.64	26.81	5.17	3.13	335.60	44.03	62
8	Ghaen	P	0.93	18.79	26.64	5.00	3.64	326.39	41.73	66
8	Ghaen	K	0.89	18.59	24.77	4.67	3.30	325.50	40.37	79
8	Ghaen	Control	0.69	16.53	25.55	4.52	2.94	327.07	38.75	85
8	Mashhad	Fe	1.34	21.08	31.09	6.14	3.59	355.42	44.03	30
8	Mashhad	P	1.20	22.86	27.72	5.12	3.18	339.18	41.73	60
8	Mashhad	K	1.07	20.72	25.11	4.94	3.12	338.82	40.37	72
8	Mashhad	Control	1.17	20.17	26.88	5.73	3.05	339.00	38.75	64
8	Torbat-Heidarieh	Fe	1.45	21.81	30.06	6.14	3.99	355.42	43.09	26
8	Torbat-Heidarieh	P	1.31	21.01	29.69	5.70	3.45	346.29	40.81	45
8	Torbat-Heidarieh	K	1.11	20.52	26.41	5.33	3.54	331.39	41.85	58
8	Torbat-Heidarieh	Control	1.09	19.41	29.72	5.75	3.40	339.32	41.84	46
8	Gonabad	Fe	1.04	18.62	26.09	5.43	3.82	341.00	43.46	49
8	Gonabad	P	0.99	18.54	26.44	5.00	3.61	327.22	41.51	68
8	Gonabad	K	0.77	18.27	25.82	4.31	3.84	333.36	43.10	67
8	Gonabad	Control	0.66	17.43	24.43	4.75	3.35	319.38	39.52	84

Table 5. Continue

Weight	Ecotype	Fertilizer	DSY	FN	FSW	DSW	SL	FFW	DFW	Rank
10	Birjand	Fe	1.70	29.52	30.00	5.50	4.14	366.41	44.54	14
10	Birjand	P	1.57	27.85	29.69	5.66	4.14	354.84	42.45	28
10	Birjand	K	1.44	28.84	26.87	5.05	3.88	352.45	44.24	38
10	Birjand	Control	1.14	26.20	27.84	4.62	3.19	362.84	42.77	50
10	Torbat-Jam	Fe	1.58	28.84	29.61	5.40	4.14	366.41	46.74	13
10	Torbat-Jam	P	1.66	29.94	26.99	5.24	4.14	354.84	45.72	29
10	Torbat-Jam	K	1.25	26.91	28.98	4.90	3.88	356.33	44.93	36
10	Torbat-Jam	Control	1.24	26.74	28.83	4.86	3.19	358.97	43.35	48
10	Ghaen	Fe	1.53	27.26	26.98	5.40	3.89	369.52	46.24	27
10	Ghaen	P	1.54	28.40	27.06	5.24	3.65	356.70	43.85	37
10	Ghaen	K	1.29	25.36	26.19	4.90	3.45	350.36	42.37	61
10	Ghaen	Control	1.19	23.47	26.92	4.86	3.54	359.62	40.75	56
10	Mashhad	Fe	1.89	28.93	32.24	6.24	3.89	372.15	46.24	9
10	Mashhad	P	1.86	31.64	28.99	5.58	3.65	351.73	43.85	31
10	Mashhad	K	1.57	28.18	27.35	5.20	3.45	358.56	42.37	42
10	Mashhad	Control	1.44	26.61	30.10	5.62	3.54	351.76	40.75	39
10	Torbat-Heidarieh	Fe	2.28	33.22	31.34	6.43	4.41	369.52	46.74	5
10	Torbat-Heidarieh	P	1.90	29.43	30.32	6.24	4.17	356.70	45.22	11
10	Torbat-Heidarieh	K	1.69	28.99	27.61	5.62	4.01	350.99	42.09	35
10	Torbat-Heidarieh	Control	1.52	26.47	29.41	5.78	3.48	358.98	46.69	25
10	Gonabad	Fe	1.53	26.98	26.98	5.50	4.41	372.15	44.54	20
10	Gonabad	P	1.54	26.87	27.06	5.66	4.17	354.60	42.45	33
10	Gonabad	K	1.29	25.40	26.19	5.05	3.74	360.24	44.24	41
10	Gonabad	Control	1.01	25.80	26.92	4.62	3.75	347.21	42.77	51
12	Birjand	Fe	2.79	48.55	31.80	5.49	3.74	362.15	46.10	10
12	Birjand	P	2.28	44.13	30.22	4.95	4.01	354.81	44.87	18
12	Birjand	K	2.25	40.46	28.96	5.49	3.36	352.51	44.30	32
12	Birjand	Control	1.72	37.23	26.90	4.47	3.21	354.96	43.70	47
12	Torbat-Jam	Fe	2.89	47.06	32.16	5.88	4.26	373.18	46.10	3
12	Torbat-Jam	P	3.11	45.88	29.58	5.98	4.16	360.13	44.40	7
12	Torbat-Jam	K	2.25	39.70	30.46	5.46	3.94	358.65	43.66	17
12	Torbat-Jam	Control	2.32	38.94	32.91	5.81	3.78	373.18	44.81	8
12	Ghaen	Fe	1.97	38.85	28.56	5.07	3.74	359.68	46.38	23
12	Ghaen	P	2.25	42.01	29.11	5.26	4.01	354.30	44.38	19
12	Ghaen	K	2.23	41.00	26.93	5.17	3.36	352.30	42.66	43
12	Ghaen	Control	1.55	34.99	27.64	4.49	3.21	354.81	41.32	54
12	Mashhad	Fe	2.95	46.14	31.80	6.19	4.18	385.48	46.38	1
12	Mashhad	P	2.48	45.04	29.25	5.56	3.69	366.35	44.38	12
12	Mashhad	K	2.79	44.08	27.85	5.81	3.41	372.98	42.66	21
12	Mashhad	Control	2.27	41.06	28.99	5.87	3.55	369.63	41.32	24
12	Torbat-Heidarieh	Fe	3.12	49.29	32.16	5.95	4.26	373.18	45.43	2
12	Torbat-Heidarieh	P	3.12	44.78	31.22	6.43	4.16	372.01	44.45	4
12	Torbat-Heidarieh	K	2.67	45.58	29.41	5.84	3.94	354.13	43.36	15
12	Torbat-Heidarieh	Control	2.82	41.53	32.31	6.78	3.78	365.81	45.37	6
12	Gonabad	Fe	2.16	40.63	28.56	5.19	4.18	360.80	46.02	16
12	Gonabad	P	2.62	42.73	29.11	5.72	3.69	339.48	44.28	22
12	Gonabad	K	2.11	43.28	26.93	4.92	3.41	360.01	46.41	34
12	Gonabad	Control	1.67	35.21	27.64	5.20	3.55	332.46	41.90	44
HSD (5%)			2.37	38.57	6.66	1.60	1.32	31.30	7.23	---

Table 6. Correlation coefficients between flowering characteristics in saffron.

	DSY					
FN[♣]	0.98	**FN**				
FSW	0.61	0.55	**FSW**			
DS	0.63	0.52	0.77	**DSW**		
SL	0.51	0.47	0.57	0.55	**SL**	
FFW	0.61	0.60	0.66	0.56	0.53	**FFW**
DF	0.48	0.47	0.57	0.53	0.50	0.68

♣- Abbreviations are described in materials and methods

Overall, it was concluded that traits FN, DSW and DFW were the most effective traits on DSY. Therefore, any increasing or improvement in these traits will increase DSY. Previous researches have shown that the correct saffron farm management including proper planting date, optimal use of fertilizer, appropriate planting density, selection of larger corms and timely irrigation saffron farms especially in the final stages of the growing season play the most effective roles in markedly improving yield components and subsequently increasing saffron in the first year and the coming years. Molina, et al., (2005) and Vurdu, et al., (2002) pointed out that mature and bigger corms gave more flowers and daughter corms. Therefore, one of the objectives in production of saffron is to obtain bigger corms (Omidbaigi, 2005).

Table 7. Regression coefficients of saffron yield using stepwise method.

Year	Model	Unstandardized Coefficients B	Unstandardized Coefficients Std.Error	Standardized Coefficients Beta	t	R2	Durbin Watson
2013	Intercept	-1.512	0.242	---	-6.253 **		
	FN	0.036	0.005	0.494	6.665 **		
	DSW	0.207	0.029	0.424	7.149 **		
	SL	0.117	0.026	0.179	4.582 **		
	FSW	-0.022	0.008	-0.186	-2.91 **		
	FFW	0.003	0.001	0.215	2.813 **		
						0.91	1.367
2014	Intercept	-0.732	0.295	---	-2.480 *		
	FN	0.058	0.001	0.934	54.293 **		
	DSW	0.398	0.023	0.214	17.256 **		
	DFW	-0.019	0.005	-0.044	-3.416 **		
	FFW	-0.002	0.001	-0.041	-2.177 *		
						0.99	1.718
Total	Intercept	-0.901	0.145	---	-6.218 **		
	FN	0.056	0.001	0.911	91.323 **		
	DSW	0.304	0.017	0.181	17.486 **		
	DFW	-0.017	0.004	-0.046	-4.556 **		
						0.99	1.96

[a] - Abbreviations are described in materials and methods

CONCLUSIONS

Results showed significant differences between years, nanofertilizers, maternal corm weights, saffron ecotypes and their interactions in terms of all flowering traits. Furthermore, it was clear that nanofertilizers (especially Fe) compared to the control increased the production of all these traits. Meteorological data illustrated that Torbat-Heydarieh, Urmia and Mashhad regions have the same climatic. However, it was the ecotypes of Torbat-Heydarieh and Mashhad regions that had the highest saffron yield in the Urmia region. On the other hand, it was highlighted that maternal corm weight had more positive effects on flowering traits rather than nanofertilizers and saffron ecotypes. Increasing maternal corm weight from 6 to 12 grams conferred an increase in all traits, especially dry saffron yield, at an increase of 5.17 times and flower number, an increase of 4.4 times. From the results of the stepwise regression and correlation coefficients, it was determined that the following traits: flowers number, dry stigma weight and dry flower weight, were the most effective indicators of dry saffron yield. In general, it can be concluded that the yield and flowering traits of saffron are strongly influenced by environmental conditions and farm management. Therefore, the correct choice of saffron ecotypes, nanofertilizer and maternal corm weight can be considered the most crucial factors in farm management and economic production of saffron.

LITERATURE CITED

Abdullaev, F.I. 2003. Crocus sativus against cancer. Archives of Medical Research 34(4)354.

Anastasaki, E., C. Kanakis, C. Pappas, L. Maggi, C.P. del Campo, M. Carmona, G.L. Alonso and M.G. Polissiou. 2010. Differentiation of saffron from four countries by mid-infrared spectroscopy and multivariate analysis. *Eur. Food. Res. Technol.* 230:571–577.

Azarpour, E., J. Asghari, H.R. Bozorgil and G. Kamalpour. 2013. Foliar Spraying of Ascophyllum nodosum Extract, Methanol and Iron Fertilizers on Fresh Flower Cover yield of Saffron plant (Crocus sativus L.). *International Journal of Agriculture and Crop Sciences*. 5 (17): 1854-1862.

Azizbekova, N.S.H. and E.L. Milyaeva. 1999. Saffron in cultivation in Azerbaijan. In: Negbi, M. (Ed.), Saffron: *Crocus sativus* L. Harwood Academic Publishers, Australia, pp. 63–71.

Baruah, S. and J. Dutta. 2009. Nanotechnology applications in sensing and pollution degradation in agriculture. *Env. Chem. Letters J.* 7: 191-204.

Bateman, A.S. and S.D. Kelly. 2007. Fertilizer nitrogen isotope signatures. *Isotopes in Environmental and Health Studies.* 43(3) 237–247.

Behnia, M.R., A. Estilai and B. Ehdaie. 1999. Application of fertilizers for increased saffron yield. *J. Agro. Sci.* 182 (1) 9-15.

Behzad, S., M. Razavi and M. Mahajeri. 1992. The effect of mineral nutrients (N.P.K.) on saffron production. International Symposium on Medicinal and Aromatic Plants, *Acta Horti.* 306:426-430. P. 75.

Botella, O., J.A. de Juan, M.A.R. Munoz, A. Moya and H. López Córcoles. 2002. Descripción morfológica y ciclo anual del azafrán (*Crocus sativus* L.). *Cuadernos de Fitopatología* 71, 18–28.

Castillo, R., J.A. Fernandez and L. Gomez-Gomez. 2005. Implications of Carotenoid Biosynthetic Genes in Apocarotenoid Formation during the Stigma Development of *Crocus sativus* and Its Closer Relatives. *Plant Physiology.* 139:674-689.

De Juan, A., A. Moya, S. López, O. Botella, H. López and R. Mu˜noz. 2003. Influence of the corm size and the density of plantation in the yield and the quality of the production of corms of *Crocus sativus* L. *ITEA.* 99:169–180.

De Mastro, G. and C. Ruta. 1993. Relation between corm size and saffron (Crocus sativus L.) flowering. *Acta Hortic.* 334:512–517.

Fernandez, J.A. 2004. *Recent Res. Dev. Pl. Sci.* 2:127–159.

Gómez-Gómez, L., A. Trapero-Mozosa, M. Dolores Gómezc, A. Rubio-Moragaa and O. Ahrazem. 2012. Identification and possible role of a MYB transcription factor from saffron (*Crocus sativus*). *Journal of Plant Physiology.* 169:509–515.

Gresta, F., G. Avola, G.M. Lombardo, L. Siracusa and G. Ruberto. 2009. Analysis of flowering, stigmas yield and qualitative traits of saffron (*Crocus sativus* L.) as affected by environmental conditions. *Sci. Hortic.* 119(3) 320–324.

Jalali-Heravi, M., H. Parastar and H. Ebrahimi-Najafabadi. 2010. Self-modeling curve resolution techniques applied to comparative analysis of volatile components of Iranian saffron from different regions. *Analytica Chimica Acta.* 662:143–154.

Juana, J.A.D., H.L. Córcolesb, R.M. Munozb and M.R. Picornella. 2009. Yield and yield components of saffron under different cropping systems. *Industrial Crops and Products.* 30:212–219.

Kafi, M., 2006. Saffron ecophysiology. In: Kafi, M., Koocheki, A., Rashed, M.H., Nassiri, M. (Eds.), Saffron (*Crocus sativus*) Production and Processing. Science Publishers, Enfield, pp. 39–58.

Kole, C., P. Kole, K.M. Randunu, P. Choudhary, R. Podila, P.C. Ke, A.M. Rao and R.K. Marcus. 2013. Nanobiotechnology can boost crop production and quality: first evidence from increased plant biomass, fruit yield and phytomedicine content in bitter melon (Momordica charantia). *BMC Biotechnology.* 13(37): 1472-6750.

Lage, M. and C.L. Cantrell. 2009. Quantification of saffron (*Crocus sativus* L.) metabolites crocins, picrocrocin and safranal for quality determination of the spice grown under different environmental Moroccan conditions. *Scientia Horticulturae.* 121 (3) 366–373.

Liu, X.M., F.D. Zhang, S.Q. Zhang, X.S. He, R. Fang, Z. Feng and Y. Wang. 2010. Effects of nano-ferric oxide on the growth and nutrients absorption of peanut. *Plant Nutr Fert Sci.* 11:14–18.

Ma, X.Q., D.Y. Zhu, C.S.P. Li, T.T.K. Dong, K.W.K. Tsim. 2001. Authentic identification of stigma croci (stigmas of Crocus sativus) from its adulterants by molecular genetics analysis. *Planta Med.* 67:183–186.

Maggi, L., M. Carmona, S.D. Kelly, N. Marigheto and G.L. Alonso. 2011. Geographical origin differentiation of saffron spice (*Crocus sativus* L. stigmas) Preliminary investigation using chemical and multi-element (H, C, N) stable isotope analysis. *Food Chemistry.* 128:543–548.

McGimpsey, J., M. Douslas and A. Wallace. 1997. Evaluation of saffron (*Crocus sativus* L.) production in New Zealand. *N. Z. J. Crop Hortic. Sci.* 25:159-168.

Molina, R.V., M. Valero, Y. Navarro, A. Garc´ıa-Luis and J.L. Guardiola. 2004. The effect of time of corm lifting and duration of incubation at inductive temperature on flowering in the saffron plant (Crocus sativus L.). Scientia Horticulturae. 103:79–91.

Molina, R.V., M. Valero, Y. Navarro, A. Garci´a-Luis, J.L. Guardiola. 2005. Low temperature storage of corms extends the flowering season of saffron (*Crocus sativus* L.). J. Hort. Sci. Biotechnol. 80:319–326.

Mollafilabi, A., 2004. Experimental findings of production and echophysiological aspects of saffron (*Crocus sativus* L.). In: Fernández, J., Abdullaev, F. (Eds.), Proceeding of the 1st on Saffron, Albacete, Spain. *Acta Hortic.* 650:195–200.

Naderi, M.R. and A. Danesh-Shahraki. 2013. Nanofertilizers and their roles in sustainable agriculture. *International Journal of Agriculture and Crop Sciences.* 5(19): 2229-2232.

Negbi, M., 1999. Saffron cultivation: past, present and future prospect. In: Negbi, M. (Ed.), Saffron: Crocus sativus L. *Harwood Academic Publishers*, Australia, pp. 1–18.

Omidbaigi, R., 2005. Effects of corm weight on quality of Saffron (Crocus sativus Linn.), Nat. Prod. Radinace, 4:193-194.

Ordoudi E, Tsimidou M. 2004. Saffron quality: Effect of agricultural practices, processing and storage. *In Production Practices and Quality Assessment of Food Crops Edited* by: Dris R, Jain SM. Dordrecht The Netherlands Kluwer Academic Publishers. 209-260.

Rees, A.R., 1992. Ornamental Bulbs, Corms and Tubers. C.A.B. International,Wallingford, UK.

Renau-Morata, B., S.G. Nebauer, M. Sánchez and R.V. Molina. 2012. Effect of corm size, water stress and cultivation conditions on photosynthesis and biomass partitioning during the vegetative growth of saffron (*Crocus sativus* L.). *Industrial Crops and Products.* 39:40– 46.

Rico, C.M., S. Majumdar, M. Duarte-Gardea, J.R. Peralta-Videa and J.L. Gardea-Torresdey. 2011. Interaction of nanoparticles with edible plants and their possible implications in the food chain. *J Agric Food Chem.* 59:3485–3498.

Rubio-Moraga, A., R. Castillo-López, L. Gómez-Gómez and O. Ahrazem. 2009. Saffron is a monomorphic species as revealed by RAPD, ISSR and microsatellite analyses. *BMC Research Notes.* 2:1-5.

Salinger, J.P., 1991. Producción comercial de flores. Butterworths of New Zealand, Wellington, NZ.

Sampathu, S.R., Shivashankar, S., Lewis, Y.S., 1984. Saffron (*Crocus sativus* L.): cultivation, processing, chemistry and standardization. *Critical Reviews in Food Science and Nutrition.* 20:123–157.

Schmidt, M., G. Betti and A. Hensel. 2007. Saffron in phytotherapy: Pharmacology and clinical uses. *Wiener medizinische Wochenschrift.* 157:315-319.

Siracusa, L., F. Gresta, G. Avola, G.M. Lombardo and G. Ruberto. 2010. Influence of corm provenance and environmental condition on yield and apocarotenoid profiles in saffron (*Crocus sativus* L.). *Journal of Food Composition and Analysis*. 23: 394–400.

Turhan, H., F. Kahriman, C.O. Egesel and M.K. Gul. 2007. The effects of different growing media on flowering and corm formation of saffron (*Crocus sativus* L.). *African Journal of Biotechnology*. 6(20): 2328-2332.

Unal, M. and A. Cavusoglu. 2005. The effect of various nitrogen fertilizers on saffron (*Crocus sativus* L.) yield. *Akdeniz University Journal of Agriculture Faculty* (in Turkish). 18(2) 257-260.

Vurdu, H., Z. Laltu and S. Ayan. 2002. *Crocus sativus* L. culture technique. *Gazi Uni. Journal of Forest Faculty* (in Turkish). 2:175-187.

Zdemür, C., P. Baran and Y. Akyol. 2006. The Morphology and Anatomy of Crocus flavus Weston subsp. flavus (Iridaceae). *Turk J Bot*. 30:175-180.

EFFECT OF LOCATION, GENOTYPE AND THEIR INTERACTIONS FOR ANTHOCYANINS AND ANTIOXIDANT ACTIVITIES OF PURPLE WAXY CORN COBS

Suchittra KHAMPAS[1], *Kamol LERTRAT*[1, 3], *Khomsorn LOMTHAISONG*[2], *Sakunkan SIMLA*[4], *Bhalang SURIHARN*[1, 3*]

[1]*Khon Kaen University, Faculty of Agriculture, Department of Plant Science and Agricultural Resources, Khon Kaen, THAILAND*
[2]*Khon Kaen University, Faculty of Science, Department of Biochemistry, Khon Kaen, THAILAND*
[3]*Khon Kaen University, Plant Breeding Research Center for Sustainable Agriculture, Khon Kaen, THAILAND*
[4]*Mahasarakham University, Faculty of Technology, Department of Agricultural Technology, Mahasarakham, THAILAND*
**Corresponding author: bsuriharn@gmail.com*

ABSTRACT

Selection of purple waxy corn genotypes for high and stable anthocyanin content in corn cobs is important for breeding programs and anthocyanin production. The objectives of this study were to evaluate the effects of location, genotype and their interaction on anthocyanin content and antioxidant activities and to identify purple waxy corn genotypes with high and stable anthocyanin content. Five purple waxy corn and a white waxy corn genotypes were evaluated in a randomized complete block design with three replications at four locations with different elevations in Thailand. Location (L), genotype (G) and GxL interaction were significant for all characters. Waxy corn grown in Nakhon Ratchasima had the highest total anthocyanin content (TAC), cyanidin 3-glucoside (C3G), pelargonidin 3-glucoside (Pg3G), and peonidin 3-glucoside (Pn3G), DPPH radical scavenging activity of phenolics (DPPH) and Trolox equivalent antioxidant potential (TEAC). Genotype KNDM4 had the highest TAC, Pg3G, Pn3G and DPPH. Its regression coefficient (b_i) was close to one but it had the highest Sd^2, indicating specific adaptation to favorable environments. KNDM4 genotype performed better than other genotypes at unfavorable environments for all studied traits. This information is useful for breeding programs and anthocyanin production from purple waxy corn.

Keywords: Anthocyanin production, Different elevations, Favorable environments, Regression coefficient, Specific adaptation

INTRODUCTION

Waxy corn (*Zea mays* L. var. *ceratina*) is an important vegetable crop in East and Southeast Asia including China, Vietnam, Cambodia, Taiwan, Korea and Thailand. The world production of vegetable waxy corn is low compared to field corn and sweet corn. It is popular among people in these countries because of its glutinous starch in endosperm due to high amylopectin. Unlike field corn, waxy corn has received little interest for genetic improvement through breeding. The increase in demand for waxy corn and market potential arouses more attention from breeders to improve waxy corn. There are numerous special cultivars that contain colored pigments and give rise to numerous varieties of black and purple corn. Purple corn produced the anthocyanin pigment throughout the plant, especially high in the husk and cob regions, although anthocyanin levels varied significantly among different plant parts (Li et al., 2008).

Anthocyanins have health benefits in reducing the incidence of cardiovascular disease, cancer, hyperlipidemias and other chronic diseases (de Pascual-Teresa and Sanchez-Ballesta, 2008; Wang and Stoner, 2008). Purple waxy corn genotypes had the highest total anthocyanin and antioxidant activity in seed at both immature and harvesting stages in comparison with white waxy corn, super sweet corn and field corn (Khampas et al., 2013). Recently, a study showed that purple waxy corn is a source of anthocyanins with a protective property against diabetic cataract (Thiraphatthanavong et al., 2014). Kernels and cobs of purple corn possess an excellent antioxidant activity, and the application of these natural food colorants by the food industry would be increased considerably (Yang and Zhai, 2010). Corn cob contributed the most in phenolic compounds and antioxidant capacity, and can be considered as excellent novel sources of natural antioxidants for the functional food and dietary supplement markets (Cevallos-Casals and Cisneros-Zevallos, 2003). This makes it more attractive for nutraceutical and functional food market.

High and stable yield across a range of environments is the primary goal of plant breeding programs. Recently, breeding for high anthocyanins is a priority of waxy corn breeding program at Khon Kaen University as this character defines it possible use as colorants and functional food. Most previous studies on stability and genotype by environment interaction in agronomic traits of cereal crops have focused on yield and the studies on quality traits such anthocyanin content and antioxidant activity are still lacking (Altay, 2012; Mitrović et al., 2012; Mohammadi et al., 2012). In wheat, cultivars Kate A1 and Mufitbey were found to be the most stable genotypes for all the environments whereas cultivar Gerek-79 was found to be the best performer for under poor soil and weather conditions (Altay, 2012), and significant genotype×environment interactions were observed among wheat genotypes (Altay, 2012; Mohammadi et al., 2012). In maize, environment accounted for the largest portion (77.83%) of total variation in grain yield, whereas genotype and genotype by environment accounted for 30% of total variation (Mitrović et al., 2012).

Several factors such as genetic, agronomic, pH, temperature, light intensity, type of light, and metallic ions, processing and storage affect anthocyanin stability (Bordignon-Luiz et al., 2007; de Rosso and Mercadante 2007; Mollov et al., 2007; Veigas et al., 2007; de Pascual-Teresa and Sanchez-Ballesta, 2008). Anthocyanin was mainly affected by pH and temperature (Cevallos-Casals and Cisneros-Zevallos, 2004). However, Zhao et al. (2008) found that anthocyanin in five Chinese purple corn hybrids was rather stable across a wide range of temperatures and times. Soil potassium concentration and population density did not significantly affect anthocyanin concentration in cobs, but growing location affected anthocyanin levels and the percentage of anthocyanins to total phenolics (Jing et al., 2007).

However, effects of various factors such as genotypes, environments and their interaction on purple waxy corn anthocyanin have not been adequately studied. The objectives of this study were to evaluate the effects of location, genotype and their interaction on anthocyanin content and antioxidant activities and to identify purple waxy corn genotypes with high and stable anthocyanin content. The information obtained from this study is important for breeding of waxy corn for high anthocyanin content and antioxidant activity and selection of locations for production of waxy corn for use as raw material for functional food products.

MATERIALS AND METHODS

Plant Materials

Five purple waxy corn genotypes (KGW1, KGW2, KGW3, KNDM4 and Fancy 111) and one white waxy corn genotype (Fancy 121) were used as plant material in the study (Table 1). Most of them are F_1 hybrids except for KNDM4, which is an open-pollinated variety (Hussanun et al., 2014).

Table 1. The waxy corn varieties used in this study.

Genotypes	Seed color	Cob color	Types
1. KGW1	Purple	Purple	Hybrid
2. KGW2	Purple	Purple	Hybrid
3. KGW3	Purple	Purple	Hybrid
4. KNDM4	Purple	Purple	Opened-pollinated
5. FANCY 111[1/]	Purple	Purple	Hybrid
6. FANCY 121[1/]	White	White	Hybrid

[1/] Commercial variety

Field Experiment

Six genotypes of waxy corn were evaluated in a randomized complete block design with three replications at four locations (Saraburi, Khon Kaen, Nakhon Ratchasima and Chiang Rai) with differences in elevations from 120 to 380 meters above sea level in the dry season 2012/13. These locations represented waxy corn growing areas in Thailand from the central plain to the North of Thailand. The details of locations were presented in Table 2.

Table 2. The testing locations used in this study.

Locations	Geographical coordinates	Elevations (masl)	Planting date
Saraburi	N 14° 43′ E 100° 47′	120	Dec 2012
Khon Kaen	N 16° 28′ E 102° 49′	200	Nov 2012
Nakhon Ratchasima	N 14° 30′ E 101° 30′	356	Dec 2012
Chiang Rai	N 20° 24′ E 99° 53′	380	Feb 2013

The genotypes were grown in four-row plots with 5 m long and spacing of 0.80 m between rows and 0.25 m between plants within rows. Recommended cultural practices were used for all locations. The crop at all location was well-irrigated for optimum growth and yield. Sprinkler irrigation was available at the locations in Khon Kaen, Nakhon Ratchasima and Chiang Rai, whereas furrow irrigation was performed in Saraburi.

15-15-15 Fertilizer of N-P-K as basal dose at the rate of 171 kg ha^{-1} was incorporated into the soil during soil preparation. Two splits of the fertilizer at the rate of 93.75 kg ha^{-1} plus urea (46-0-0) at the rate of 93.75 kg ha^{-1} for first split and the fertilizer at the rate of 125 kg ha^{-1} plus urea at the rate of 62.5 kg ha^{-1} for second split were applied to the crop at 14 days after planting (DAP) and 30 DAP, respectively. At flowering stage, 13-13-21fertilizer was applied at the rate of 156.25 kg ha^{-1}. Therefore, total dose of fertilizers consisted of 150.65 kg ha^{-1} nitrogen, 78.78 kg ha^{-1} phosphorus and 91.27 kg ha^{-1} potassium, respectively. Weather and soil data for all locations were presented in Table 3 and Table 4, respectively.

Table 3. Weather data at four locations.

Days after planting	Average Temperature (°C)	Solar radiation ($MJ\ m^{-2}day^{-1}$)	Total rainfall (mm)	Relative humanity (%)
Saraburi				
0-30	28.5	15.3	49	68.7
31-60	27.1	17.8	0	62.7
61-90	29.7	18.9	3	66.8
mean	28.4	17.3		66.1
Khon Kaen				
0-30	26.9	17.1	23	60.0
31-60	23.7	17.6	0	52
61-90	26.7	17.1	6	52
mean	25.8	17.3		78.6
Nakhon Ratchasima				
0-30	25.1	16.9	5	68.4
31-60	25.0	18.2	46	62.4
61-90	26.7	19.7	42	71.1
mean	25.6	18.3		67.3
Chiang Rai				
0-30	25.2	16.9	53	74.3
31-60	27.1	19.8	8	71.4
61-90	28.8	20.8	84	62.3
mean	27.0	19.2		69.4

Table 4. Soil physical and chemical properties.

Locations	Soil types	pH (1:1 H_20)	EC (1:5 H_20) (dS/m)	Organic Matter (%)	Total nitrogen (%)	Available Phosphorus (mg/kg)	Available potassium (mg/kg)	Available calcium (mg/kg)
Saraburi	Sandy loam	7.4	0.09	1.22	0.06	234	155	2,432
Khon Kaen	Sandy	4.9	0.01	0.73	0.03	307	139	54
Nakhon Ratchasima	Loamy sand	7.1	0.13	2.49	0.13	52	162	3,137
Chiang Rai	Loamy sand	5.6	0.08	3.40	0.09	45	149	593

Corn cobs were collected at harvest maturity stage (R6) or 90 days after planting and stored at room temperature for analysis of anthocyanin and antioxidant activity. Data were recorded for total anthocyanin content (TAC), cyanidin 3-glucoside (C3G), pelargonidin 3-glucoside (Pg3G), and peonidin 3-glucoside (Pn3G), DPPH radical scavenging activity of phenolics (DPPH) and the Trolox equivalent anti-oxidant potential (TEAC) as described by Yang et al. (2008).

Sample extraction

The portions of 0.5 g cob powder samples were weighted and pounded to fine powder using mortar and pestle. The extraction method used for TAC and antioxidant activity analysis was modified from the method reported previously (Yang et al., 2008). The samples were extracted in 25 ml of methanol containing 1% 1 M citric acid in falcon tubes. The mixtures were vortexed and stored at 4 °C for 24 hours. The homogenates were centrifuged at 5000 rpm at 4 °C for 15 minutes. After centrifugation, the mixtures were acquired by filtration through Whatman No.1 filter papers, adjusted to 25 ml and stored in dark room at -20 °C until analysis for TAC and antioxidant activity.

Analysis of total anthocyanin content (TAC)

TAC was quantified by pH differential method (Giusti and Wrolstad, 2001). The absorbance of diluted extracts was measured to detect anthocyanins level at 510 nm and 700 nm with UV-Visible Spectrophotometer (model Specord 250 plus, Analytik jena, Germany) against a blank cell filled with distilled water. The absorbance was calculated as Abs = (Abs_{510} - A_{700}) pH 1.0-(Abs_{510}-A_{700}) pH 4.5. Then monomeric anthocyanin pigment in samples was calculated as;

Monomeric anthocyanin pigment = (Abs x MW x DF X 1000)/(ε x 1),

Where MW is the molecular weight of cyanidin 3-glucoside (MW of 449.2 g/mol), DF is dilution factor of the sample and ε is the molar extinction coefficient (Equal to 26900 M^{-1} cm^{-1}).

Anthocyanin levels were presented in mg of cyanidin 3-glucoside equivalents (CGE) per 100 g of dry weight (DW).

HPLC analysis of anthocyanin components

Total anthocyanin content was determined by the method described by Kim et al. (2007) with minor modification, and individual anthocyanin components were quantified with high-performance liquid chromatography (HPLC), Shimadzu LC-20AC pushes (Shimadzu Co., Kyoto, Japan), SPD-M20A with a photodiode array detector (Waters, USA). A XselectCHS C-18 line (4.6 mm x 250 mm, i.d. 5 μm). The extracts of cob samples were filtered through a 0.22 μm filter at 30 °C. The anthocyanins were eluted at 1 mL/min using gradient system consisting of two solvents: (A) acidified methanol (methanol 0.1% HCl, 85:15, v/v) and (B) 10% formic acid. The gradient elutions of solvent B were performed as follows: for 0.0-0.5 min (80%), 0.5-9.5 min (15%), 9.5-10.0 min (5%), 10.0-15.0 min (80%), and 15.0-20.0 min (80%). The divided solutions were consequently detected and identified at 520 nm for chromatographic retention times. Three purified anthocyanins consist of cyanidin 3-glucoside, pelargonidin 3-glucoside, and peonidin 3-glucoside and one anthocyanidin (cyanidin chloride) (Sigma, USA) were used for quantification.

The purified anthocyanins extractions were utilized to aid the identification of individual anthocyanins. The standard solutions were prepared in acidified methanol (pH 1) by weighing exactly 200-300 μg in 200 μL; 20 μL of standard solutions and then diluted to 500 μL to prepare working standard. The standard anthocyanins revealed a linear relationship with a HPLC optimum peak area with a concentration range of 0.0-1.0 μg. The coefficient of dedication (r^2) varied from 0.9933 to 0.9998 for a mixture of purified anthocyanins, which were divided by HPLC using 5, 10, and 15 μL of operating standard solution.

Analysis of antioxidant activity

DPPH radical scavenging activity of methanolic extracts was determined according to Xu et al. (2010). Briefly, 0.5 mL of sample extract was added and mixed with 4.5 mL of 60 μM DPPH demolished in methanol. The mixture was shaken thoroughly and allowed to stand for half an hour in the dark room. The absorbance was detected at 517 nm against the solvent blank. Trolox solution (100-1000 μM) was used as a reference conventional. The outcomes were provided in micromoles Trolox equivalents (TE) per g of DW.

Analysis of Trolox equivalent antioxidant capacity (TEAC) measured by the reduction in radical cation of ABTS by anti-oxidants was carried out as previously described by Jemai et al. (2009). Briefly, ABTS+ radical cation was conducted with a result of 7 mmol/L ABTS and 2.45 mmol/L potassium per sulfate. The combination was allowed to stay in dark room at ambient temperature 16-24 hours prior to use and should be used within 2 days. The ABTS solution was diluted with methanol for absorbance of 0.700 ± 0.050 at 734 nm. All samples were diluted appropriately to provide 20-80% inhibition of the blank absorbance. 50 μL of the diluted extract was taken by a pipette and combined with 1.9 mL of diluted ABTS solution. The analysis of the combination was performed in three replications. The combination was allowed to stand for 6 min at ambient temperature and the absorbance was instantly measured at 734 nm. Trolox solution (100-1000 μM) was used as a reference point. The outcomes were indicated as micromoles Trolox equivalents (TE) per g of DW.

Statistical Analysis

Data for each location were analyzed separately and error variances were tested for variance homogeneity. As variances were homogeneous and the differences of error variances were not larger than 3 folds, then combined analysis of variance was performed across the locations. Linear regression was carried out to study stability of corn genotypes by plotting cultivar's means against environmental means for each genotype as described by Eberhart and Russell (1966). Means were compared statistically using least significant difference (LSD) at $P \leq 0.05$.

RESULTS AND DISCUSSION

Variations in anthocyanin and antioxidant activity

Location (L), genotype (G) and GxL interaction were significant ($P \leq 0.01$) for total anthocyanin content (TAC), cyanidin 3-glucoside (C3G), pelargonidin 3-glucoside (Pg3G), and peonidin 3-glucoside (Pn3G), DPPH radical scavenging activity of phenolics (DPPH) and the Trolox equivalent anti-oxidant potential (TEAC) (Table 5). Location contributed to large portions of total variations in TAC (44%), C3G (56%), Pg3G (60%), Pn3G (62%), DPPH (52%) and TEAC (57%).

Genotype contributed to relatively large portion of total variation in TAC (42%), but genotype contributed to rather small portions of total variations in C3G (18%), Pg3G (20%), Pn3G (2%), DPPH (24%) and TEAC (11%). G x L interactions contributed to medium to small portions of total variations in TAC (13%), C3G (25%), Pg3G (20%), Pn3G (15%), DPPH (22%) and TEAC (32%).

Anthocyanins are phytochemicals that commonly occur in many cereal and fruit crops, and the trait is quantitatively inherited (Vizzotto et al., 2007). In corn, anthocyanins are present in all tissues and found at high concentrations in kernel skin and cobs (Moreno et al., 2005). In fruit crops such as grape, grapes may be either white or colored, ranging from the lightest pink to the darkest purple tones according to the amount of anthocyanin accumulated in the berry skin, which is a crucial trait for both wine quality and human nutrition (Fournier-Level et al., 2009). In peach, anthocyanin content, phenolic content, and antioxidant activity were higher in red-flesh than in light-colored flesh peaches (Vizzotto et al., 2007). Genetic variation in anthocyanins has been reported in maize (Chander et al., 2008), grape (Fournier-Level et al., 2009) and wheat (Knievel et al., 2009). The results indicated that selection for high TAC would be possible among purple waxy corn genotypes. Variation among locations also indicated the importance of selection for the best location for anthocyanin production in corn.

Table 5. Mean squares for total anthocyanin content (TAC), cyanidin 3-glucoside (C3G), pelargonidin 3-glucoside (Pg3G), and peonidin 3-glucoside (Pn3G), DPPH radical scavenging activity of phenolics (DPPH) and the Trolox equivalent anti-oxidant potential (TEAC) of six waxy corn lines/cultivars.

SOV	df	TAC mg/100 gDW	C3G mg/100 gDW	Pg3G mg/100 gDW	Pn3G mg/100 gDW	DPPH µmol/gDW	TEAC µmol/gDW
Location (L)	3	1,004,062**	60,464**	6,320**	1,564**	179.4**	86.1**
		(44)	(56)	(60)	(62)	(52)	(57)
Replication (R)/L (a)	8	432	50	6	2	1.4	0.1
Genotype (G)	4	15,975**	14,582**	1,542**	411**	61.4**	12.3**
		(42)	(18)	(20)	(2)	(24)	(11)
G x L	12	75,107**	6,596**	512**	96**	18.5**	12.0**
		(13)	(25)	(20)	(15)	(22)	(32)
Error (b)	32	2,726	67	5	3	0.4	0.1
C.V. (%) (a)		4.1	15.6	14.0	12.5	8.0	7.5
C.V. (%) (b)		10.3	18.1	12.0	15.2	4.1	5.4

** Highly significant difference at $p \leq 0.01$
Values in parenthesis are percentages of sum squares.

Effects of location and genotype on anthocyanins and antioxidant activity

Location was the main source of variation in TAC, C3G, Pg3G, Pn3G, DPPH and TEAC of five purple waxy corns and one white waxy corn. Nakhon Ratchasima had the highest TAC of 730 mg/100 gDW followed by 418 mg/100 gDW in Saraburi, whereas Khon Kaen and Chiang Rai had rather low TAC, accounting for 275 and 271 mg/100 gDW, respectively (Table 6). Similar patterns were also observed for C3G, Pg3G, Pn3G and DPPH. For these parameters, Nakhon Ratchasima was still highest followed by Saraburi, whereas Khon Kaen and Chiang Rai were similar and lower than Nakhon Ratchasima and Saraburi. Nakhon Ratchasima was also highest for TAEC followed by Khon Kaen, Saraburi and Chiang Rai, respectively.

In general, the highest TAC and anthocyanin components in corn cobs were found in Nakhon Ratchasima compared to the nearest locations in Khon Kaen and Saraburi. Nakhon Ratchasima is located at higher alleviation and had lower temperature during growing season and lower phosphorus. Low temperature and low phosphorus might be the cause of the high TAC and its components.

Low temperature induced anthocyanin synthesis in maize seedlings (Christie et al., 1994). Anthocyanin concentration increased rapidly during low temperature in apple and peach shoots (Leng et al., 2000). Anthocyanins and phenolics contents were enhanced with higher light intensity and lower temperature in potato tubers (Reyes et al., 2004). Visible and UVB radiation, cold temperatures and water stress has been shown to induce anthocyanin synthesis in plant tissue (Chalker-Scott, 1999).

However, low temperatures in the absence of either visible light or UVB prevent anthocyanin biosynthesis (Janda et al., 1996; Oren-Shamir and Levi-Nissim, 1997). Wang and Zheng (2001) also found that high temperature (30/22 day/night °C) condition enhanced phenolic acid, flavonols, and anthocyanins in strawberry.

Phosphorus deficiency induces anthocyanins accumulation in maize (Cobbina and Miller, 1987). Antocyanin content was increased and growth was decreased on nitrogen and phosphorus deficiency in Cruciferous (Hodges and Nozzolillo 1996). However, Chiang Rai also had low phosphorus but it had the lowest anthocyanins possibly due to high temperature and high solar radiation during reproductive periods. It should be note that stresses of low temperature and low phosphorus were the factors enhancing anthocyanin accumulation.

Nakhon Ratchasima (low temperature and low phosphorus) also had the highest antioxidant activities (DPPH and TEAC) but Chiang Rai (high temperature and low phosphorus) had the lowest highest antioxidant activities. These results were in agreement with previous study of Lachman et al. (2008) who reported that low temperature during vegetative period showed high antioxidant activity in potatoes.

Table 6. Means for total anthocyanin content (TAC), cyanidin 3-glucoside (C3G), pelargonidin 3-glucoside (Pg3G), and peonidin 3-glucoside (Pn3G), DPPH radical scavenging activity of phenolics (DPPH) and the Trolox equivalent anti-oxidant potential (TEAC) of six waxy corn lines/cultivars evaluated in four locations 2012/13.

	Saraburi	Khon Kaen	Nakhon Ratchasima	Chiang Rai
		TAC (mg/100gDW)		
KGW1	247^{j}	671^{m}	609^{e}	129^{kl}
KGW2	449^{g}	341^{h}	982^{c}	373^{h}
KGW3	328^{hi}	263^{ij}	547^{ef}	330^{hi}
KNDM4	$1{,}132^{ab}$	826^{d}	$1{,}166^{a}$	490^{fg}
FANCY111	346^{h}	148^{k}	$1{,}076^{b}$	302^{hij}
FANCY121	4^{m}	6^{m}	3^{m}	2^{m}
Means	418^{B}	275^{C}	730^{A}	271^{C}
		C3G (mg/100g DW)		
KGW1	5^{fg}	4^{fg}	37^{d}	1^{g}
KGW2	27^{de}	9^{fg}	246^{a}	2^{g}
KGW3	16^{ef}	7^{fg}	93^{b}	3^{g}
KNDM4	96^{b}	23^{e}	246^{a}	5^{fg}
FANCY111	9^{f}	2^{g}	72^{c}	1^{g}
FANCY121	1^{g}	1^{g}	1^{g}	1^{g}
Means	26^{B}	8^{C}	116^{A}	2^{D}
		Pg3G (mg/100g DW)		
KGW1	3^{hi}	3^{hi}	9^{f}	2^{hi}
KGW2	11^{f}	5^{gh}	54^{c}	3^{hi}
KGW3	10^{f}	5^{ghi}	29^{d}	3^{hi}
KNDM4	30^{d}	15^{e}	86^{a}	7^{fg}
FANCY111	16^{e}	5^{gh}	63^{b}	3^{hi}
FANCY121	2^{hi}	2^{i}	2^{hi}	2^{hi}
Means	12^{B}	6^{C}	41^{A}	3^{D}
		Pn3G (mg/100g DW)		
KGW1	3^{ghi}	3^{ghi}	5^{fg}	2^{i}
KGW2	15^{c}	7^{ef}	35^{a}	3^{ghi}
KGW3	9^{de}	5^{fg}	26^{b}	3^{ghi}
KNDM4	26^{b}	11^{d}	35^{a}	5^{fgh}
FANCY111	9^{de}	3^{ghi}	28^{b}	2^{hi}
FANCY121	2^{i}	1i	2^{i}	2^{i}
Means	11^{B}	5^{C}	22^{A}	3^{D}
		DPPH (µmol/gDW)		
KGW1	11^{i}	10^{i}	18^{cde}	8^{jk}
KGW2	17^{ef}	14^{h}	18^{cd}	10^{i}
KGW3	17^{de}	13^{h}	19^{cd}	9^{j}
KNDM4	22^{a}	18^{cd}	15^{g}	16^{fg}
FANCY111	19^{c}	13^{h}	21^{b}	11^{i}
FANCY121	5^{l}	5^{l}	4^{l}	7^{k}
Means	15^{A}	12^{B}	16^{A}	10^{C}
		TEAC (µmol/gDW)		
KGW1	0.8^{mn}	2.8^{h}	5.3_{e}	1.3^{kl}
KGW2	6.4^{d}	3.8^{g}	8.1^{b}	1.3^{lk}
KGW3	7.3^{c}	3.7^{g}	6.9^{c}	2.2^{ij}
KNDM4	1.1^{lm}	5.7^{e}	8.2^{b}	4.2^{f}
FANCY111	1.4^{k}	2.5^{hi}	10.2^{a}	2.1^{j}
FANCY121	0.3^{op}	0.0^{p}	0.0^{p}	0.5^{no}
Means	2.9^{B}	3.1^{B}	6.4^{A}	1.9^{C}

Means of varieties in the same column and means of locations in the same row with the same letter(s) are not significantly different at $P \leq 0.05$ by LSD

As the interactions between location and genotype were significant for all parameters, identification of the best genotypes for each parameter was difficult. The varieties that showed good performance in one location did not perform well in other locations. However, KNDM4 showed consistently the highest values for TAC, C3G, Pg3G and Pn3G across locations. High values for these parameters in KNDM4 would be due to the fact that it is an open-pollinated variety and other hybrids had one parent with white kernels. For DPPH, KNDM4 was highest in Saraburi, Khon Kaen and Chiang Rai, whereas FANCY111 was highest in Nakhon Ratchasima. For

TEAC, KNDM4 was the winner in two locations in Khon Kaen and Chiang Rai. FANCY111 was still highest in Nakhon Ratchasima, whereas KGW3 was highest in Saraburi. FANCY111, which is the waxy corn, showed consistently the lowest values for all parameters across locations.

Stability analysis for anthocyanins and antioxidant activity

The genotypes with stable yield should have high mean yield across locations, regression coefficient close to one (b_i=1) and deviation from regression (Sd^2) close to zero (Eberhart and Russell, 1966). KNDM4 had the highest TAC (903 mg/100 gDW) and regression coefficient close to one (0.9), but it showed high deviation from regression (254) (Table 7). KNDM4 was the highest variety for TAC across locations. It did not performed well under unfavorable environment in Chiang Rai but it was still better than others (Table 6). However, anthocyanins of KNDM4 was rather fluctuated across environments as it has high Sd^2. KNDM4 was suitable for favorable environments. KGW2 had the second rank for TAC (536 mg/100g DW), regression coefficient close to one (1.1) and low deviation from regression (67). This genotype was most stable for TAC across environments.

Table 7. Stability parameters for total anthocyanin content (TAC), cyanidin 3-glucoside (C3G), pelargonidin 3-glucoside (Pg3G), and peonidin 3-glucoside (Pn3G) of six waxy corn lines/cultivars.

Genotypes	TAC (mg/100 gDW)			C3G (mg /100 gDW)			Pg3G (mg /100 gDW)			Pn3G (mg /100 gDW)		
	Mean	b[1/]	Sd^2	Mean	b	Sd^2	Mean	b	Sd^2	Mean	b	Sd^2
Purple waxy corn												
KGW1	263[e]	0.9	34	12[e]	0.3**	2	4[e]	0.2**	1	3.5d	0.1**	0.4
KGW2	536[b]	1.1	67	71[b]	1.8	14	18[c]	1.2	2	15.1b	1.4**	0.3
KGW3	367[d]	0.5	45	30[c]	0.7	4	11[d]	0.6**	1	10.9c	1.0	1.7
KNDM4	903[a]	0.9	254	93[a]	1.7	24	35[a]	1.7**	3	19.2a	1.3	5.4
FANCY111	468[c]	1.6	120	21[d]	0.5	4	22[b]	1.4**	1	10.4c	1.1	1.9
White waxy corn												
FANCY121	4			1			2			1.9		

** Significant different from 1 at $P \leq 0.01$

Means in the same column with the same letter(s) were not significantly different at $P \leq 0.05$ probability level by LSD.

[1/]Regression coefficients were analyzed without white waxy corn.

KNDM4 had the highest C3G (93 mg /100 gDW) followed by KGW2 (71 mg /100 gDW) and their regression coefficients were not different from one (1.7 and 1.8, respectively). However, their deviations from regression were high, indicating that they performed well under favorable conditions but performed poorly under unfavorable conditions (Chiang Rai). KGW1 had the lowest C3G (12 mg/100 gDW) and regression coefficient (0.3**) was significantly lower than one, indicating that this genotype had low C3G and stability for C3G across environments.

Regression coefficients for Pg3G for most genotypes were significantly different from one except KGW2, indicating that these genotypes were not stable for this trait. KNDM4 had the highest Pg3G of 35 mg/100 g DW followed by FANCY111 (22 mg/100g DW) but regression coefficients (1.7** and 1.4**, respectively) were significantly higher than one, indicating that these genotypes performed well under favorable conditions but they had low Pg3G under unfavorable conditions.

KNDM4 had the highest Pn3G (19.2 mg/100 gDW), regression coefficient close to one (1.3) and high deviation from regression. However, it performed best for all locations but it had low Pn3G under unfavorable conditions (Chiang Rai) (Table 5). KGW2 showed high Pn3G (15.1 mg/100 gDW) but it had regression coefficient higher than one (1.4**) and low deviation from regression, indicating that it performed well for Pn3G under favorable conditions but performed poorly under unfavorable conditions.

KNDM4 had the highest DPPH (18.0 μmol/gDW), regression coefficient smaller than one (0.3**) and high deviation from regression (13.6) (Table 8). This genotype showed good adaptation to favorable conditions but it had poor adaptation to unfavorable conditions. KGW3 and KGW2 had high TEAC and regression coefficient close to one with high deviation from regression. Therefore, they may be suitable for some specific conditions.

Table 8. Stability parameters for DPPH radical scavenging activity of phenolics (DPPH) and the Trolox equivalent anti-oxidant potential (TEAC) of six waxy corn lines/cultivars.

Genotypes	DPPH (µmol / gDW)			TEAC (µmol / gDW)		
	Mean	b	Sd^2	Mean	b	Sd^2
Purple waxy corn						
KGW1	11.8^{d}	1.0	3.0	2.6^{d}	0.8	0.9
KGW2	14.6^{c}	1.1	0.3	4.9^{ab}	1.0	2.0
KGW3	14.3^{c}	1.3**	0.3	5.0^{a}	0.6	2.4
KNDM4	18.0^{a}	0.3**	13.6	4.8^{b}	0.9	2.5
FANCY111	15.9^{b}	1.3	1.3	4.1^{c}	1.6**	1.5
White waxy corn						
FANCY121	5.2			0.2		

** Significant different from 1 at *P≤0.01*

Means in the same column with the same letter(s) were not significantly different at *P≤0.05* probability level by LSD.

Khampas et al. (2013) and Harakotr et al. (2014) reported that kernels of purple waxy corn had high TAC and antioxidant activity (DPPH, FRAP and TEAC). In this study, cobs of purple waxy corn also had high TAC and antioxidant activity, and cobs can be used for production of natural colorant with health benefits.

CONCLUSION

Location (L), genotype (G) and GxL interaction were highly significant for all characters. Location contributed to large portion of total variation in anthocyanin content and antioxidant activity in corn cobs, whereas genotype and genotype by location interaction contributed smaller portions of total variations in these traits. Nakhon Ratchasima, which was characterized by low temperature and low phosphorus, was the highest location for TAC and C3G. KNDM4 gave the highest TAC and C3G, and its regression coefficients (b) were close to one. However, KNDM4 also had the highest Sd^2, indicating that it had specific adaptation to favorable environments. KNDM4 performed well under unfavorable environments for all traits. This information is useful for breeding programs in multi-location trails and production of anthocyanins from purple waxy corn.

ACKNOWLEDGEMENTS

The project was funded by the Thailand Research Fund, the commission for High Education and Khon Kaen University through the Distinguished Research Professor Grant of Professor Dr. Aran Patanothai and the Plant Breeding Research Center for Sustainable Agriculture, Faculty of Agriculture, Khon Kaen University, Thailand. Acknowledgement is extended to Khon Kaen University and the Faculty of Agriculture for providing financial support for manuscript preparation activities.

LITERATURE CITED

Altay, F. 2012. Yield stability of some Turkish winter wheat (*Triticum aestivum* L.) genotypes in the western transitional zone of Turkey. Turk. J. Field Crops 17:129-134.

Bordignon-Luiz, M.T., C. Gauche, E.F. Gris and L.D. Falcao. 2007. Colour stability of anthocyanins from Isabel grapes (*Vitis labrusca* L.) in model systems. LWT-Food Sci. Technol. 40:594-599.

Cevallos-Casals, B.A. and L. Cisneros-Zevallos. 2003. Stoichiometric and kinetic studies of phenolic antioxidants from Andean purple corn and red-fleshed sweet potato. J. Agric. Food Chem. 51:3313-3319.

Cevallos-Casals, B.A. and L. Cisneros-Zevallos. 2004. Stability of anthocyanin-based aqueous extracts of Andean purple corn and redfleshed sweet potato compared to synthetic and natural colorants. Food Chem. 86:69-77.

Chalker-Scott, L. 1999. Environmental significance of anthocyanins in plant stress responses. J. Photochem. Photobiol. 70:1-9.

Chander, S., Y. Meng, Y. Zhang, J. Yan and J. Li. 2008. Comparison of nutritional traits variability in selected eighty-seven inbreds from Chinese maize (*Zea mays* L.) germplasm. J. Agric. Food Chem. 56:6506-6511.

Christie, P.J., M.R. Alfenito and V. Walbot. 1994. Impact of low-temperature stress on general phenylpropanoid and anthocyanin pathways: enhancement of transcript abundance and anthocyanin pigmentation in maize seedlings. Planta. 194:541-549.

Cobbina J. and M.H. Miller. 1987. Purpling in maize hybrids as influenced by temperature and soil phosphorus. Agron. J. 79:576-582.

de Pascual-Teresa S. and M.T. Sanchez-Ballesta. 2008. Anthocyanins: from plant to health. Phytochem. Rev. 7(12):281-299.

de Rosso, V.V. and A.Z. Mercadante. 2007. Evaluation of colour and stability of anthocyanins from tropical fruits in an isotonic soft drink system. Innovative Food Sci. Emerging Technol. 8:347-352.

Eberhart, S.A. and W.A. Russel. 1966. Stability parameters for controlling variation. Crop Sci. 6:36-40.

Fournier-Level, A., L. Le Cunff, C. Gomez, A. Doligez, A. Ageorges, C. Roux, Y. Bertrand, J.M. Souquet, V. Cheynier and P. This. 2009. Quantitative genetic bases of anthocyanin variation in grape (*Vitis vinifera* L. ssp. *sativa*) Berry: A Quantitative trait locus to quantitative trait nucleotide integrated study. Genetics 183:1127-1139.

Giusti, M.M. and R.E. Wrolstad. 2001. Characterization and measurement of anthocyanins by UV-visible spectroscopy. In: Current Protocol in Food Analytical Chemistry, ed. Wrolstad, R.E., 5:1-13, John Willy & Sons Inc., New York.

Harakotr, B., B. Suriharn, R. Tangwongchai, M.P. Scott and K. Lertrat. 2014. Anthocyanins and antioxidant activity in coloured waxy corn at different maturation stages. J. Funct. Foods 9:109-118.

Hodges D.M. and C. Nozzolillo. 1996. Anthocyanin and anthocyanoplast content of cruciferous seedlings subjected to mineral nutrient deficiencies. J. Plant Physiol. 147:749-754.

Hussanun, S., B. Suriharn, K. Lertrat. 2014. Yield and early maturity response to four cycles of modified mass selection in purple waxy corn. Turk. J. Field Crops 19:84-89.

Janda, T., G. Szalai and E. Paldi. 1996. Chlorophyll fluorescence and anthocyanin content in chilled maize plants after return to a non-chilling temperature under various irradiances. Biol. Plant. 38:625-627.

Jemai, H., M. Bouaziz and S. Sayadi. 2009. Phenolic composition, sugar contents and antioxidant activity of Tunisian sweet olive cultivar with regard to fruit ripening. J. Agric. Food Chem. 57:2961-2968.

Jing, P., V. Noriega, S.J. Schwartz and M.M. Giusti. 2007. Effects of growing conditions on purple corncob (*Zea mays* L.) anthocyanins. J. Agric. Food Chem. 55:8625-8629.

Khampas, S., K. Lertrat, K. Lomthaisong, and B. Suriharn. 2013. Variability in phytochemicals and antioxidant activity in corn at immaturity and physiological maturity stages. Inter. Food Res. J. 20:3149-3157.

Kim, M.J., J.N. Hyun, J.A. Kim, J.C. Park, M.Y. Kim, J.G. Kim, S.J. Lee, S.C. Chun and I.M. Chung. 2007. Relationship between phenolic compounds, anthocyanins content and antioxidant activity in colored barley germplasm. J. Agric. Food Chem. 55:4802-4809.

Knievel, D.C., E.S.M. Abdel-Aal, I. Rabalski, T. Nakamura and T. Hucl. 2009. Grain color development and the inheritance of high anthocyanin blue aleurone and purple pericarp in spring wheat (*Triticum aestivum* L.). J. Cereal Sci. 50:113-120.

Lachman, J., K. Hamouz, M. Orsák, V. Pivec and P. Dvořák. 2008. The influence of flesh colour and growing locality on polyphenolic content and antioxidant activity in potatoes. Sci. Hortic. 117:109-114.

Leng, P., H. Itamura, H. Yamamura and X.M. Deng. 2000. Anthocyanin accumulation in apple and peach shoots during cold acclimation. Sci. Hortic. 83:43-50.

Li, C.Y., H.W. Kim, S.R. Won, H.K. Min, K.J. Park, J.Y. Park, M.S. Ahn and H.I. Rhee. 2008. Corn husk as a potential source of anthocyanins. J. Agric. Food Chem. 56:11413-11416.

Mitrović, B., D. Stanisavljevi, S. Treski, M. Stojaković, M. Ivanović, G. Bekavac and M. Rajković. 2012. Evaluation of Experimental Maize Hybrids Tested in Multi-Location Trials Using AMMI and GGE Biplot Analyses. Turk. J. Field Crops 17:35-40.

Mohammadi, M., R. Karimizadeh, N. Sabaghnia and M.K. Shefazadeh. 2012. Genotype x environment interaction and yield stability analysis of new improved bread wheat genotypes. Turk. J. Field Crops 17:67-73.

Mollov, P., K. Mihalev, V. Shikov, N. Yoncheva and V. Karagyozov. 2007. Colour stability improvement of strawberry beverage by fortification with polyphenolic copigments naturally occurring in rose petals. Innovative Food Sci. Emerging Technol. 8:318-321.

Moreno, Y.S., G.S. Sánchez, D.R. Hernández, N.R. Lobato. 2005. Characterization of anthocyanin extracts from maize kernels. J. Chromatogr. Sci., 43:483-487.

Oren-Shamir, M. and A. Levi-Nissim 1997. UV-light effect on the leaf pigmentation of *Cotinus coggygria* 'Royal Purple'. Sci. Hort. 71:59-66.

Reyes, L.E., J.C. Miller, and L. Cisneros-Zevallos. 2004. Environmental conditions influence the content and yield of anthocyanins and total phenolics in purple- and red-flesh potatoes during tuber development. Amer. J. of Potato Res. 81:187-193.

Thiraphatthanavong, P., J. Wattanathorn, S. Muchimapura, W. Thukham-mee, P. Wannanon, T. Tong-un, B. Suriharn, and K. Lertrat. 2014. Preventive Effect of *Zea mays* L. (Purple Waxy Corn) on Experimental Diabetic Cataract. Biomed. Res. Int., 1-8.

Veigas, J.M., M.S. Narayan, P.M. Laxman and B. Neelwarne. 2007. Chemical nature, stability and bioefficacies of anthocyanins from fruit peel of *syzygium cumini* Skeels. Food Chem. 105:619-627.

Vizzotto, M., L. Cisneros-Zevallos and D.H. Byrne. 2007. Large variation found in the phytochemical and antioxidant activity of peach and plum germplasm. J. Amer. Soc. Hort. Sci. 132(3):334-340.

Wang, L. and G.D. Stoner. 2008. Anthocyanins and their role in cancer prevention. Cancer Lett. 269:281-290.

Wang, S.Y. and W. Zheng. 2001. Effects of plant growth temperature on antioxidant capacity in strawberry. J. Agric. Food Chem. 49:4977-4982.

Xu, J.G., Q.P. Hu, X.D. Wang, J.Y. Luo, Y. Liu and C.R. Tian. 2010. Changes in the main nutrients, phytochemicals, and antioxidant activity in yellow corn grain during maturation. J. Agric. Food Chem. 58:5751-5756.

Yang, Z., G. Fan, Z. Gu, Y. Han and Z. Chen. 2008. Optimization extraction of anthocyanins from purple corn (*Zea mays* L.) cob using tristimulus colorimetry. Eur. J. Lipid Sci. Tech. 227:409-415.

Yang, Z. and W. Zhai. 2010. Identification and antioxidant activity of anthocyanins extracted from the seed and cob of purple corn (*Zea mays* L.). Innovative Food Sci. Emerging Technol. 11:169-176.

Zhao, X., M. Corrales, C. Zhang, X. Hu, Y. Ma and B. Tauscher. 2008. Composition and thermal stability of anthocyanins from chinese purple corn (*Zea mays* L.). J Agric Food Chem. 56:10761-10766.

GENETIC ANALYSIS OF THE GRAIN PROTEIN CONTENT IN SOFT RED WINTER WHEAT (*Triticum aestivum* L.)

Jinbao YAO, Hongxiang MA, Xueming YANG, Miaoping ZHOU, Dan YANG*

Provincial Key Laboratory of Agrobiology, Jiangsu Academy of Agricultural Sciences, Nanjing, CHINA
**Corresponding author: yaojb@jaas.ac.cn*

ABSTRACT

Improvement of grain protein content is the main objective in soft red wheat breeding programs. The main objective of this study was to investigate the mode of inheritance and combining ability of grain protein content in wheat. Seven genetically diverse soft red wheat cultivars were crossed in an incomplete 7×7 diallel fashion. F1 and F2 progenies along with their parents were evaluated in two separate experiments to do F1 and F2 diallel analysis. The results indicated significant differences among the parents for general combining ability and crosses for specific combining ability. Among the parents Ningmai 9 was the best combiner and could be utilized for developing soft wheat cultivars with low protein content. The specific crosses for grain protein content were Ningmai 8×Ningmai 9 and Ningmai 8×Yangmai 5. The additive-dominance model was adequate for grain protein content in F1, while showed partially adequate in F2. Greater value of additive (D) over dominance (H1 and H2) demonstrated additive nature of genes for grain protein content in both generations. Grain protein content exhibited high value of narrow sense heritability. The genetic analysis suggested that grain protein content could be improved through pedigree and progeny selection.

Key words: Combining ability, Diallel cross, Grain protein content, Triticum aestivum L.

INTRODUCTION

Soft wheat flours are most suitable for making cakes and biscuits. Good quality soft wheat flour produces large spread cookies with large diameter, low thickness, and tender texture (Moiraghi et al., 2011). Cookie quality is associated with soft wheat flour of low protein content (Leno et al., 1996). Significant negative correlation between grain protein content and cookie diameter has been reported (Ram and Singh, 2004; Guttieri et al., 2004). Moreover, the significant negative correlation of wet gluten content, solvent retention capacities (SRCs), water soluble pentosan, with cookie diameter has also been observed (Guttieri and Souza, 2003; Zhang et al., 2007; Geng et al., 2012). To develop the wheat cultivars with low protein content is one of the main breeding objectives in the Yangtze River basin, where it is the largest area of soft wheat production in China.

Genetic differences and environmental effects on grain protein content have been reported previously (Miezan et al., 1977; Kramer, 1979; Baenziger et al., 1985). In most of the diallel studies of wheat, grain protein content seemed to be controlled by the partial dominance with additive gene effects (Huo et al., 1996; Akram et al., 2007). However, over dominance type of gene action was also noted for grain protein content (Hsu and Sosulski, 1969; Halloran, 1975). Rahman et al. (2003) reported that grain protein content could result from additive and dominant genes with the possibility of epistatic genetic effects. Heritability estimate is a valuable parameter for determining the magnitude of genetic gain from selection. Low, medium, and high narrow sense heritability estimates were reported for grain protein content (Mckendry et al., 1988; Wang et al., 1991; Ekiz et al., 1998; Rong et al., 2001; Bnejdi et al., 2010).

The development of the unique, low grain protein content, high grain yield genotype in the Yangtze River basin soft wheat breeding program afforded the authors with an excellent opportunity to study the inheritance of grain protein content using diallel crosses. Accordingly, the objectives of this study were: (i) to estimate GCA, SCA and other genetic parameters by using F1 and F2 progenies in two sets of diallel crosses, (ii) identify the best general and specific combiners among different cultivars and crosses, (iii) perform graphical analysis of gene actions for grain protein content. The result of this study can be used in the selection of desirable parents for an effective breeding program to develop the new soft wheat varieties with low grain protein content.

MATERIALS AND METHODS

Plant material and experimental design

The research work comprised of parents, F1 and F2 population experiments of wheat was conducted during the crop seasons of 2010-2012 at the Jiangsu Academy of

Agricultural Sciences, Nanjing, China. Seven diverse soft red wheat cultivars viz. Ningmai 8, Ningmai 9, Yangmai 5, Yangmai 9, Yangmai 11, Sumai 3 and Wangshuibai varied by pedigree, morphological characters, yield potential and quality traits were sown in Oct. 2010. Plots consisting of 3 rows, each 2 m in length, with a plant and row spacing of 5 cm and 30 cm, respectively were used to facilitate hand emasculation and crossing. All the cultivars were crossed following incomplete 7×7 diallel fashion. The F1 and F2 population having 21 hybrids along with 7 parents were sown in a Randomized Complete Block Design with three replications during Oct. 2011 and 2012, respectively. In F1's, each genotype occupied a plot of two 2-m rows spaced 30 cm apart, while in F2's, each genotype consisted of seven rows spaced 30 cm apart. Enough seeds were planted, and then plants thinned to have 20 plants per meter of row for harvesting. For proper growth and development of the plants, agronomic practices and plant protection measures recommended for wheat crop were adopted during the growing season. The experimental field was fertilized with 210 kg ha-1 of N, 105 kg ha-1 of P2O5, and 105 kg ha-1 of K2O. Both the phosphorus and potassium were totally applied as base fertilizer during land reparation. N was applied at preplant, the five-leaf stage, and booting stage at the ratio of 6:2:2. Harvesting was performed after the grain was mature and the plants were dried, at which time grain moisture was determined.

Trait measurement and statistical analysis

The data were recorded for grain protein content, tested by near infrared reflectance (NIR) after one month of harvesting. The data were subjected to analysis of variance using the SAS software (SAS Institute Inc. version 9.1). The general and specific combining ability values were estimated using Method II, Model I of Griffing's (1956). Two scaling tests (Mather and Jinks, 1982) were applied to test the validity of the additive-dominance model. Further, the Hayman (1954) method was used for estimation of gene actions.

RESULTS

Combining ability analysis

Analysis of variance revealed highly significant differences among the genotypes in both F1 and F2 diallel experiments for grain protein content (Table 1). These results permitted further analysis of combining abilities. The mean squares of general combining ability (GCA) and specific combining ability (SCA) in F1 and F2 diallel experiments were highly significant (Table 1), indicating the importance of both additive and non-additive gene effects. However, the magnitudes of GCA variances were several times than SCA (9.10 in F1 and 11.75 in F2). This indicates the preponderance of additive gene effects in the genetic control of grain protein content.

Table 1. Analysis of variance for combining ability of grain protein content in the F_1 and F_2 generations of wheat

Source of variation	df	F_1			F_2		
		SS	MS	F	SS	MS	F
Replications	2	0.052	0.026	0.28	0.222	0.111	1.00
Genotypes	27	27.185	1.007	10.61**	24.970	0.925	8.33**
Error	54	5.125	0.095		6.009	0.111	
GCA	6	6.552	1.092	34.52**	6.416	1.069	28.89**
SCA	21	2.509	0.120	3.78**	1.908	0.091	2.46**
Error	54	1.708	0.032		2.003	0.037	

**, Significant at 1% probability level.

Table 2. Mean values and general combining ability (GCA) effect for grain protein content in F_1 and F_2 generations

Parents	F_1		F_2	
	Protein content (%)	GCA	Protein content (%)	GCA
Wangshuibai	13.55 aA	0.508 aA	13.66 aA	0.430 aA
Sumai 3	13.08 bAB	0.314 bAB	13.40 aAB	0.241 bAB
Yangmai 11	12.71 bcBC	0.110 cBC	12.77 bcBC	0.022 cBC
Ningmai 8	12.34 cdC	− 0.147 deDE	12.75 bcBC	− 0.094 cdC
Yangmai 5	12.22 dC	− 0.250 eE	12.44 cC	− 0.200 dC
Yangmai 9	12.20 dC	− 0.011 cdCD	13.31 abAB	0.214 bAB
Ningmai 9	11.46 eD	− 0.524 fF	11.63 dD	− 0.613 eD
Correlation coefficient [a]	0.975**		0.977**	

The values followed by different capital or small letters within the same column are significantly different at 1% and 5% probability levels, respectively.

*, **, Significant at 5% and 1% probability level, respectively

a, correlation coefficient (r) between the mean value for grain protein content and the value for GCA of seven genotypes

Performance of parents and combining ability

Mean value and GCA effects of the seven parents for grain protein content in both F1 and F2 generations were given in Table 2. Significant differences were found among parents. Grain protein content ranged from 11.46% and 11.63% for Ningmai 9 to 13.55% and 13.66% for Wangshuibai, with an overall mean of 12.51% and 12.85% in F1 and F2 generations, respectively (Table 2).

Estimates for GCA effects varied between □ 0.524 (Ningmai 9) and 0.508 (Wangshuibai) in F1, and between □ 0.613 (Ningmai 9) and 0.430 (Wangshuibai) in F2 diallel analysis for grain protein content. The greatest negative GCA effects, i.e. contribution to lower grain protein, were exhibited by Ningmai 9 in both F1 and F2. This result indicated that Ningmai 9 was the best combiner for decreasing grain protein content in its progeny. Yangmai 5 and Ningmai 8 showed negative GCA in both F1 and F2, and they may also be recommended as superior parents for breeding programs aimed at reducing the grain protein content. Yangmai 9 showed negative GCA effects in F1 but positive GCA effects in F2, suggesting that Yangmai 9 was not stable in performance over generation. The parents that consistently showed high and positive GCA effects in both F1 and F2 were Wangshuibai, Sumai 3 and Yangmai 11.

The correlations between GCA and parental performance were positive and significant with values of 0.975** and 0.977** in F1 and F2, respectively. The result suggested that GCA could be predicted from parental performance in different generations.

Unlike GCA-effects, the SCA-effects (Table 3) are the result of non-additive gene action. Number of crosses showing negative SCA-effects was a little high in F2 than F1. Crosses, such as Ningmai 8 × Ningmai 9, Ningmai 8 × Yangmai 5, Ningmai 9 × Wangshuibai, and Yangmai 9 × Yangmai 11 exhibited strong negative SCA effects for grain protein content in both F1 and F2 generations. These crosses were lower grain protein content (data not shown) and could be utilized as low grain protein content hybrids.

Table 3. Effects of specific combining ability for grain protein content in 21 crosses

Cross	F_1	F_2
Ningmai 8×Ningmai 9	- 0.366	- 0.265
Ningmai 8×Yangmai 5	- 0.356	- 0.508
Ningmai 8×Yangmai 9	0.018	0.035
Ningmai 8×Yangmai 11	- 0.213	0.113
Ningmai 8×Sumai 3	0.329	- 0.072
Ningmai 8×Wangshuibai	- 0.408	- 0.114
Ningmai 9×Yangmai 5	0.157	- 0.169
Ningmai 9×Yangmai 9	0.089	0.251
Ningmai 9×Yangmai 11	0.161	- 0.237
Ningmai 9×Sumai 3	- 0.573	- 0.056
Ningmai 9×Wangshuibai	- 0.224	- 0.165
Yangmai 5×Yangmai 9	- 0.278	- 0.075
Yangmai 5×Yangmai 11	- 0.220	0.003
Yangmai 5×Sumai 3	- 0.204	- 0.152
Yangmai 5×Wangshuibai	- 0.282	0.292
Yangmai 9×Yangmai 11	- 0.172	- 0.277
Yangmai 9×Sumai 3	0.037	- 0.089
Yangmai 9×Wangshuibai	0.127	- 0.548
Yangmai 11×Sumai 3	- 0.238	- 0.194
Yangmai 11×Wangshuibai	- 0.032	0.210
Sumai 3×Wangshuibai	0.011	- 0.212
$SE(S_{ij}-S_{ik})$	0.474	0.514
$SE(S_{ij}-S_{kl})$	0.444	0.480

Adequacy tests for additive-dominance model

The adequacy of data to additive dominance (AD) model was tested by two scaling tests i.e. joint regression analysis and analysis of variance for Wr+ Vr and Wr− Vr (Table 4). The regression analysis revealed that regression coefficient (b) departed significantly from zero but not from unity in both generations, suggesting the absence of non-allelic interactions in genetic behavior of grain protein content which attested grain protein content valid for AD model in both generations. The suitability of the model data analysis was also tested with the analysis of variance of (Wr+ Vr) and (Wr− Vr) (Table 4). The lack of significant variation in the (Wr− Vr) arrays for grain protein content in both generations suggested that any kind of epistasis was not involved in the phenotypic expression of the trait. Although the value of regression coefficient (b) proved the fitness of the data of grain protein content (0.901) in F2 generation for AD model, mean square values of (Wr+ Vr) for grain protein content in F2 generation indicated no significant deviation, thus emphasizing partial adequacy of AD model for grain protein content in F2. However, grain protein content exhibited full adequacy for AD model in F1.

Table 4. Adequacy test of additive-dominance model for grain protein content

Parameters	Grain Protein Content (%)	
	F_1	F_2
Joint regression (b)	0.901±0.134	0.784±0.170
Test for b=0	6.724**	4.612**
Test for b=1	0.739 NS	1.271 NS
Mean squares of Wr+ Vr between arrays	0.399 **	0.056 NS
Mean squares of Wr− Vr between arrays	0.012 NS	0.003 NS
Fitness of the data to Additive-Dominance model	Full	Partial

**, Significant at 1% probability level, NS, Non-significant.

Components of genetic variation

Additive (D) and non-additive (H1 and H2) components were all highly significant in F1 and F2, indicating that both additive and dominance effects were important components of genetic variation for grain protein content (Table 5). However, additive (D) effects were greater than dominance (H1 and H2) in both generations, suggesting that additive gene action played a predominant role in controlling the genetic mechanism of the trait. This advocated that selection can be helpful for the improvement of the trait. The magnitudes of H2/4H were closer to 0.25 in both generations, indicating a symmetrical distribution of positive and negative alleles among the parents demonstrated by non-significant F value determining the relative frequency of dominant and recessive alleles. Degree of dominance (H1/D)0.5 was less than unity in both generations, thus confirming the presence of partial dominance with additive effects. The mean dominance effect of the heterozygote locus (h2) was significant, suggesting that high heterotic effect values would be expected for grain protein content among crosses. Significant environmental component (E) indicated that the grain protein content was highly affected by environmental factors. Narrow sense heritability (h2N) was estimated to be 70.76% and 71.70% in F1 and F2, respectively.

Table 5. Components of genetic variation for grain protein content in wheat

Components of variation	Grain Protein Content (%)	
	F_1	F_2
D	0.429**±0.027	0.439**±0.021
F	− 0.044±0.064	− 0.010±0.051
H_1	0.308**±0.065	0.185**±0.052
H_2	0.286**±0.057	0.197**±0.045
h^2	0.729**±0.038	0.479**±0.031
E	0.031**±0.009	0.037**±0.008
$(H_1/D)^{1/2}$	0.847	0.649
$H_2/4H_1$	0.232	0.266
h^2_N (%)	70.76	71.70
r[(Wr+Vr), Yr]	0.869*	0.101

*, **, Significant at 5% and 1% probability level, respectively

Graphical Analysis

The Vr/Wr graph (Fig.1 and Fig.2) showed that the regression line cut the Wr-axis above the point of origin in both generations which indicated the presence of partial dominance in the inheritance of grain protein content. The distribution of array points along the regression line (Fig.1) showed that Ningmai 9 and Yangmai 5 contained maximum dominant genes as they were located nearer to the point of origin. Wangshuibai and Sumai 3 were furthest away from the origin and thus they possessed maximum recessive genes. The Vr/Wr graph (Fig. 2) exhibited that Yangmai 9 was located near the origin, indicating that it carried maximum dominant genes. In contrast, Ningmai 8 and Yangmai 5 were located further away from the origin, indicating a relative abundance of recessive genes in these two parents.

Positive r-value between parental values (Pr) with (Wr + Vr) in F1 and F2 (Table 5) indicated a tendency for low grain protein content to be associated with dominance and high grain protein content with recessive genes.

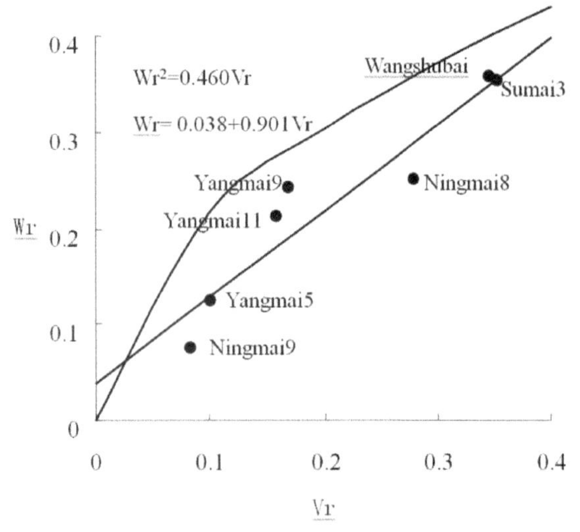

Fig 1. Vr versus Wr graph for grain protein content in F_1

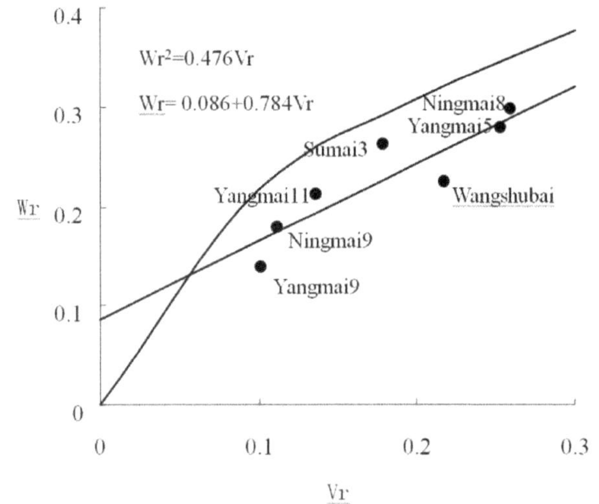

Fig 2. Vr versus Wr graph for grain protein content in F_2

DISCUSSION

In the present investigation grain protein content was found to be controlled by additive and non-additive gene effects, however, additive effects were more important in the genetic control of grain protein content. The present findings thus supported the results of Wang and Lu (1991), Rong et al. (2001), Joshi et al. (2004), Lysa (2009) and Akram et al. (2011), which also showed that additive genetic variance as the main component of genetic variance of grain protein content in wheat. However, preponderance of non-additive gene effect was reported by Kraljevic-Balalic et al. (1982). In this study, due to presence of additive effects, the estimates of narrow sense heritability were higher for grain protein content. Such higher heritable value for grain protein content was also reported by other researchers (Ekiz et al., 1998; Bnejdi et al., 2010). Saranga et al. (1992) stated that higher magnitude of heritability in F2 may have been due to greater recombination of genes or low environmental component. However, Falconer (1989) stated that the estimates of heritability were subjected to environmental variation, and therefore these must be reported and used with great care while making selection from segregating material. Nonetheless, the study offers a lot of scope for improvement of grain protein content in soft wheat breeding program through individual plant selection in early generations.

Based on GCA of parental lines (Table 2), it can be found that Ningmai 9 was the best combiner parents for improvement of grain protein content in soft wheat breeding program, as with the highest negative and significant GCA effects for grain protein content in both F1 and F2 diallels. In addition, the lower mean grain protein content in both F1 and F2 diallels was obtained for Ningmai 8×Ningmai 9 and Ningmai 9×Yangmai 5, respectively (data not shown). Accordingly, to improve grain protein content in soft red wheat, parents like Ningmai 9 can be used in breeding programs as the sources of desirable genes. In fact, the parent Ningmai 9 has been proven to be a highly useful genetic source of low protein content for soft wheat breeding in the Mid lower reaches of the Yangtze River, China. The five soft red wheat cultivars, namely, Ningmai 13, Ningmai 18, Shengxuan 6, Yangmai 18 and Yangmai 21 with low protein content, high yield potential and nine wheat lines with larger cookie diameter than Ningmai 9 have been developed, using Ningmai 9 as a direct crossing parent since 2006 (Yao et al., 2010; Yao et al., 2012). Therefore, the use of the cultivar Ningmai 9 as a parent in the crosses is recommended by which one selects segregates for lower grain protein content.

In general, there was considerable consistency for the results obtained from analysis of F1 and F2 progenies in terms of GCA, SCA, H2N (Joshi et al., 2004; Golkar et al., 2011). It is expected that the results of analysis for F1 and F2 progenies in a diallel set of crosses should not be different in terms of theoretical aspects (Mather and Jinks, 1982). The results of this study and the others in safflower (Pahlavani et al., 2007; Golkar et al. 2011) and in cotton (Khan et al., 2009) indicated that there are some consistencies between results of F1 and F2 diallel analyses. Therefore, it seems that F2 diallels may provide similar results as F1 diallels in other crops. Verhalen and Murray (1969) used F2 data in cotton. Cho and Scott (2000) used F2 data in soybean. Amiri-Oghana et al. (2009) used F2 data in oilseed rape. These studied all reported that F2 analysis provided reliable and better information than F1 generation. One of the most important advantages of the F2 diallel is to have sufficient seeds to overcome the difficulties regarding artificial crossing and F1 seed production in many hermaphroditic plants such as wheat. Also, sufficient F2 seeds provide the opportunity to have more replication or environments in the experiments.

ACKNOWLEDGMENTS

This work was partially supported by the research projects (BE2013439, CX122026) funded by the Department of Science and Technology, Jiangsu, and Public Finance Office, Jiangsu, respectively.

LITERATURE CITED

Akram, Z., S.U. Ajmal, A.A. Kiani and M. Jamil. 2007. Genetic analysis of protein, lysine, gluten and flour yield in bread wheat (Triticum aestivum L.). Pak. J. Boil. Sci. 10:1990-1995.

Akram, Z., S.U. Ajmal, K.S. Khan, R. Qureshi and M. Zubair. 2011. Combining ability estimates of some yield and quality related traits in spring wheat (Triticum aestivum L.). Pak. J. Bot. 43:221-231.

Amiri-Oghana, H., M.H. Fotokianb, F. Javidfar and B. Alizadeh. 2009. Genetic analysis of grain yield, days to flowering and maturity in oilseed rape (Brassica napus L.) using diallel crosses. Int. J. Plant Prod. 3:19-26.

Baenziger, P.S., R.L. Clements, M.S. McIntosh, W.T. Yamazaki, T.M. Starling, D.J. Sammons and J.W. Johnson. 1985. Effect of cultivar, environment, and their interaction and stability analyses on milling and baking quality of soft red winter wheat. Crop Sci 25:5-8.

Bnejdi, F. and M.E. Gazzah. 2010. Epistasis and genotype-by-environment interaction of grain protein content in durum wheat. Genet. Mol. Biol. 33:125-130.

Cho, Y. and R.A. Scott. 2000. Combining ability of seed vigor and grain yield in soybean. Euphytica. 112:145-150.

Ekiz, H., A. Kiral, A. Akçin and L. Simsek. 1998. Cytoplasmic effects on quality traits of bread wheat (Triticum aestivum L.). Euphytica. 100:189-196.

Falconer D.S. 1989. Introduction of quantitative genetics, 2nd edition, Longman New York, USA.

Geng, Z.M., P.P. Zhang, J.B. Yao, D. Yang, H.X. Ma and P. Rayas-Duarte. 2012. Physicochemical and rheological properties of Chinese soft wheat flours and their relationships with cookie-making quality. Cereal Chem. 89:237-241

Golkar, P., A. Arzani and A.M. Rezaei. 2011. Genetic analysis of oil content and fatty acid composition in safflower (Carthamus tinctorius L.). J. AM. Oil Soc. 88:975-982.

Griffing, B. 1956. Concept of general and specific combining ability in relation to diallel crossing system. Aust. J. Biol. Sci. 9:463-493.

Guttieri, M.J., C. Becker and E.J. Souza. 2004. Application of wheat meal solvent retention capacity tests within soft wheat breeding populations. Cereal Chem. 81:261-266.

Guttieri, M.J. and E.J. Souza. 2003. Sources of variation in the solvent retention capacity test of wheat flour. Crop Sci. 43:1628-1633.

Halloran, G.M. 1975. Genetic analysis of grain protein percentage in wheat. Theor. Appl. Genet. 46:79-86.

Hayman, B.I. 1954. The theory and analysis of diallel crosses. Genetics. 39:789-809.

Hsu, C.S. and F.W. Sosulski. 1969. Inheritance of protein content and sedimentation value in diallel crosses of spring wheat (Triticum aestivum). Can. J. Genet. Cytol. 11:967-976.

Huo, Q.T., D.B. Lu, D.Q. Cui and B.G. Xia. 1996. Studies on the genetic model of main quality characters in wheat. Acta Agritulturae Boreali-Sinica. 11:8-14.

Joshi, S.K., S.N. Sharma, D.L. Singhania and R.S. Sain. 2004. Combining ability in the F1 and F2 generations of diallel cross in hexaploid wheat (Triticum aestivum L. em. Thell). Hereditas. 141:115-121.

Khan, N.U., K.B. Marwat, , G. Hassan, M.B. Kumbhar, F. Ullah, Z.A. Soomro, N. Khan, A. Parveen and Umm-E-Aiman. 2009. Study of fiber quality traits in upland cotton using additive-dominance model. Pak. J. Bot. 41:1271-1283.

Kraljevic-Balalic, M., D. Stajner and O. Gasic. 1982. Inheritance of grain proteins in wheat. Theor. Appl. Genet. 63:121-124.

Kramer, T.H. 1979. Environmental and genetic variation for protein content in winter wheat (Triticum aestivum L). Euphytica. 28:209-218.

Leno, A.E., A. Rubiolo and M.C. Anon. 1996. Use of triticale flours in cookies: Quality factors. Cereal Chem. 73:779-784.

Lysa, L.L. 2009. Identification of the genetic controlling system of the protein content in the grain of winter wheat. Cytol. Genet. 43:258-261.

Mahter, K. and J.L. Jinks. 1982. Biometrical Genetics. 3rd Edition. Chapman and Hall, London, UK.

Mckendry, A.L., P.B.E. Mcvetty, and L.E. Evans. 1988. Inheritance of grain protein concentration, grain yield, and related traits in spring wheat (Tnticum aestivum L.). Genome. 30:857-864.

Miezan, K., E.G. Heyne and K.F. Finney. 1977. Genetic and environmental effects on the grain protein content in wheat. Crop Sci. 17:591-593.

Moiraghi, M., L. Vanzetti, C. Bainotti, M. Helguera, A. Leon and G. Perez. 2011. Relationship between soft wheat flour physicochemical composition and cookie-making performance. Cereal Chem 88:130-136.

Pahlavani, M.H., G. Saeidi and A.F. Mirlohi. 2007. Genetic analysis of seed yield and oil content in safflower using F1 and F2 progenies of diallel crosses. Int. J. Plant Prod. 1:129-140.

Rahman, M.A., N.A. Siddquie, M.R. Alam, A.S. Khan and M.S. Alam. 2003. Genetic analysis of some yield contributing and quality characters in spring wheat (Tnticum aestivum L.). Asian J. Plant Sci. 2:277-282.

Ram, S. and R.P. Singh. 2004. Solvent retention capacities of Indian wheats and their relationship with cookie-making quality. Cereal Chem. 81:128-133.

Rong, G.Z., L.Q. Xie and J.T. Gu 2001. Heredity study on grain protein content of different type winter wheat varieties. Journal of Agricultural University of Hebei 24:9-12.

Saranga, Y., D. Zamir, A. Marain and J. Pudich. 1992. breeding tomatoes for plant tolerance: inheritance of salt tolerance and related traits in interspecific populations. Theor. Appl. Genet. 84:390-396.

Verhalen, L.M. and J.C. Murray. 1969. A diallel analysis of several fiber property traits in upland cotton (Gossypium hirsutum L.) Ⅱ. Crop Sci 9:311-315.

Wang, Z.Y. and S.Y. Lu. 1991. Genetic analysis of quality and yield characters of wheat. Journal of Agricultural University of Hebei 14:1-5.

Yao, J.B., H.X. Ma, P.P. Zhang, G.C. Yao, X.M. Yang, L.J. Ren, P. Zhang and M.P. Zhou. 2012. Research of wheat elite parent Ningmai 9 and its utilization. Acta Agriculturae Nucleatae Sinica 26:17-21.

Yao, J.B., E. Souza, H.X. Ma, P.P. Zhang, G.C. Yao, X.M. Yang, L.J. Ren and P. Zhang. 2010. Relationship between Quality Traits of Soft Red Winter Wheat and Cookie Diameter. Acta Agronomica Sinica 36:695-700.

Zhang, Q.J., Y. Zhang, Y. Zhang, Z.H. He and R.J. Peña. 2007. Effects of solvent retention capacities, pentosan content, and dought rheological properties on sugar snap cookie quality in Chinese soft wheat genotypes. Crop Sci. 47:656-664

EVALUATION OF THE QUALITY OF TWO DIFFERENT ARTIFICIAL PASTURES AND THEIR EFFECTS ON THE PERFORMANCE OF GRAZING BEEF ANIMALS

*Mevlut TURK**, *Sebahattin ALBAYRAK, Yalcin BOZKURT*

Suleyman Demirel University, Faculty of Agriculture, Department of Field Crops, Isparta, TURKEY
**Corresponding author: mevlutturk@sdu.edu.tr*

ABSTRACT

This research was conducted to determine the feeding quality of different artificial pastures and their effects on animal performance from the years 2010 to 2012. For this purpose, two different artificial pastures were established, each one covered 1.5 ha area during the first year of the research. The mixtures of the pastures used were as follows: Pasture 1 (P1): *Medicago sativa* L. (20%) + *Bromus inermis* L. (40%) + *Agropyron cristatum* L. (30%) + *Poteriumsanguisorba* (10%); Pasture 2 (P2): *Medicago sativa* L. (15%) + *Onobrychis sativa* Lam. (15%) + *Agropyron cristatum* L. (35%) + *Bromus inermis* L. (35%). The grazing was conducted in the second and the third years of the research. Twenty Holstein male calves with 6 months old were assigned to the experimental areas randomly, each pasture containing 10 animals. Forage samples were collected from grazing and non-grazing areas once every 15 days during the grazing seasons. Concurrently with forage sampling, liveweight gains of animals were also obtained. The dry matter (DM) yield, crude protein (CP), acid detergent fibre (ADF), neutral detergent fibre (NDF) contents, *in vitro* dry matter digestibility (IVDMD) were determined on the forage samples collected from grazing and non-grazing areas. The results showed that as grazing season advanced forage quality was declined in both pastures. The CP content and IVDMD were decreased with advancing growth stages in grazing and non-grazing areas of two pastures. The ADF and NDF contents were increased with advancing growth stages. There were no significant differences in the animal performances grazed on both pastures in terms of final weights, total weight gains and daily live weight gains.

Keywords: Animal performance; dry matter yield; forage quality; grazing; liveweight gain.

INTRODUCTION

The most important problems of raising livestock are not only the shortage of feed stuff but also the quality of forage that can be supplied in many countries with continental climate. Rangelands and harvest residue have a significant role in extensive animal husbandry in Turkey, due to the shortage of rangelands. Therefore, rangelands have substantial importance for animal production, especially during crop growing season because there are no artificial pastures or feed resources for extensive animal husbandry during this period.

In order to establish artificial pasture to produce sufficient and good quality feed in completely damaged vegetation is a suitable choice (Vallentine, 1989; Altin et al., 2005). In most situations, dryland pastures best comprised a simple mixture containing two and three species having similar palatability, season of growth, grazing tolerance, drought tolerance and rare cases of regrowth (Holzworth et al., 2003). Alfalfa (*Medicago sativa* L.), sainfoin (*Onobrychis sativa* Lam.), or other legumes planted in mixtures with grasses, provide nitrogen to increase yield and nutritive values of the entire mixture. However, it is sometimes difficult to keep legumes in the mixture because of their high palatability (Holzworth and Weisher, 2010). Alfalfa, crested wheatgrass and smooth brome cultivar would be suited for use in binary grass-alfalfa mixtures for dryland hay production (Berdahl et al., 2001).

Forage quality can be defined as the relative performance of animals (Buxton et al., 1996). In general, higher levels of cell-soluble, crude protein and minerals are considered as criteria for higher nutritive quality. These components of forage decline substantially with the advanced plant growth and reach the lowest level when plants become dormant (Koc and Gokkus, 1994) in all steppe vegetation. The changing trend of nutritive component of forage shows great differences among range types because the timing and length of growing season differ among them due to climate (Holechek et al., 2004). Most plants show a similarity in declining nutrient composition with advancing development towards maturation (Rao et al., 1973; Stubbendieck and Foster, 1978; Rebole et al., 2004; Turk et al., 2014).

Bozkurt (2012) reported about the superior performance of Holstein cattle compared to other some local and European breed cattle and concluded that under the Mediterranean conditions Holstein cattle were better suited to the feedlot beef systems than other local and some European type cattle. The live weight gain of grazing animals decreases parallel to the declining forage quality. Therefore, it is essential to give supplemental feed to grazing animals in order to maintain live weight gain after plants begin drying due to summer dormancy (Koc et al., 2014).

The objective of this research was to determine the feeding quality of different artificial pastures and monitor their effects on animal performance in the Mediterranean climatic conditions.

MATERIALS AND METHODS

Experimental location

This study was carried out at Suleyman Demirel University Research Farm in Isparta Province (37°45′N, 30°33′E, elevation 1035 m) located in the Mediterranean region of Turkey on three consecutive years of 2010 and 2012. The total precipitation and average temperature data for the experimental area are given in Figures 1 and 2. The major soil characteristics of the research area, determined based on the method described by Rowell (1994), were as follows: The soil texture was clay loam, the organic matter was 1.3% as determined using the Walkley–Black method, the lime was 7.1% as determined using a Scheibler calcimeter, the total salt was 0.29%, the exchangeable K was 122 mg kg^{-1} by 1 N NH_4OAc, the extractable P was 3.3 mg kg^{-1} by 0.4 N $NaHCO_3$ extraction, and the pH of a soil-saturated extract was 7.7. The soil type was calcareous fluvisol (Akgul and Basayigit, 2005).

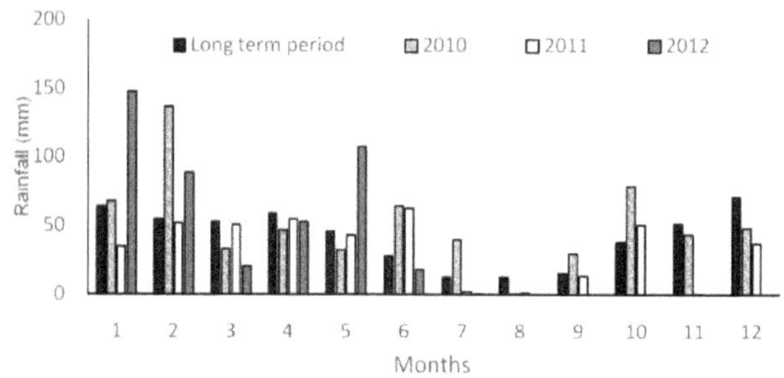

Figure 1. Rainfall values for individual experimental years and over the long term.

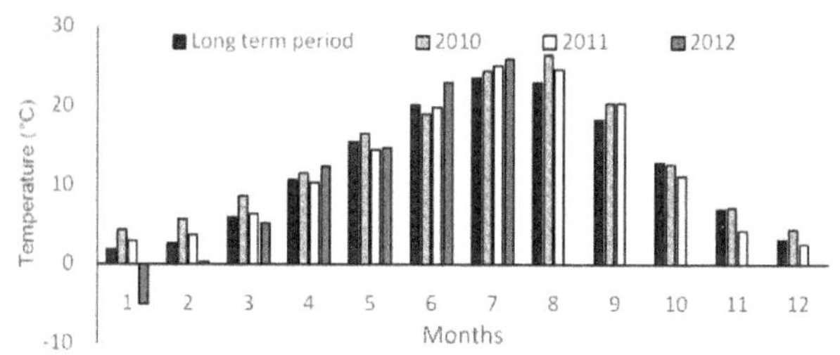

Figure 2. Temperature values for individual experimental years and over the long term.

Animal Management

In order to monitor animal performance, a total of 20 Holstein male calves with an average 6 months old were included and allocated evenly to two artificially established pastures in the experiment which lasted for 90 days in 2011 and 2012. The same breed and the same number of animals were used in both years. Animals were initially weighed at the beginning of the experiments and were randomly divided according to their weights into two grazing groups. Each group was weighed and monitored on a fortnightly basis, using electronic weighing scale (True-Test 2000 SmartUnit). The animals were turned out to pasture for grazing on the 1st of May and the grazing was terminated on the 1st of August each year. The free access of the animals to water and salt was available throughout the experimental period.

Pasture Management

For the establishment of artificial grazing land, 3 ha pasture land was chosen adjacent to the university farm and cultivated in March 2010 with two different botanical compositions. Pasture 1 (P1) was composed of *Medicago sativa* L. (20%) + *Bromus inermis* L. (40%) + *Agropyron cristatum* L. (30%) + *Poterium sanguisorba* (10%); and Pasture 2 (P2) had mixtures of *Medicago sativa* L. (15%) + *Onobrychis sativa* Lam. (15%) + *Agropyron cristatum* L. (35%) + *Bromus inermis* L. (35%), respectively. In first year, cutting and maintenance applications were made. Pastures were harvested twice during the end of June and beginning of October in 2010. Animal grazing applications were performed in the second and the third year of the study since the first year covered only the establishment of the artificial pastures.

In order to monitor chemical composition changes in pastures, 4 non-grazed areas within each pasture were established and fenced with wires by 4×3m size and grass samples were collected by using 0.5m^2 (0.5x1 m) quadrats fortnightly from May to August each year. The dry matter (DM) yield, crude protein (CP), acid detergent fibre (ADF), neutral detergent fibre (NDF) contents, *in vitro* dry matter digestibility (IVDMD) were determined as well.

After the harvest, samples were dried at 70 °C for 48 h, and weighed. The dried samples were reassembled and ground to pass through a 1-mm screen. The crude protein (CP) content was calculated by multiplying the Kjeldahl nitrogen concentration by 6.25 (Kacar and Inal, 2008). The acid detergent fiber (ADF) and neutral detergent fiber (NDF) concentrations were measured according to methods from Ankom Technology. Tilley and Terry's (1963) methods were used to determine *in vitro* dry matter digestibility (IVDMD) of samples.

Statistical analysis

A randomized complete block experimental design was used in this study. General Linear Model (GLM) procedure was applied for the statistical analysis of the data by using Minitab16 statistical software programme and for the data obtained from animal experiment, initial weight and age were taken as covariates to eliminate the weight and age differences at the start of the experiment (Minitab,2010).In statistical analysis, the average of two-year data was taken into account since the data related to animal and pastures were collected after the establishment of the artificial pastures.

RESULTS AND DISCUSSION

Dry Matter Yield and Forage Quality

The results of the variance analysis showed that the effects of the years, pasture types, grazing and sampling times on all investigated traits were significant (Table 1). The highest DM yields of non-grazed areas were determined in May 30 and June 15 in P1 and P2, respectively (Figure 3). The DM yields of grazed areas in both pastures decreased linearly during the grazing season. There was a quadratic relationship between DM yield and sampling times in non-grazed areas as opposed to expectation that the concentration of DM linearly increases with advancing stages of maturity (McDonald et al., 1995; Sankhyan et al., 1999). The DM yield increased until the end of May and reached the highest value (11.78 t ha^{-1}) in May 30 in non-grazed area of P 1 (Figure 3). This may be due to the continued growth of the plant as expected. The DM yield started to decrease in June 15 due to drying plants. In the non-grazed area of P 2, increase in DM yield lasted until June 15, after June 30 it began to decrease due to drying plants. However, DM yield was decreased linearly throughout the grazing season in the grazed areas of P 1 and P 2 due to the consumptions by animals. Those variations caused significant interaction between pasture types × grazing × sampling times for DM yield. Decreases in DM yields due to grazing have been reported in numerous studies (Karslı et al., 2003; Gokkus et al., 2005). DM yields of non-grazed areas were higher than those of grazed areas in the present study as expected. The DM yield of P2 was higher than that of P1. The reason for this difference is *Onobrychis sativa* in mixture of P2. Alfalfa, crested wheatgrass and smooth brome which types of both pastures are already suitable for arid rangelands (Altin et al., 2005). However, although sainfoin is not very suitable for long-term artificial pastures, it showed better performance than burnet.

Table 1. Results of analysis of variance and mean squares of the traits determined

Sources of variations	df	Dry Matter (DM) Yield	Crude Protein (CP)	NDF	ADF	IVDMD
Block (year)	6	11231**	0.015 ns	3.16**	1.798**	0.82 ns
Year	1	258766**	88.812**	280.59**	56.754**	102.09**
Pasture type (P)	1	2897919**	1.051**	103.44**	328.509**	409.54**
Grazing (G)	1	14069012**	3.745**	23.21**	2.870**	198.76**
Sampling Times (ST)	6	1294804**	291.814**	1278.55**	780.733**	1424.31**
P x G intr.	1	185985**	0.047 ns	24.21**	7.959**	3.72 ns
P x ST intr.	6	106535**	0.227**	0.97ns	0.559 ns	1.06 ns
G x ST intr	6	675914**	0.464**	5.81**	2.759**	8.57**
P x Year	1	76671**	3.874**	16.50**	0.081 ns	1.53 ns
G x Year	1	23961**	1.335**	13.85**	0.101 ns	1.57 ns
ST x Year	6	15638**	0.545**	0.22 ns	0.255 ns	0.13 ns
P x G x ST intr.	6	28726**	0.031 ns	1.99 ns	0.866 ns	1.18 ns
P x G x Year intr.	1	63 ns	0.007 ns	1.97 ns	0.039 ns	0.05 ns
P x ST x Year intr.	6	10636**	0.042 ns	0.06 ns	0.156 ns	0.00 ns
G x ST x Year intr.	6	18267**	0.046 ns	0.51 ns	0.150 ns	0.05 ns
P x G x ST x Year int.	6	1376 ns	0.001 ns	0.10 ns	0.160 ns	0.00 ns
Error	162	1144	0.027	0.75	0.346	0.77 ns

df: degrees of freedom, ns: not significant, *P<0.05 and **P<0.01.

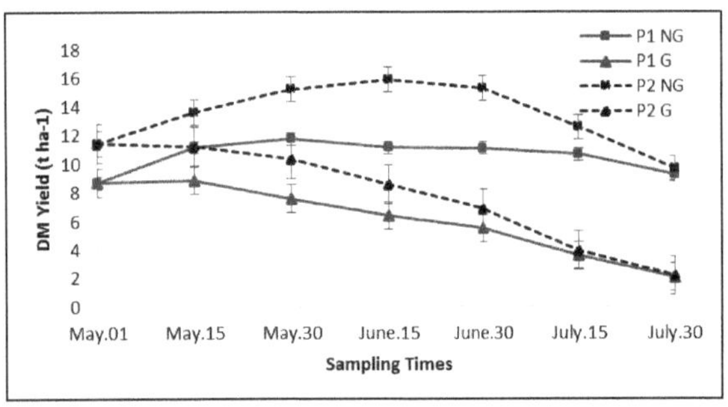

Figure 3. Seasonal variation of DM yields of two different artificial pastures (P1:Pasture 1, P2:Pasture 2, NG: Non grazing, G:Grazing).

The effects of the years, pasture types, grazing and sampling times on CP contents were significant (Table 1). The CP contents decreased linearly in grazed and non-grazed areas in both pastures. The highest CP contents were obtained from beginning of the grazing season while the lowest CP contents were determined at end of the grazing season (Figure 4). Statistically significant interactions between pasture types × sampling times and grazing × sampling times were found for CP ratios (Table 1). These interactions indicated that harvesting stage affected CP ratios differently according to the different pastures and grazing. The CP contents were decreased linearly throughout the grazing season in grazed and non-grazed areas of P1 and P2 (Figure 4). Maturity stage at harvest is the most important factor determining forage quality. Besides N, and hence protein, most minerals, decline with advancing plant development. Other reports also support that the CP contents decreases by advancing stage of maturity (Koc et al., 2000; Rebole et al., 2004), suggesting that animals should be supplemented with protein sources, especially towards the end of the grazing season. As a result of this process, forage quality lessens substantially towards the end of growing season. The CP ratios of the grazed areas were higher than that of non-grazed areas in the present study. This could be associated with the continued re-growth of the plants in grazed areas because young plant tissues are more nutritious than dead or mature plant (Lyons et al., 1996). The CP ratio of P2 was higher than that of P1. The reason for this was that legume rates of pastures were different. Legumes are rich in terms of protein concentration (Altin et al., 2011). Pasture 1 have 20% legume ratio while Pasture 2 have 30% legume ratio in the present study.

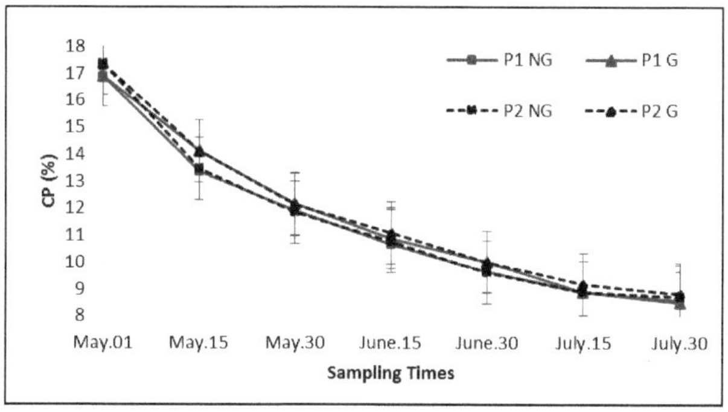

Figure 4. Seasonal variation of CP ratios yield of two different artificial pastures (P1:Pasture 1, P2:Pasture 2, NG: Non grazing, G:Grazing).

The effects of the years, pasture types, grazing and sampling times on ADF and NDF contents were significant (Table 1). The NDF and ADF contents increased linearly in grazed and non-grazed areas in both pastures. The lowest NDF and ADF contents were obtained from beginning of the grazing season while the highest NDF and ADF contents were determined at end of the grazing season (Figures 5 and 6). The effects of pasture types × grazing and grazing × sampling times interactions on ADF and NDF contents were found statistically significant ($P< 0.05$). The reason for this, differences are the changes in the ADF and NDF rates. Acid detergent fiber and NDF contents were increased linearly during the grazing season (Figures 5 and 6). This could be explained by the decrease in proportion of leaves and the increase of the stems proportion with advanced maturity. Because, ADF and NDF contents of stems is higher than the leaves. Similar results were reported by Karslı et al. (2003), Kaya et al. (2004), Erkovan et al. (2009), Turk et al. (2007), Albayrak et al. (2009). The trends in ADF and NDF contents with increasing maturity are normally the reverse of protein (Rebole et al., 2004). Young plant cells have the primary cell wall, but also the secondary cell wall occurs with maturing. This causes being the more fibrous of mature plants (Arzani et al., 2004). ADF and NDF contents of non-grazed areas were higher than those of grazed areas in the present study. This could be explained by the continued re-growth of the plants in the grazed areas. The ADF and NDF contents of P2 were higher than those of P1. Plants with mature structural carbohydrates (cellulose, hemicellulose, lignin) increases (Arzani et al., 2001). As a result, decrease in proportion of leaves and the increase of the stems proportion with advanced maturity causes a decrease in feed quality (Lacefield et al., 1999; Linn and Martin, 1999).

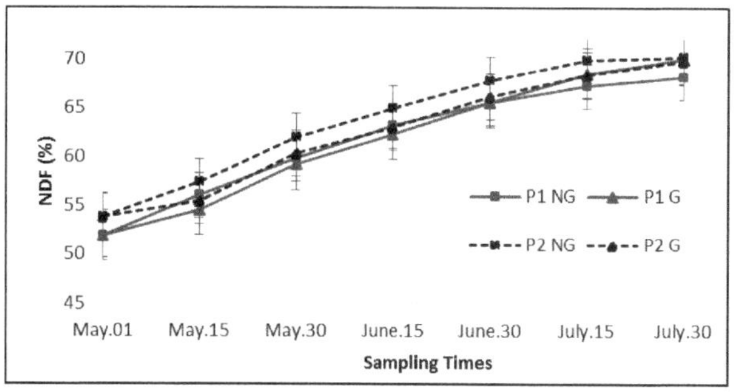

Figure 5. Seasonal variation of NDF contents of two different artificial pastures (P1:Pasture 1, P2:Pasture 2, NG: Non grazing, G:Grazing).

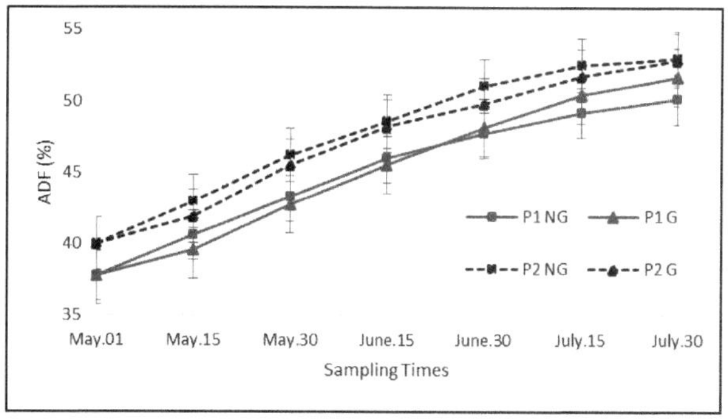

Figure 6. Seasonal variation of ADF contents of two different artificial pastures (P1:Pasture 1, P2:Pasture 2, NG: Non grazing, G:Grazing).

The effects of the years, pasture types, grazing and sampling times on IVDMD were significant (Table 1). The IVDMD was decreased throughout the grazing season in grazed and non-grazed areas of P 1 and P 2 (Figure 7). Similar results were reported by Horney et al. (1996), Hitz and Russell (1998), Karsli et al., (1999). The reason for this decrease in digestibility of plants is the increase in the lignin content (Jung and Vogel, 1992; Van Soest, 1982). The decrease in IVDMD resulted from the increase structural tissues in stems (Arzani et al., 2004). Pinkerton (1996) stated that there is a close relationship between digestibility and cell wall structure. Fiber content is increased as the plants grow, the digestibility decreases (Erfanzadeh, 2001). Overall, IVDMD varies depending on crude cellulose levels in its structure. Digestibility of plenty of leafy green forage is very high, whereas the increase of the stems proportion with advanced maturity causes a decrease digestibility (Aksoy and Yilmaz, 2003).

The IVDMD of legumes are higher than others. Therefore, IVDMD content of P2 was higher than that of P1.

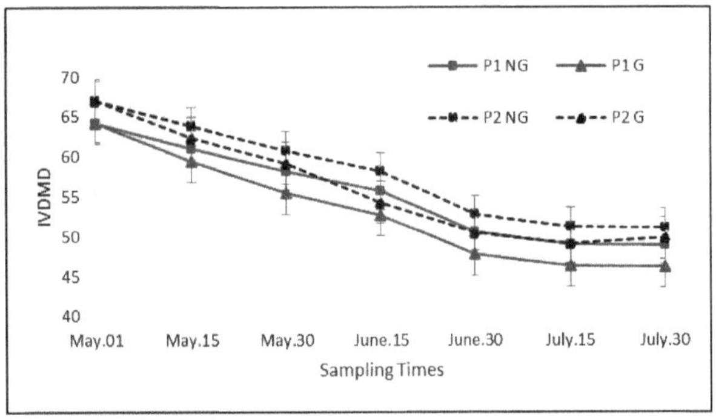

Figure 7. Seasonal variation of IVDMD of two different artificial pastures (P1:Pasture 1, P2:Pasture 2, NG: Non grazing, G:Grazing).

Animal Performance

Animal live weights were increased during the grazing season in both pastures and years (Figure 8). It was observed that these increases were greater until July 15, then these increases slowed down in both years.

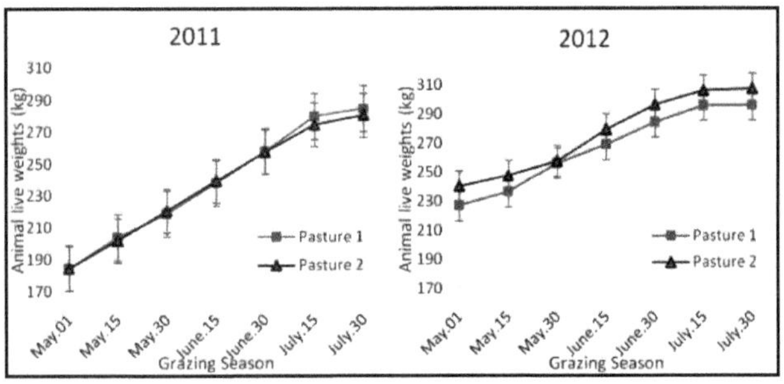

Figure 8. Animal live weights throughout the grazing season determined in 2011 and 2012 years.

There were no significant (P >0.05) differences between pasture types in final weight (FW), total weight gain (TWG) and daily live weight gain (DLWG) (Figure 8). In respect to performance of animals in pasture types, the final weights were 284.7 and 280.5 kg for P1 and P2 in 2011, 296.2 and 307.4 kg for P1 and P2 in 2012, respectively. The average total weight gains 100.1 and 95.9 kg for P1 and P2 in 2011, 69.1 and 68.1 kg for P1 and P2 in 2012 and finally daily live weight gains of 1.112 and 1.066 kg for P1 and P2 in 2011, 0.768 and 0.746 kg for P1 and P2 in 2012, respectively (Table 2).

Table 2. Overall performance comparisons of animals by years and pasture types

	2011			2012		
	Pasture 1	**Pasture 2**	**Mean**	**Pasture 1**	**Pasture 2**	**Mean**
IW	184.6	184.6	184.6	227.1	240.3	233.7
FW	284.7	280.5	282.6	296.2	307.4	301.8
TWG	100.1	95.9	98.0	69.1	67.1	68.1
DLWG	1.112	1.066	1.089	0.768	0.746	0.757

IW= Initial weight, FW= Final weight, TWG= Total weight gain, DLWG= Daily live weight gain.

There were significant (P <0.05) differences between years in Final weight (FW), Total weight gain (TWG) and DLWG. The animals in 2011 tended to perform better than the cattle in 2012 in all parameters observed. As it is presented in Table 2, final weights of the animals in the years 2011 and 2012 were 282.6 and 301.8 kg; the average total weight gains 98.0 and 68.1 kg and finally daily live weight gains of 1.089 and 0.757 kg respectively. The reason for the TWG and DLWG differences between years could be due to higher DM yield of pastures in 2011.

In literature, there are no many published studies on performance of different breeds and comparison on different beef production systems in the Mediterranean conditions. However, Bozkurt (2012) reported about the superior performance of Holstein cattle compared to other some local and European breed cattle and concluded that under the Mediterranean conditions Holstein cattle were better suited to the feedlot beef systems than other local and some European type cattle.

Keane et al. (1989) and Keane and More O'Ferrall (1992) reported some results on breed comparisons indicating that differences in factors such as production systems, slaughter weights and climate conditions are of great importance. Similarly, it was stated that breeds and crosses of beef cattle show distinctive differences in performance in different production systems (Bozkurt and ApDewi, 1996). Performance potential vary greatly between different breeds of cattle and different production systems. While there are certainly differences between performance of animals in growth rate, the live weight gain which can be achieved from a given area of grass or quantity of feed is similar for most breeds of animals, provided that animal is fed and managed according to its own particular requirements in its own environment (Wilkinson, 1985).

CONCLUSIONS

It can be concluded thatthe highest DM yields of non-grazed areas were determined in May 30 and June 15 in P1 and P2, respectively, while the DM yields of grazed areas in both pastures decreased linearly during the grazing season. The harvesting at the late stages caused a reduction in forage quality. The CP content and IVDMD decreased throughout the grazing season, while ADF and NDF contents increased in grazing and non-grazing areas in both pastures. The effects of pastures on animal performance were not significant. Animal performances were similar during both grazing seasons in both pastures and years.

At the end of a three-year research conducted in the Mediterranean type of climatic conditions, pasture 2 (*Medicago sativa* L. (15%) + *Onobrychis sativa* Lam. (15%) + *Agropyron cristatum* L. (35%) + *Bromus inermis* L. (35%)) can be recommended in respect for high yield and quality for the regions with similar type of climatic conditions and especially for the inner parts of the region with an altitude higher than a thousand meter.

ACKNOWLEDGMENTS

This work was financially supported by TUBITAK (Turkish Scientific and Technological Research Council, project no. 109O162).

LITERATURE CITED

Akgul M, L Basayigit.2005. Detailed soil survey and mapping of Suleyman Demirel University farming land. In: Bulletin (Fen Bilimleri Enstitusu Dergisi) 9 (3), Suleyman Demirel Universitesi, Isparta, Turkey, 54–63, available from URL: http://edergi.sdu.edu.tr/index.php/fbed/article/viewFile/822/1119

Aksoy A, A. Yilmaz.2003. Metabolizable energy values and dry matter and organic matter digestibilities of some alfalfa varieties. Journal of Agricultural Sciences, 9 (4): 440–444.

Albayrak, S., M. Turk and O. Yuksel. 2009. Effects of phosphorus fertilization and harvesting stages on forage yield and quality of woolypod vetch. Turkish Journal of Field Crops, 14(1):30-40.

Altin M, A. Gokkus, AKoc. 2005. Range and Meadow Improvement and Development. The Ministry of Agriculture and Rural Affairs, Republic of Turkey, Ankara, 1–468.

Arzani H, J. Torkan, M. Jafari, A. Nikkhah. 2001. Investigation on effects of phenological stages and environmental factors (soil and climate) on forage quality of some important range species. Iran J AgricSci 32:385 –397.

Arzani H, M. Zohdi, E. Fish, G.H. ZahediAmiri, A. Nikkhah, D. Wester.2004. Phenological effects on forage quality of five grass species. J Range Manage 57:624 –629.

Berdahl J.D., J.F.Karn, J.R. Hendrickson. 2001. Dry matter yields of cool-season grass monocultures and grass–alfalfa binary mixtures. Agron J 93: 463–467.

Bozkurt Y. 2012. Seasonal performance of different breeds of feedlot beef cattle grown under the Mediterranean conditions. Bulgarian J AgricSci 18 (3):443–445.

Bozkurt Y., I. Ap Dewi.1996. Effect of breed type, sex, birth year and season of birth and their interactions on live weight change in beef cattle. In: Bulletin 10 (13) Faculty of Agriculture, Selcuk University, Aleaddin Keykubat, Turkey, 125–140.

Buxton R.D., D.R. Mertens, D.S. Fisher.1996. Forage quality and ruminant utilization. In: Cool-Season Forage Grasses (Eds Moser, LE, Buxton DR, Casler MD), American Society of Agronomy, Crop Science Society of America & Soil Science Society of America, Madison, Wisconsin, USA, 229–267.

Erfanzadeh R. 2001. Investigation on quality indices changes of rangeland species during two phenological stages. In: Proceedings of the 1st National Livestock and Rangeland Management Congress, Research Institute of Forests and Rangelands of Iran, Tehran, 60.

Erkovan H.I., M.K. Gullap, M. Dasci, A. Koc. 2009. Changes in leaf area index, forage quality and above-ground biomass in grazed and ungrazed rangelands of Eastern Anatolia Region. Journal of Agricultural Sciences,15 (3): 217–223.

Gokkus A., B.H. Hakyemez, I.Y. Yurtman, T. Savas. 2005. The effects of different stocking rates on pasture hay yield and milk yield of goats in different types of pastures. Akdeniz University Journal of the Faculty of Agriculture 18: 207–212.

Hitz A.C., J.R. Russell. 1998. Potential of stockpiled perennial forages in winter grazing systems for pregnant beef cows. J AnimSci 76:404–415.

Holechek J.L., R.D.Preper, C.H. Herbel. 2004. Range Management: Principles and Practices, 5th edn. Prentice Hall, New Jersey, 1–607.

Holzworth L, J. Mosley, D. Cash, D. Koch, K. Crane. 2003. Dryland pastures in Montana and Wyoming. Montana State University Extension Service, Montana, available from URL: http://msuextension.org/publications/AgandNaturalResources/ EB0019.pdf .

Holzworth L.K., L.E. Wiesner. 2010. Dryland seeded pastures: management of seedings - principles of establishment and seedling year management. Publication & Information Animal & Range Sciences, Extension Service, Montana State University, Montana, available from URL: http://animalrangeextension.montana.edu/articles/forage/Chpt3/3-mgt-dry.htm

Horney M.R., T. DelCurto, M.M. Stamm, R.K. Bailey, S.D. Brandyberry. 1996. Early-vegetative tall fescue hay vs alfalfa hay as a supplement for cattle consuming low-quality roughages. J AnimSci 74: 1959–1966.

Jung H.J.G., K.P. Vogel. 1992. Lignification of switcgrass (*Panicum virgatum*) and big bluestem (*Andropogon gerardii*) plant parts during maturation and its effect on fibre degradability. J Sci Food Agric 59:169–176.

Kacar B, A. Inal. 2008. BitkiAnalizleri. Nobel Akademik Yayıncılık, Ankara, 1–892.

Karsli M.A., J.R.Russell, M.J. Hersom. 1999. Evaluation of berseem clover in diets of ruminants consuming corn crop residues. J AnimSci 77:2873–2882.

Karsli M.A., S. Deniz, H. Nursoy, N. Denek, H. Akdeniz. 2003. The effects of maturity on range quality and animal performance. Turk J Vet Anim Sci 27:117–124.

Kaya I, A. Oncuer, Y. Unal, S. Yıldız. 2004. Nutritive value of pastures in Kars district. I. Botanical and nutrient composition at different stages of maturity. Turk J Vet Anim Sci 28:275–280.

Keane M.G., G.J. More O'ferrall.1992. Comparison of Friesian, Canadian Hereford × Friesian and Simmental × Friesian steers for growth and carcass composition. AnimProd 55: 377–387.

Keane M.G., G.J. More O'ferrall, J. Connolly. 1989. Growth and carcass composition of Friesian, Limousin × Friesian and Blonde d'Aquitaine × Friesian steers. Anim Prod 48: 353–365.

Koc A, A. Gokkus.1994. Determination of basal area, botanical composition and suitable stubble height of the rangelands of Guzelyurt district, Erzurum. Turk J Agric For 18: 495–500.

Koc A, A. Gokkus, A. Bakoglu, A. Ozaslan.2000.The changing in some chemical properties of grass samples received from different sections of Palandoken pastures in grazing season, In: Proceedings of the International Animal Nutrition Congress, 4–6 September 2000, Isparta, 471–478.

Koc, A., A.Kaya, M.K.Gullap, H.I.Erkovan, M.Macit, M.Karaoğlu, 2014. The effect of supplemental concentrate feed on live weight gain of yearling heifers over grazing season in subirrigated rangelands of East Anatolia. Turk J Vet Anim Sci. 38: 278-284.

Lacefield G., J.C. Henning, M. Collins, L. Swetnam.1999. Quality hay production. Cooperative Extension Service,

University of Kentucky, Collage of Agriculture, Lexington, available from URL: http://www2.ca.uky.edu/agc/pubs/agr/agr62/agr62.pdf

Linn J.G., N.P. Martin. 1999. Forage Quality Tests and Interpretations. Publication No. 02637, University of Minnesota Extension Service, Minnesota, 1–4.

Lyons R.K., R. Machen, T.D.A. Forbes.1996. Why range forage quality changes. Texas A&MAgriLife Extension Service, Texas, USA, available from URL: http://gillespie.agrilife.org/files/2013/02/WhyRangeForageQualityChanges.pdf

McDonald P., R.A. Edwards, J.F.D. Greenhalgh, C.A. Morgan.1995. Animal Nutrition, 5th edn. Longman Scientific and Technical, Harlow, UK, 1–607.

Minitab.2010. Minitab Reference Manual: Release 16 for Windows. Minitab Inc., State College, Pennsylvania, USA.

Pinkerton B. 1996. Forage Quality. Crop and Soil Environment Science Department, College of Agriculture, Forest, and Life Science, Clemson University, Clemson, South Carolina, 1–998.

Rao M.R., L.H. Harbers, E.F. Smith, 1973. Seasonal change in nutritive value of bluestem pastures. J Range Manage 26: 419–422.

Rebole A., C.Alzueta, L.T.Ortiz, C. Barro, M.L. Rodriguez, R. Caballero. 2004. Yields and chemical composition of different parts of the common vetch at flowering and at two seed filling stages. Span J Agric Res 2: 550–557.

Rowell D.L. 1994. Soil Science: Methods and Applications. Longman Scientific and Technical, Harlow, UK, 1–350.

Sankhyan S.K., A.K. Shinde, R. Bhatta, S.A. Karim.1999. Comparison of diet and faecal collection methods for assessment of seasonal variation in dry matter intake by sheep maintained on a Cenchrusciliaris pasture. Anim Feed SciTechnol 82: 261–269.

Stubbendieck J., M.A. Foster.1978. Herbage yield and quality of threadleaf sedge. J Range Manage 31: 290–292.

Tilley J.M.A., R.A. Terry.1963. A two-stage technique for the in vitro digestion of forage crops. J Br GrasslSoc 18: 104–111.

Turk M, S. Albayrak, O. Yuksel. 2007. Effects of phosphorus fertilization and harvesting stages on forage yield and quality of narbon vetch. N Z J Agric Res 50:457–462.

Turk M, S. Albayrak, Y.Bozkurt.2014. Seasonal trends in chemical composition of different artificial pastures. Turkish Journal of Field Crops, 19(1):53-58.

Vallentine J.F. 1989. Range Development and Improvements, 3rd edn. Academic Press, Inc., San Diego, California, 1–524.

Van Soest P.J.1982.Nutritional Ecology of the Ruminant. O & B Books Inc., Corvallis, Oregon, 1–374.

Wilkinson J.M. 1985. Beef Production from Silage and Other Conserved Forage. Longman Group Limited, London and Newyork, 1–152.

RESISTANCE TO POWDERY MILDEW AND LEAF RUST IN WHEAT LINES DERIVED FROM A *Triticum aestivum/Aegilops variabilis* CROSS

Penko SPETSOV[1,], Nadia DASKALOVA[2], Dragomir PLAMENOV[2], Todor MORALIYSKI[2]*

[1]Dobroudja Agricultural Institute, General Toshevo, BULGARIA
[2]Faculty of Marine Sciences and Ecology, Technical University – Varna, BULGARIA
**Corresponding author: pspetsov@abv.bg*

ABSTRACT

Lines with resistance to powdery mildew in adult plant stage have been selected from two groups of F_1 seeds produced by crossing *T. aestivum* with *Aegilops variabilis* (2n=28, UUSS). Two lines were derived from non-irradiated F_1 seeds and another five lines originated from irradiated hybrid seeds. Rusalka and Pliska bread wheat varieties were used as backcrossing parents. Two lines selected from the irradiated group, showed different resistance pattern in comparison to 18 wheat cultivars/lines possessing known powdery mildew resistance genes. Line 11-8 was previously identified with C-banding as a 1U(1B) substitution. It expressed a distinctive disease pattern, different from the parents and differential set of lines. It is suggested that the alien 1U chromosome is a carrier of resistance to powdery mildew, as evidenced by the disease pattern of doubled haploid lines derived from cv. Charodeika x line 11-8 cross. The DH lines expressed resistance to 8 from 11 isolates used. Additionally, the lines obtained from the original cross were subjected to a test for leaf rust resistance and compared to 11 cultivars possessing documented leaf rust resistance genes. At least four lines were postulated to carry unidentified genes that could not be resolved by the leaf rust isolates used.

Key words: *Aegilops variabilis*, 1U chromosome, Introgression, Leaf rust, Powdery mildew, *T. aestivum*

INTRODUCTION

Introgression of genetic variability into wheat from distantly or closely related species has been connected to economically important traits such as insect and disease resistance (Friebe et al., 1996; Hsam and Zeller, 2002; Xu et al., 2009). *Aegilops* species belong to the primary, secondary and tertiary gene pool of wheat and gene transfers into cultivated wheat can be achieved by direct hybridization, homologous and homoeologous recombination, irradiation, backcrossing and selection (Sears, 1956; Fedak, 1998; De Pace et al., 2001).

Aegilops variabilis Eig (syn. *Aegilops peregrina* 2n=28, genome UUSS) is an annual allotetraploid species inclusive in the tertiary gene pool of wheat. Genes for resistance to root knot and cereal cyst nematodes (Yu et al., 1990; Barloy et al., 2007), kernel bunt (Williams and Mujeeb-Kazi, 1996), eyespot (Thiele et al., 2002), powdery mildew (Spetsov et al., 1997), stripe rust (Liu et al., 2011) and leaf rust (Marais et al., 2008) from this species have been successfully transferred into common wheat. U genome is involved in diploid (*Ae. umbellulata*), tetraploid (*Ae. columnaris*, *biuncialis*, *geniculata*, *neglecta*, *triuncialis*, *kotschyi* and *peregrina*) and hexaploid (*Ae. neglecta* 6x and *Ae. juvenalis*) species of genus *Aegilops* (van Slageren, 1994). Several attempts succeeded in transferring genes for powdery mildew resistance from *Aegilops* species with U genome, *Ae. umbellulata* (Frauenstein and Hammer, 1985), *Ae. ovata* (syn. *geniculata*) (Zeller et al., 2002; Stoilova and Spetsov, 2006), and *Ae. triuncialis* (Harjit-Singh et al., 2000). Specifying the U chromosomes, Zhu et al. (2006) were the first to transfer powdery mildew resistance gene from *Ae. umbellulata* to wheat as selected and studied 2U(2B) substitution line. The new gene was temporarily designated *PmY39*.

Leaf rust resistance genes in wheat have been derived from species of *Triticum*, *Secale*, *Thinopyrum* and *Aegilops* (*Ae. ventricosa*, *kotschyi*, *sharonensis*, *peregrina*, *triuncialis*, *geniculata* and *neglecta*) (McIntosh et al., 2009). Chhuneja et al. (2008) indicated two new genes (*LrU1* and *LrU2*) that have been introgressed into wheat from *Ae. umbellulata*, but the alien chromosome/s was not identified. During backcrossing a chromosome segment containing a leaf rust resistance gene from *Ae. peregrina*, designated *Lr59*, was transferred and incorporated in wheat 1A chromosome (Marais et al., 2008). Thus, distantly related genera and species provide useful genetic resources of disease resistance for wheat breeding.

Following the aim to exploit the genes for powdery mildew resistance in *Aegilops* species through irradiation treatment, backcrossing and selection, we investigated the disease response of selected lines from a *T. aestivum*/*Ae. variabilis* cross with emphasis on their resistance to powdery mildew and leaf rust in seedlings, and possible donors of the transferred resistance in genotypes deriving from a group of hybrid seeds treated with gamma rays and backcrossing.

MATERIALS AND METHODS

Plant Materials

Lines with 2n=42 have been produced from a *T. aestivum* cv. Rusalka/*Ae. variabilis* cross. The alien species is an accession from the collection at the Dobroudja Agricultural Institute – General Toshevo. F_1 plants produced were divided into four groups depending on the variety of backcrossing and application of irradiation (Spetsov et al., 1997). Part of the hybrid seeds was irradiated with 100 Gy gamma rays of Co^{60}. Following one backcross to wheat and by 6 to 10 generations of selfing with selection for resistance in each generation, several hexaploid lines resistant to the mildew pathogen *Blumeria graminis* (DC.) E.O. Speer f.sp. *tritici* were isolated. All genotypes used in the study are presented in Table 1. Out of the seven hexaploid lines studied from the original cross, two lines, 28-1 and 28-7 were selected from the group of F_1 non-irradiated (normal) seeds and the hybrid plants were backcrossed once to another cultivar that resembling Rusalka. The remaining five lines (11-8, 37-3, 37-7, 104-4 and 104-5) were developed from the group of irradiated F_1 seeds and mixed pollen from Rusalka and Pliska varieties were used for obtaining BC_1 seeds. The selected substitution line 11-8 (1U/1B) was cross to common wheat cv. Charodeika to obtain several doubled haploid lines. Two of them, lines 51 and 131 did not possess 1U chromosome but were hexaploids and expressed resistance to powdery mildew.

Table 1. Pedigree and satellited chromosomes (sat) of genotypes (2n=42) derived from a *T. aestivum*/*Aegilops variabilis* cross

Line No.	Cross/grouping	Sat chromosomes
	Rusalka/*Ae. variabilis*	
28-1	F_1 backcrossed with Rusalka*	1B, 6B
28-7	-//- -//-	1B, 6B
11 8	Irradiated F_1 backcrossed with Rusalka-Pliska	1U, 6B
37-3	-//- -//-	1B, 6B
37-7	-//- -//-	1B, 6B
104-4	-//- -//-	1B, 6B
104-5	-//- -//-	1B, 6B
	Charodeika/line 11-8	
51	DH line	6B
131	-//- -//-	6B

Rusalka*, a wheat cultivar resembling Rusalka;
Rusalka-Pliska, mixed pollen from the two varieties;
DH line, doubled haploid line.

Powdery Mildew Test

The near-isogenic lines of Chancellor with known powdery mildew resistance genes and TP114/2*Starke possessing gene *Pm6* were supplied by R. A. McIntosh, Sydney, Australia. The eleven *Blumeria graminis* f.sp. *tritici* (Bgt) isolates used for the differentiation of the known major resistance genes were collected from different parts of Europe and selected from single spore progenies. The test for mildew resistance were conducted on primary leaf segments cultured on 0.6% w/v agar and 35mg/l benzimidazole in plastic boxes. The methods of inoculation, conditions of incubation and disease assessment were according to Hsam et al. (2003). Three major classes of host reactions were distinguished: r = resistant, i = intermediate, s = susceptible.

Leaf Rust Test

Resistance test was also conducted on primary leaf segments in petri dishes on 6 g/l agar and a final benzimidazole concentration of 35 mg/l following the method of Felsenstein et al. (1998). Uredospores of race specific *Puccinia triticina* (*Pt*) isolates were applied to leaf segments of the cultivars and assessed for disease response eleven days after inoculation using the 0 to 4 scale of Stakman et al. (1962). Infection types ; and 1 were regarded as resistant (R), types 2 and 3 as moderately resistant and moderately susceptible (MR and MS), respectively, and infection type 4 as susceptible (S). Most of the Pt isolates used were collected from Europe and Mediterranean regions and selected from single spore progenies. In the two experiments each line was tested in five replications. Pathological tests were carried out in Technical University of Munich, Institute of Plant Breeding, D-85350 Freising-Weihenstephan, Germany, with the isolates classified and maintained by F.J. Zeller's group.

Cytogenetical Analysis

Chromosome counts were made from squashes of root-tip cells pretreated with mono-bromonaphtalene and stained by the Feulgen method. 15 plants per line were checked for chromosome number and the presence of the satellited (sat) chromosome, starting from BC_1F_6 generation. The alien 1U chromosome is shorter than the wheat 1B and 6B chromosomes, with a secondary constriction that is visually differentiated in mitotic metaphases. In the course of 3 years (totally 45 plants per line used for chromosome counting) these plants were grown in greenhouse and estimated for plant morphology and resistance to powdery mildew.

RESULTS

Plant Morphology and Cytogenetical Analysis

Several hexaploid derivatives from a *T. aestivum* cv. Rusalka/*Ae. variabilis* cross with powdery mildew resistance in adult plant stage, were isolated. Only line 11-8 expressed different plant morphology compared to the wheat parent Rusalka and the other 6 lines by having short stem, short spike and brown coloured glumes. It was cytologically confirmed to have a 1U chromosome pair from *Ae. variabilis*, a carrier of genes for powdery mildew resistance and brown glumes in wheat (Spetsov,

1998). The variety Rusalka and *Ae. variabilis* exhibited two pairs of sat chromosomes, but differing in arm ratios, size and satellite expression. 1B and 6B chromosomes in common wheat have a secondary constriction, and can be visually differentiated in mitotic metaphases. Line 11-8 possessed 42 chromosomes with four sat chromosomes: one pair is 6B from the wheat Rusalka and the other is 1U chromosome pair from the alien species (Table 1). All the remaining lines had 42 chromosomes and exhibited the two sat chromosomes, 1B and 6B, a general characteristic for any cultivar of hexaploid wheat.

Line 11-8 was crossed to wheat variety Charodeika. From several doubled haploid lines produced with resistance to powdery mildew, two lines 51 and 131 were found not carrying the alien 1U chromosome. The analysis showed also that from the wheat sat chromosomes, only 6B chromosome pair was visualized.

Powdery Mildew Resistance

Five lines (28-1, 28-7, 37-3, 37-7 and 104-4) including Rusalka, showed susceptible response at seedlings stage to the eleven differential isolates of *Bgt*, while *Ae. variabilis* expressed a complete resistance response to all of them (Table 2). Line 104-5 expressed a little different pattern giving a resistant reaction to isolate 12 only, and the alien species showed the same reaction to this isolate too. The best interesting line is 11-8 expressing a unique resistant pattern, different from parents and lines with known pm resistance genes. Variety Pliska as a possible parent in five of the lines, including line 11-8, expressed a different response in comparison to Rusalka, *Ae. variabilis* and derivatives selected from the cross. Parents of Pliska are Rusalka and 5517-A-5-5-1P3, and the former variety is susceptible to wheat powdery mildew, when for the latter line originating from a complicate cross involving four cultivars, information is not available.

Table 2. Postulated powdery mildew resistance (pmr) genes of common wheat cultivars Rusalka, Pliska, Charodeika and their derivatives from crosses with *Ae. variabilis* in comparison with reactions after inoculation with 11 isolates of *Blumeria graminis* f.sp. *tritici* (Bgt)

Bgt isolates Cultivar/line	2	5	6	9	10	12	13	14	15	16	17	*Pm* gene
Aegilops variabilis	r	r	r	r	r	r	r	r	r	r	r	*u*
Rusalka	s	s	s	s	s	s	s	s	s	s	s	-
Pliska	s	i	r	s	r	s	i	r	i	i	i	*u*
11-8[1]	r	r	r	r	s	i	s	r	r	s	s	*u*
28-7, 28-1, 37-3, 37-7, 104-4[2]	s	s	s	s	s	s	s	s	s	s	s	-
104-5	s	s	s	s	s	r	s	s	s	s	s	*u*
Charodeika	s	s	i	s	r	s	s	r	s	s	s	*Pm5*
51, 131[3]	r	r	r	r	r	r	s	r	r	s	s	*u*
Axminister/8*Cc[4]	r	s	r	i,s	r	s	s	s	s	s	s	*Pm1*
Ulka/8*Cc	s	s	r	s	r	s	s	s	s	s	s	*Pm2*
Asosan/8*Cc	r	s	r	r	r	s	r	r	s	s	r	*Pm3a*
Chul/8*Cc	r	s	s	r	r	r	r	r	s	r	i,s	*Pm3b*
Sonora/8*Cc	r	s	s	i	r	s	r	i,s	s	s	s	*Pm3c*
Kolibri	s	s	s	r	s	r	s	r	r	s	r	*Pm3d*
W150	s	i,s	i,s	i	r	i,s	r	r,s	s	s	s	*Pm3e*
Michigan Amber/8*Cc	r	s	s	i	r	s	r	i,s	s	s	s	*Pm3f*
Khapli/8*Cc	s	r	s	r	i	r	s	s	i	s	i	*Pm4a*
Armada	s	r	s	r	r	r	s	s	s	s	s	*Pm4b*
Hope	s	s	s	s	r	s	s	r	s	s	s	*Pm5*
TP114/2*Starke[5]	s	r,i	r,i	r	r,i	s	r,i	r,i	r,i	i	s	*Pm 6*
Disponent	r	s	s	r	s	r	s	s	s	s	r	*Pm8*
BRG 3N[6]	r	r	r	r	r	r	r	r	r	r	r	*Pm16*
Amigo	i	i	i,s	i	i	i	r	s	i	r	r	*Pm17*
XX 186[7]	s	s	r	i	r	r	i	i	s	i	r	*Pm19*
Virest	r	i	r	r	r	i	i,r	i	r	i,s	i,s	*Pm22*
Chiyacao	r	r	r	r	r	r	r	r	r	r	r	*Pm24*

[1]Lines 11-8, 37-3, 37-7, 104-4 and 104-5, selected from irradiated F_1 (Rusalka/*Ae. variabilis*) and backcrossed with mixed pollen from Rusalka and Pliska

[2]Lines 28-7, 28-1, obtained from F_1 (Rusalka/*Ae. variabilis*) and backcrossed once to cultivar resembling Rusalka; [3]Lines 51 and 131, doubled haploids developed from Charodeika/line 11-8 cross

[4], Seven times backcrossed to Chancellor

[5], Once backcrossed to Starke; r-resistant, s-susceptible, i-intermediate; u-unknown

[6], BRG 3N/76-F_2-205, a wheat-*T. turgidum* var. *dicoccoides* derivative;

[7], XX 186, a *T. durum*/*Ae. tauschii* hexaploid synthetic wheat line;

The disease response of variety Charodeika fitted the expression of Hope from the differential set of lines, and postulated to carry a gene *Pm5*. The two DH lines, derived from a Charodeika/line 11-8 cross, exhibited resistance pattern different from all NILs used in this study. Both lines expressed resistance to 8 and susceptibility to 3 isolates, and in this term they surpassed Pliska, but stood between line 11-8 and *Ae. variabilis*. The resistance of all new derived lines were tentatively characterized as unknown (u) resistance, because their response patterns did not correspond to any of the lines possessing major single resistance genes.

Leaf Rust Resistance

Eight isolates of *Puccinia triticina* (*Pt*) were used to check the resistance of lines derived from the original cross in seedlings in comparison with a differential set of wheat lines possessing individual documented genes (NILs) resistant to leaf rust. Both parents, Rusalka and *Ae. variabilis*, showed different resistance responses to four isolates (S12, S16, S29 and S48) (Table 3). All the six lines possessed specific resistance patterns to the *Pt* isolates used. Their reactions were different from the parents as well as the NILs. Lines 28-1 and 28-7 were infected in higher extent giving susceptible reaction to all isolates, and thus differentiated from Rusalka. Lines 37-7 and 104-5 possessed a wide spectrum of resistance to the eight *Pt* isolates employed. Their reactions were distinguishable from the parents on at least five isolates (S12, S16, S47, S48 and S71). Lines 37-3 and 104-4 behaved also differently from parents (reactions to S28, S47, S48 and S71). As compared to other lines, they showed an intermediate response to at least three isolates, possibly indicating the presence of quantitatively inherited resistance genes. Results showed that at least four lines (37-3, 37-7, 104-4 and 104-5) are very different from Pliska regarding the resistance pattern, and the probability for this cultivar to donate some resistance genes in the selected lines during backcrossing remains very low. Non of the lines and parents fitted any of the 11 NILs with known resistance genes.

Table 3. Disease response patterns of six lines derived from a cv. Rusalka/*Ae. variabilis* cross and their parents in comparison to lines with documented resistance genes to isolates of the leaf rust pathogen

Parents	*Puccinia triticina* isolates								*Lr*
Lines	S12	S16	S27	S28	S29	S47	S48	S71	genes
Ae. variabilis	i	s	r	r	i	s	s	s	*u*
Rusalka	s	r	r	r	s	s	r	s	*u*
Pliska	r,s	i	i	i	s	r,s	i	r,s	*u*
28-1	s	s	i	s	s	s	s	s	*u*
28-7	s	s	s	s	s	s	s	s	-
37-3	r	r	i	s	i	s	s	r	*u*
37-7	r	r	r	r	i	r	r	r	*u*
104-4	i	i	r	r	i	r	r	r	*u*
104-5	r	r	r	r	s	r	r	r	*u*
Centenario	r	r	r	r	r	s	r	r	*Lr1*
Corina	s	s	r	s	r	s	r	i	*Lr2b*
Democrat	r	r	r	r	s	r	r	s	*Lr3*
Lee	i	r	r	s	i	s	r	i	*Lr10*
Kenya E19	r	r	r	r	s	s	i	s	*Lr15*
Kl. Lucero	i	i	s	i	s	s	i	r	*Lr17*
Ae. tauschii	s	s	s	s	s	i	s	s	*Lr21*
Ae. speltoides	r	r	r	r	r	s	i	s	*Lr28*
Terenzio	s	s	r	r	s	r	s	i	*Lr30*
PI58458	s	s	s	s	s	i	i	i	*Lr33*
T. spelta	i	i	s	s	s	i	s	s	*Lr44*

Lines 28-7 and 28-1 derived from F_1, backcrossed once to a cultivar resembling Rusalka;
Lines 37-3, 37-7, 104-4 and 104-5 were originated from irradiated F_1 and backcrossed once with mixed pollen from Rusalka and Pliska; r - resistant, i - intermediate, s – susceptible.

The variety Pliska, parents of which are Rusalka and line 5517-A-5-5-1P3, showed intermediate disease response as a whole. Rusalka expressed a disease pattern that is different from all lines used, and is characterized as unknown (u) resistance. The latter parent originated from a cross: Genesee/4/(54117A3)Genesee/3/Yorkwin/ /Brevor/Norin10 (T. Payne, personal information). Genesee was considered to be susceptible to both powdery mildew and leaf rust based on infection type following inoculation (Ellingboe, 1975), and the same information was published for Yokwin (Jensen and Kent, 1952). Norin 10 carried *Pm14* and *Pm15* genes for resistance to wheatgrass powdery mildew, but was highly susceptible to wheat mildew fungus (Tosa and Sakai, 1990).

DISCUSSION

Ionizing radiation and somaclonal variation have been recently utilised to induce wheat mutants resistant to powdery mildew and rusts (Duggal et al., 2000; Kinane and Jones, 2001; Rakszegi et al., 2010). Kinane and Jones (2001) reported thirteen mutants with increased resistance to powdery mildew and eleven from them exhibited grain yields at least as high as that of the wheat cv. Guardian, while one mutant line yielded significantly higher than the wheat parent. According to Brevis et al. (2008), wild *Triticeae* species have been extensively used to expand the pool of resistance genes in cultivated wheat and account for approximately half of the 55 named *Lr* resistance genes. One of the alien resistance gene *Lr47*, located on the short arm of *T. speltoides* chromosome 7S#1, has probably originated from a population obtained by irradiation of the hybrid CI15092/*T. speltoides*//Fletcher/3/5*Centurk with fast neutrons (Wells et al., 1982).

The wheat-*Aegilops* lines 37-7, 37-3, 104-4 and 104-5 in this experiment, originating from F_1 seeds irradiated with gamma rays, have showed increased resistance to powdery mildew and leaf rust in seedlings due to various wheat-alien chromosome rearrangements and/or mutations occurred during the selection process. This is supported by the waxy bloom character in these lines in contrast to lines 28-1 and 28-7, and cv. Rusalka. Additionally, lines 104-4 and 104-5 produced grain yield higher than Rusalka wheat (data not shown). Line 11-8 used in this study was reported as disomic substitution line with a 1U chromosome pair from *Ae. variabilis* substituted for 1B chromosome and causing brown glume character in maturity (Spetsov, 1998). All other lines selected from the Rusalka/*Ae. variabilis* cross possessed white glumes and absence of the alien 1U chromosome.

DHL 51 and 131 did not contain the 1U chromosome too, but were verified by checking to have only 6B satellited chromosomes. The introgression between 1B and 1U chromosomes might be occurred in these genotypes giving rise to resistance to 8 from the 11 pm isolates used. Noticeable the Charodeika itself has postulated *Pm5* gene which is widespread resistance gene in Russia (Zeller and Hsam, 1998) and Bulgaria (Petrova et al., 2000), , while Rusalka is susceptible. Two lines in this study, 28-1 and 28-7, that originated from a group of plants once backcrossed to variety resembling Rusalka, were morphologically similar to the wheat parent.

Among the 37 cultivars studied from Bulgaria, cv. Pliska was determined as intermediate in response, while Rusalka expressed resistance response pattern to nine isolates of the leaf rust pathogen (Petrova et al., 2000). Here, Rusalka showed four resistance reactions (to S16, 27, 28 and 48) and four susceptible patterns (to S12, 29, 47 and 71), and expressed better resistance than *Ae. variabilis* and Pliska. The resistance of parents and lines derived from a Rusalka/*Ae. variabilis* cross, except for line 28-7, was characterized as unknown (u) resistance, because their response patterns did not correspond to any of the lines possessing major single resistance genes.

CONCLUSION

The selected lines manifested different disease pattern from the parents and lines possessing known powdery mildew and leaf rust resistance genes. Line 11-8 as a 1U(1B) substitution and the two doubled haploid lines that are probably introgressed genotypes, exhibited different resistance to powdery mildew from any of the NILs and characterized tentatively as unknown resistance. Results showed that at least four lines (37-3, 37-7, 104-4 and 104-5) were very different from Pliska regarding the resistance to leaf rust, and these facts do not support this variety as a potential donor of resistance. Non of the lines derived from Rusalka/*Ae.variabilis* and Charodeika/11-8 crosses fitted any of the differential set possessing documented resistance genes to powdery mildew and leaf rust used in the study, and should be further investigated as promising genetic stocks in wheat breeding.

ACKNOWLEDGEMENTS

This research has been financed by grants made by the International Atomic Energy Agency–Vienna (No. 10755/R) and the project BG051PO001-3.3.06-0005, Program 'Human Resources Development'. We thank Prof. F.J. Zeller for providing laboratory facilities and research guidance, and Dr. S. Hsam and N. Petrova for their help in the pathological tests.

LITERATURE CITED

Barloy, D., J. Lemoine, P. Abelard, A.M. Tanguy, R. Rivoal, J. Jahier, 2007. Marker-assisted pyramiding of two cereal cyst nematode resistance genes from *Aegilops variabilis* in wheat. Mol. Breed. 20: 31-40.

Brevis, J.C., O. Chicaiza, I.A. Khan, L. Jackson, C.F. Morris, J. Dubcovsky, 2008. Agronomic and quality evaluation of common wheat near-isogenic lines carrying the leaf rust resistance gene *Lr47*. Crop Sci. 48: 1441-1451.

Chhuneja, P., S. Kaur, R.K. Goel, M. Aghaee-Sarbarzeh, M. Prashar, H.S. Dhaliwal, 2008. Transfer of leaf rust and stripe rust resistance from *Aegilops umbellulata* Zhuk. to bread wheat (*Triticum aestivum* L.). Gen. Res. Crop Evol. 55: 849-859.

De Pace, C., D. Snidaro, M. Ciaffi, D. Vittori, A. Ciofo, A. Cenci, O.A. Tanzarella, C.O. Qualset, G.T. Scarascia Nugnozza, 2001. Introgression of *Dasypyrum villosum* chromatin into common wheat improves grain protein quality. Euphytica 117: 67-75.

Duggal, V., G.J. Jellis, T.W. Hollinst, R. Stratford, 2000. Resistance to powdery mildew in mutant lines of the susceptible wheat cultivar Hobbit 'sib'. Plant Pathol. 49: 468-476.

Ellingboe, A.,1975. Horizontal resistance: an artifact of experimental procedure ? Austr. Plant Pathol. Soc. Newsl. 4: 44-46.

Fedak, G., 1998. Procedures for transferring agronomic traits from alien species to crop plants. In: Slinkard, A.E. (ed.), Proc. 9th Int Wheat Genet Symp, vol.1:1-7. University of Saskatchewan, Canada.

Felsenstein, F.G., R.F. Park, F.J. Zeller, 1998. The use of detached seedling leaves of *Triticum aestivum* to study pathogenicity in *Puccinia recondita* f. sp. *tritici*. J. Phytopathol. 146: 115-121.

Frauenstein, K., K. Hammer, 1985. Pruefung von *Aegilops*-Arten auf Resistenz gegen Echten Mehltau, *Eyisiphe graminis* DC.,

Braunrost, *Puccinia recondita* Rob. Ex Desm., und Spelzenbraeune, *Septoria nodorum* Berk. Kulturpflanze 33: 155-163.

Friebe, B., J. Jiang, W.J. Raupp, R.A. McIntosh, B.S. Gill, 1996. Characterization of wheat-alien translocation conferring resistance to disease and pests: current status. Euphytica 91: 59-87.

Harjit-Singh, H. Tsujimoto, P.K. Sakhuja, T. Singh, H.S. Dhaliwal, 2000. Transfer of resistance to wheat pathogens from *Aegilops triuncialis* into bread wheat. Wheat Inf. Serv. 91: 5-10.

Hsam, S.L.K., F.J. Zeller, 2002. Breeding for powdery mildew resistance in common wheat (*Triticum aestivum* L.). In: Belanger, R.R., W.R. Bushnell, A.J. Dik, T.L.W. Carver (eds), The Powdery Mildews (A comprehensive treatise), The American Phytopathological Society, St. Paul, Minnesota, pp.219-238.

Hsam, S.L.K., I.F. Lapochkina, F.J. Zeller, 2003. Chromosomal location of genes for resistance to powdery mildew in common wheat (*Triticum aestivum* L. em Thell.). 8. Gene *Pm32* in a wheat-*Aegilops speltoides* translocation line. Euphytica 133: 367-370.

Jensen, N.F., G.C. Kent, 1952. Disease resistance from a wheat x rye cross. J. Hered. 43: 242.

Kinane, J.T., P.W. Jones, 2001. Isolation of wheat mutants with increased resistance to powdery mildew from small induced variant populations. Euphytica 117: 251-260.

Liu, D., Z. Xiang, L. Zhang, Y. Zheng, W. Yang, G. Chen, C. Wan, H. Zhang, 2011. Transfer of stripe rust resistance from *Aegilops variabilis* to bread wheat. Afr. J. Biotechn. 10: 136-139.

Marais, G.F., B. McCallum, A.S. Marais, 2008. Wheat leaf resistance gene *Lr59* derived from *Aegilops peregrina*. Plant Breed. 127: 340-345.

McIntosh, R.A., J. Dubcovsky, W.J. Rogers, C. Morris, R. Appels, X.C. Xia, 2009. Catalogue of gene symbols for wheat: 2009 supplement. Wheat Genetic Resource Database Komugi-Natl. BioResource Res. Project, Japan.

Petrova, N., S.L.K. Hsam, P. Spetsov, F.J. Zeller, 2000. Identification of powdery mildew and leaf rust resistance genes in common wheat (*Triticum aestivum* L. em. Thell.) cultivars grown in Bulgaria and Russia. Plant Gen. Res. Newsl. 122: 32-35.

Rakszegi, M., B.N. Kisgyörgy, K. Tearall, P.R. Shewry, L. Láng, A. Phillips, Z. Bedő, 2010. Diversity of agronomic and morphological traits in a mutant population of bread wheat studied in the Healthgrain program. Euphytica 174: 409-421.

Sears, E.R., 1956. The transfer of leaf rust resistance from *Aegilops umbellulata* to wheat. Brookhaven Symp. Biol. 9: 1-22.

Spetsov, P., D. Mingeot, J.M. Jacquemin, K. Samardjieva, E. Marinova, 1997. Transfer of powdery mildew resistance from *Aegilops variabilis* into bread wheat. Euphytica 93: 49-54.

Spetsov, P., 1998. Introgression from *Aegilops species* (2n=28, UUSS) of powdery mildew resistance and other agronomic traits into common wheat. D.Sc. Thesis, Agricultural Academy, Sofia (in Bulg. with resume in English).

Stakman, E.C., D.M. Stewart, W.Q. Loegering, 1962. Identification of physiological races of *Puccinia recondita* var. *tritici*. Miss. Agric. Expt. Stn. Sci. J. Ser. Paper 4691.

Stoilova, T., P. Spetsov, 2006. Chromosome 6U from *Aegilops geniculata* Roth bearing powdery mildew resistance in bread wheat. Breed. Sci. 56: 351-357.

Thiele, A., E. Schumann, A. Peil, W.E. Weber, 2002. Eyespot resistance in wheat x *Aegilops kotschyi* backcross lines. Plant Breed. 121: 29-35.

Tosa, Y., K. Sakai, 1990. The genetics of resistance of hexaploid wheat to the wheatgrass powdery mildew fungus. Genome 33: 225-230.

Van Slageren, M.W., 1994. Wild wheats: a monograph of *Aegilops* L. and *Amblyopyrum* (Jaub. & Spach) Eig (*Poaceae*). Wageningen Agricultural University Papers, Wageningen, The Netherlands.

Wells, D.G., R.S. Kota, H.S. Sandhu, W.A.S. Gardner, K.F. Finney, 1982. Registration of one disomic substitution line and five translocation lines of winter wheat germplasm resistant to wheat streak mosaic virus. Crop Sci. 22: 1277-1278.

Williams, M.D.H., A. Mujeeb-Kazi, 1996. Development of genetic stocks and biochemical markers to facilitate utilization of *Aegilops variabilis* in wheat improvement. Cytologia 61: 7-13.

Xu, S.S., Y. Jin, D.L. Klindworth, R.R-C. Wang, X. Cai, 2009. Evaluation and characterization of seedling resistances to stem rust Ug99 races in wheat-alien species derivatives. Crop Sci. 49: 2167-2175.

Yu, M.Q., F. Person-Dedryver, J. Jahier, 1990. Resistance to root knot nematode, *Meloidogyne naasi* (Franklin), transferred from *Aegilops variabilis* Eig to bread wheat. Agronomie 6: 451-456.

Zeller, F.J., S.L.K. Hsam, 1998. Progress in breeding for resistance to powdery mildew in common wheat (*Triticum aestivum* L.). In: Slincard, A.E. (ed), Proceedings of the 9th International Wheat Genetics Symposium, University Extension Press, Saskatoon, Canada, v. 1: 178-180.

Zeller, F.J., L. Kong, L. Hartl, V. Mohler, S.L.K. Hsam, 2002. Chromosomal location of genes for resistance to powdery mildew in common wheat (*Triticum aestivum* L. em Thell.) 7. Gene *Pm29* in line Pova. Euphytica 123: 187-194.

Zhu, Z., R. Zhou, X. Kong, Y. Dong, J. Jia, 2006. Microsatellite marker identification of a *Triticum aestivum* - *Aegilops umbellulata* substitution line with powdery mildew resistance. Euphytica 150: 149-153.

SOYBEAN (*Glycine max* L.) RESPONSE TO MICRO-NUTRIENTS

Faisal ZAHOOR[1], Mansoor AHMED[1], M. Azim MALIK[2], Khuram MUBEEN[2], M. H. SIDDIQUI[1], Muhammad RASHEED[2], Ramzan ANSAR[2], Khalid MEHMOOD[1]*

[1]*Department of Agronomy, Faculty of Agriculture, The University of Poonch, Rawalakot, Azad Jammu Kashmir (AJK, Pakistan).*
[2]*Department of Agronomy, PMAS-Arid Agriculture Rawalpindi.*
[*]*Corresponding author's e-mail: khuram.mubeen@gmail.com*

ABSTRACT

Soybean (*Glycine max* L.) generally respond to micro-nutrients by enhanced nodulation and more seed yield. A field trial was carried out to evaluate the response of soybean to micronutrients viz. iron (Fe), molybdenum (Mo) and cobalt (Co). Results indicated that iron applied at the rate of 400 g ha^{-1} and molybdenum @ 20 g ha^{-1} had a significant effect on shoot length, shoot dry weight, number of nodules plant^{-1}, nodules fresh weight and thousand seed weight. The highest seed yield 42.28% over control, dry matter yield , seed nitrogen and seed protein content was recorded by combined application of Fe at the rate of 400 g ha^{-1} and Mo @ 20 g ha^{-1}.

Key Words: Cobalt, Iron, Molybdenum, nodulation, soybean, yield.

INTRODUCTION

Soybean (*Glycine max* L.) is herbaceous, erect grown, annual, warm climate and legume plant. It is considered as highly nutritive food for human, livestock (Berglund, 2002; Akparobi, 2009). Soybean enhances the content of soil organic matter when rotated in farming systems (O'Hara et al., 2002). But poor availability of plant nutrients limits yield in many world's crop production areas (Heitholt et al., 2003).

Micronutrients deficiency is wide spread in calcareous soils (Ahmad et al., 2012) and all over the world (Gurmani et al., 2012). Metal such as iron is very important for normal growth of soybean (Fageria, 2007). It also plays an important role in the formation of some nodule proteins like nitrogenase and leghaemoglobin (Moran et al., 1997). Iron is necessary for plant growth because of its participation in most important metabolic processes such as ribonucleotides and molecular nitrogen reduction, and energy yielding electron transfer reactions of respiration and photosynthesis (Guerinot and Yie, 1994). Iron deficiency in soybean results in chlorosis (Rotaru and Sinclair, 2009). It was assumed that a single unit increase in chlorosis score resulted in 20% yield losses in soybean (Froehlich and Fehr, 1981). Similarly Molybdenum plays a role as a co-factor of proteins, responsible for electron transfer in synthesis of nitrogenase enzyme and conversion of N_2 into ammonia in nitrogen fixation process (Martens and Westerman, 1991). Deficiency of molybdenum reduces chlorophyll content in soybean and corn (Liu and Yang, 2003). Along with iron and molybdenum, cobalt is an essential element for legume crops because it is utilized by micro-organisms for the process of atmospheric nitrogen fixation (Evan and Kliwer, 1964). Growth processes such as coleoptiles and stem elongation, bud development, leaf disc expansion and opening of hypocotyle hooks are enhanced by cobalt application in soybean (Klein 1959). High concentration of cobalt becomes toxic and has an adverse effect on physiological and biochemical functions of plant (Parmar and Chanda, 2005).

Little information about the effects of these micronutrients (Fe, Mo and Co) on growth, yield and quality of soybean is available. Hence a need was felt to devise an experiment with the aim to evaluate the interactive effect of application of various doses of iron, molybdenum and cobalt on growth, nodulation, yield and quality of soybean.

MATERIALS AND METHODS

The present study was carried out under field conditions at the experimental area of the Department of Agronomy, Faculty of Agriculture, Rawalakot, Azad Jammu Kashmir, Pakistan during 2010 and 2011. The study was aimed to find out the effect of cobalt, Iron and molybdenum application on nodulation, growth, yield and N nutrition of the soybean plants. The experimental area was between 33−36° N and 73−75° E with an elevation of 1633 m above sea level. Sub-mountainous topography of valleys and extended plains with temperate sub-humid climate are the feature of the area. Average rainfall range between 500–2000 mm annually. Summer is mild with mean annual temperature of 27°C while moderately cold temperature even below freezing point prevails in winter (Tahir et al., 2009).

The trial was laid out in Randomized Complete Block Design (RCBD) having three replications with a net plot size of 2.5m x 2.5m. Main plot treatments consist of three Fe rates; 0, 200 and 400 g Fe ha^{-1}. Subplot treatments were two Mo rates; 0 and 20 g Mo ha^{-1}. The sub-subplot treatments were two rates of Co; 0 and 20 g Co ha^{-1}. Fe was applied as foliar spray, each consisting of 1.82 kg or 3.64 kg ha^{-1} of FeEDTA (5.5% Fe and 2% EDTA) and 50 ml of surfactant in 200 ml ha^{-1} of water, using a knapsack sprayer. The physico-chemical properties of soil are as follows.

Table 1. Soil physico-chemical properties before sowing of soybean during 2010 and 2011

Year	Soil Bulk density	Soil pH	Soil total N	Soil organic matter	Gravimetric soil moisture content	Fe (mg/kg)	Mo (mg/kg)	Co (mg/kg)
2010	1.33 (g cm^{-3})	7.65	0.08 (%)	1.20 (%)	0.07 (%)	2.99	0.09	0.115
2011	1.27 (g cm^{-3})	7.55	0.11 (%)	1.29 (%)	0.09 (%)	3.21	0.11	0.235

Half of the Fe doses were applied at the 1–2 trifoliolate stage and application was repeated at the 4–5 trifoliolate stage as recommended by Goos and Johnson (2000). Plant growth stages were determined by the methods of Fehr and Caviness (1977). Mo and Co was applied as sodium molybdate ($Na_2MoO_4.2H_2O$) and cobalt chloride ($CoCl_2$) as seed treatment. Seeds were coated with $Na_2MoO_4.2H_2O$ and $CoCl_2$ by using 10% sucrose solution. A basal dose of phosphorous and potassium was applied in all plots including control at sowing time at the rate of 90 kg P_2O_5 ha^{-1} and 60 kg K_2O ha^{-1} as single super phosphate (SSP) and sulphate of potash (SOP), correspondingly by broadcast method. The various treatment combinations were:

Soybean cultivar NARC-1 at the rate of 80 kg ha^{-1} was planted as test crop on May 15th and 5th, (2010 and 2011 respectively) at 40 and 10 cm inter- and intra-row distance, respectively to obtain an appropriate population density ha^{-1}. Observations on shoot length (cm) , shoot dry weight (g), number of nodules $plant^{-1}$, nodules fresh weight (g), 1000 seed weight (g), seed yield (kg ha^{-1}), dry matter yield (kg ha^{-1}), seed nitrogen content (%) and seed protein content (%) were taken to determine the effect of treatments on the growth and yield of crop characteristics, analysis of variance (ANOVA) and least significant difference (LSD) tests was carried out for each trait individually using a MSTAT-C. Comparison of means for each treatment was done at the 5% probability level based on the F-test of the analysis of variance (Steel et al., 1997).

RESULTS AND DISCUSSION

The interaction of various micro-nutrients (Fe, Mo, Co) showed a positive and significant response to growth and yield parameters of soybean. The highest shoot length (76.70) was recorded in $Fe_2Mo_1Co_0$ where Fe was applied @ 400 g ha^{-1}, Mo @ 20 g ha^{-1} and Co 0 g ha^{-1}. The percent increase over control was 25.06 %. While the lowest shoot length (61.33) was observed in control ($Fe_0Mo_0Co_0$). The significant increase in shoot length in $Fe_2Mo_1Co_0$ can be attributed to Fe and Mo application as Fe played a vital role in nitrogen fixation and photosynthesis (Bennett, 1993) that ultimately increased shoot length (Ahmad et al., 2002). Goos and Johnson (2000) and Rashid and Ryan (2004) from their studies reported that application of Fe correct Fe chlorosis and improve vegetative growth in soybean. Abd El-Hai (1995) observed significant response of Mo on plant height.

Likewise the lowest shoot dry weight (0.80) was recorded in control ($Fe_0Mo_0Co_0$) where no micro-nutrition was added and the maximum (1.50) was obtained in $Fe_2Mo_1Co_0$ (Table 2). Maximum shoot dry weight was probably due to integrated use of Fe+Mo+Co because these results favours the findings of Guerinot and Yie (1994); Bhuiyan et al., (2008) and Caliskan et al., (2008) who observed a significant increase in shoot and root dry weight in mungbean with Mo application and Fe that improves photosynthesis activity that contributed for better plant dry biomass.

Regarding number of nodules $plant^{-1}$ the maximum number of nodules $plant^{-1}$ (106.7) were observed in $Fe_2Mo_1Co_0$. The increase in number of nodules $plant^{-1}$ in $Fe_2Mo_1Co_0$ was 48.87 % over control. It was at par with $Fe_2Mo_1Co_1$, $Fe_1Mo_1Co_1$ and $Fe_1Mo_1Co_0$ where number of nodules $plant^{-1}$ were 103.3, 99 and 94.67 respectively and Fe was applied at the rate of 400 g ha^{-1}, 200 g ha^{-1} while Mo and Co at the rate of 20 g ha^{-1} each. Data also showed the lesser number of nodules $plant^{-1}$ (71.67) in control. Highest number of nodules $plant^{-1}$ might be attributed to Mo and Co application as Mo is utilized by specific plant enzymes which play a role in oxidative and reduction reactions and Co being a part of cobalamin (vit. B12 and its derivaties), a component of different enzymes in nitrogen fixing micro-organisms (Hegazi et al., 2011). Further more, application of Fe enhance number of nodules $plant^{-1}$ as Fe is an important element for rhizobia nutrition, nodulation, nodule activity and biological nitrogen fixation (O' Hara, 2001). Balachandar et al., (2003) and Subasinghe et al., (2003) reported that Mo and Co significantly enhance nodules number $plant^{-1}$ in black gram and cowpea, respectively.

Similar trend was observed for nodules fresh weight. Maximum nodules fresh weight (0.067) were noted in $Fe_2Mo_1Co_0$ and was at par with $Fe_2Mo_1Co_1$ and $Fe_1Mo_1Co_1$ where nodule fresh weight were 0.050 g each respectively while the lowest nodules fresh weight (0.023) were recorded in control. The maximum nodules fresh weight in $Fe_2Mo_1Co_0$ (Table 2) can be attributed to Fe and Mo application because these results supports the findings of Balachander et al., (2003) and Subasinghe et al., (2003) who reported that nodules fresh weight can be increased by Mo and Co application in black gram and cowpea. Also Krouma and Abdelly (2003) from their studies suggested that Fe is necessary for overall growth of nodules.

Table 2. Effect of micronutrients on growth, yield and nodulation of soybean (*Glycine max* L.) during 2010-11 (pooled over years)

Treatments	Shoot length (cm)	Shoot dry weight (g)	Number of nodules plant^{-1}	Nodules fresh weight (g)	1000 seed weight (g)	Seed yield (kg ha^{-1})	Dry matter yield (kg ha^{-1})	Seed nitrogen content (%)	Seed Protein content (%)
$Fe_0Mo_0Co_0$	61.33 c	0.80 d	71.67 f	0.023 c	151.3 d	1213 d	1741 e	2.50 e	34.00 e
$Fe_0Mo_0Co_1$	65.00 bc	1.04 cd	76.67 ef	0.023 c	161.7 cd	1283 cd	1843 e	2.70 de	35.00 de
$Fe_0Mo_1Co_0$	66.73 bc	1.08 bcd	77.67 ef	0.023 c	163.0 bcd	1289 cd	1973 de	2.80 de	35.40 cde
$Fe_0Mo_1Co_1$	67.03 bc	1.17 abc	81.67 def	0.030 c	165.7 bc	1348 bcd	2074 cde	2.90 cde	36.00 cd
$Fe_1Mo_0Co_0$	67.57 bc	1.19 abc	85.00 cdef	0.030 c	166.7 bc	1415 bcd	2150 bcde	3.10 bcde	36.00 cd
$Fe_1Mo_0Co_1$	68.10 bc	1.21 abc	88.67 bcde	0.033 bc	167.3 abc	1449 bc	2227 bcde	3.30 bcde	36.00 cd
$Fe_1Mo_1Co_0$	71.27 ab	1.38 abc	94.67 abcd	0.037 bc	173.3 abc	1788 a	2611 bc	3.70 abc	37.00 abc
$Fe_1Mo_1Co_1$	72.40 ab	1.40 ab	99.00 abc	0.050 ab	174.7 ab	1801 a	2688 ab	3.80 ab	38.00 ab
$Fe_2Mo_0Co_0$	69.13 abc	1.22 abc	88.67 bcde	0.033 bc	169.7 abc	1542 b	2483 bcd	3.40 abcd	36.33 cd
$Fe_2Mo_0Co_1$	70.23 ab	1.28 abc	89.00 bcde	0.037 bc	171.0 abc	1547 b	2586 bc	3.50 abcd	36.40 bcd
$Fe_2Mo_1Co_0$	76.70 a	1.50 a	106.7 a	0.067 a	179.3 a	1834 a	3226 a	4.20 a	38.40 a
$Fe_2Mo_1Co_1$	73.10 ab	1.40 ab	103.3 ab	0.050 ab	179.0 a	1821 a	3226 a	3.90 ab	38.17 a
LSD($P \leq 0.05$)	**8.15**	**0.35**	**16.82**	**0.02**	**12.13**	**230.90**	**545.30**	**0.87**	**1.633**

Means not sharing same letter with in column differ significantly from each other at 5% level of probability by LSD test

$Fe_0Mo_0Co_0$ (Control), $Fe_0Mo_0Co_1$ (Fe = 0, Mo = 0, Co = 20 g ha^{-1}), $Fe_0Mo_1Co_1$ (Fe = 0, Mo = 20 g ha^{-1}, Co = 20 g ha^{-1}), $Fe_0Mo_1Co_0$ (Fe = 0, Mo = 20 g ha^{-1}, Co = 0), $Fe_1Mo_1Co_0$ (Fe = 200 g ha^{-1}, Mo = 20 g ha^{-1}, Co = 0), $Fe_1Mo_1Co_1$ (Fe = 200 g ha^{-1}, Mo = 20 g ha^{-1}, Co = 20 g ha^{-1}), $Fe_1Mo_0Co_1$ (Fe = 200 g ha^{-1}, Mo = 0, Co = 20 g ha^{-1}), $Fe_1Mo_0Co_0$ (Fe = 200 g ha^{-1}, Mo = 0, Co = 0), $Fe_2Mo_0Co_0$ (Fe = 400 g ha^{-1}, Mo = 0, Co = 0), $Fe_2Mo_0Co_1$(Fe = 400 g ha^{-1}, Mo = 0, Co = 20 g ha^{-1}), $Fe_2Mo_1Co_1$(Fe = 400 g ha^{-1}, Mo = 20 g ha^{-1}, Co = 20 g ha^{-1}), $Fe_2Mo_1Co_0$ (Fe = 400 g ha^{-1}, Mo = 20 g ha^{-1}, Co = 0).

The perusal of data given in Table 2 indicated that interaction between various micronutrients viz iron (Fe), molybdenum (Mo) and cobalt (Co) for yield parameters were statistically significant. It revealed that the highest thousand seed weight (179.3) was recorded in $Fe_2Mo_1Co_0$ and the percent increase over control was 10 % that was at par with $Fe_0Mo_1Co_0$ and $Fe_1Mo_0Co_0$ where thousand seed weight was 163.0 and 166.7 respectively. Whereas the lowest thousand seed weight was recorded control. Similarly the lowest seed yield (1213 kg ha^{-1}) was noted in control ($Fe_0Mo_0Co_0$) whereas the highest seed yield (1834 kg ha^{-1}) was recorded in $Fe_2Mo_1Co_0$ (Table 2). Similarly the significant increase in thousand seed weight (Table 2) can also be attributed to Fe and Mo application as both are the important components of enzyme nitrogenase and nitrate reductase that enhance bacterial growth and symbiotic efficiency resulting in improved plant growth and yield (Solaiman, 1999). These results favors the findings of Rahman et al., (2008) and Heidarian et al., (2011).

The percent increase in seed yield was 42.28 % over control. It was at par with $Fe_2Mo_1Co_1$, $Fe_1Mo_1Co_1$ and $Fe_1Mo_1Co_0$ where seed yield were 1821, 1801 and 1788 kg ha^{-1}, respectively. The seed yield and number of nodules plant^{-1} of soybean showed a linear correlation (Fig. 1). It revealed that with an increase in number of nodules plant^{-1} the seed yield can also be increased. The highest response in seed yield (Table 2) can be attributed to Fe and Mo application as Fe plays an important role in plant growth and crop yield (Guerinot and Yie, 1994) and Mo is responsible for increasing process of nitrogen fixation by nodule tissues formation (Sharma et al., 1988) that ultimately resulted in better plant growth and yield. Caliskan et al., (2008) and Adesoji et al., (2009) reported highest yield with Fe and Mo application.

Also the highest dry matter yield (3226 kg ha^{-1}) was recorded in $Fe_2Mo_1Co_0$ (Table 2) and the percent increase was 85.29 % over control. It was at par with $Fe_2Mo_1Co_1$ and $Fe_1Mo_1Co_1$ where dry matter yield was 3226 and 2688 kg ha^{-1} respectively. While the lesser dry matter yield (1741 kg ha^{-1}) were observed in control. The significant increase in dry matter yield in $Fe_2Mo_1Co_0$ might be due to Mo and Fe application for activation of nitrogenase enzymes. Similarly, Fe has a role in formation of nodule protein nitrogenase and leghaemoglobin (Moran et al., 1997) thus reduces iron deficiency chlorosis that results in excellent growth and yield. These results are confirmed by Goos and Johnson (2001); Goos et al., (2004) and Adesoji et al., (2009) who concluded that Fe and Mo application resulted in better dry matter yield of soybean.

Micro-nutrients viz iron (Fe), molybdenum (Mo) and cobalt (Co) produce significant effect on seed nitrogen content. It was highest (4.20 %) in $Fe_2Mo_1Co_0$ where Fe was applied at the rate of 400 g ha^{-1} and Mo at the rate of 20 g ha^{-1}(Table 2).Whereas the lowest was produced by control. The percent increase over control was 68 %. When correlation between seed nitrogen content and number of nodules plant^{-1} was developed, a positive correlation was observed between seed nitrogen content and number of nodules plant^{-1} of soybean (Fig. 2).

Significant effect of micronutrients on seed nitrogen content can be attributed to Fe and Mo application as Fe and Mo are the key components of nitrogenase enzyme and nitrate reductase that enhance bacterial growth and symbiotic efficiency (Solaiman, 1999). Campo et al., (2009) noticed significant effect of applied micronutrients on seed nitrogen content.

Likewise the lowest seed protein content (34 %) was recorded in control ($Fe_0Mo_0Co_0$) while the highest seed protein content (38.40 %) was noted in $Fe_2Mo_1Co_0$ where Fe was applied at the rate of 400 g ha^{-1} while Mo at the rate of 20 g ha^{-1}. The percent increase in seed protein content in this treatment was 12.94 % over control. It was at par with $Fe_2Mo_1Co_1$, $Fe_1Mo_1Co_1$ and $Fe_1Mo_1Co_0$ where seed protein content were 38.17 %, 38 % and 37 % respectively. The significant increase in seed protein content can be attributed to Mo, Co and Fe application, as Mo take a part in increasing photosynthetic rates (Pang et al., 2001; Liu and Yang, 2003; Sun et al., 2006) and Co due to its utilization by micro-organisms to continue the process of atmospheric nitrogen fixation (Evan and Kliwer, 1964) that enhance growth, yield and mineral contents of plant. Along with Co and Mo, Fe significantly improved photosynthesis activity that contributed for overall plant growth, mineral content and yield (Guerinot and Yie, 1994; Caliskan et al., 2008). Campo et al., (2009) and Gad (2010) concluded from their studies that Mo and Co significantly enhance protein contents of seed.

CONCLUSION

From the results it may be inferred that the plots fertilized with Fe, Mo and Co @ 400 g, 20 g and 20 g ha^{-1} respectively revealed similar results to plots with same level of Fe and Mo but without Co application. Hence keeping in view the cost effectiveness for the farmer, it could be recommended that better growth, nodulation, yield and quality of soybean could be achieved by the application of Fe @ 400 g ha^{-1} and Mo @ 20 g ha^{-1}.

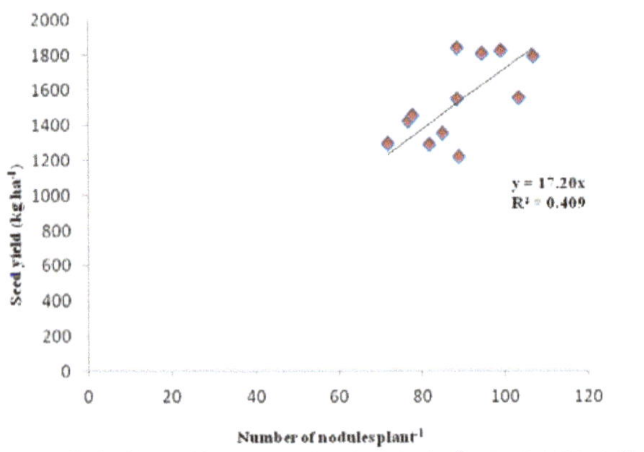

Fig.01: Relationship between number of nodules plant^{-1} and seed yield (kg ha^{-1})

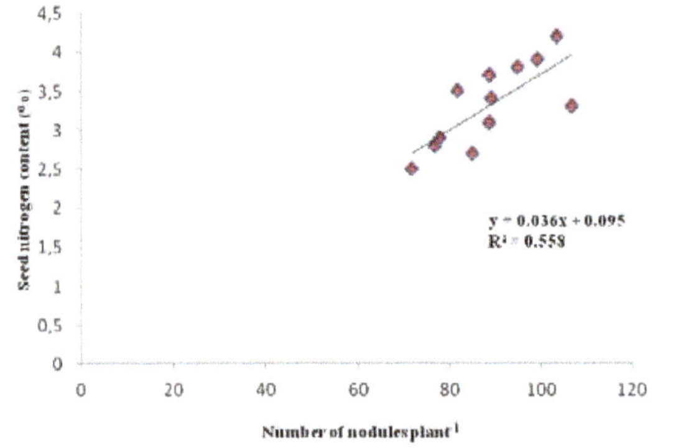

Fig.02: Relationship between number of nodules plant^{-1} and seed nitrogen content (%)

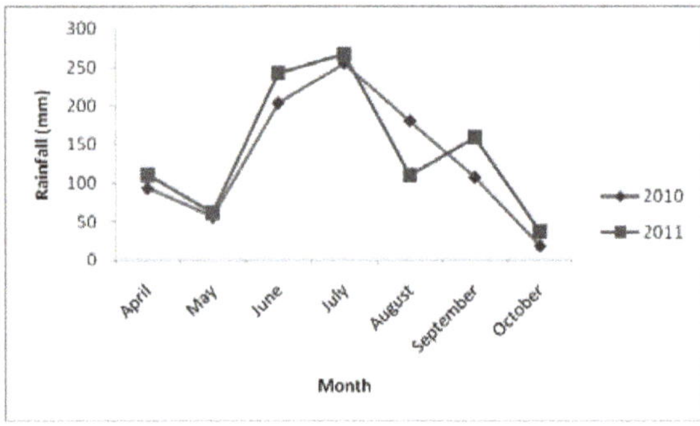

Fig. 03 Monthly rainfall (mm) of experimental area during the growing period of soybean

LITERATURE CITED

Abd El-Hai, K. M., 1995. Effect of certain growth substances and some microelements on soybean Plants. M.Sc. Thesis, Faculty of Agriculture, Mansoura University, Egypt.

Adesoji, A. G., I. U. Abubakar, D. B. Ishaya, 2009. Performance of soybean (*Glycine max* (L.) Merrill) as influenced by method and rate of molybdenum application in Samaru, Northern Guinea Savanna of Nigeria. American Eurasian J. Sustain. Agric., 3: 845-849.

Ahmad, H.R., T. Aziz, S. Hussain, M. Akraam, M. Sabir, S.U.R. Khashif, M.M. Hanafi, 2012. Zinc-enriched farm yard manure improves grain yield and grain zinc concentration in rice grown on a saline-sodic soil. Int. J. Agric. Biol., 14: 787-792

Ahmad, M., M. Hussain, M. Shafique, 2002. Important macro and microelements in chickpea and lentil. Nucleus, 39: 101-105.

Akparobi, S. O., 2009. Evaluation of six cultivars of soybean under the soil of rainforest agro-ecological zones of Nigeria. Middle-East J. Sci. Res., 4: 06-09.

Balachandar, D., P. Nagarajan, S. Gunasekaran, 2003. Effect of organic amendments and micronutrients on nodulation and yield of blackgram in acid soil. Legume Res., 26: 192-195.

Bennett, W. F., 1993. Plant nutrient utilization and diagnostic plant symptoms.*In: Nutrient deficiencies and toxicities in crop plants*. Bennett WF (Ed.) St. Paul, MN: The APS Press. American Phytopathol. Soc., pp: 1-7.

Berglund, D. R., 2002. Soybean production field guide for North Dakota and Northwestern Minnesota. Published in cooperative and with support from the North Dakota Soybean Council, pp: 136.

Bhuiyan, M. M. H., M. M. Rahman, F. Afroze, G. N. C. Sutradhar, M. S. I. Bhutyan, 2008. Effect of phosphorus, molybdenum and rhizobium inoculation on growth and nodulation on mungbean. J. Soil Nature, 2: 25-30.

Caliskan, S., I. Ozkaya, M. E. Caliskan, M. Arslan, 2008. The effects of nitrogen and iron fertilization on growth, yield and fertilizer use efficiency of soybean in a mediterranean type soil. Field Crops Res., 108:126-132.

Campo, R. J., R. S. Araujo, M. Hungria, 2009. Molybdenum-enriched soybean seeds enhance nitrogen accumulation, seed yield and seed protein content in Brazil. Field Crop Res., 110: 219-224.

Evan, H. J., M. Kliwer, 1964. Vitamin B compounds in relation to the requirements of cobalt for higher plants and nitrogen fixing organisms. Ann. New York Acad. Sci., 2: 732-755.

Fageria, N. K., 2007. Soil fertility and plant nutrition research under field conditions: Basic principles and methodology. J. Plant Nutr., 30 (2) :203-223.

Fehr, W. R., C. E. Caviness, 1977. Stages of soybean development. Special report 80. Iowa State University, Am, IA, USA.

Froehlich, D. M., W. R. Fehr, 1981. Agronomic performance of soybeans with different levels of iron deficiency chlorosis on calcareous soil. Crop Sci., 21: 438-441.

Gad, N., 2010. Improving quantity and quality of canola oil yield through cobalt nutrition. American J. Agric.Biol.,1: 1090-1097.

Goos, R. J., B. E. Johnson, 2000. A comparison of three methods for reducing iron deficiency chlorosis in soybean. J. Agron., 92: 1135-1139.

Goos, R. J., B. E. Johnson, 2001. Seed treatment, seeding rate and cultivar effects on iron deficiency chlorosis of soybean. J. Plant Nutr., 24: 1255-1268.

Goos, R. J., B. Johnson, G. Jackson, G. Hargrove, 2004. Greenhouse evaluation of controlled release iron fertilizers for soybean. J. Plant Nutr., 24: 1255- 1268.

Guerinot, M. L., Y. Yie, 1994. Iron: nutritious, noxious, and not readily available. Plant Physiol., 104: 815-820.

Gurmani, A.R., J.U. Din, S.U. Khan, R. Andaleep, K. Waseem, A. Khan, Hadyat-Ullah, 2012. Soil Application of zinc improves growth and yield of tomato. Int. J. Agric. Biol., 14: 91–96

Hegazi, A., M. A. Mohamed, A. G. S. Sayed, M. H. Elsherif, N. Gad, 2011. Reducing N doses by enhancing nodule formation in groundnut plants via Co and Mo. *Australian.* J. Basic Appl. Sci., 5: 2568-2577.

Heidarian, A. R., H. Kord, K. Mostafavi, A. P. Lak, F. A. Mashhadi, 2011. Investigating Fe and Zn foliar application on yield and its components of soybean (*Glycine max* (L) Merr.) at different growth stages. J. Agric. Biotech. Sustain. Dev., 3: 189-197.

Heitholt, J. J., J. J. Sloan, C. T. MacKown and R. I. Cabrera, 2003. Soybean growth on calcareous soil as affected by three iron sources. J. Plant Nutr., 4: 935-948.

Klein, W. H., 1959. Interaction of growth factors with photoprocess in seedling growth. *In:* Photoperiodism and related phenomena in plants and animals. Withrow, R. B. (Ed.). *American Assoc. Adv. Sci*., Washington, D. C, pp: 207-215.

Krouma, A., C. Abdelly, 2003. Importance of iron use efficiency of nodules in common bean (*Phaseolus vulgaris* L.) for iron deficiency chlorosis resistance. J. Plant Nutr. Soil Sci., 166: 525-528.

Liu, P., Y. A. Yang, 2003. Effect of molybdenum and boron on photosynthetic efficiency of soybean. Plant Nutr.Fertil. Sci., 9: 456–461.

Martens, D. C., D. T. Westermann, 1991. Fertilizers applications for correcting micronutrient deficiencies. *In: Micronutrients in Agriculture* 2nd edn. Mortvedt, J. J., Cox, F. R., Shuman, L. M. and Welch, R. M. (Eds.). Soil Sci. Soc. America. Madison, WI. pp: 90-112.

Moran, J. F., R. V. Klucas, R. J. Grayer, J. Abian, M. Becana, 1997. Complexes of iron with phenolic compounds from soybean nodules and other leguminous tissues: prooxidant and antioxidant properties. Free Radic. Biol. Med., 22: 861-870.

O'Hara, G. W., 2001. Nutritional constrains on root nodule bacteria affecting symbiotic nitrogen fixation: a review. Australian. J. Exp. Agric., 41: 417- 433.

O'Hara, G., R. Yates, J. Howieson, 2002. Selection of strains of root nodule bacteria to improve inoculant performance and increase legume productivity in stress full environments.In: Inoculants and nitrogen fixation of legumes in Vietnam. Herridge, D. (Ed.). ACIAR, Proceedings, 109: 75-80.

Pang, J., C. X. Hu, Y. H. Wang and W. X. Wei, 2001. The mechanism of molybdenum effects on carbon metabolism of winter wheat in acid yellow brown earth. Huazhong J. Agric. Univ., 20: 33–35.

Parmar, G. and V. Chanda, 2005. Effects of Mercury and Chromium on Peroxidase and IAA Oxidase Enzymes in the Seedlings of Phaseolus vulgaris Turk. J. Biol., 29: 15-21.

Rahman, M. M., M. M. H., Bhuiyan, G. N. C., Sutradhar, M. M. Rahman, A. K., Paul, 2008. Effect of phosphorus, molybdenum and rhizobium inoculation on yield and yield attributes of mungbean. Int. J. Sustain. Crop Prod., 3: 26-33.

Rashid, A., J. Ryan, 2004. Micronutrients constrains to crop production in soils with Mediterranean type characteristic: a review. J. Plant Nutr., 27: 959-975.

Rotaru, V., T. R. Sinclair, 2009. Influence of plant phosphorus and iron concentrations on growth of soybean. J. Plant Nutr., 32: 1513-1526.

Sharma, M. S., M. S. Upadhyay, S. S. Tomar, 1988. Water use efficiency of some rainfed crop on a vertisol as influenced by soil micronutrients and straw mulching. Indian J. Soil Sci., 33: 387-390.

Solaiman, A. R. M., 1999. Effects of Bradyrhizobiumjaponicum inoculation and molybdenum on soybean. Bangladesh J. Bot., 28: 181-183.

Steel, R. G. D., J. H. Torrie, D. A. Dickey, 1997. *Principles and procedures of statistics*: A Biometrical Approach. 3rd ed. McGraw Hill Book Co. Inc. New York.

Subasinghe, S., G. A. Dayatilake, R. Senaratne, 2003. Effect of B, Co and Mo on nodulation, growth and yield of cowpea (*Vigna unguiculata*). (Special issue: Sustainable land and water management in the 21stcentury). Trop. Agric. Res. Ext., 6: 108-112.

Sun, X. C., C. X. Hu, Q. L. Tan, Q. Q. Gan, 2006. Effects of molybdenum on photosynthetic characteristics in winter wheat under low temperature stress. China. Agric. Sci., 32: 1418–1422.

Tahir, M. M., M. K. Abbasi., N. Rahim., A. Khaliq, M. H. Kazmi, 2009. Effect of Rhizobium inoculation and NP fertilization on growth, yield and nodulation of soybean (*Glycine max* L.) in the sub-humid hilly region of Rawlakot Azad Jammu and Kashmir, Pakistan. African. J. Biotechnol., 8: 6191-6200.

INDUCTION AND INHERITANCE OF COMPOUND LEAF AND CONE STIPULE IN THE CULTIVATED CHICKPEA (*Cicer arietinum* L.)

Cengiz TOKER[*], *Fatma Oncu CEYLAN*

Akdeniz University, Faculty of Agriculture, Department of Field Crops, Antalya, Turkey,
** Corresponding Author: toker@akdeniz.edu.tr*

ABSTRACT

The cultivated chickpea (*Cicer arietinum* L.) has multipinnate, simple and normal leaves, caused by mutations. About 1500 air-dried seeds (500 seeds per dose) of chickpea genotype ICC 6119 were irradiated with 200, 300 and 400 Gy gamma rays. Two new induced characteristics, compound leaf and cone stipule on internodes, were isolated in a single mutant chickpea in the M_3 generation. Inheritance of the new characteristics was studied in the segregating populations of the M_4 and the results indicated that the new characteristics were governed by two recessive genes designed as *clcl* and *cscs* symbols. These characteristics could be linked since they were jointly segregated. The new genes identified may be useful for gene mapping of *Cicer* species.

Keywords: Chickpea, *Cicer arietinum*, mutagenesis, mutant, leaf types

INTRODUCTION

The cultivated chickpea (*Cicer arietinum* L.) is a diploid species with 2n = 16 chromosomes. It is self-pollinated species due to its cleistogamic flowers (van der Maesen, 1972; Cubero, 1987). Rachis length of the cultivated chickpea is 25-75 mm ending a leaflet. Stipules are ovate to oblique-triangular incised with 2-4 teeth. Leaves are often pseudo-imparipinnate (Cubero, 1987) and typically bear 11-15 leaflets, with teeth covering nearly 2/3 of the foliar blade (van der Maesen, 1972). This shape of the leaves in the cultivated chickpea is referred to as normal leaf (Pundir et al., 1990). However, different leaf derivatives have been reported in the cultivated chickpea. Rao et al. (1980) illustrated four different leaf shapes; normal, bipinnate, simple and tiny leaves. Muehlbauer and Singh (1987) indicated five different leaves and leaflets; normal leaf, narrow leaflets, simple leaf, tinny leaflets and bipinnate leaf. After that, Toker and Cagirgan (2004) reported an additional different leaf shape, isolated from an induced mutant chickpea. Despite of these reports, Pundir et al. (1990) showed that the cultivated chickpea consisted of three major leaf shapes (e.g., normal, simple and multipinnate).

Normal leaf in the cultivated chickpea is referred to as imparipinnate, alternifolia, bunchy leaves, filicoid, and narrow leaf. Multipinnate leaf is called tiny, fasciculifoliate, bipinnate and decompound leaf. Simple leaf is known as chrysanthefolia leaf (Pundir et al., 1990). These leaves were reported to be under the control of genetic factors (Muehlbauer and Singh, 1987). Similarly, Toker et al. (2012b) reported genetics of three kinds of leaves in the cultivated chickpea. Chickpea with the normal leaf is predominantly grown in the world (van der Maesen, 1972; Cubero, 1987).

Variation in the leaves of the cultivated chickpea has been induced by spontaneous or induced mutations (Muehlbauer and Singh, 1987; Salimath et al., 2007; Toker et al., 2012b). Induced mutations are one of the most effective approaches to create variation and new genes in the chickpea (Yildirim et al., 2013). The purpose of the present study was to study the inheritance of a new leaf shape and stipule in the cultivated chickpea induced by gamma rays.

MATERIALS AND METHODS

About 1500 air-dried seeds (500 seeds per dose) of chickpea genotype ICC 6119 were irradiated with 200, 300 and 400 Gy of gamma rays from a ^{60}Co source in the Turkish Atomic Energy Agency (TAEK), Ankara, Turkey (Toker et al., 2005). Morphological characteristics of ICC 6119 as well as mutant were given in Table 1.

Table 1. Morphological characteristics of the parent chickpea genotype ICC 6119 and the mutants, ACC 3204, and ACC 3204-1 with compound leaf and cone stipule on internodes

Characteristics	ICC 6119	ACC 3204	ACC 3204-1
Selected generation	-	M_2	M_3
Flower color	Pink	White	White
Flower per peduncle	1	2	1
Flower shape	Cleistogamic	Cleistogamic or open	Open
Seed color	Brown	Cream	Cream
Pigmentation	Pigmented	Non-pigmented	Non-pigmented
Stipule shape	Small with 2-3 teeth	Small with 2-3 teeth	Giant with 2-6 teeth
Stipule in internodes	Absent	Absent	Cone-stipulated
Leaf shape	Multipinnate	Multipinnate	Compound
Petioles	Short	Short	Too long
Fe-deficiency	Fe-deficient	Fe-efficient	Fe-efficient

In 2005, the M_1 plants were grown at Antalya location (30° 38' E, 36° 53' N, 32 m from sea level) in the spring of 2005 and harvested individually in the beginning of summer. In 2006, the M_2 was grown as single plant progeny rows at the same location. In 2007, the selected mutants and their sibs were grown as the M_3 at the same location. Certain and additional mutant chickpeas were confirmed and selected in the M_3 generation, respectively. In 2008 and in 2009, the progenies of the mutant plants were grown in the separate rows as single plant progeny in the M_4 and the M_5 at the same location. The mutant 'ACC 3204-1' was confirmed in the M_4 and the M_5.

Genetic materials and generations were grown in plot with 45 cm row and 5 cm plant spacing. Materials were fertilized with N, P and K at the rate of 20 kg per ha prior to sowing. Weed control was done by hand during the seedling stage.

Chi-squares ($\chi 2$) test was performed for goodness of fit of 15 multipinnate leaves to 1 compound leaf and cone stipule on internodes ratio in the segregating population in the M_4 generation using the formula:

$$\chi 2 = \Sigma (O-E)^2/E,$$

where O and E are observed and expected values, respectively (Steel and Torrie, 1980).

RESULTS AND DISCUSSION

The parent genotype 'ICC 6119' produced pigment on the aerial part of plants with pink flowers and brown seeds, while the mutant selected in the M_2 generation (ACC 3204) was free of pigmentation with white flowers and cream seeds. The parent genotype 'ICC 6119' had single flower per peduncle, while 'ACC 3204' had double flowers per peduncle. The parent genotype 'ICC 6119' and the mutant 'ACC 3204' had multipinnate leaves. The mutant 'ACC 3204-1' selected from the mutant 'ACC 3204' in the M_3 had cone stipules, open flowers and compound leaves with long petioles (Figure 1). As seen in Table 1, the mutant 'ACC 3204-1' had considerably different characteristics from the parent genotype 'ICC 6119' and the mutant 'ACC 3204'. It had compound leaves with long petioles like the genus *Lupinus* L (Cowling, 1998a, 1998b). Also, it had stipules like cone on internodes (Figure 1). A joint segregation of the compound leaf and cone stipules in the mutant 'ACC 3204-1' indicated that these characteristics could be linked. Leaves of the mutant 'ACC 3204-1' were different from those described in following reports (Rao et al., 1980; Muehlbauer and Singh, 1987; Pundir et al., 1990; Toker et al., 2012b). According to our knowledge, stipules and leaves of the mutant 'ACC 3204-1' were the first report in the cultivated chickpea. The mutated characteristics in the mutant 'ACC 3204-1' proved that mutations could affect more than one locus. The similar and useful mutations in the progenitor (*C. reticulatum* Ladiz.) were also reported (Toker et al., 2012a).

Figure 1. A: The mutant 'ACC 3204-1' with compound leaves with long petioles and cone stipules on internodes. B: Leaf derivatives in the cultivated chickpea (LtoR: multipinnate, simple and normal leaves. C: LtoR: simple, compound and normal leaves).

In the M_4 generation, the segregation ratio of the mutant 'ACC 3204' to the mutant 'ACC 3204-1' was found as 90 : 9 (Table 2). The chi-square test indicated that segregation in the M_4 generation fit well to the expected ratio of 15 : 1 (i.e. 15 multipinnate leaves : 1 compound leaf). The results showed that the compound leaf and cone stipule on internodes were governed by two recessive genes, which show supplementary gene action. Pundir et al. (1990) studied inheritance of leaf shapes in the cultivated chickpea. Multipinnate, simple and normal leaves were occurred $ml^{+}sl/.sl$, $ml./ml.$ and $ml^{+}sl^{+}/..$, respectively (Pundir et al., 1990). The gene symbols *clcl* and *cscs* were designed for compound leaves and cone stipules, respectively.

Table 2. Inheritance of compound leaf and cone stipule in an induced mutant chickpea

No. of filials	M_4				
	Observed		Expected	χ2	*P*
	Leaf and stipule type	No. of filials			
99	Multipinnate leaf and normal stipule	90	15 : 1	1.35	0.25-0.10
	Compound leaf and cone stipule	9			

In conclusions, two new genes were the first report of a discovery by using gamma rays. These genes will be useful for the gene mapping studies in *Cicer* species.

ACKNOWLEDGEMENTS

The work was supported by the Scientific and Technical Research Council of Turkey (TUBITAK) and Akdeniz University Scientific Research Projects Coordination Unit. We thank Drs. B.V. Rao and H.D. Upadhyaya (ICRISAT, Patencheru, Hyderabad, India) for kindly supplying the plant material. The gene symbols dedicated to the first authors' daughter, T. Pelin Toker and son, A. Karahan Toker.

LITERATURE CITED

Cowling, W.A., B. J. Buirchell, M.E. Tapia, 1998a. Lupin, *Lupinus* L., Promoting the conservation and use of underutilized and neglected crops. 23. Institute of Plant Genetics and Crop Plant Research, Gatersleben/International Plant Genetic Resources Institute, Rome, Italy.

Cowling, W.A., C. Hughe, W. Swiecicki, 1998b. Lupin breeding. In: J.S. Gladstones, C. Atkins, J. Hamblin (ed.), Lupins as Crop Plants, Biology, Production and Utilization, CAB International, Wallingford, pp. 93-120.

Cubero, J.I., 1987. Morphology of chickpea. In: M.C. Saxena, K.B. Singh (ed.), The Chickpea, CAB International, Wallingford, pp. 157-170.

Muehlbauer, F.J., K.B. Singh, 1987. Genetics of chickpea. In: M.C. Saxena, K.B. Singh (ed.), The Chickpea, CAB International, Wallingford, pp. 99-125.

Pundir, R.P.S., M.H. Mengesha, G.V. Reddy, 1990. Leaf types and their genetics in chickpea (*Cicer arietinum* L.). Euphytica 45, 197-200.

Rao, N.K., R.P.S. Pundir, L.J.G. van der Maesen, 1980. Inheritance of some qualitative characters in chickpea (*Cicer arietinum* L.). Plant Science (Proc. Indian Acad. Sci.) 89, 497-503.

Salimath, P.M., C. Toker, J.S. Sandhu, J. Kumar, B. Suma, S.S. Yadav, P.N. Bahl, 2007. Conventional breeding methods. In: S.S. Yadav, B. Redden, W. Chen, B. Sharma (ed.), Chickpea Breeding and Management, CAB International, Wallingford, pp. 369-390.

Steel, R.G.D., J.H. Torrie, 1980. Principles and Procedures of Statistics: A Biometrical Approach. McGraw-Hill Book Company, New York.

Toker, C., M.I. Cagirgan, 2004. Spectrum and frequency of induced mutations in chickpea. International Chickpea and Pigeonpea Newsletter 11, 8-10.

Toker, C., B. Uzun, H. Canci, F.O. Ceylan, 2005. Effects of gamma irradiation on the shoot length of *Cicer* seeds. Rad Phy Chem 73, 365-367.

Toker, C., H. Canci, N.E. Inci, F.O. Ceylan, 2012a. Improvement in imidazolinone resistance in *Cicer* species by induced mutation. Plant Breeding 131, 535-539.

Toker, C., F.O. Ceylan, H. Canci, N.E. Inci, T. Yildirim, M.I. Cagirgan, 2012b. Inheritance of leaf shape in the cultivated chickpea (*Cicer arietinum* L.). Turkish Journal of Field Crops 17, 16-18.

Van der Maesen, L.J.G. 1972. *Cicer* L., a monograph of the genus, with special reference to the chickpea (*Cicer arietinum* L.), its ecology and cultivation. Mededelingen Landbouwhogeschool (Communications Agricultural University) Wageningen. The Netherlands 72-10.

Yildirim, T., H. Canci, N.E. Inci, F.O.C. Baloglu, C. Ikten, C. Toker, 2013. Inheritance of female sterility in induced *Cicer* species. Turkish Journal of Field Crops 18, 78-81.

GENETIC DIVERSITY OF DIPLOID AND TETRAPLOID COTTONS DETERMINED BY SSR AND ISSR MARKERS

Adem BARDAK[1*], *Yuksel BOLEK*[2]

[1] *Bingol University, Faculty of Agriculture, Department of Field Crops, Bingol, TURKEY*
[2] *Kahramanmaras Sutcu Imam University, Faculty of Agriculture, Department of Field Crops, Kahramanmaraş, TURKEY*
**Corresponding author: adembardak@ksu.edu.tr*

ABSTRACT

Cotton as an annual crop is mainly grown for its fiber and oil in the seed. Determining genetic diversity in the germplasm is the first step of plant breeding. This study aimed to determine genetic diversity for diploid and tetraploid cotton genotypes grown in different parts of the world. SSR (Simple Sequence Repeats) and ISSR (Inter Simple Sequence Repeats) markers were used to determine genetic relationships among *Gossypium* species and genotypes. Using 39 SSR and 5 ISSR markers, 173 alleles (averaging 3.93 alleles per locus) were produced from 25 cotton genotypes. Out of 173 alleles, 155 (89.60%) were polymorphic among the genotypes and polymorphic information content (PIC) values were between 0.0040 and 0.9993 averaging 0.4396. Genetic diversity ranged from 0.04 to 0.58 among all the genotypes inspected. This ratio was 0.04-0.23 within *G. hirsutum* L. and 0.07-0.26 within *G. barbadense* L. species. Out of these two species, genetic diversity ranged from 0.23 to 0.57 among other diploid and tetraploid species. Genetic diversity was low within commercial cultivars that are also frequently used in breeding programs (0.08-0.20). It is advisable to use wild type cottons to increase present genetic diversity in germplasm pools to have a better chance for the selection of the desired traits.

Key Words: cotton, *Gossypium*, genetic diversity, molecular markers, diploid, tetraploid

INTRODUCTION

Cotton as an annual crop is mainly grown for its fiber and oil in the seed. It is worth to determine genetic diversity in the germplasm for the success of the breeding process. Cotton has a lot of diseases and pests which affect growth and development of the plant. Especially, diploid cottons have resistance genes for diseases, pests and biotic stress conditions. Their characteristics can be revealed at DNA levels, and the information for these pests, diseases and biotic stress resistant genes can be used in cotton breeding programs.

Information about the degree and distribution of genetic diversity and relationships among breeding materials has a significant effect on crop improvement. Selection of suitable parents is a priority for the most promising crosses and increases the efficiency of breeding programs. Molecular markers increasingly play an important role in crop improvement programs. They have been used to predict genetic variance among inbred lines (Manjarrez-Sandoval et al., 1997), estimate genetic diversity in crops (Iqbal et al., 2001; Gutierrez et al., 2002; Zhang et al., 2005; Chen et al., 2007; Abdurakhmonov et al., 2006; Zhikun et al., 2008; Kantartzi et al., 2009; Nas et al., 2011), protect plant variety rights (Smith and Smith, 1992), classify heterotic groups (Dudley et al., 1991; Senior et al., 1998), study phylogenetic relationships among crops and their wild relatives (Li and et al., 2000), analyze pedigrees (Smith et al., 1997), and select desired traits (Young, 1999).

The hypothesized narrow genetic base of upland cotton germplasm used in breeding has been considered as one of the reasons contributing to the lack of progress in the improvement of cotton growers and industry in the world during the last 15 years (Lewis, 2001). Multivariate analysis of agronomic and fiber traits of ancestral cultivars detected high similarity (Van Esbroeck et al., 1999), supporting the conclusion that modern cotton cultivars have narrow genetic base when evaluated with isozyme and DNA markers (Wendel et al., 1992). This suggested that pedigree analysis may overestimate genetic distance among modern cultivars (Van Esbroeck et al., 1999).

Microsatellites are regions of short, tandemly repeated DNA sequences of 1 to 6 base pairs in eukaryotic genomes. Two different marker strategies have been used based on microsatellites: SSR (simple sequence repeats) and ISSR (inter-simple sequence repeats). SSRs are highly reproducible co-dominant markers, in which the repeated sequences produce polymorphic patterns among alleles, depending on the length of the repeats. Although these markers are generally highly polymorphic, the initial cost of developing them is relatively high (Reddy et al., 2001).

ISSR is a different microsatellite-based method that does not need prior knowledge of the genome, cloning or primer design (Zietkiewicz et al., 1994). While the SSR protocol relies on the amplification of the repeated region using two locus specific primers, in ISSR, a single primer composed of a microsatellite sequence anchored at the 3'or 5' end by 2 to 4 arbitrary, often degenerate nucleotides, is used to amplify the DNA between two opposed microsatellites of the same type. Allelic polymorphisms occur whenever one genome is missing the sequence repeated or have a deletion or insertion. For 5' anchored primers, polymorphisms also occur due to differences in the length of the microsatellite. The sequences of repeats and anchored nucleotides are randomly selected. ISSRs are dominant markers but they have the advantage of analyzing multiple loci in a single reaction.

The goal of this study was to estimate genetic diversity among commercial cultivars, genetic stock materials and wild cotton species using SSR and ISSR markers.

MATERIALS AND METHODS

Twenty-five cotton genotypes were used in the experiment. Of these, 13 were belong to *G. hirsutum* L., 2 to *G. barbadense* L., and 10 were wild accessions (4 diploid and 6 tetraploid species) (Table 1). Genotypes and wild accession were planted in pots in greenhouse to get leaf samples for DNA extraction. Genomic DNA was extracted from approximately 0.5 g of young leaves according to a slight modification of the procedure described by Zhang and Steward (2000).

Table 1. Genotypes used to estimate genetic diversity

Genotypes	Species	Accession number	Origin	Genome
*Çukurova 1518	*G. hirsutum* L.	N/A	Turkey	AD
*Nazilli 84S	*G. hirsutum* L.	N/A	Turkey	AD
*Ersan 92	*G. hirsutum* L.	N/A	Turkey	AD
*Maras 92	*G. hirsutum* L.	N/A	Turkey	AD
*Gürelbey Ms34/1	*G. hirsutum* L.	N/A	Turkey	AD
*Suregrow 125	*G. hirsutum* L.	N/A	USA	AD
*Sayar 314	*G. hirsutum* L.	N/A	Turkey	AD
Ekstrem Okra (Brown)	*G. hirsutum* L.	N/A	Turkey	AD
Albania 6172	*G. hirsutum* L.	N/A	Albania	AD
Lifsiz	*G. hirsutum* L.	N/A	Turkey	AD
Siocra	*G. hirsutum* L.	N/A	Australia	AD
Taşkent 6	*G. hirsutum* L.	N/A	Uzbekistan	AD
Acala Maxa	*G. hirsutum* L.	N/A	USA	AD
Askabat 91	*G. barbadense* L.	N/A	Turkmenistan	AD
Bahar 82	*G. barbadense* L.	N/A	Uzbekistan	AD
G. herbaceum L..	-	A_1-20,PI 408778	Afghanistan	A
G. sturtianum var. *nandewarense*	-	C_{1-n}-3,PI 530754	Australia	C
G. harknesii Brandg	-	D22-8	Mexico	D
G. incanum (schwartz) Hill	-	E_4-4	Pakistan	E
G. mustelinum Miers ex Watt	-	$(AD)_4$-9	Brazil	AD
G. hirsutum var. *yucatanense*	-	$(AD)_1$.1469	USA	AD
G. hirsutum var. *marie galante*	-	$(AD)_1$.1607	USA	AD
G. lanceolatum Tod	-	$(AD)_1$-1 (P)	Mexico	AD
G. darwinii Watt	-	$(AD)_5$-14	Galapagos Island	AD
G. barbadense L.. (GB-4)	-	$(AD)_{2-23}$	Egypt	AD

*: Commercial varieties; N/A: Not Available

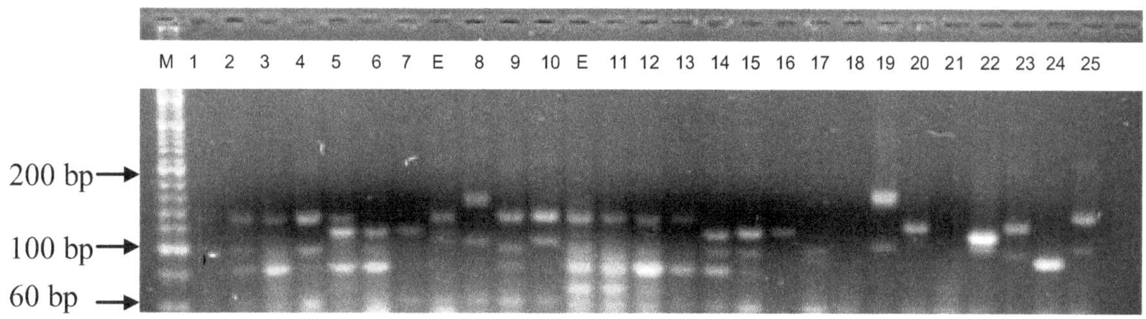

Figure. 1. Microsatellite polymorphsm (locus JESPR 153), M= 20 base pair marker size; 1=Suregrow 125; 2=Cukurova 1518; 3=Nazilli 84S; 4=Gurelbey MS34/1; 5=Ekstrem Okra (Brown); 6=Albania 6172; 7=Ersan 92; 8=Askabat 91; 9=Lifsiz; 10=Bahar 82; 11=Sayar 314; 12=Siocra; 13=Maras 92; 14=Taskent 6; 15=Acala Maxa; 16=G. mustelinum Miers ex Watt; 17=G. harknesii Brandg; 18=G. herbaceum L.; 19=G. barbadence L. (GB-4); 20=G. hirsutum var. yucatanence; 21=G. incanum (Schwartz) Hill; 22=G. lanceolatum Tod; 23=G. hirsutum var. marie galante; 24=G. sturtianum var. nandewarence; 25=G. darwinii Watt; E= Excluded

Extracted genomic DNA was PCR-amplified using 39 SSR and 5 ISSR primer pairs (Table 2). The primers were synthesized by Iontek, Turkey. PCR reactions were performed in a 20 µl volume. The reaction mixture contained 0.25 mM of each dNTP, 2 µl of 10X PCR buffer (Favorgen), 5 µM of each primer, 1 unit of Taq DNA polymerase (Fermentas) and 60 ng of genomic DNA. The PCR-amplification program consisted of one cycle at 95 °C for 3 min, then 35 cycles of [94°C for 1 min, 55 – 60 °C (depending on primer annealing temperatures) for 1 min, 72 °C for 1 min], a final cycle at 72 °C for 5 min. The PCR reactions were carried out in a 96-well block Eppendorf Mastercycler. Amplified PCR products were separated by electrophoresis using 2.5% (w/w) Metaphor agarose (Lonza, USA) + 1.5% (w/w) low melting agarose (Sigma, A5093) gel, stained with ethidium bromide, visualized and photographed under UV light using an AlphaImager Gel Documentation and Analysis System (Figure 1), and fragment lengths were calculated by molecular weight comparison with 20 bp DNA step ladder mobility (Promega).

The products of SSR and ISSR amplification were recorded as present (1) or absent (0). Polymorphic information content (PIC) values provide an estimate of the discriminatory power of a marker by taking into account not only the number of alleles at a locus, but also the relative frequencies of those alleles in the population under study (Pei et al., 2010). The PIC for each SSR and ISSR locus was determined as described by Kalivas et al. (2011) the frequency of alleles per locus was calculated using the following formula:

$$PIC = 1 - \sum_{j=1}^{n} P_{ij}^{2}$$

where P_{ij} is the frequency of the *jth* allele for primer *i* The level of genetic distance (GD) between pair of genotypes was estimated using pairwise comparison (Nei, 1972). GDs were used for cluster analysis with the software Popgen version 3.2, and the dendrogram was drawn with Mega 3.1 software (Kumar et al., 2004) using Neighbor-Joining (NJ) method. The data were also visualized with the principal components analysis module of NTSYSpc.

RESULTS

The 39 SSR and 5 ISSR primer pairs produced a total of 173 unambiguous scorable fragments, and 155 (89.60%) fragments were polymorphic in at least one *Gossypium* species analyzed. The average number of alleles per SSR and ISSR markers was 3.93, ranging from 1 to 8 alleles (Table 2). The mean of the PIC value over the 39 SSR and 5 ISSR markers averaged 0.4396, ranging from 0.0040 for JESPR 135 to 0.9993 for BNL1673.

Table 2. Locus, repeat motif, chromosome, allele size (bp) and number, and polymorphism information content (PIC) for the 39 SSR and 5 ISSR loci used in the analysis of the 25 cotton cultivars

Primer Name*	Repeat Type	Chromosome	No. of Alleles	Allele size (bp)	PIC
BNL448	(CT)13	AD20/AD22	2	100-200	0.4358
BNL852	(CA)13	AD18/19	5	90-110	0.6610
BNL1047	(CA)12	AD25	1	100-220	0.2944
BNL1053	(AC)16	AD21	3	90-220	0.3200
BNL1317	(AG)14	AD9/23	4	100-180	0.8031
BNL1414	(AG)16	AD9/24	6	100-200	0.7059
BNL1611-L1	(AG)12	AD19	3	100-200	0.2470
BNL1611-L2			1		0.2256
BNL1673	(AG)24	A12	3	100-200	0.9994
BNL1679	(AG)17	A12	4	100-200	0.5679
BNL226	(GA)16	AD3	1	100-200	0.0000
BNL2646	(GA)3+G+A2+(AG)4+ (GA)4. (TC)4+(CT)17	AD7/9/15	5	100-200	0.2546
BNL2652	(TC)32	AD13/18	2	100-300	0.5328
BNL2847	(GA)17	AD9	4	150-600	0.3020
BNL285	(GA)12. (GA)3+GC+ (GA)12+A+(AG)2	AD19	4	180-200	0.9576
BNL2960	(GA)10	AD10	4	100-200	0.9010
BNL2986	(AG)10	A7/AD16	5	120-500	0.0389
BNL3103	(GA)13. (TC)14	AD25	1	100-200	0.8976
BNL3171	(GA)26	AD21	2	180-200	0.8640
BNL3511	(AC)11	AD23	4	150-500	0.0044
BNL3816	(TG)15. (TG)5TA(TG)15	AD12/AD26	2	100-200	0.0624
CM13	(AG)23	-	3	80-120	0.9516
CM23	(CTCA)4. (CT)14	AD11	5	80-120	0.4138
CM3	(AG)22	-	2	180-200	0.9872
CM71	(TC)19	AD10	5	70-120	0.9485
CM76	(TC)13(AC)11. (AAAC)4	AD1	1	200	0.9936
CIR084	CA_8TA_7	AD14	1	100-150	0.2944
CIR398	$ACAT_5CA_7$	AD21	3	80-500	0.4838
JESPR135	$(CT)_{11}$	AD11/AD23	3	100-200	0.0040
JESPR153	$(CTA)_{18}$	AD18	8	80-160	0.1594
JESPR169	$(GA)_5(CTT)_{10}$	-	1	100-200	0.0784
JESPR224	$(GA)_{22}$	A6/AD25	7	80-200	0.1896
JESPR232	$(CT)_{18}$	AD8	3	120-180	0.8249
JESPR292	$(CTT)_7$	AD16	2	80-200	0.1536
JESPR292	$(CTT)_7$	AD16	2	80-200	0.6400
JESPR50	$(CAA)_5$	A4/AD5/AD25	3	100-250	0.2316
JESPR56	$(GAA)_{23}$	AD18/AD10/AD1/AD20/AD9	4	80-220	0.9928
Meghes16		-	4	80-180	0.3153
Meghes16		-	1	80-180	0.0784
MUcs0570	(AAT)4	AD17	1	80-300	0.9936
NAU1032	(ATC)6	-	5	100-200	0.0540
NAU1032	(ATC)6	-	1	100-200	0.1536
NAU1369	(AGGCGG)3	AD8/AD24/AD25	2	180-220	0.2240
NAU1369	(AGGCGG)3	AD8/AD24/AD25	1	180-220	0.2256
UDC811**	(GA)8 C	-	4	200-100	0.2203
ISSR12**	(GA)9	-	7	200-1000	0.2962
UDC817 L1**	(CA)8A	-	1	200-1000	0.0784
UDC817 L2**	(CA)8A	-	4	200-1000	0.3409
UDC827 L1**	(AC)8+G	-	1	100-500	0.1536
UDC827 L2**	(AC)8+G	-	5	100-500	0.3936
UDC826**	(AC)8	-	5	200-1000	0.4723

* Microsatellite primers producing multiple loci are indicated by "L".

** ISSR primers

Genetic diversity (GD) ranged from 0.04 to 0.58 among 25 cotton genotypes and wild accessions (Table 3). The lowest GD (0.04) examined was between Siocra and Nazilli 84S genotypes. The highest GD (0.58) examined was between *G. sturtianum* var. *nandewarense* belonging to C genome and Erşan92 (*G. hirsutum*) belonging to AD genome.

Table 3. Pairwise genetic distance coefficients [minimum. maximum and (mean) respectively] within and between cotton species, accessions and genotypes.

Genetic Distance	Wild Type Accessions	Commercial Cultivars	*G. hirsutum*	*G. barbadense*
Wild Type Accessions	0.23 - 0.57 (0.40)			
Commercial Cultivars	0.20 - 0.58 (0.35)	0.08 - 0.20 (0.15)		
G. hirsutum	0.18 - 0.58 (0.34)	0.04 - 0.23 (0.14)	0.04 - 0.23 (0.13)	
G. barbadense	0.23 - 0.49 (0.32)	0.13 - 0.32 (0.23)	0.13 - 0.31 (0.21)	0.07 - 0.26 (0.19)

On the other hand, GD among wild type accessions ranged from 0.23 to 0.57. The lowest GD (0.23) among wild type accessions was between *G. hirsutum* var. *yucatanense* and *G. mustelinum* genotypes while the highest GD (0.57) was among *G. lanceolatum* and *G. sturtianum* var. *nandewarense*.

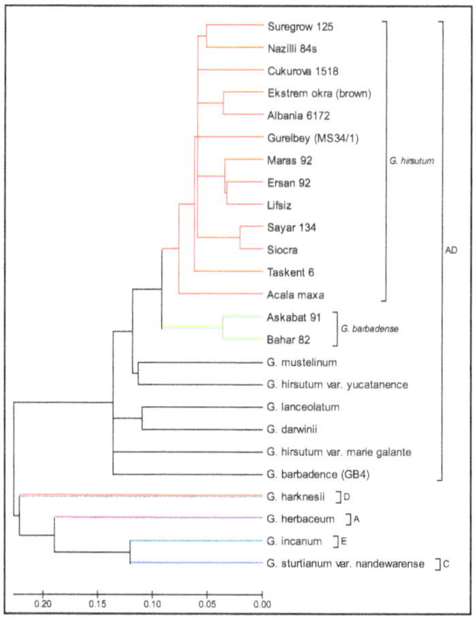

Figure. 2. Dendrogram obtained by Neighbor-Joining (NJ) method cluster analysis based on genetic distances (0.50) among cotton genotypes and wild accessions

The principal components analysis also showed similar relationships between diploid and the other tetraploid cottons as the NJ method shows (Figure 3).

GD among commercial cottons grown in Turkey ranged from 0.08 to 0.20. The lowest GD was between Sayar314 and Erşan92, Sayar314 and Maraş92, and Erşan 92 and Maraş 92 genotypes. The highest GD obtained were between Suregrow 125 and Erşan92 genotypes. GD among genotypes belonging to *G. hirsutum* L. ranged from 0.04 to 0.23. The lowest GD among *G. hirsutum* genotypes were observed between Siocra and Sayar314 genotypes. The highest GD examined between Acala Maxa and Nazilli84S genotypes.

Cluster analysis clearly discriminated diploid wild type cotton from other tetraploid wild types, accessions and genotypes. The dendrogram separated diploid cottons and the other tetraploid cottons into two major clusters. One of the clusters formed by tetraploid cotton was also separated into two groups; one included the *G. hirsutum* genotypes and the other one had the *G. barbadense* genotypes (Figure 2).

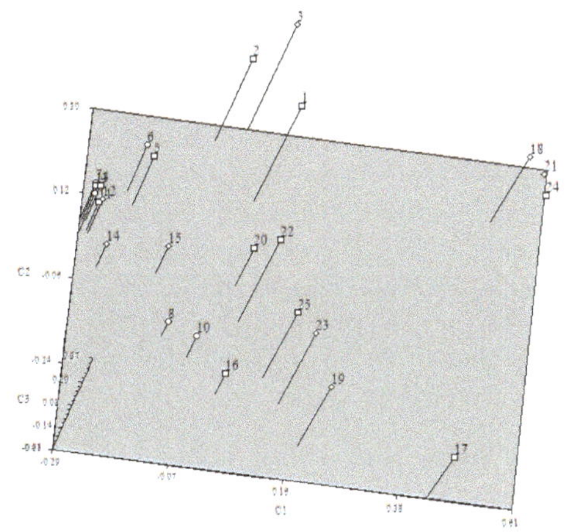

Figure. 3. Principal components analysis of 173 marker (SSR and ISSR) loci for *Gossypium* sp. Genotypes. (1=Suregrow 125; 2=Cukurova 1518; 3=Nazilli 84S; 4=Gurelbey MS34/1; 5=Ekstrem Okra (Brown); 6=Albania 6172; 7=Ersan 92; 8=Askabat 91; 9=Lifsiz; 10=Bahar 82; 11=Sayar 314; 12=Siocra; 13=Maras 92; 14=Taskent 6; 15=Acala Maxa; 16=*G. mustelinum* Miers ex Watt; 17=*G. harknesii* Brandg; 18=*G. herbaceum* L.; 19=*G. barbadence* L. (GB-4); 20=*G. hirsutum* var. *yucatanence*; 21=*G. incanum* (Schwartz) Hill; 22=*G. lanceolatum* Tod; 23=*G. hirsutum* var. *marie galante*; 24=*G. sturtianum* var. *nandewarence*; 25=*G. darwinii* Watt)

DISCUSSION

Primers produced a total of 173 alleles averaging 3.93 per locus (Table 2). Bertini et al. (2006) found 2.13 alleles per locus while Gutierrez et al. (2002) found 139 alleles in a total of 69 loci, averaging 2 alleles per locus. On the other hand, Liu et al. (2000) used 62 primers that produced total 325 alleles, averaging 5 alleles per locus. Working with 47 wild *Gossypium* species (Lacape et al., 2007) obtained total 1128 alleles for 201 SSR loci and averaged 5.61 alleles per locus. Our results for the number of alleles per locus were higher than Gutierrez et al. (2002) and Bertini et al. (2006) but lower than Liu et al. (2000) and Lacape et al. (2007). This may be due to comparing wild and cultivated cottons together or differences in the amplified regions on the DNA.

PIC values averaged 0.44 and ranged from 0.0040 to 0.9993. Liu et al. (2000) found that PIC values ranged from 0.05 to 0.82 and averaged 0.31 in a work using wild *G. hirsutum* accessions. Different cotton genotypes may yield differences in the PIC values. In the work by Bertini et al. (2006), PIC values ranged from 0.18 to 0.62 and averaged 0.40. The lower PIC values for this experiment were due to the genotypes that came from a breeding program as explained by the author. Average value is very close to what we found but the difference in the range probably comes from the wild cotton accessions used in the present study..

GD values among *G. hirsutum* cotton genotypes were between 0.04 and 0.23. Similar results were previously obtained by Gutierrez et al. (2002) that genetic diversity belonging to *G. hirsutum* L. ranged from 0.06 to 0.34 among 11 parental lines. On the other hand, Zhang et al. (2005) found that GD was between 0.06 and 0.38 among Acala 1517 genotypes with SSR molecular markers. Multani and Lyon (1995) observed that GD ranged from 0.01 to 0.08 among nine Australian cultivars which also showed a lack of genetic diversity. Iqbal et al. (1997) found very high genetic similarity ranging from 0.82 to 0.93 among 17 *G. hirsutum* cultivars on the basis of random amplified polymorphic DNA (RAPD) markers. Khan et al. (2009) used SSR markers to determine genetic differences among *G. hirsutum* genotypes and found GD between 0.19 and 0.36.

Van Esbroeck et al. (1998) have pointed out that the monoculture of some successful cultivars and their extensive use as progenitors in breeding programs has limited the genetic diversity of cultivated cotton cultivars. They found genetic variation as 0.13 among cultivated cotton cultivars.

Comparisons of Aşkabat 91 and Bahar 82 to commercial cultivars yielded genetic diversity as 0.21. Genetic differences among wild cottons ranged from 0.23 to 0.57. Higher genetic variability among wild cotton species shows the usefulness of including them in breeding programs to increase variability within the germplasm pools. Wild type cotton can especially be used to increase biotic and abiotic stresses in cotton and have high fiber quality and yield under stress conditions.

ACKNOWLEDGEMENTS

This work was supported by Kahramanmaras Sutcu Imam University (BAP: 2005/4-8). Authors also would like to thank Dr. Ziya DUMLUPINAR for PIC calculations.

LITERATURE CITED

Abdurakhmonow, I.Y., R.J. Kohel, S. Saha, A.E. Pepper, J.Z. Yu, Z.T. Buriev, S.E. Shermatov, A.A. Abdullaev, F.N. Kushanov, J.N. Jenkins, B. Scheffler, A. Abdukarimov, 2006. Molecular Genetic diversity of *G. hirsutum* accessions from Uzbek cotton germplasm revealed by core set and chromosome specific microsatellite markers. Plant & Animal Genomes XIV Conference, P131.

Bertini, C.H.C.M., I. Schuster, T. Sediyama, E. G. De Barros, M. A. Moreira, 2006. Characterization and genetic diversity analysis of cotton cultivars using microsatellites. Genet. Mol. Biol. 29(2):321-329.

Chen, G., X.M. Du, 2006. Genetic diversity of source germplasm of upland cotton in China as determined by SSR marker analysis. Acta Genetica Sinica 33(8):733-745.

Gutierrez, O. S., S. Basu, S. Saha, J. N. Jenkins, D. B. Shoemaker, C. L. Cheatham, J. C. Jr. Mccarty, 2002. Genetic distance among selected cotton genotypes and its relationship with F_2 performance. Crop Sci. 42:1841-1847.

Iqbal, M. J., N. Aziz, N.A. Saeed, Y. Zafar, K.A. Malik, 1997. Genetic diversity evaluation of some elite cotton varieties by RAPD analysis. Theor. Appl. Genet. 94:139-144.

Iqbal, K. J., O. U. K. Reddy, K. M. El-Zik, A. E. Pepper, 2001. A Genetic bottleneck in the 'evolution under domestication' of upland cotton *Gossypium hirsutum* L. examined using DNA fingerprinting. Theor. Appl. Genet. 103(4):547-554.

Kalivas, A., F. Xanthopoulos, O. Kehagia, A. S. Tsaftaris, 2011. Nonanchored inter simple sequence repeat (ISSR) markers: reproducible and specific tools for genome fingerprinting. Genetics and Molecular Research 10 (1): 208-217.

Kantartzi, S. K., M. Ulloa, E. Sacks, J. Mc D. Stewart, 2009. Assessing genetic diversity in *Gossypium arboreum* L. cultivars using genomic and EST-derived microsatellites. Genetica 136:141–147.

Khan, A. I., Y. B. Fu, I. A. Khan, 2009. Genetic diversity of Pakistani cotton cultivars as revealed by simple sequence repeat markers. Communications in Biometry and Crop Sci. 4(1), 21-30.

Lacape, J. M., D. Dessauw, M. Rajab, J. L. Noyer, B. Hau, 2007. Microsatellite diversity in tetraploid *Gossypium* germplasm: assembling a highly informative genotyping set of cotton SSRs. Mol. Breed. 19:45–58.

Lewis, H., 2001. A review of yield and fiber quality trends and components in American upland cotton. Pp. 1447-1453. In Proc. Beltwide Cotton Res. Conf., Anaheim, Ca. 9-13 Jan. Natl. Cotton Council Am., Memphis, Tn.

Liu, S., R. G. Cantrell, J. C. J. R. Mccarty, J. Mc.D. Stewart, 2000. Simple sequence repeat based assessment of genetic diversity in cotton race stock accessions. Crop Sci. 40:1459-1469.

Manjarrez-Sandoval, P., T. E. Jr. Carter, D. M. Webb, J. W. Burton, 1997. RFLP genetic similarity estimates and coefficient of parentage as genetic variance predictors for soybean yield. Crop Sci. 37:698-703.

Multani, D.S., B.R. Lyon, 1995. Genetic fingerprinting of Australian cotton cultivars with RAPD markers. Genome 38:1005-1008.

Nas, M. N., Y. Bolek, A. Bardak, 2011. Genetic diversity and phylogenetic relationships of Prunus microcarpa C.A. Mey. subsp. tortusa analyzed by simple sequence repeats (SSRs). Sci. Hortic. 127(3) 220-227.

Nei, M., 1972. Genetic distance between populations. Am. Nat., 106:283-292.

Pei, Z., J. Gao, Q. Chen, J. Wei, Z. Li, F. Luo, L. Shi, B. Ding, S.Sun, 2010. Genetic diversity of elite sweet sorghum genotypes assessed by SSR markers. Biologia Plantarum 54 (4): 653-658.

Reddy, O. U. K., A. E. Pepper, I. Y. Abdurakhmonov, S. Saha, J. H. Jenkins, T. Brooks, Y. Bolek, K. M. El-Zik, 2001. New dinucleotide and trinucleotide microsatellite marker resources for cotton genome research. J Cotton Sci. 5:103-113.

Senior, L. M., J. P. Murphy, M. M. Goodman, C. V. Stuber, 1998. Utility of SSRs for determining genetic similarities and relationships in maize using an agarose gel system. Crop Sci. 38:1088-1098.

Smith, J. S. C., E. C. L. Chin, H. Shu, O. S. Smith, S. J. Wall, M. L. Senior, S. E. Mitchell, S. Kresovich, J. Ziegle, 1997. An evaluation of the utility of SSR loci as molecular markers in maize *(Zea Mays* L.) comparisons with data from RFLPs and pedigree. Theor Appl Genet. 95:163-173.

Smith, J. S. C., and O. S. Smith, 1992. Fingerprinting crop varieties. Adv. Agron. 47:85-140.

Van Becelaere, G., E. L. Lubber, A. H. Paterson, P. W. Chee, 2005. Pedigree- Vs. DNA marker-based genetic similarity estimates in cotton. Crop Sci. 45: 2281-2287.

Van Esbroeck, G.A., D. T. Bowman, O. L. May, D. S. Calhoun, 1999. Genetic similarity indices for ancestral cotton cultivars and their impact on genetic diversity estimates of modern cultivars. Crop Sci. 39:976-984.

Young, N. D., 1999. A cautiously optimistic version for marker-assisted breeding. Mol. Breed. 5:505-510.

Zhang, J. F., Y. Lu, H. Adragna, E. Hughs, 2005. Genetic improvement of New Mexico Acala cotton germplasm and their genetic diversity. Crop Sci. 45:2363-2373.

Zhang, J.F., and J. M. Stewart, 2000. Economical and rapid method for extracting cotton genomic DNA. J. Cotton Sci. 4:193–201.

Zietkiewicz, E., A. Rafalski, D. Labuda, 1994. Genome fingerprinting by simple sequence repeat (SSR)-anchored polymerase chain reaction amplification. Genomics 20:176-183.

THE EFFECT OF WATER STRESS ON SPIKELET POSITION WEIGHT VARIATION IN WHEAT

Mustafa YILDIRIM

Kelkit Aydin Dogan Vocational High School, Gumushane University, Gumushane, TURKEY.
Corresponding author: myildirimkm@gmail.com

ABSTRACT

This study was conducted in Eskisehir during the 2004-05 and 2005-06 growing seasons, and aimed to evaluate the response of six bread wheat varieties to rainfed and irrigated conditions in terms of grain yield, heading date, plant height and position weight variation (PWV). Of the six varieties tested, three (Sultan, Yildiz, Bezostaja) are drought-sensitive cultivars and three (Gerek, Altay, Sonmez) are drought-tolerant cultivars. Spikes from each variety were divided into five spikelet groups in order to study the effect of spikelet position upon grain weight (S1, S2, S3, S4 and S5, with S1 the spikelet closest to the base). This study focused on position weight variation (PWV) in order to better understand how 1000-kernel weight is affected by spikelet position under both rainfed and irrigated conditions and thus gains a better understanding of wheat resistance to drought. Linear increases in grain yield, heading date and plant height were observed in response to irrigated conditions, whereas a linear increase in PWV was observed in response to rainfed conditions for both seasons. Grain yield was negatively associated with heading date ($r=-0.746$). PWV was positively associated with grain yield and negatively associated with heading date ($p<0.01$; $r=0.804$ and $r=-0.860$, respectively). The observed differences in grain weight between spikelet positions have important implications for drought tolerance in wheat and can be used as a new parameter for predicting drought tolerance.

Key Words; spikelet position weight, water stress, wheat

INTRODUCTION

An increase in yield per unit area is required to feed an ever-expanding world population which was estimated at 7 billion as of 2011 (Rosenberg, 2012). According to the FAO (2009), annual cereal production will need to rise to about 3 billion tons from 2.1 billion. However, in recent years, drought has sprung up unpredictably in a world experiencing climate change (Brown, 2008). In order to support climate-related decisions such as which varieties to plant and when, how much water is needed for irrigation, when and where disease outbreaks are likely to occur, or whether to reduce livestock numbers in case of drought, monthly, seasonal and decadal climate predictions are needed (Bernardi, 2011).

The greater adaptation of bread wheat relative to other cultivated species has made it one of the most important global food crops (Rajaram, 2005) cultivated across different environments. In the process of screening wheat genotypes for breeding, grain yield is considered the most important parameter (Yildirim, 2005; Forgone, 2009). Rain is the most important factor affecting grain yield (Altay, 2006), the improvement of which is the primary goal of most wheat-breeding programs. Under drought conditions, however, actual yield is dependent upon genotype phenology as well as yield potential (Acevedo, 1991). Unfortunately, plant response to moisture stress has a negative effect on grain yield (Schmidt, 2003), which decreases significantly as a result of drought stress (Madani et al., 2010). In temperate climates, grain yields are limited by drought conditions occurring in early summer, which coincides with the grain-filling period of wheat (Forgone, 2009). Given the polygenic character of drought tolerance, a single drought-tolerance parameter may not be very meaningful on its own (Sadiq, 1994; Tirado and Cotter, 2010); rather, breeders must work with a number of parameters to achieve drought tolerance (Cekic, 2007), and considerable variations can be found among varieties developed for rainfed conditions versus those designed for irrigated conditions (Cekic, 2007). Drought conditions often involve not only water, but also heat stress, which reduces the duration of grain filling (Blum, 1998) and negatively affects the translocation of dry matter after flowering (Garcia del Moral et al., 2003; Ilker et al., 2011). This process occurs at different levels: tillers, spikelets and even grains within spikelets (Ozturk and Aydin, 2004).Breeders believe that tall varieties tolerate drought stress better than shorter varieties (Blum et al., 1999; Atay, 2006). Flowering is another parameter used for measuring drought tolerance, with late flowering resulting in reductions in grain yield under drought conditions (Rajaram et al., 1995; Blum et al., 1999).

Grain yield is determined by the number of spikes per plant, number of grains per spike and grain weight. Grain weight, which is one of the most important traits in terms of yield and has been an important criterion in the selection of higher-yielding plants (Roderet al., 2008), is affected by differences in kernel weight, which varies according to spike position (Bremner, 1972). Grain filling rates and durations for individual kernels are affected by environmental factors such as drought stress (Li et al., 2001). Wheat flowering as well as grain filling begin at mid-spike and continue upwards and downwards (Peterson, 1965), and it is possible to distinguish between varieties developed for rainfed and irrigated conditions according to the order of spikelet grain filling within the spike. This study is based on the assumption that the effect of water stress on kernel weight varies according to spikelet position; thus, the aim of the study was to examine the effects of drought on plant height, grain yield, heading date and position weight variation (PWV) of six bread wheat varieties over two growing seasons.

MATERIALS AND METHODS

This study was carried out at Anatolian Agricultural Research Institute in Eskisehir, Turkey during the 2004-2005 and 2005-2006 growing seasons. The experiment station is located at 39° 45' N latitude, 30° 33' E longitude at an altitude of 801 m above sea level (Anon., 2012a). The trial area was divided into two areas, one for each year of the experiment, and each of these was sub-divided into two separate growing areas, one for irrigated and one for rainfed conditions. Wheat varieties included Sultan, Yildiz and Bezostaja, which are bread wheat cultivars registered in Turkey for cultivation under irrigated conditions, and Gerek, Altay and Sonmez, which were registered in Turkey for cultivation under rainfed conditions (Anon., 2003). The experimental design and agronomic treatments from planting to harvest for both growing seasons are presented in Table 1 (Cekic, 2007). Experimental plots consisted of 5-m long with six rows 20-cm apart on each row.

Table 1. Experimental design and agronomic treatments from planting to harvest.

Material and Methods		2004-2005 Crop Season		2005-2006 Crop Season	
		Irrigated	Rainfed	Irrigated	Rainfed
Cultivars		6	6	6	6
Replications		3	3	3	3
Sowing density		550 seeds m^{-2}	550 seeds m^{-2}	550 seeds m^{-2}	550 seeds m^{-2}
Sowing date		14 Oct., 2004	14 Oct., 2004	12 Oct., 2005	12 Oct., 2005
Irrigation	1st	14 Oct., 2004	14 Oct., 2004	12 Oct., 2005	12 Oct., 2005
	2nd	8 Apr., 2005	-	3 Apr., 2006	-
	3rd	4 May 2005	-	8 May 2006	-
	4th	7 June 2005	-	5 June 2006	-
Fertilization	**Sowing date**	P_2O_5 ;70 kg ha^{-1}	P_2O_5;70 kg ha^{-1}	P_2O_5 ;70 kg ha^{-1}	P_2O_5;70 kg ha^{-1}
		DAP; 27 kg ha^{-1}	DAP; 27 kg ha^{-1}	DAP; 27 kg ha^{-1}	DAP; 27 kg ha^{-1}
	Zadoks 24 †	AN; 40 kg ha^{-1}	AN; 70 kg ha^{-1}	AN; 40 kg ha^{-1}	AN; 70 kg ha^{-1}
	Zadoks 50 †	AN; 30 kg ha^{-1}	-	AN; 30 kg ha^{-1}	-
Weed control		Herbicide	Herbicide	Herbicide	Herbicide

†Zadoks, 1974.

The soil in the experimental area was characterized as loamy with mild alkalinity, no salinity problem, medium amounts of organic matter and calcium, poor in terms of available phosphorus and rich in terms of available potassium (Anon., 2004).

Monthly rainfall data over two growing seasons and long-term are shown in Figure 1 (Anon., 2006). As the figure shows, rainfall was below average in both the 2004-05 and 2005-06 growing seasons (287.1 and 280.3 mm, respectively). Early summer drought limits grain yield and yield components. Precipitation amount in April, May and June are very important in terms of total rainfall. Therefore, precipitation in these months was low enough in terms of drought in these years.

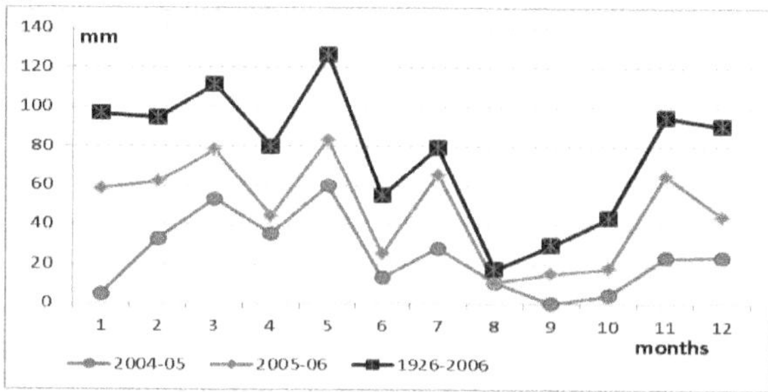

Figure 1 Average monthly rainfall (mm) in Eskisehir during the 2004-05 and 2006-06 experimental years and over the long-term.

Growing parameters measured in this study included grain yield, heading date, plant height and 1000-kernel weight by spikelet position. Heading date was calculated from May 1 to heading stage. Plant height was measured in centimeters during the pre-harvest period. Plot yields were measured in kilograms after harvesting and were converted to hectare yields. For each irrigated and rainfed plots, 40 main stem of plants with the same heading date were marked and cut pre-harvest. In order to study the effect of spikelet position upon grain weight, fertile spikelets from each spikes were distributed equally among 5 spikelet regions (S1, S2, S3, S4 and S5, starting at the spike base and proceeding upwards (Figure 2), with remaining spikelets distributed from the center spikelet (S3) outwards. (For example, in grain spikes with 22 spikelets, 4 spikelets each were included in S1, S2, S4 and S5 and 6 spikelets were included in S3, whereas in grain spikes with 23 spikelets, 4 spikelets each were included in S1 and S5 and 5 spikelets each were included in S2, S3 and S4.) Sterile spikelets were not counted. Spikelets from 40 spikes were divided according to spikelet position, and spikelets from the same position were combined for measurement in the same cup, with seeds separated from spikelets by hand to prevent grain loss. Grain was weighed on a precision scale with an accuracy of 200g/0.001g, and 1.000-kernel weight was calculated by spikelet position.

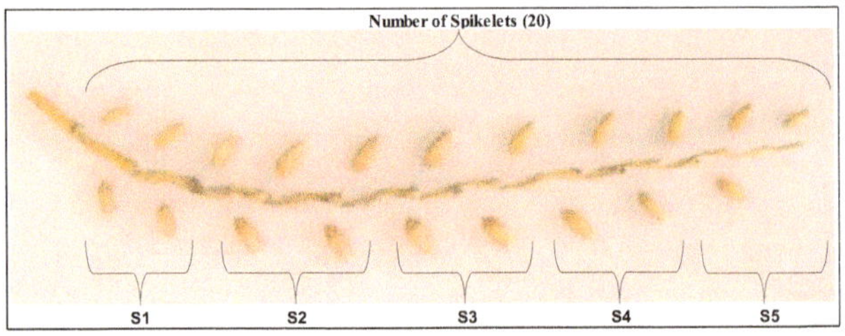

Figure 2 A grain spike divided into 5 regions of fertile spikelets.

Severity of Water Stress and Drought Sensitivity Index were calculated as follows (Fischer and Maurer, 1978:

$$\text{Severity of Water Stress} \quad \frac{(\text{Yield of Irrigated Condition}^* - \text{Yield of Rainfed Condition}^*)}{\text{Mean Yields of Irrigated}}$$

* The mean of yields of all varieties for each condition.

$$\text{Drought Sensitivity Index} \quad \frac{[(\text{Yield of Irrigated}^\dagger - \text{Yield of Rainfed}^\dagger) / \text{Yield of Irrigated}^\dagger]}{\text{Severity of Water Stress}}$$

† The grain yield of variety.

PWV was calculated using the following formulas;

$$PWV^{\ddagger} = P_{S3} - \frac{(P_{S1}+P_{S5})}{2}$$

PWV : Position weight variation (g),
P_{S1} :1.000-kernel weight of bottom spikelets (g),
P_{S3} :1.000-kernel weight of middle spikelets (g),
P_{S5} :1.000-kernel weight of top spikelets (g),
‡ In these calculations, S1, S3 and S5 were used. S2 and S4 were not included in the calculation.

DTKW: Differences in 1.000-kernel weight of spikelet positions between irrigated and rainfed conditions (g),

$$DTKW_{(sn)} = (P_{sn} \text{ of Irrigated}\S - P_{sn} \text{ of Rainfed}\S)$$

sn: Calculated 1.000-kernel weight of spikelet position (g),
§ Calculated using the overall mean of all plots for both years

Mean values of grain yield, plant height, heading date and PWV for each variety and each year were calculated according to the following formula:

$$D_{(n)} = (Irrigated_{(n)}\P - Rainfed_{(n)}\P)$$

¶ *the mean of all plots of a variety in a treatment*

$D_{(n)}$: Data for the parameter being calculated,

Degree of freedom (SD) was calculated using the following formula:

$$SD = (2 \text{ years} \times 6 \text{ varieties}) - 1 = 11$$

All data analysis was performed using the MSTAT statistical package for analysis of variance (ANOVA) (Anon., 1986). Correlation values were calculated and graphs drawn using MS-Excel 2010 (Anon., 2012b).

RESULTS AND DISCUSION

This study found a wide range of heading dates, grain yields and PWVs for irrigated and rainfed plants. Genotype and treatment (irrigated vs. rainfed) had the greatest effect on the parameters tested ($p<0.01$). Treatment had the highest effect ($p<0.01$) on grain yield and PWV in both 2004-05 and 2005-06. Whereas heading date, grain yield and plant height all increased with irrigation, PWV decreased with irrigation. A mean increase of 51.2% in grain yields and a mean decrease of 42.9% in PWV values of drought-sensitive cultivars were observed for the two growing seasons studied (Table 2). Earliest mean heading dates were obtained for Sonmez, Gerek and Altay varieties (17, 17 and 20 days, respectively) under rainfed conditions. When the effects of irrigation treatment on grain yields are examined, Altay was found to have the highest grain yields under rainfed conditions (3.410 $kgha^{-1}$), whereas Sultan performed best under irrigated condition (6.546 $kg\ ha^{-1}$).

Table 2 Mean heading dates, grain yields, plant heights and PWV values for combined experimental years by treatment (irrigated vs. Rainfed conditions) and variety.

Treatment	Varieties	Heading (day)	Grain Yield (ha^{-1})	Height (cm)	PWV (g)
Rainfed*	Gerek	17 E	2.791 B	73.7 A	17.2 BC
	Sonmez	17 E	3.321 A	71.0 AB	20.6 A
	Altay	20 D	3.410 A	66.3 C	18.3 AB
	Sultan	29 A	2.184 C	60.5 D	7.7 D
	Yildiz	28 B	2.017 D	69.0 BC	9.6 D
	Bezostaja	24 C	2.034 D	71.3 AB	14.7 C
LSD (P < 0.05)		0.5	106	3.2	2.8
Irrigated#	Gerek	23 F	5.142 D	109.8 A	9.3 D
	Sonmez	23 E	5.388 C	109.2 A	13.6 B
	Altay	25 D	5.161 D	110.6 A	11.1 C
	Sultan	30 A	6.546 A	101.0 C	12.2 C
	Yildiz	29 B	5.752 B	99.1 C	9.1 D
	Bezostaja	26 C	4.467 E	105.4 B	16.3 A
LSD (P < 0.05)		0.6	207	2.5	1.2

#Mean heading dates, grain yields, plant heights and PWV values for combined experimental years.

The PWV values of Sonmez, Altay and Gerek (20.6%, 18.3% and 17.2%, respectively) were higher than the PWV values of Bezostaja, Yildiz and Sultan (14.7%, 9.6% and 7.7%, respectively) under rainfed conditions (Figure 3). The PWV value of Gerek was decreased by 7.9 percentage points to 9.3% under irrigated conditions in comparison to rainfed conditions, indicating that under drought conditions, central spikelets of Gerek spikes receive more nutrients, with top and bottom spikelets receiving fewer nutrients. A similar phenomenon was observed in Sonmez and Altay, two-other drought-tolerant cultivars tested, whereas irrigation did not have a significant effect on the PWV values of any of the drought-sensitive cultivars tested.

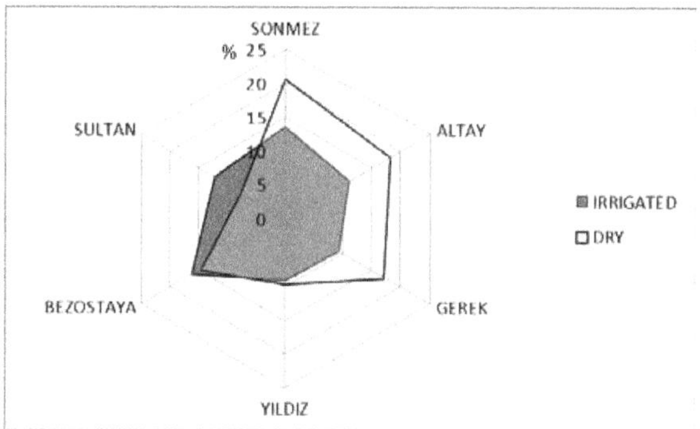

Figure 3 Position weight variations (PWV) of six wheat varieties under irrigated and Rainfed conditions.

The water stress severities of irrigated and rainfed treatments were 49% and 54% (2004-2005 and 2005-2006 growing seasons, respectively). Drought sensitivity index values of the varieties tested are given in Table 3. The values showed that grain yields of Bezostaja, Yildiz and Sultan (1.06, 1.26 and 1.29, respectively) were higher than those of Altay, Sonmez and Gerek (0.66, 0.75 and 0.89, respectively). (It should be noted that in the formula used here, lower PWV values represent greater sensitivity to drought, and vice versa).

Table 3 Drought sensitivity indexes for the parameters heading date, grain yield, plant height and PWV, by wheat variety and experimental year.

		Varieties					
Parameters	**Years**	**Gerek**	**Sonmez**	**Altay**	**Sultan**	**Yildiz**	**Bezostaja**
Heading	2005	0.44	0.53	0.40	0.00	0.00	-0.05
	2006	0.54	0.58	0.40	0.09	0.11	0.28
	Mean	0.49	0.56	0.40	0.04	0.05	0.11
Yield	2005	0.87	0.74	0.67	1.31	1.27	1.07
	2006	0.91	0.75	0.65	1.28	1.26	1.05
	Mean	0.89	0.75	0.66	1.29	1.26	1.06
Height	2005	0.73	0.71	0.81	0.84	0.65	0.66
	2006	0.55	0.65	0.75	0.72	0.53	0.60
	Mean	0.64	0.68	0.78	0.78	0.59	0.63
PWV	2005	-1.46	-1.11	-1.22	0.77	-0.19	0.25
	2006	-1.87	-0.92	-1.28	0.68	0.04	0.14
	Mean	-1.66	-1.02	-1.25	0.73	-0.08	0.19

Of all varieties tested, Sultan, a cultivar developed for irrigated conditions, showed the highest sensitivity to drought stress in terms of drought sensitivity index values both 2004-05 and 2005-06 (1.31 and 1.28, respectively), whereas Altay, which was developed for rainfed conditions, showed the highest tolerance in terms of drought sensitivity index values (0.67 and 0.65, respectively, for 2004-05 and 2005-06). Sonmez showed the highest tolerance to drought in terms of heading date in both 2004-05 and 2005-06. In terms of plant height, Sultan was the most drought-tolerant in 2004-2005, whereas Altay was the most drought-tolerant in 2005-06. According to PWV values, Gerek showed the highest drought tolerance in both 2004-2005 and 2005-2006 (-1.46 and -1.87, respectively), whereas Sultan showed the highest sensitivity to drought (0.77 and 0.68 in 2004-2005 and 2005-2006, respectively).

Grain yield was negatively associated with heading date (r=-0.746) (Figure 4). PWV was positively associated with grain yield and negatively associated with heading date (r=0.804 and r=-0.860; p<0.01, respectively); PWV was not associated with plant height.

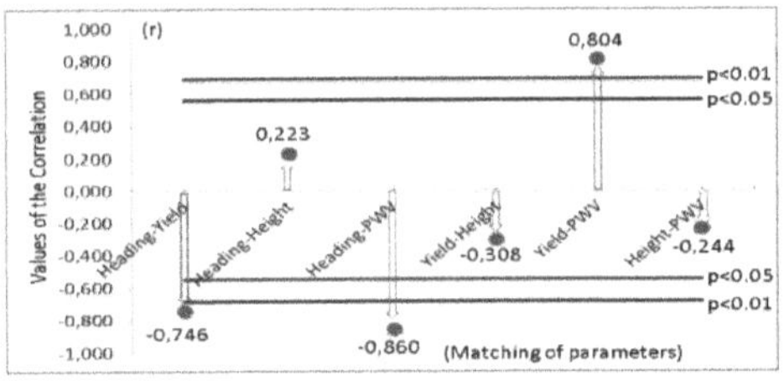

Figure 4 Relationships among heading date, grain yield, plant height and PWV.

Heading dates showed greater variation under rainfed and irrigated-conditions among drought-tolerant varieties (23.9%) in comparison to drought-sensitive varieties (4.7%) (Table 4). Plant height showed greater variation under rainfed and irrigated-conditions among drought-tolerant varieties (36%) in comparison to drought-sensitive varieties (34.3%). Grain yield showed greater variation under rainfed and irrigated-conditions among drought-sensitive varieties (39.3%) in comparison to drought- tolerant varieties (62.8%). PWV values showed greater variation under rainfed and irrigated-conditions among drought-tolerant varieties (65%) in comparison to drought-sensitive varieties (14.9%).

Gerek showed the lowest variation in PWV rates (-84.9%), followed by Altay (-64.9%) and Sonmez (-51.5%). Sultan showed the highest variation in grain yield rates (66.6%), followed by Yildiz (64.9%) and Bezostaja (54.6%).

Polynomial graphs of mean differences in 1.000-kernel weight by spikelet position between drought-sensitive and drought-tolerant cultivars exhibited inverted symmetry (Figure 5). Whereas polynomials of drought-tolerant varieties (Gerek, Sonmez and Altay) exhibited upward-curving graphs, those of drought-sensitive varieties (Sultan, Yildiz and Bezostaja) exhibited downward-curving graphs.

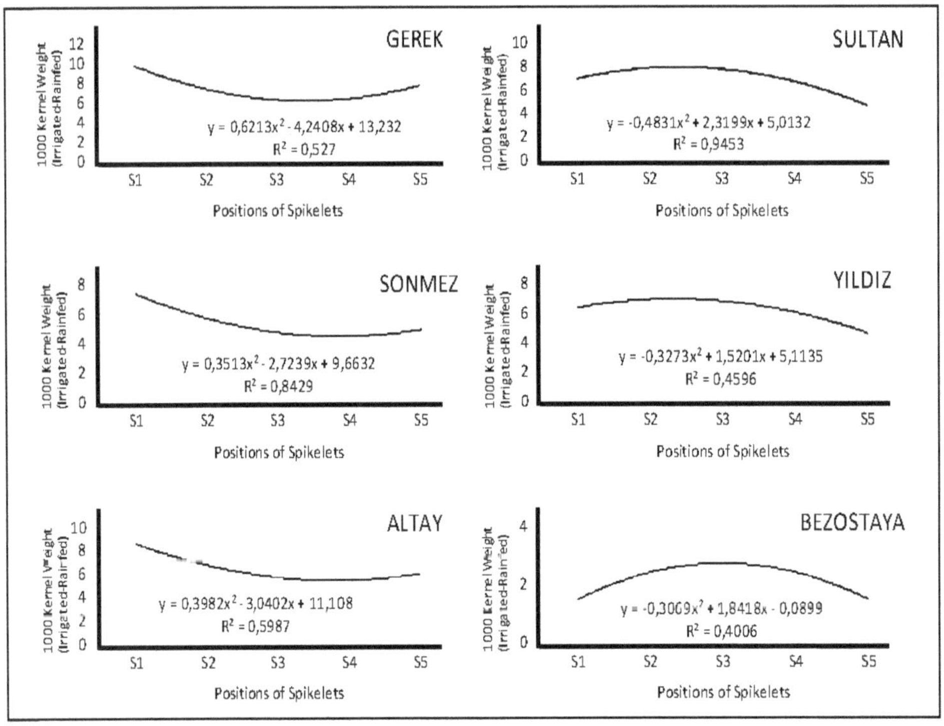

Figure 5 Differences in 1.000-kernel weight (DTKW) of spikelet positions between irrigated and Rainfed conditions for six wheat varieties.

Table 4 Mean differences in heading date, grain yield, plant height and PWV under irrigation vs rainfed conditions, by wheat variety.

	Heading Date		Grain Yield		Plant Height		PWV	
Varieties	I-R*	%#	I-R*	%#	I-R*	%#	I-R*	%#
Gerek (G)	5.7	24.8%	2.351	45.7%	36.1	32.9%	-7.9	-84.9%
Sonmez (Sn)	6.6	28.7%	2.067	38.4%	38.2	35.0%	-7.0	-51.5%
Altay (A)	5.2	20.8%	1.751	33.9%	44.3	40.1%	-7.2	-64.9%
Sultan (Sl)	1.0	3.3%	4.362	66.6%	40.5	40.1%	4.5	36.9%
Yildiz (Y)	0.8	2.8%	3.735	64.9%	30.1	30.4%	-0.5	-5.5%
Bezostaja (B)	1.6	6.2%	2.433	54.5%	34.1	32.4%	1.6	9.8%
Rainfed Character Varieties	5.7	23.9%	2.056	39.3%	39.5	36.0%	-7.4	-65.0%
Irrigated Character Varieties	1.3	4.7%	3.510	62.8%	34.9	34.3%	1.9	14.9%

* Irrigated – Rainfed

(Irrigated – Rainfed) x 100 / Irrigated

Polynomial graphs of mean differences in spikelet positions of drought-sensitive and drought-tolerant cultivars exhibited inverted symmetry (Figure 6). Data for mean differences in 1.000-kernel weight by spikelet position is in line with drought sensitivity index data.

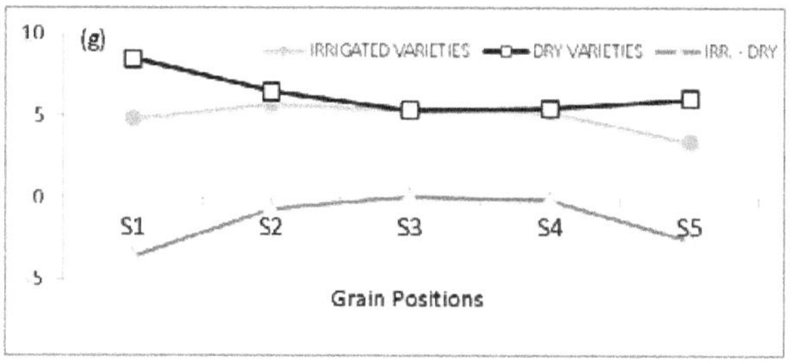

Figure 6 Mean differences in 1.000-kernel weight by spikelet positions between drought-tolerant (Gerek, Sonmez and Altay) and drought-sensitive varieties (Sultan, Yildiz and Bezostaja).

CONCLUSION

Water is the most important factor to influence wheat grain yields. Individual yield components such as fertile spikelets per spike and grain weight per spikelet differ in their responses to drought conditions and are used as indicators of water stress in breeding for drought tolerance. The grain filling rates of drought-tolerant varieties are optimal under drought conditions. Position weight variation (PWV) values of drought-tolerant varieties differ significantly from those of drought-sensitive varieties due to variations in the distribution of photosynthetic products among spikelets. In comparison to drought-sensitive cultivars, drought-tolerant cultivars distribute less photosynthetic products towards the ends of the spike under drought conditions. Drought-tolerant varieties maintain acceptable grain size in the middle of the spike; therefore, PWV can be used as a parameter in developing drought-tolerant wheat cultivars.

LITERATURE CITED

Acevedo, E., M. Nachit and G. Ortiz-Ferrara. 1991. Effects of heat stress on wheat and possible selection tools for use in breeding for tolerance. In: D.A Saunders, (ed) Wheat for the Nontraditional Warm Areas. Mexico, D.F.: CIMMYT. pp. 401-421.

Altay, F. 2006. Winter wheat yield statistics. Eskisehir Anatolian Agricultural Research Institute.

Anonymous. 1986. MSTAT-C Statistical Program. Michigan State. East Lansing, Michigan State University, USA.

Anonymous. 2003. Wheat varieties (irrigated & rainfed) and durum wheat varieties presentation catalog. Anatolian Agriculture Research Institute, Eskisehir.

Anonymous. 2004. Soil analysis report. Soil and Water Resources Research Institute, Eskisehir.

Anonymous. 2006. Weather report data from Eskisehir Anatolian Agricultural Research Institute.

Anonymous. 2009. Food and Agriculture Organization of the United Nations. Rome, www.fao.org

Anonymous. 2012a. Eskisehir Province latitude, longitude and altitude, and the General Directorate of State Meteorological, www.dmi.gov.tr.

Anonymous. 2012b. Microsoft Excel 2010. Microsoft Corporation. USA.

Atay, A.T. 2006. History of the Turkish seed breeding. Chapter 2, Seed File 4, the republic of seed breeding period. Agriculture and Engineering. 78-79: 45-50.

Bernardi, M. 2011. Understanding user needs for climate services in agriculture. Bulletin, Volume 60(2). http://www.wmo.int/pages.

Blum, A. 1998. Improving wheat grain filling under stress by stem reserve mobilization. In: H. J. Braun et al. (eds). Wheat: Prospects for Global Improvement. Kluwer Academic Puplishers. pp. 135-141.

Blum, A., G. Golan, J. Mayer, B. Sinmena, L. Shpiler and J. Bura. 1999. The drought response of landraces of wheat from the Northern Negev Desert in Israel. Euphytica. 43: 87-96.

Bremner, P.M. 1972. The accumulation of dry matter and nitrogen by grains in different position of the wheat ear, as influenced by shading and defoliation. Aust. J. Biol. Sci. 25: 657–668.

Brown, O. 2008. Migration and climate change. IOM Migration Research Series, No: 31. International Organization for Migration. ISSN 1607-338X.

Cekic , C. 2007. A study on physilogical parameters which can be used as selection criteria in breeding wheat (*Triticum aestivum* L.) for drought resistance. Ankara Univ. Institute of Natural Science. Phd Thesis.

Fischer, R.A. and R. Maurer. 1978. Drought resistance in spring wheat cultivars. I. grain yield responses. Austr. J. Agric. Res. 29: 897-912.

Forgone, A.G. 2009. Physiological indicators of drought tolerance of wheat. Biology PhD Program. University of Szeged Faculty of Science and Informatics Department of plant Biology, Szeged.

Garcia del Moral, L.F., Y. Rharrabti, D. Villegas and C. Royo. 2003. Evaluation of grain yield and its components in durum wheat under Mediterranean conditions: An onto genic approach. Agronomy Journal. 95: 266-274.

Gunes, A.M. 2010. Framework convention on climate change and the Kyoto Protocol which are legal developments in the international arena against global warming. Journal of the Bar Association of Turkey, 87:43.

Ilker, E., O. Tatar, F.A. Tonk and M. Tosun. 2011. Determination of tolerance level of some wheat genotypes to post-anthesis drought. Turkish Journal of Field Crops. 16: 59-63.

Li, A., Y. Hou and A. Trent. 2001. Effects of elevated atmospheric CO_2 and drought stress on individual grain filling rates and durations of the main stem in spring wheat. Agricultural and Forest Meteorology. 106: 289-301.

Madani, A., A. Shirani Rad, A. Pazoki, G. Nourmohammadi and R. Zarghami. 2010. Wheat (*Triticum aestivum* L.) grain filling and dry matter partitioning responses to source: sink modifications under post-anthesis water and nitrogen deficiency. Acta Scientiarum. Agronomy. Maringá. 32(1): 145-151.

Ozturk, A. and F. Aydin. 2004. Effect of water stress at various growth stages on some quality characteristics of winter wheat. J. Agronomy and Crop Sci. 190: 93-99.

Peterson, R.F. 1965. Wheat: Botany, cultivation and utilisation. Leonard Hill, London, pp. 448.

Rajaram, S., H.J. Braun, M. Van Ginkel and P.M.A. Tigerstedt. 1995. CIMMYT'S Approach to breed wheat for drought tolerance. Euphytica. 92(1-2): 147-153.

Rajaram, S. 2005. Role of conventional plant breeding and biotechnology in future wheat production. Turkish Journal Agric. Forestry. 29: 105-111.

Roder, M.S., X.Q. Huang and A. Borner. 2008. Fine mapping of the region on wheat chromosome 7D controlling grain weight. Funct. Integr. Genomics, 8: 79-86.

Rosenberg, M. 2012. Current world population and world population growth since the year One. http://geography.about.com/od/obtainpopulationdata/a/worldpopulation.htm

Sadiq, M.S., K.A. Siddiqui, C.R. Arain and A.R. Azami. 1994. Wheat breeding in water stressed environment. First delineation of drought tolerance and susceptibility. Plant Breeding. 113: 36-46.

Schmidt, J.W. 2003. Drought resistance and wheat breeding. Paper No. 6974, Journal series, Nebraska Agricultural Experiment Station. Department of Agronomy, University of Nebraska, USA.

Tirado, R. and J. Cotter. 2010. Ecological farming: Drought-resistant Agriculture Greenpeace Research Laboratories. GRL-TN 02/2010. www.greenpeace.org.

Yildirim, M. 2005. A study on heredity of some agronomical, physiological and quality characteristics in diallel F_1 off springs of selected six bread wheat (*Triticum aestivum* L.) cultivars. Natural Science of Cukurova Univ. PhD Thesis, register no: 893 pp.57.

Zadoks, J.C., T.T. Chang and C.F. Konzak. 1974. "A decimal code for the growth stages of cereals". Weed Res. 14: 415-421.

GENETIC RELATIONSHIPS BETWEEN CULTIVARS OF *Cicer arietinum* AND ITS PROGENITOR GROWN IN TURKEY DETERMINED BY USING THE SSR MARKERS

Sertan CEVIK[1], Serpil UNYAYAR[1], Ali ERGUL[2]*

[1]Mersin University, Faculty of Art and Sciences, Mersin, TURKEY
[2]Ankara University, Institute of Biotechnology, Ankara, TURKEY
[*]Corresponding author: srtncvk@gmail.com

ABSTRACT

Genetic characterization plays a key role in chickpea breeding programs since genetic base of the cultivated chickpea is narrow. This study aims to determine the genetic relationships among 23 cultivated chickpea (*Cicer arietinum* L.) and 2 genotypes of *Cicer reticulatum* Ladizusing simple sequence repeat (SSR) molecular markers with 10 SSR primer pairs. A total of 58 alleles were detected at 10 loci. The number of alleles per locus ranged from 2 to 11, with an average of 5.8. The polymorphic information (PI) ranged from 0.074 (NCPGR42) to 0.806 (NCPGR28) with an average of 0.42. The genetic similarity among all accessions ranged from 0.10 to 0.90. Genotypes of *C. reticulatum* were considerably different from the cultivated chickpea. The cultivars of chickpea were divided into four groups while Er99 and Dikbas cultivars were genetically the closest each other. ICC 4958, 'microsperma' or desi' chickpea was found genetically the closest cultivar to *Cicer reticulatum*. It was suggested that different groups should be crossed in breeding programs to increase variations in chickpeas.

Key words: Chickpea, *Cicer arietinum*, *Cicer reticulatum*, Genetic relationship, SSR.

INTRODUCTION

Chickpea (*Cicer arietinum*L.) is the only cultivated species belonging to the *Cicer* genus. *Cicer* species are predominantly self-pollinated, and the chromosome number of the species is 2n = 16. It is one of the most important pulse crop in the whole World and Turkey as well (Sepetoglu et al., 2008; Ozalkan et al., 2010; Cagirgan et al., 2011, 2012; Toker et al., 2012; Cevik et al., 2014)and is presently grown in 53 countries. Major producing countries include India, Australia, Pakistan, Turkey, and Myanmar (FAOSTAT 2013). Chickpea is globally produced about 13.1 million t from 13.5 million ha area with an average yield of 968 kg per ha (FAOSTAT 2015). However, the average annual yield world-wide (968 $kgha^{-1}$) is considered to be somewhat lower than its potential yield (Canci and Toker, 2009). Therefore, many chickpea breeding programs are focused on improving the genetic potential to increase yield (Sudupak et al., 2002).

Genetic characterization can be made by different methods. It is generally believed that the use of molecular markers is more reliable and repeatable as compared to other methods. Molecular markers provide highly discriminatory information and, therefore, are frequently used for genetic studies. DNA-based markers such as restriction fragment length poylmorphisms (Serret et al., 1997), amplified fragment length polymorphisms (Nguyen et al., 2004; Singh et al., 2008), randomly amplified polymorphic DNAs (Sudupak et al., 2002; Rao et al., 2007) and inter simple sequence repeats (Sudupak et al., 2004) were used to study genetic diversity and relationships in chickpea. Most of these studies reported narrow genomic variability in chickpea. Genetic variation is very limited in *Cicer arietinum*, and this causes difficulty in breeding efforts aimed at developing new cultivars resistant to stress conditions (Atalay and Babaoglu, 2012).In order to enhance genetic potential, there must be a comprehensive understanding of the amount and pattern of genetic variation that exists within and between the available cultivated and wild *Cicer* accessions (Rao et al., 2007).SSRs, also known as microsatellites, short tandem repeats or sequence-tagged microsatellite sites, are PCR-based markers. They are randomly tandem repeats of short nucleotide motifs (2-6 bp/nucleotides long). Di-, tri- and tetra-nucleotide repeats, e.g. (GT)n, (AAT)n and (GATA)n, are widely distributed throughout the genomes of plants and animals (Jiang, 2013). SSR markers are very useful for the analysis of genetic diversity, genomic mapping and marker-assisted selection in many plant species compared to many other marker systems (Atalay and Babaoglu, 2012).

Turkey has a large number of chickpea germplasm (Sudupak, 2013). There are several investigations about genetic relationships of chickpea germplasm which are grown in Turkey, but most of them have not been characterized at molecular levels. The present study aims to determine the genetic relationships of some chickpea cultivars and the closest wild species (*C. reticulatum* Ladiz.) of the cultivated chickpea in Turkey through SSR molecular markers.

MATERIALS AND METHODS

Plant materials

A total of 25 chickpea consisting of 23 cultivars of *Cicer arietinum* and 2 genotypes of *Cicer reticulatum* Ladiz. (Table 1) were used for SSR analyses. *Cicer arietinum* cultivars were developed in Turkey, and most of them were grown by farmers. The seeds of the genetic material were obtained from Akdeniz University Faculty of Agriculture and the Ministry of Agriculture, Field Crops Central Research Institute, Ankara. The plants were grown at 26/22 °C (day/night) temperature and 65±5% relative humidity in a growth chamber with 480 μmolm^{-2}s^{-2} light (day/ night 16/8 h) for 21 days.

SSR analysis

Young and healthy leaves of genotypes were used for DNA extraction (Table 1). DNA was extracted using the procedure described by Lefort et al. (2001). Concentration and purity of the extracted DNA were analyzed by using a NanoDrop®ND-1000 spectrophotometer and 1% agarose gel. PCR reactions were performed in 15 μL volume containing 15 ng of template DNA, 5 pmol each of forward and reverse primers, 0.5 mM of each of dNTPs, and 0.5 U of *Taq* polymerase. The PCR program was run on a Bio-Rad thermocycler. The PCR conditions consisted of an initial cycle of 3 min at 94°C, followed by 35 cycles of 1 min at 94°C, 1 min at 58-66°C and 2 min at 72°C, with a final extension at 72°C for 10 min. To examine the amplification of PCR fragments, PCR products were resolved on ethidium bromide stained with 2% agarose gel in 1 X TBE buffer at 80 V.

Table 1. The Chickpea cultivars/genotypes used as a plant material in SSR analyses.

Cultivars	Kabuli/Desi/Wild	Donor Institutes or Universities
Canitez 87	Kabuli	Anatolian Agricultural Research Institute
Yasa 05	Kabuli	Anatolian Agricultural Research Institute
Isik 05	Kabuli	Anatolian Agricultural Research Institute
Sari 98	Kabuli	Aegean Agricultural Research Institute
Cevdetbey 98	Kabuli	Aegean Agricultural Research Institute
İzmir 92	Kabuli	Aegean Agricultural Research Institute
Menemen 92	Kabuli	Aegean Agricultural Research Institute
Aydin 92	Kabuli	Aegean Agricultural Research Institute
Gokce	Kabuli	Central Research Institute For Field Crops
Akcin 91	Kabuli	Central Research Institute For Field Crops
Dikbas	Kabuli	Central Research Institute For Field Crops
Kusmen 99	Kabuli	Central Research Institute For Field Crops
Uzunlu 99	Kabuli	Central Research Institute For Field Crops
Er 99	Kabuli	Central Research Institute For Field Crops
Cagatay	Kabuli	Black Sea Agricultural Research Institute
Damla 89	Kabuli	Black Sea Agricultural Research Institute
Inci	Kabuli	East Mediterranean Sea Agricul. Research Institute
Aziziye	Kabuli	Eastern Anatolian Agricultural Research Institute
ILC482	Kabuli	GAP Int. Agricultural Research and Training Center
Diyar 95	Kabuli	GAP Int. Agricultural Research and Training Center
TAEK Sagel	Kabuli	Turkish Atomic Energy Authory
ILC 8617	Kabuli	Faculty of Agriculture, Akdeniz University
ICC 4958	Desi	Faculty of Agriculture, Akdeniz University
AWC 611	Wild	Faculty of Agriculture, Akdeniz University
AWC 612	Wild	Faculty of Agriculture, Akdeniz University

Genetic analysis of SSR

The 10 SSR primers selected in a consequence of a literature review were used for SSR analysis (Table 2). PCR products were diluted in sample loading solution (SLS) and standards from the Genome lab DNA standard Kit-400 were included. The amplified fragments were analyzed using a CEQ 8800 XL capillary DNA analysis system (Beckman Coulter, Fullerton, CA, USA) to ensure reproducibility. Allele sizes were determined for each SSR locus using the Beckman CEQ fragment analysis software. The number of alleles (n), allele frequency, expected (He) and observed (Ho) heterozygosity, estimated frequency of null alleles (r), probability of identity (PI), and presence of identical genotypes were determined for each locus using IDENTITY version 1.0 software. Microsat version 1.5 was used to calculate the proportion of shared alleles with the ps option [i.e. option 1-(ps)] selected to assess genetic dissimilarity. Data were then converted to a similarity matrix, and a dendrogram was constructed via the unweighted pair-group with arithmetic mean (UPGMA) method, using the Numerical Taxonomy and Multiware Analysis System (NTSYSpc) software, version 2.0.

Table 2. SSR primers information used in this study.

Locus	Primer sequence (5'–3')	Repeat motif	Length (bp)	Tm
NCPGR21F	TCTACCTCGTTTTTCGTGCC	$(CT)_{15}$	137	64
NCPGR21R	TTGCTCCTTCAACAAAACCC			
NCPGR27F	ACCCCATTTTTGGGTTTTCT	$(CA)_{12}$	278	63
NCPGR27R	TGCATCCAACTTGTTGTCTTG			
NCPGR28F	TGATGGAAGGTGATGTGGAA	$(AT)_6(GT)_5gc(GT)_7$	224	64
NCPGR28R	GAGGGGGAAACGTTTTCTTT			
NCPGR33F	ACATCTTGAAGTGCCCCAAC	$(GA)_{20}$	248	64
NCPGR33R	TGCAAGCAGACGGTTACAAG			
NCPGR34F	TGGAAGGTGTTTTAGTGGGTG	$(CT)_{17}$	240	64
NCPGR34R	GACTAACTGGCCCCCAAAA			
NCPGR36F	GTGGAGCCAAAAATCGACAT	$(GT)_{12}$	204	64
NCPGR36R	AACTTTATTTTCATTTGTCCATCAA			
NCPGR40F	TGAACGAATCATGGCAAGAG	$(GA)_{12}gt(GA)ca(GA)_3gtaagt$ $(GA)_4gtgg(GA)_{10}gt(GA)gt(GA)_6$	193	64
NCPGR40R	GCCCTCCTTCTTGCTTACAA			
NCPGR42F	CCCCTAGTAGCAAATATTTTGACC	$(CT)_{27}$	170	62
NCPGR42R	TTTGAATGCATTTCTTCATAGCA			
NCPGR93F	CAAAGTTTGTTGCTAGGATTC	$(CA)_2$ $(CT)_{24}$ $(CA)_{13}$	299	58
NCPGR93R	GAAGATCTCCGACGATGATA			
NCPGR100F	CCATTTTCTACAATCTCATGTCT	$(CT)_{15}N42(CT)_2cc(CT)_5tt$ $(CT)_6at(CT)_7$	263	59
NCPGR100R	GTAGAAAGAGCCAAGAGGCA			

RESULTS AND DISCUSSION

Ten SSR markers were used for 23 chickpea cultivars and 2 genotypes of *C. reticulatum* to determine genetic variation. A total of 58 alleles were detected at 10 loci. The number of alleles per locus ranged from 2 (NCPGR28-the least informative locus, PI: 0,806) to 11(NCPGR42-the most informative locus, PI: 0,074), with an average of 5.8 (Table 3). Some researchers have previously demonstrated genetic variation in chickpea using SSR markers. For instance, Choudhary et al. (2012) used 48 chickpea cultivars and detected 2 to 11 alleles and an average of 4.8 alleles per locus using 100 SSR markers. Saeed et al. (2011) obtained 2 to 13 alleles among 44 chickpea genotypes using 16 SSR markers while Singh et al. (2008) analyzed 21 chickpea cultivars using 18 SSR markers and detected 2 to 5 alleles. Huttel et al. (1999) used 22 SSR markers to determine genetic diversity in 4 chickpea genotypes and obtained 2 to 4 alleles whereas Singh et al. (2003) detected 1 to 4 alleles among 13 chickpea cultivars using 12 SSR markers. Data of allele numbers per locus, obtained during these studies, were either close to or less than our data. On the other hand, Naghavi et al. (2012), Seferaet al. (2011), Atalay and Babaoglu (2012) and Udupa et al. (1999) used SSR markers within chickpea genotypes and found 19.31, 10.5, 13.4 and 14.08 average alleles per locus, respectively. These values were higher than our findings. Utilization of different population sizes and also from different geographical origins and use of different SSR markers might be the reason of these differences. (Naghavi et al., 2012).

Table 3. Simple sequence repeat (SSR) loci, number of alleles (n), expected heterozygosity (He), observed heterozygosity (Ho), probability of identity (PI), and the frequency of null alleles (r) of 25 cultivars analyzed at 10 SSR markers.

SSR locus	n	He	Ho	PI	r
NCPGR36	4	0.36	0.36	0.482	0.006
NCPGR40	6	0.77	0.92	0.160	-0.082
NCPGR93	7	0.69	0.12	0.188	0.338
NCPGR28	2	0.11	0.04	0.806	0.065
NCPGR33	5	0.70	0.08	0.265	0.365
NCPGR34	5	0.62	0.04	0.366	0.358
NCPGR21	10	0.69	0.20	0.208	0.290
NCPGR42	11	0.84	0.60	0.074	0.134
NCPGR100	3	0.51	1.00	0.577	-0.316
NCPGR27	5	0.56	0.04	0.297	0.337
Total	58				
Average	5.8	0.58	0.34		

The mean heterozygosity (Ho) and expected heterozygosity (He) were determined to be 0.344 and 0.594, respectively (Table 3). Genetic information is quite similar in chickpea genotypes (Atalay and Babaoglu, 2012) and this may be the cause of low heterozygosity. Saeed et al. (2011), Atalay and Babaoglu (2012), Sethy et al. (2006) reported similar results to our data. Chickpea is a self-pollinated crop with low frequency of outcrossing

(Toker et al. 2006). The heterozygosity in self-pollinated crop is mainly resulted in the low level.

In the dendrogram from SSR analysis, it was observed that genotypes of *C. reticulatum* were found to be clearly different from cultivars of *C. arietinum*, and 23 cultivars of *Cicer arietinum* branched in four main groups (Figure 1). According to the dendogram, most of the genetic similarity values were between 0.55 and 0.85. These results may demonstrate that cultivated chickpea cultivars are closely related among themselves. Most of the cultivars may share the same genetic gene pools. This explains why yield improvement and increased tolerance to various biotic and abiotic stresses have been slow in chickpea. (Choudhardy et al., 2012)

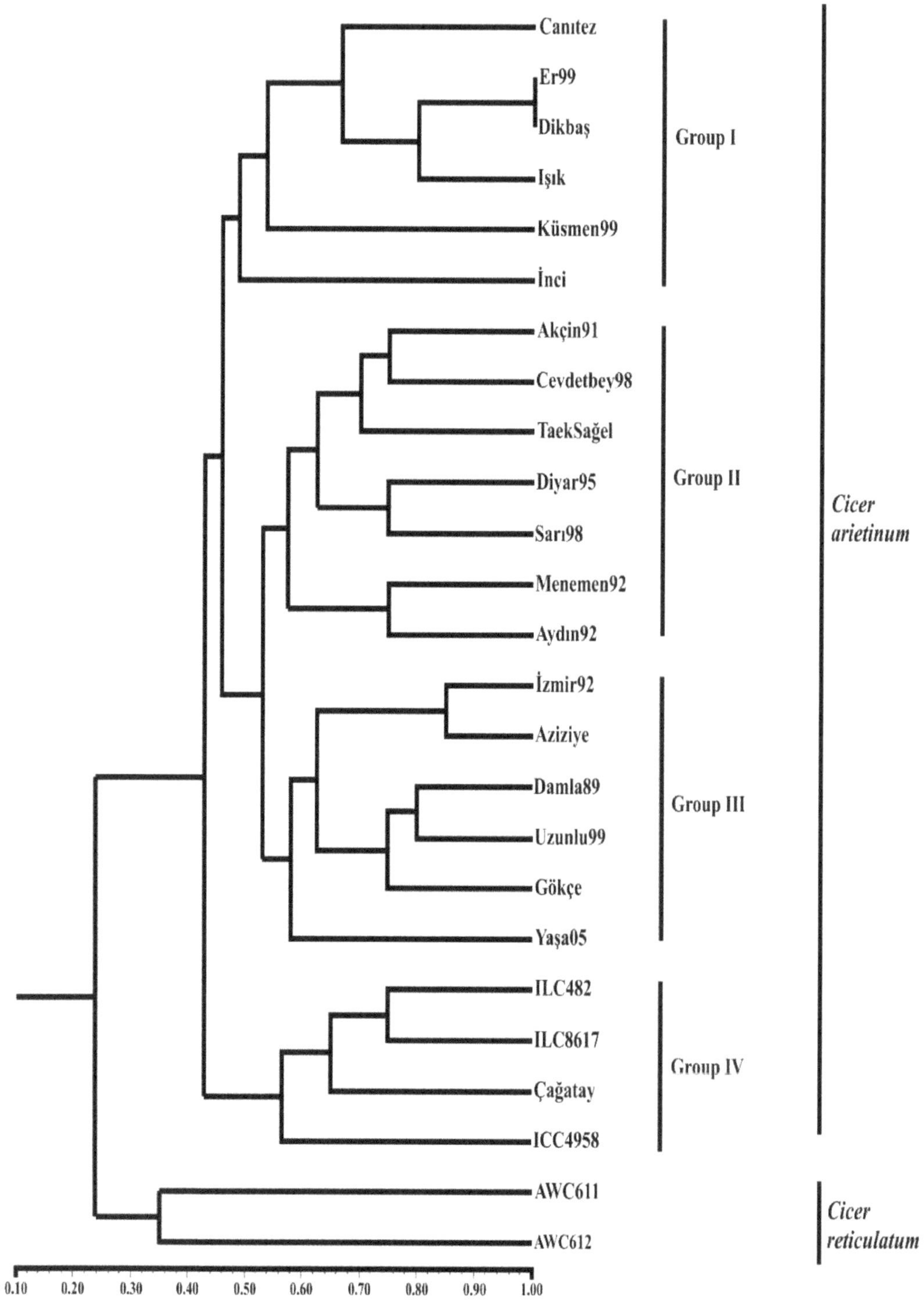

Figure 1. Genetic similarity (%) dendogram of chickpea cultivars used in this study.

For a long time, cultivated species have been selected and hybridized naturally and artificially; the wild species have almost not been used (Weiguo et al., 2007). Resistance to biotic and abiotic stresses has been found in wild *Cicer* species (Singh, 1998); therefore the use of resistant cultivars will help to stabilize chickpea yield (Tar'an et al., 2007). Selection for resistance to biotic and abiotic stresses has become important. The knowledge of genetic relationships between the cultivated chickpea and its wild relatives can be used for introgression of useful traits into the cultigen in plant breeding programs (Singh et al., 2008; Toker et al. 2013). Chickpea improvement programs are underway, but traditional breeding approaches are slow, and possibilities to introgress useful traits from wild germplasm are limited by crossability barriers (Smykal et al. 2014). Molecular markers have proven to be powerful tools in the assessment of genetic variation and in the elucidation of genetic relationships within and among species. Our result demonstrated that ICC 4958 (*C. arietinum*) which Radhika et al. (2007) indicated it as a drought tolerant cultivar was genetically the closest cultivar to ACW611 and AWC612 (*C. reticulatum*). Toker et al. (2013) reported that the hybrid vigor for yield and yield components in F_1s was higher in interspecific crosses than those of intraspecific crosses. Also, transgressive segregations in $F_{2\text{-}3}$s were higher in interspecific crosses than those of intraspecific ones. Their results have proven that genetic differences among *Cicer* species based on SSR markers have been powerful predictor for hybrid vigor in F_1s and fruitful segregations. Therefore, genetically different cultivars should be undertaken in breeding programs.

ACKNOWLEDGEMENTS

The authors would like to thank Prof. Dr. CengizToker and Dr. AbdulkadirAydogan for providing the plant material. This work was supported by the University of Mersin; Project number is BAP-FBE BB (SC) 2008-8 YL.

LITERATURE CITED

Atalay, E. and M. Babaoglu. 2012. Determination of genetic relationship in Turkish chickpea (*Cicer arietinum* L.) genotypes using SSR molecular markers and capillary electrophoresis. J Anim Plant Sci 22: 369-375.

Cagirgan, M.I., C. Toker, M. Karhan, M. Aksu, S. Ulger and H. Canci. 2011. Assessment of endogenous organic acid levels in Ascochyta blight (Ascochyta rabiei (Pass.) Labr.) susceptible and resistant Chickpeas (Cicer arietinum L.) Turk J Field Crops 16: 121-124.

Cagirgan, M.I., C. Toker, S. Ulger and M. Karhan. 2012. Determination of endogenous hormone levels in *Ascochyta blight* [*Ascochyta rabiei* (Pass.) Labr.] susceptible and resistant chickpeas (*Cicer arietinum* L.) Turk J Field Crops. 17: 5-9.

Canci, H. and C. Toker. 2009. Evaluation of yield criteria for drought and heat resistance in chickpea (*Cicer arietinum* L.). J Agron Crop Sci 195: 47–54.

Cevik, S., A. Yildizli, G. Yandim, H. Goksu, M.S. Gultekin, A. Guzel Deger, A. Celik, N. Simsek Kus and S. Unyayar. 2014. Some synthetic cyclitol derivatives alleviate the effect of water deficit in cultivated and wild-type chickpea species. J Plant Phys. 171: 807-816.

Choudhary, S., R. Gaur, S. Gupta and S. Bhatia, 2012. EST-derived genic molecular markers: development and utilization for generating an advanced transcript map of chickpea. TeorAppl Gen 124: 1449-1462.

Huttel, B., P. Winter, K. Weising, W. Choumane, F. Weigand and G. Kahl, 1999. Sequence-tagged microsatellite site markers for chickpea (*Cicer arietinum* L.). Genome 42: 210-217.

Jiang, G.L.2013. Molecular markers and Marker-assisted breeding in plants. In: Sven Bode Andersen, editor. Agricultural and Biological Sciences: Plant Breeding from Laboratories to Fields. 1st edition. Chapter 3. Brookings, USA.

Lefort, F. and K.A.R. Angelakis.2001. Genetic comparison of Greek cultivars of *Vitis vinifera* L. by nuclear microsatellite profiling. Am J EnolVitic 52: 101–108.

Naghavi, M.R., S, Rashidi Monfared and G. Humberto.2012. Genetic diversity in Iranian Chickpea (*Cicer arietinum* L.) landraces as revealed by microsatellite markers. Czech J Gen Plant Breed 48: 131–138.

Nguyen, T.T., P.W.J. Taylor, R.J. Redden and R. Ford. 2004. Genetic diversity estimates in *Cicer*using AFLP analysis. Plant Breed 123: 173-179.

Ozalkan, C., H.T. Sepetoglu, I. Daur and O.F. Sen, 2010. Relationship between some plant growth parameters and grain yield of chickpea (Cicer arietinum L.) during different growth stages. Turkish J Field Crops, 15 (1): 79-83.

Radhika, P., S.J.M. Gowda, N.Y. Kadoo, L.B. Mhase, B.M. Jamadagni, M.N. Sainani, S. Chandra and V.S. Gupta. 2007. Development of an integrated intraspecific map of chickpea (*Cicer arietinum* L.) using two recombinant inbred line populations. Theor Appl Gen115: 209-216.

Rao, L.S., P.U. Rani, P.S. Deshmukh, P.A. Kumar and S.K. Panguluri.2007. RAPD and ISSR fingerprinting in cultivated chickpea (*Cicer arietinum*L.) and its wild progenitor *Cicer reticulatum* Ladizinsky. Genet Res Crop Evol54: 1235-1244.

Saeed, A,, R. Darvishzadeh and A. Basirnia. 2013. Simple sequence repeat markers associated with agro-morphological traits in chickpea (*Cicer arietinum* L.). Zem-Agri 100: 433–440.

Sefera, T., B.Abebie, P.M. Gaur, K. Assefa and K.V. Rejeev. 2011. Characterization and genetic diversity analysis of selected chickpea cultivars of nine countries using simple sequence repeat (SSR) markers. Crop Pasture Sci62: 177-187.

Sepetoglu, H.T., M. Altinbas and I. Daur. 2008. Uptake of some essential nutrients in chickpea during different growth stages in relation to biomass yield. Turk J Field Crops 13: 1-11.

Serret, M.D., S.M. Udupa and F. Weigand.1997. Assessment of genetic diversity of cultivated chickpea using microsatellite-drived RFLP markers: Implications for origin. Plant Breed 116: 573-578.

Sethy, N.K., B. Shokeen, K.J. Edwards and S. Bhatia.2006. Development of microsatellite markers and analysis of intraspecific genetic variability in chickpea (*Cicer arietinum* L.). Theor Appl Gen 112: 1416-1428.

Singh, K.B., B. Ocampo and L.D. Robertson. 1998. Diversity for abiotic and biotic stress resistance in the wild annual Cicer species. Gen. Resources and Crop Evolution 45: 9-17.

Singh, R., C.D. Prasad, V. Singhal and G.J. Randhawa. 2003. Assessment of genetic diversity in chickpea cultivars using RAPD, AFLP and STMS markers. J Ge Breed 57: 165-174.

Singh, R., V. Singhal and G.J. Randhawa. 2008. Molecular analysis of chickpea (*Cicer arietinum*L.) cultivars using AFLP and STMS markers. J P Bioch Biotech 17: 167-171.

Sudupak, M.A., M.Akkaya and A. Kence. 2002. Analysis of genetic relationships among perennial and annual *Cicer* species growing in Turkey using RAPD markers. TheorAppl Gen 105: 1220-1228.

Sudupak, M.A., M. Akkaya and A. Kence. 2004. Genetic relationships among perennial and annual *Cicer* species growing in Turkey assessed by AFLP fingerprinting. TheorAppl Gen 108: 937-944.

Sudupak, M.A. 2013. SSR-based genetic diversity assessment of Turkish Chickpea varieties. Biotechnology & Biotechnological Equipment 27: 4087-4090.

Tar'an, B., T. Warkentin, A. Tulu and A. Vandenberg. 2007. Genetic relationships among Chickpea (*Cicer arietinum* L.) genotypes based on the SSRs at the quantitative trait Loci for resistance to Ascochyta Blight. Eur J of Plant Patho 51: 119-139.

Toker, C., H. Canci and F.O. Ceylan. 2006. Estimation of outcrossing rate in chickpea (*Cicer arietinum* L.) sown in autumn. Euphytica 151: 201-205.

Toker, C., H. Canci, N.E. İnci, F.O. Ceylan, B. Uzun, S. Sonmez, S. Citak, and C. Ikten. 2012. Pyramiding of the resistance to fe-deficiency chlorosis and leaf miner (*Liriomyza cicerina* ROND.) in chickpea (*Cicer arietinum* L.) by mutation breeding. Turk J Field Crops 17: 41-45.

Toker, C., C. Ikten, F.O. Ceylan, E. Bolucek and B. Uzun. 2013. Genetic relationship between transgressive segregations and genetic distance based on SSR markers in cicer species. Current opinion in biotechnology 24: 38-39.

Udupa, S.M., L.D. Robertson, F. Weigand, M. Baum and G. Kahl. 1999. Allelic variation at $(TAA)_n$ microsatellite loci in a world collection of chickpea (*Cicer arietinum*L.) germplasm. Mol Gen Genet 261: 354-363.

Weiguo, Z., Z. Zhihua, M. Xuexia, Z. Yong, W. Sibao, H. Jianhua, X. Hui, P. Yile and H. Yongping.2007. A comparison of genetic variation among wild and cultivated *Morus* Species (Moraceae: *Morus*) as revealed by ISSR and SSR markers. Biodi and Con 16: 275-290.

38

HIGH LEVELS OF GENETIC VARIATION AS DETECTED BY AFLP IN *Sideritis tmolea* FROM WESTERN TURKEY

Seda NEMLI[1], *Umit SUBASI*[2], *Volkan EROGLU*[2], *Serdar Gokhan SENOL*[2], *Muhammed Bahattin TANYOLAC*[*1]

[1]*Department of Bioengineering, Ege University, Bornova-Izmir, TURKEY*
[2]*Botanical Garden & Herbarium Research and Application Center, Ege University, Bornova-Izmir, TURKEY*
[*]*Corresponding author: bahattin.tanyolac@ege.edu.tr*

ABSTRACT

The objective of this study was to evaluate the level of genetic diversity among 29 *S. tmolea* genotypes collected from Bozdag, Turkey, using the Amplified Fragment Length Polymorphism (AFLP) technique. A total of 392 AFLP markers were detected using 14 primer combinations. The number of polymorphic bands per AFLP primer combination ranged from 16 to 42, with an average of 28. The Polymorphism Information Content (PIC) values varied from 0.25 (M-CTC/E-AGC) to 0.72 (M-CAC/E-AAG) among 14 selective primers. The genetic dissimilarity that was detected using the NTSYS-PC software ranged from 0.09 to 0.82. According to the dissimilarity results, a high level of genetic diversity existed among the studied genotypes. A model-based structural analysis revealed the presence of 2 populations. The defined population structure was helpful when studying the *S. tmolea* genotypes for diversity and classification.

Keywords: AFLP, genetic diversity, population structure, *Sideritis tmolea*

INTRODUCTION

Sivri çay *(Sideritis tmolea* L.) belongs to the Lamioideae (Stachyoideae) subfamily (Ryding, 1994; Hickey and King, 1997). Although taxa belonging to Lamiaceae Lindl. (Labiatae Juss.) are spread over a large area in habitat ranging from the North Pole to the Himalayas and from South eastern Asia to Hawaii, Australia, Africa and America, they are mainly distributed in the Mediterranean basin (Heywood, 1996). This family, which has 224 types and almost 5600 species, is divided into eight subfamilies according to Cantinoya. *Sideritis* L. is distributed especially in Mediterranean basin, is represented by more than 150 species in a wide area from the Bahamas to China and from Germany to Morocco and is divided into two subgenera. While Lamiaceae in the flora of Turkey is represented by 565 genera and 735 taxa (Guner et al., 2000), *Sideritis* is represented by 3 sections: Hesiodia (Moench) Bentham, Burgsdorfia (Moench) Briquet and Empedoclia (Rafin) Bentham. In Turkey, according to Duman (2013), *Sideritis* is represented by 53 taxa belonging to 45 species. While 40 of these taxa are endemic, the endemism ratio is approximately 75%. In Turkey, there are 31 *Sideritis* taxa, of which 25 are endemic and used in domestic and foreign trade, as well as *S. sipylea*, *S. tmolea* and *S. trojona*, which are endemic and among the first 50 species that are in the danger of extinction due to unconscious collecting (Ozhatay et al., 1998).

There are a number of studies of the pharmaceutical compounds in *Sideritis* species and these studies especially have focused on essential oils, diterpenes and flavonoids (Akcos et al., 1999; Kirimer et al., 2001; Topcu et al., 2002). Various anti-inflammatory, analgesic, diuretical, antiulcer, antidepressant, antimicrobial and insect repellent effects of *Sideritis* species that are grown in Turkey have been studied (Ozturk et al., 1996; Akcos et al., 1999; Bondi et al., 2000). *Sideritis* species have an important place among other plants both as an herbal tea and household remedy. While *Sideritis* species are known under various names in Anatolia, they are usually referred to as "Mountain Tea", "Plateau Tea" and "Sage". *S. tmolea* is known as "Acute Tea".

An analysis of genetic variation between individuals within a species or between different species or populations is useful in identifying the level of genetic variation (Brummer et al., 1995). These diversity studies provide useful information for understanding the genetic bases of various gene pools, for describing and conserving their germplasm and for correctly identifying the varieties in facilitating the selection of sources for new genes for improved yield and quality improvements (Cho et al., 2008). Traditionally, morphological traits, karyotypes and protein and isozyme markers have been widely used to assess genetic diversity (Mudibu et al., 2011). Therefore, the measurement of genetic variation based on morphological characteristics is time-consuming and incoherent and requires extensive field trials and evaluation (Astarini et al., 2004). Different DNA techniques (RFLP, RAPD, AFLP, SSR, CAPS, ISSR and SNP) that are not affected by the environment (Maciel et al., 2003) are necessary for genetic screening among categories of closely related species in germplasm banks (Lapitan et al., 2007). The major applications of DNA markers for plant genetics include the assessment of phylogeny in selecting suitable parents for the generation of heterosis (Biton et al., 2012). Thus, several researchers have developed a wide range of molecular markers that have been used in several genetic diversity studies (Shimomura and Hirashima, 2006; Martins-Lopes et al., 2008; Cui et al., 2010). Among the different molecular marker approaches, AFLP is one of the most widely used in various plant species with different degrees of relatedness (Witkowicz et al., 2003). AFLP is suitable for studying diversity. The AFLP technique is based on the amplification of short restriction endonuclease-digested genomic DNA fragments onto which adaptors have been ligated at both ends using Polymerase Chain reaction (PCR) (Saiki et al., 1988). The presence or absence of these selective nucleotides in the genomic fragments being amplified provides the polymorphism. Alternately, the AFLP technique relies on a system of dominant markers that allow for the simultaneous analysis of a large number of markers in the genome. This method is highly reproducible and can be used to survey the overall genetic differences in the nuclear genome in a single assay without the need for primary sequence knowledge to design primers (Vuylsteke et al., 1999).

The objective of this research was to investigate the level of molecular diversity and phylogenetic relationships at the AFLP marker level among epidemic *Sideritis* that were collected from Bozdag.

MATERIALS AND METHODS

The collected S. tmolea species

The material of the current study is *S. tmolea* P.H. Davis by Lamiaceae. *S. tmolea* is distributed on the West side of Bozdag and on the slopes of the North and Northwest sides of Kangal Mount, which is a part of Bozdağlar at an altitude of 1450-2100 m. DNA samples were collected from the Bozdag Mountains; a population of individuals from different points was sampled in at least 10 - m intervals and GPS data for all of the samples was recorded (Table 1).

Table 1. Geographic distribution of *S. tmolea* landraces sampled in this study.

Genotype number	Altitude (m*)	GPS*	Genotype number	Altitude (m*)	GPS*
1	1654	38°16' 06" N* 028°04'52" E*	**16**	1564	38°201' 14" N 028°07'22" E
2	1695	38°19' 12" N 028°04'58" E	**17**	1615	38°20' 12" N 028°07'21" E
3	1747	38°19' 16" N 028°04'58" E	**18**	1566	38°20' 47" N 028°08'27" E
4	1651	38°19' 44" N 028°06'53" E	**19**	1566	38°20' 47" N 028°08'27" E
5	1703	38°19' 37" N 028°06'49" E	**20**	1555	38°20' 49" N 028°08'27" E
6	1857	38°19' 29" N 028°06'42" E	**21**	1980	38°19' 27" N 028°05'39" E
7	1967	38°19' 26" N 028°06'23" E	**22**	1960	38°01' 24" N 028°05'36" E
8	1967	38°19' 26" N 028°06'23" E	**23**	1960	38°01' 24" N 028°05'36" E
9	1709	38°19' 48" N 028°06'30" E	**24**	1920	38°19' 23" N 028°05'29" E
10	1669	38°19' 40" N 028°06'32" E	**25**	1920	38°19' 23" N 028°05'29" E
11	1597	38°20' 34" N 028°06'34" E	**26**	1786	38°19' 46" N 028°07'20" E
12	1597	38°20' 34" N 028°06'34" E	**27**	1751	38°19' 48" N 028°07'17" E
13	1641	38°20' 59" N 028°05'47" E	**28**	1664	38°19' 49" N 028°07'12" E
14	1534	38°21' 19" N 028°05'39" E	**29**	1588	38°19' 52" N 028°07'60" E
15	1534	38°21' 19" N 028°05'39" E			

*N: North, E: East, m: meter, GPS: Global Positioning System

DNA isolation

The young leaves from *Sideritis* were ground to a fine powder with liquid nitrogen in a TissueLyser (Technogen Co. Izmir Turkey). The total genomic DNA was isolated according to the protocol of Doyle and Doyle (1987) with some modifications. The purified DNA was resuspended in 100 µL of 10 mM Tris-HCl (pH 8.0) buffer containing 1 mM EDTA for use in the analysis. The DNA concentration of the sample was determined spectrophotometrically by measuring the absorbance at 260 and 280 nm using a NanoDrop ND-1000 (Thermo Sci. Co) spectrophotometer and the quality was checked on 1% agarose gels. The stock DNA was diluted to a working solution of 200 ng/µL and the samples were stored at -20 °C for the PCR reactions.

AFLP analysis

An amplified fragment length polymorphism analysis was performed using the Li-Cor AFLP Kit (catalog number: 830-06195 AFLP2-DYE Selective Amplification Kit, Lincoln, NE, USA). followed the manufacturer's instructions. According to the kit, genomic DNA (200 ng/µL) was digested with the *Eco*RI and *Mse*I restriction enzymes at 37 °C for 2 h followed by 15 min at 70 °C. The digested DNA fragments were then ligated with specific enzyme adaptors in the presence of T4 DNA ligase. A ligation reaction was performed at 37 °C for 3 h. Pre-amplification PCR was performed after diluting the ligated DNA ten-fold with double deionized water and then pre-amplifying using *Eco*RI and *Mse*I non-selective primers with one additional selective nucleotide at the 3′ end of the *Mse*I primer (*Mse*I + C) and the *Eco*RI primer (*Eco*RI + A). The primer combinations were named according to the restriction enzyme initials, such as M-CAC/E-GGA: "M" stands for *Mse*I, "CAC" stands for the nucleotide extensions and "E" stands for the *Eco*RI enzyme. After the pre-selective PCR, the reaction mixture was diluted 1:40 and used as a template for selective amplification. Selective amplifications were performed with the selective primer combinations of *Eco*RI (labeled with two different fluorescent dyes (IRD 700 or 800) at the 5′ ends) and *Mse*I (unlabeled) with three selective nucleotides. Fourteen primer combinations were used to screen for polymorphisms among the samples (Table 2). The AFLP products were separated electrophoretically in denaturing 8% polyacrylamide gels (19:1 ratio acrylamide:bisacrylamide, 7,5 M urea and 1X TBE buffer) at 45 W constant power for approximately 4 h 30 min and run on a Li-Cor 4300s DNA Analyzer. The size of the products was determined using the Li-Cor® IRDye® as an internal size standard (50 to 700 bp).The imaging of the AFLP fragments was performed using the SAGA software.

Table 2. Number of polymorphic bands and PIC for each AFLP primer pair that was used in the analysis of the 29 *S. tmolea* genotypes

Primer Number	Primer Pairs	Number of polymorphic bands	PIC
1	M-CAA / E-AAG	19	0.54
2	M-CAA / E-ACT	20	0.56
3	M-CAC / E-AAG	19	0.72
4	M-CAC / E-ACT	23	0.57
5	M-CTT / E-AAC	31	0.41
6	M-CTT / E-AGG	29	0.34
7	M-CTC / E-ACA	23	0.42
8	M-CTC / E-AGC	16	0.25
9	M-CTC / E-AAG	27	0.43
10	M-CTC / E-ACT	22	0.39
11	M-CTG / E-ACA	40	0.56
12	M-CTG / E-AGC	41	0.52
13	M-CTG / E-AAG	42	0.64
14	M-CTG / E-ACT	40	0.48
	TOTAL	**392**	
	AVERAGE	**28**	**0.49**

Data analysis

The AFLP profiles were assessed individually for each primer pair combination. The polymorphic AFLP bands were manually scored as a binary data matrix with presence as "1" and absence as "0" across all of the 29 genotypes. Only the clear and strong bands were recorded and used for the analysis. Genetic dissimilarity estimates were calculated between the genotypes using Jaccard's coefficient of dissimilarity. The calculations and analyses were generated with the help of NTSYS-PC software version 2.02 (Rohlf, 1998). The PIC of each marker was calculated using $PIC = 1 - \Sigma pi^2$, where P_i is the band frequency of the ith allele (Smith et al., 1997).

Assessment of population structure with AFLP

All of the AFLP markers that were developed in this study were used to infer the population structure in the set of 29 genotypes. For the analysis of population structure, we used the STRUCTURE version 2.3.4 software, which implements a Bayesian, model-based clustering algorithm (Pritchard et al., 2000). The individuals in the sample are assigned to populations (clusters) or jointly to additional populations if their genotypes indicate that the populations are admixed. The optimal number of populations (K) was determined under a burn-in period of 100,000 with 100,000 Markov Chain Monte Carlo iterations. The membership of each genotype was run for the range of genetic clusters from value of K = 1 to 10 with the admixture model and for each K, the run was replicated 10 times. For each K value, the runs showing the highest posterior probability of data were considered. The true value of K and the number of populations were detected by an ad hoc quantity based on the second order rate of change of the likelihood function with respect to K (ΔK) (Evanno et al., 2005).

RESULTS AND DISCUSSION

In the current study, we detected the genetic diversity and phylogenetic relationships of *S. tmolea* at the AFLP marker level. The size range between the smallest and largest alleles that were observed for a given AFLP varied from 50 to 500 bp, which was able to discriminate between the 29 *Sideritis* genotypes. Similar results were observed by Murtaza (2006) and Strikic et al. (2010), who found band sizes ranging from 50 to 500 bp using cotton and olive genotypes, respectively. AFLP marker systems were highly effective in discriminating the 29 *S. tmolea* genotypes that were analyzed. The total number of assays included 14 primer combinations for AFLP, as listed in Table 2. In total, the 14 selective primer combinations amplified 392 polymorphic products for the 29 analyzed genotypes (Table 2). Previously, AFLP marker systems were used to characterize various genotypes (Maras et al., 2008; Strikic et al., 2010; Nemli, 2013; Panahi et al., 2013). For example, Grati-Kamoun et al. (2006) found 172 polymorphic bands from 9 primer combinations in olive genotypes. Khalighi et al. (2008) obtained 387 bands from 14 primer combinations. Colomba and Gregorini (2011) studied durum wheat using 8 AFLP primer pairs that generated 137 amplification products and 17 polymorphic bands per primer combination. Farah Fazwa et al. (2013) detected a total of 170 AFLP fragments among 62 accessions for two varieties of *L. pumila* (*L. pumila* var. *pumila* and *L. pumila* var. *alata*). From the studies that are mentioned above, the authors concluded that AFLP was a very sensitive technique for detecting markers for genetic studies. Based on the results of our study, 14 AFLP primer pairs amplified in the 29 *S. tmolea* genotypes from Turkey, the number of polymorphic bands was higher than that of previous studies (Colomba and Gregorini, 2011; Farah Fazwa et al., 2013). These findings indicate a high level of genetic variation among the studied genotypes. The number of polymorphic bands per AFLP primer combination ranged from 16 (M-CTC / E-AGC) to 42 (M-CTG / E-AAG). On average, 28 polymorphic bands were amplified by each primer combination (Table 2). These results are similar to those of this study in terms of the average number of scored polymorphic bands. Khalighi et al. (2008) found the average number of 27 in *Triticum* spp. and *Aegilops* spp. and Zhang et al. (2012) reported 31 polymorphic bands per primer combination from their AFLP study in *Brachypodium* species.

The PIC is defined as the probability that two alleles that are taken at random from a population can be distinguished using the marker in question and is a measure of allele diversity at a locus. The success of AFLP markers in a variety of genotypes that were corrected with the PIC value indicates the genetic variation among the genotypes (Muse et al., 2005). The PIC was calculated to identify the most informative AFLP primers. The PIC value ranged from 0,25 (M-CTC / E-AGC) to 0,72 (M-CAC / E-AAG) among the 14 selective primers. The average PIC was calculated as 0,49, demonstrating the good discriminatory power of the identified markers (Table 2). Furthermore, except for M-CTC/E-AGC (PIC value: 0,25), all of the AFLP combinations were highly informative, with an average PIC of 0.6 (Table 2). However, higher mean PIC values were obtained by AFLP markers compared to those of other studies, such as Yuan et al. (2011) (0,33) and Zhang et al. (2012). Pecina-Quintero et al. (2012) reported a mean PIC value of 0,294 with 10 primer combinations using 41 genotypes of the sweet sorghum germplasm. The same AFLP markers were used by Pecina-Quintero et al. (2012) to produce a lower PIC value (0,27) for the M-CAC/E-AAG and M-CTG/E-ACA primer combinations. However, in this study, the PIC values were 0,72 and 0,56 for the M-CAC/E-AAG and M-CTG/E-ACA primer combinations, respectively. Therefore, the AFLP combinations that were used in this study have a high discriminatory power and are recommended for use in genetic diversity analyses of *S. tmolea*.

Understanding the genetic similarity/dissimilarity among genotypes is crucial in germplasm collection and also facilitates the thorough utility of these genetic resources in crop improvement programs (Farah Fazwa et al., 2013). The genetic distance of the improved *Sideritis* genotypes is shown in Table 3. The 0/1 matrix was used to calculate the distance among the 29 *Sideritis* genotypes according to Jaccard (1908). The matrix of genetic dissimilarity ranged from 0,09 to 0,82. Among these values, the highest degree of dissimilarity (0,82) was found between genotypes 13 and 23, indicating that these genotypes were the most distinct from each other. The lowest degree of dissimilarity (0,09) that was revealed by the AFLP analysis was found between genotypes 27 and 28. In this study, a higher AFLP polymorphism was observed among the genotypes than in several previous reports using AFLP markers. The genetic distance values demonstrated a large genetic variation among the studied genotypes in a narrow geographic region. A high level of genetic variation was found among the genotypes that may be attributed to habitat fragmentation due to anthropopression, geographic isolation and genetic drift (Qian et al., 2013). Similarly, Cinar et al. (2009) used 8 *Sideritis* species for genotyping and reported that the genetic distance varied from 0,284 (between *Sideritis arguta* 1370 and *Sideritis arguta* 778) to 0,903 (between

Table 3. Genetic dissimilarity coefficient matrices of the 29 genotypes of *S. tmolea* using AFLP markers (Jaccard's coefficient)

Genotype number	01	02	03	04	05.	06	07	08	09	10	11	12	13	14	15	16	17	18	19	20	21	22	23	24	25	26	27	28	29
01	0.00																												
02	0.29	0.00																											
03	0.40	0.23	0.00																										
04	0.48	0.59	0.56	0.00																									
05	0.33	0.49	0.49	0.36	0.00																								
06	0.33	0.24	0.26	0.46	0.45	0.00																							
07	0.31	0.24	0.30	0.49	0.44	0.23	0.00																						
08	0.55	0.50	0.45	0.52	0.42	0.43	0.45	0.00																					
09	0.40	0.35	0.39	0.48	0.36	0.36	0.40	0.41	0.00																				
10	0.43	0.47	0.56	0.46	0.45	0.47	0.44	0.56	0.36	0.00																			
11	0.52	0.33	0.39	0.71	0.62	0.41	0.43	0.51	0.34	0.44	0.00																		
12	0.45	0.30	0.35	0.67	0.55	0.35	0.37	0.51	0.32	0.37	0.15	0.00																	
13	0.56	0.37	0.39	0.71	0.62	0.41	0.44	0.56	0.39	0.45	0.11	0.18	0.00																
14	0.56	0.40	0.41	0.71	0.63	0.43	0.45	0.56	0.39	0.47	0.14	0.21	0.11	0.00															
15	0.41	0.45	0.47	0.60	0.50	0.44	0.44	0.51	0.35	0.33	0.31	0.29	0.34	0.34	0.00														
16	0.51	0.34	0.40	0.73	0.64	0.44	0.40	0.44	0.44	0.45	0.16	0.21	0.18	0.22	0.32	0.00													
17	0.47	0.37	0.43	0.69	0.58	0.43	0.44	0.53	0.41	0.45	0.16	0.20	0.18	0.18	0.31	0.13	0.00												
18	0.40	0.49	0.52	0.47	0.49	0.47	0.47	0.52	0.41	0.34	0.47	0.42	0.50	0.50	0.37	0.50	0.47	0.00											
19	0.40	0.51	0.52	0.48	0.45	0.52	0.49	0.49	0.39	0.41	0.46	0.38	0.47	0.48	0.35	0.47	0.47	0.32	0.00										
20	0.50	0.58	0.63	0.56	0.58	0.61	0.52	0.65	0.52	0.44	0.53	0.46	0.55	0.52	0.42	0.57	0.52	0.38	0.37	0.00									
21	0.49	0.68	0.72	0.45	0.44	0.64	0.57	0.58	0.60	0.49	0.79	0.66	0.81	0.80	0.54	0.77	0.72	0.49	0.38	0.44	0.00								
22	0.41	0.65	0.68	0.44	0.42	0.65	0.60	0.56	0.48	0.45	0.73	0.65	0.79	0.80	0.52	0.76	0.68	0.43	0.38	0.42	0.32	0.00							
23	0.61	0.82	0.79	0.49	0.57	0.81	0.73	0.66	0.60	0.52	0.80	0.68	0.83	0.79	0.54	0.83	0.80	0.48	0.43	0.43	0.38	0.33	0.00						
24	0.45	0.47	0.47	0.56	0.51	0.49	0.45	0.62	0.43	0.42	0.39	0.38	0.42	0.44	0.40	0.45	0.41	0.43	0.44	0.42	0.50	0.45	0.56	0.00					
25	0.49	0.42	0.49	0.65	0.56	0.44	0.41	0.55	0.42	0.41	0.31	0.32	0.33	0.35	0.33	0.28	0.28	0.41	0.45	0.51	0.63	0.60	0.71	0.29	0.00				
26	0.52	0.73	0.71	0.40	0.46	0.67	0.60	0.45	0.52	0.43	0.72	0.63	0.77	0.76	0.47	0.72	0.71	0.45	0.36	0.43	0.37	0.35	0.33	0.49	0.60	0.00			
27	0.51	0.42	0.43	0.75	0.62	0.45	0.49	0.58	0.45	0.47	0.24	0.28	0.21	0.22	0.38	0.26	0.23	0.45	0.49	0.55	0.73	0.71	0.79	0.40	0.28	0.74	0.00		
28	0.52	0.38	0.43	0.75	0.61	0.46	0.49	0.52	0.47	0.48	0.25	0.30	0.25	0.26	0.41	0.28	0.26	0.48	0.49	0.57	0.73	0.72	0.81	0.46	0.33	0.70	0.10	0.00	
29	0.48	0.42	0.49	0.52	0.52	0.45	0.44	0.57	0.40	0.37	0.34	0.32	0.37	0.35	0.34	0.41	0.33	0.38	0.41	0.42	0.59	0.55	0.61	0.38	0.38	0.53	0.31	0.33	0.00

Sideritis perfoliata 20 and *Sideritis perfoliata* 47). These results indicate a high degree of genetic diversity among different species. We calculated a high degree of genetic diversity within *S. tmolea* samples that were collected in a narrow area of existing endemic species. Furthermore, the reproductive system directly influences the degree and distribution of genetic variation and can lead to reproductive isolation and eventually to speciation (Gonza´lez-Pe´rez et al., 2008). The level of total heterozygosity in outcrossing plants was higher than that from comparable analyses of self-pollinating plants (Nybom and Bartish, 2000). The high genetic diversity may be an indication that the number of individuals of this species was larger than previously studied (Prohens et al., 2007). The species in the genus *Sideritis* are distributed throughout the world and the distribution of these species could have once been much wider than it is currently, which could explain the high levels of genetic variation that were detected in this endemic species.

Twenty-nine *Sideritis* genotypes were analyzed in order to determine the population structure using the STRUCTURE. After running the K value, the curve peaked at ΔK = 2 for the final analysis (Figure 1), indicating that the 2 population, namely Population 1 (red) and Population 2 (green), were the most capable of explaining the 29 *Sideritis* genotypes. Population 1 consists of 16 accessions (genotypes 1, 4, 5, 6, 7, 8, 9, 10, 18, 19, 20, 21, 22, 23, 24 and 26) showed admixture levels at 1555, 1703, 1857 and 1980 m. The remaining 13 accessions from seemingly similar levels, ranging from 1500-1700 (m) except for genotype 25 (1920 m), were classified as Population 2 by the STRUCTURE version 2.3.4 software (Figure 1). The genotypes collected from same location did not form into one cluster. The differences among the genotypes may be explained by the assessment of certain morphological characteristics.

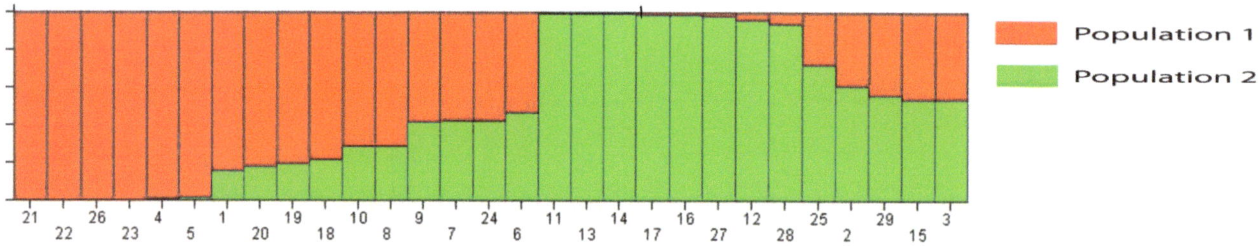

Figure 1. Bar plot showing the genetic diversity structure for the 29 *S. tmolea* genotypes using the program STRUCTURE version 2.3.4. Each population is represented by a different color as listed: red (POP1) and green (POP2).

In conclusion, the genus *Sideritis* provides a wide range of research possibilities due to its richness in compounds. This study is the first report of the use of the AFLP technique to evaluate the genetic relationships among *S. tmolea* genotypes and provides information on the population structure. A wide degree of genetic diversity is important for the characterization and development of improved varieties. Some *Sideritis* species are very difficult to distinguish morphologically due to lack of ecotype variation; therefore, this study demonstrates that the AFLP technique could play an important role in the identification of closely related taxa in *Sideritis* for future germplasm collection efforts in a cost-effective way. These preliminary results suggest that AFLP can be used in marker-assisted parental selection for future genetic diversity in genotypes while introgressing the desirable characters. The high levels of genetic variation that were observed between genotypes in *S. tmolea* indicate that management should aim to conserve as many of the small populations as possible in this study.

ACKNOWLEDGEMENTS

The authors would like to thank the Scientific and Technological Research Council of Turkey (TUBITAK Project No: 108T851) and the Ege University Science and Technology Centre (EBILTEM) for supporting this study (Project No: 2010BIL03).

LITERATURE CITED

Akcos, Y., N. Ezer, I. Calıs, R. Demirdamar, B.C. Tel. 1999. Polyphenolic compounds of *Sideritis lycia* and their anti-inflammatory activity. Pharm Biol. 36: 1-5.

Astarini, A.I., A.J. Plummer, A.R. Lancaster and G. Yan. 2004. Fingerprinting of cauliflower cultivars using RAPD markers. Aust J Agri Res. 55: 112-124.

Baser, K.H.C., G. Honda and W. Miki. 1986. Herb Drugs and Herbalists in Turkey- Studia Culturae Islamicae 27, Institute for the Study of Languages and Cultures of Asia and Africa, Tokyo.

Biton, I., S. Shevtsov, O. Ostersetzer, Y. Mani, S. Lavee, B. Avıdan and G. Ben-ari, 2012. Genetic relationships and hybrid vigour in olive (*Olea europaea* L.) by Microsatellites. Plant Breeding. 131: 767-774.

Bondi, M.L., M. Bruno, F. Piozzi, K.H.C. Baser and M.S.J. Simmonds. 2000. Diversity and antifeedant activity of diterpenes from Turkish species of *Sideritis*. Biochem Sys Ecol. 28: 299-303.

Brummer, E.C., J.H. Bouton and G. Korchert. 1995. Analysis of annual *Medicago* species using RAPD markers. Genome. 38: 362-367.

Cho, G-T., J. Lee, J-K. Moon, M-S. Yoon, H-J. Baek, J-H. Kang, T-S. Kim and N-C. Paek. 2008. Genetic diversity and population structure of Korean soybean landrace [*Glycine max* (L.) Merr.]. J Crop Sci Biotech. 11(2): 83-90.

Cinar, A., S. Elmasulu, A.G. Ince, M. Karaca, A.N. Onus and K. Turgut. 2009. A suitable DNA marker technique (minisatellites) for *Sideritis* (Lamiaceae) genotyping studies. Acta Horticulturae. 826(1): 439-446.

Colomba, M.S. and A. Gregorini. 2011. Genetic diversity analysis of the durum wheat Graziella Ra, *Triticum turgidum*

L. subsp. durum (Desf.) Husn. (Poales, Poaceae). Biodiversity Journal. 2(2): 73-84.

Cui, H., K.T. Moe, J.W. Chung, Y.I. Cho, G.A. Lee and Y.J. Park. 2010. Genetic diversity and population structure of rice accessions from South Asia using SSR markers. Korean Journal of Breeding Science. 42(1): 11-22.

Davis, P.H. 1978. Flora of Turkey and the East Aegean Islands. Vol. 6, Edinburgh.

Doyle, J.J. and J.L .Doyle. 1987. A rapid DNA isolation procedure for small quantities of fresh leaf tissue. Phytochem Bull. 19: 11–15.

Duman, H. 2013. *Sideritis*. ed. Güner, A., S. Aslan, T. Ekim, M. Vural and M.T. Babaç, Nezahat Gokyigit Botanik Bahçesi ve Flora Araştırmaları Derneği Yayını, Istanbul.

Evanno, G., S. Regnaut and Goudet, J. 2005. Detecting the number of clusters of individuals using the software STRUCTURE: A simulation study. Molecular Ecology. (14)8: 2611-2620.

Farah Fazwa, M.A., H. Siti Salwana, H. Maideen and O. Mohamad. 2013. An Assessment of Genetic Relationship Among Superior Accessions of *Labisia Pumila* Analized by Amplified Fragment Length Polymorphism (AFLP) Markers. Open Science Repository Agriculture. Online(open-access). doi:10.7392/Agriculture.70081945.

Grati-Kamoun, N., F. Lamy Mahmoud, A. Rebai, A. Gargouri, O. Panaud and A. Saar. 2006. Genetic diversity of Tunisian olive tree (*Olea europaea* L.) cultivars assessed by AFLP markers. Genet Resour Crop Ev. 53: 265–275.

Gonza´lez-Pe´rez, M., P.A Asosa and F.J. Batista. 2008. Genetic variation and conservation of the endangered endemic Anagyris latifolia Brouss. ex Willd. (Leguminosae) from the Canary Islands. Plant Syst Evol. doi 10.1007/s00606-009-0146-z.

Guner, A., N. Ozhatay, T. Ekim and K.H.C. Baser. 2000. Flora of Turkey and the East Aegean Islands. Edinburgh: Edinburgh University Press. 11(2): 35-37.

Heywood, V.H. 1996. Flowering Plant Of The World - BT Bastrford LTD, London, pp:239.

Hickey, M. and C. King. 1997. Common Families of Lowering Plants. Chambridge Univ. Press., England, pp: 119-127.

Jaccard, P. 1908. Nouvelles recherches sur la distribution florale. Bul. Soc. Vaudoise Sci. Nat. 44: 223–270.

Khalighi, M., A. Arzani and M.A. Poursiahbidi. 2008. Assessment of genetic diversity in *Triticum* spp. and *Aegilops* spp. using AFLP markers. Afri J Biotechn. 7(5): 546-552.

Kirimer, N., N. Tabanea and B. Demirci. et al. 2001. The essential oil of a new *Sideritis* species: *Sideritis ozturkii*. Chem Nat Compd. 37(3): 234-237.

Lapitan, V.C., D.S. Brar, T. Abe and E.D. Redona. 2007. Assessment of genetic diversity of Philippine rice carrying good quality traits using SSR markers. Breeding Sci. 57: 263-270.

Maciel, F.L., S. Echeverrigaray, L.T.S. Gerald and F.G. Grazziotin. 2003. Genetic relationships and diversity among Brazilian cultivars and landraces of common beans (*Phaseolus vulgaris* L.) revealed by AFLP markers. Genetic Resources and Crop Evolution. 50: 887–893.

Maras, M., J. Sustar-Vozlic, B. Javornik and V. Meglic. 2008. The efficiency of AFLP and SSR markers in genetic diversity estimation and gene pool classification of common bean (*Phaseolus vulgaris* L.). Acta Agric Slov. 91: 87–96.

Martins-Lopes, P.S., G.E. Santos and H. Guedes-Pinto. 2008. DNA markers for Portuguese olive oil fingerprinting. J Agric Food Chem. 56: 11786-11791.

Mudibu, J., K.K.C. Nkongolo, M. Mehes-Smith and A. Kanlonji-Mbuyi 2011. Genetic analysis of a soybean genetic pool using ISSR markers: Effect of gamma radiation on genetic variability. Plant Breeding Genet. 5: 235-245.

Murtaza, N. 2006. Cotton genetic diversity study by AFLP markers. Electronic Journal of Biotechnology ISSN: 0717-3458, 9(4), Issue of July 15.

Muse, S.V. and K.J. Liu. 2005. Powermarker: An integrated analysis environment for genetic marker analysis. Bioinformatics. 21: 2128- 2129.

Nemli, S. 2013. Detection of DNA Markers Controlling Some Economically Important Agronomic Characters by Association Mapping in Common Bean (*Phaseolus vulgaris* L.) - Dissertation, University of Ege, Izmir, Turkey .

Nybom, H. and I.V. Bartish. 2000. Effects of life history traits and sampling strategies on genetic diversity estimates obtained with RAPD markers in plants. Perspect Plant Ecol Evol Syst. 3: 93–114.

Ozturk, Y., S. Aydın, N. Ozturk and K.H.C. Baser. 1996. Effects of extracts from certain *Sideritis* species on swimming performance in mice. Phytother Res. 10: 70-73.

Ozhatay, N., M. Koyuncu, S. Atay and A. Byfield. 1998. The trade in wild medicinal plants in Turkey. In 'Medicinal plant trade in Europe: conservation and supply', eds Traffic Europe pp. 5-18 - Proceeding of the First International Symposium on the Conservation of Medicinal Plants in Trade in Europe. TRAFFIC Europe, Brussels.

Panahi, B., R. Afzal, M.G. Neghab, M. Mahmoodnia and B. Paymard. 2013. Relationship among AFLP, RAPD marker diversity and Agromorphological traits in safflower (*Carthamus tinctorius* L.) - Progress in Biological Sci. pp 90-99.

Pecina-Quintero, V., J.L. Anaya-López, A. Zamarripa-Colmenero, N. Montes-García, C. Nuñez-Colín, J.L. Solis-Bonilla and et al. 2012. Genetic diversity of sweet sorghum germplasm in Mexico using AFLP and SSR markers. Pesqui Agropecu Bras. 47: 1095–1102.

Pritchard, J.K., M. Stephens and P. Donnelly. 2000. Inference of population structure using multilocus genotype data. Genetics. 155(2): 945-959.

Prohens, J., G.J. Anderson, F.J. Herraiz, G. Bernardello, A. Santos-Guerra, D. Crawford and , F. Nuez. 2007. Genetic diversity and conservation of two endangered eggplant relatives (*Solanum verpertilio* Aiton, and *Solanum lidii* Sunding) endemic to the Canary Islands. Genet Resour Crop Evol. 54: 451–464

Qian, X., C-X. Wang and M. Tian. 2013. Genetic diversity and population differentiation of *Calanthe tsoongiana*, a rare and endemic orchid in China. Int J Mol Sci. 14: 20399-20413.

Rohlf, F.J. 1998. NTSYSpc Numerical Taxonomy and Multivariate Analysis System Version 2.0. User Guide. Applied Biostatistics Inc., Setauket, New York, pp.37.

Ryding, O. 1994. The importance of pericarp structure in the classification of Labiatae. Lamiales Newsletter 3: 1-3.

Saiki, R.K., D.H. Gelfand, S. Stoffel, S.J. Scharf, R. Higuchi, G.T. Horn, K.B. Mullis and H.A. Erlich. 1988. Primer-directed enzymatic amplification of DNA with thermostable DNA polymerase. Science. 239: 487-91.

Shimomura, K. and K. Hirashima, 2006. Development and characterization of simple sequence repeats (SSR) as marker to identify strawberry cultivars (Fragiria x ananassa Duch.). Journal of the Japanese Society for Horticultural Sci. 75(5): 399-402.

Smith, J.S.C., E.C.L. Chin, H. Shu, O.S. Smith, S.J. Wall, M.L. Senior, S.E. Mitchel, S. Resorich and J. Tiegle. 1997. An evaluation of the utility of SSR loci as molecular marker in maize *Zea mays*): comparisons with data from RFLP and pedigree. Theor Appl Genet. 95: 163-173.

Strikic, F., D.B. Mavsar, S. Perica, Z. Cmelik, Z. Satovic and B. Javornik. 2010. Genetic variation within the olive (*Olea europaea* L.) cultivar Oblica detected using amplified fragment length polymorphism (AFLP) markers. African Journal of Biotechnology. 9(20): 2880-2883.

Topcu, G., A.C. Goren, T. Kilic, Y.K. Yildiz and G. Tumen. 2002. Diterpenes from *Sideritis trojana.* Nat Prod Lett. 16: 33–3.

Vuylsteke, M., R. Mank, R. Antonise, E. Bastiaans, M.L. Senior, C.W. Stuber, A.E. Melchinger and et al. 1999. Two high-density AFLP® linkage maps of *Zea mays* L.: analysis of distribution of AFLP markers. Theor Appl Genet. 99: 921-935.

Witkowicz, J., E. Urbanczyk-Wochniak and Z. Przybecki. 2003. AFLP marker polymorphism in cucumber (*Cucumis sativus* L.) near isogenic lines differing in sex expression. Cell Mol Biol Lett. 8: 857-857.

Yuan, W.J., J.H. Yuan, M.F. Dong and F.D. Shang. 2011. Assessment of genetic diversity and relationships among *Osmanthus fragrans* cultivars using AFLP markers. Electronic Journal of Biotechnology. ISSN: 0717-3458 http://www.ejbiotechnology.info. doi: 10.2225/vol14-issue1-fulltext-9.

Zhang, L., Y-J. Jeon, S-Y. Kang and G-J. Lee. 2012. Genetic Diversity of Natural and Artificial Populations of Model Grass Brachypodium Species Evaluated by AFLP Markers. Hort Environ Biotechnol. 53(2): 143-150.

HERITABILITY ESTIMATES OF SOME QUANTITATIVE TRAITS IN POTATOES

Gulsum OZTURK, Zihin YILDIRIM*

Ege University, Faculty of Agriculture, Department of Field Crops, Bornova-Izmir, TURKEY
Corresponding author: gulsum.ozturk@ege.edu.tr

ABSTRACT

Broad-sense heritability (H) values of some quantitative traits in a potato population comprising 21 genotypes grown at Bornova, Izmir, Turkey during the 2013 and 2014 growing seasons were estimated by using the variance components analysis method. Based on the combined analysis of variance performed over two years, the following moderate to high level H values were found for plant height (0.77), leaf width (0.69), leaf length (0.71), single tuber weight (0.74), plant yield (0.60) and starch content (0.87).The expected genetic advance based on the selection of 25 % upper part was 8.2 cm for plant height; 0.5 for tuber number; 18.5 g for single tuber weight; 150.8 g for plant yield; 2.2 kg for plot yield and 1.9 % for starch content.

Keywords: Variance components, genotypic variance, genetic coefficient of variation, genetic advance

INTRODUCTION

In plant breeding, selection is applied to the populations having genetic variability for the desired characters. The genetic variability is generated by crossing different individuals or mutating genes of interest by induced mutagenesis. Main purpose of a selection program is to increase the mean of the selected portion of the population and thus to select individuals with high genotypic value. Selection in practice depends on the phenotypes of the individuals and in the absence of genotypic portion of the variance the genetic advance from selection cannot be estimated.

Phenotypic variance includes genotypic variance and environmental variance as well as the interactions of genotype x environment. Also genotypic variance may be divided in to additive, dominance and epistatic components. The ratio of genotypic variance to total phenotypic variance describes broad-sense heritability while the ratio of pure additive component of genotypic variance to total phenotypic variance is referred to as narrow-sense heritability. Thus, these heritability estimates may be helpful to calculate expected genetic gain from selection or response to selection being applied.

Potatoes produced worldwide are tetraploid so their genetic structures are much more complicated as compared to diploid crops. The gamete formation and the ratio of gametes in the segregation of the AAaa duplex parent are more complex and mixed up with high number of ratios of the individuals. The genotype frequencies at equilibrium are estimated as $(p+q)^4$ in autotetraploids (Li, 1955) in comparison to the Hardy-Weinberg equation ,$(p+q)^2$, of the diploid (Falconer, 1960). Therefore, most of the genetic studies in potatoeshave been conducted at the diploid level.

As described above, heritability in broad-sense has been defined as the ratio of genotypic variance (σ^2g) to the phenotypic variance (σ^2ph) and denoted by H. The narrow sense heritability (h^2) was defined as the ratio of additive variance to the phenotypic variance (σ^2ph). Although narrow sense heritability is more precise, it presents difficulties in its estimation. The simple and widely used technique in the genetic analyses of crops is the variance components analysis developed by Comstock and Moll (1963). This technic has been used widely in potatoes since the clones maintained the genetic variance.

Yildirim and Caliskan (1985) estimated broad-sense heritabilities in potatoes based on the variance component analysis for plant height (0.63), leaf area (0.78), tuber number (0.73), tuber weight (0.90), yield (0.87) and starch content (0.76).

Eraslan and Yildirim (1997) reported high values for plant height (0.92), leaf area index (0.82), tuber weight (0.86), and starch content (0.81) at Menemen, Izmir location. Desai and Jaimini (1997) reported moderate H values in a potato population. Similar broad-sense heritability estimates reported by Regassa and Basavaraj (2005), which are for plant height (0.42), stem number (0.49), leaf area (0.83), tuber number (0.73), plant yield (0.72) and dry matter content (0.71). Most recent heritability estimates obtained based on 38 genotypes grown at one location for two years for plant height, leaf area, stem number, tuber yield having high H values (Mishra et al., 2006).

However, there is no recent study of heritabilities in potato in the Aegean region. Therefore, the purpose of this study was to update the estimation of variance components and heritability values for certain physiological and agronomical traits of the potato population grown at Bornova, Izmir, Turkey for two years.

MATERIALS AND METHODS

Genetic Material

A total of 21 potato genotypes with various origins were selected to grow at Bornova-Izmir, Turkey. The information pertinent to these genotypes is shown in Table 1 (Anonymous, 2014a; Anonymous, 2014b; Anonymous, 2014c).

Table 1. The origin of potato genotypes and their pedigrees

Genotypes	Origin	Pedigree
Juwell	Germany	-
Shelford	United States	FL-1625 x Hermes
Bettina	Germany	-
Soprano	Netherlands	-
Agria	Germany	Quarta x Semlo
Pentland Dell	UK	Roslin Chania x Roslin Sasumua
Lady Olimpia	Netherlands	Agria x KW78-34-470
Musica	USA	FL1625 x Hermes
Royal	United Kingdom	Midas x 92-BUY-1
Lady Claire	United Kingdom/Netherlands	Agria x KW 78-34-470
Melody	United Kingdom	VF74-45 x W72-22-496
Agata	Netherlands	BM 52.72 x Sirco
Folva	Denmark	Mirada x Maris Piper
Brooke	USA	FL1867 x Hermes
Atlantic	Canada	Wauseon x Lenape
Granola	Germany	3333 60 x 267 04
Hermes	Netherlands	DDR5158 x SW163/55
Marabel	Netherlands	Nena x MA75364
Nif	Ege Univ.Fac. of Agri.	Cosima x R.143
Clone 122	Ege Univ.Fac. of Agri.	Cosima x R.143
Resy	Netherlands	SVP502017 x MPI 19268

Climate and soil characteristics of the Experimental area

The field trial was set up at the experimental field of Agricultural Faculty,Ege University located in Bornova, Izmir, Turkey. The native root zone of the soil in the experimental field was composed of 80.2 % sand, 18.1 % silt and 1.7 % clay. The soil was loamy sand with pH of 8.1. The long term average temperature was 17.1 ^{0}C, and the precipitation was 596.3 mm. A typical Mediterranean type climate conditions with hot and dry summers and rainy winters prevail in the region (Kavut et al., 2014).

Experimental Design

The field trial was arranged in a Randomized Complete Blocks Design with 3 replications. One plot consisted of 2 rows, 3 m long and 70 cm apart. The spacing between the plants on the rows was 30 cm, having 20 plants.

The field trial was run during the 2013 and 2014 growing seasons. The seed tubers were planted on 6 March 2013 and on 13 February 2014 by hand and the required standard agronomical practices were applied. The yield trial was harvested on July 8, 2013 and June 16-17, 2014.

Plant and tuber characteristics were measured during the growing season and at the harvest. The following morphological and yield characteristics were measured on randomly selected 3 plants in each plot: plant height (cm), stem number, leaf number, leaf width and length, tuber number, single tuber weight (g), plant yield (g). At the harvest, plot yield (kg), tuber width (cm) and length (cm) and starch content (%) were measured as described by Yildirim and Caliskan, (1985).

Statistical Analyses

The analysis of variance suitable for the experimental design was run on the data combined over two years. The means of traits studied were compared by using the Standard LSD test as described by Steel et al. (1997).

Estimations of variance components and genetic advance

The random model was accepted in order to construct the expected mean squares for the genotype, genotype x year and environmental variances. The expected mean squares given by Comstock and Moll (1963) are shown in Table 2.

Table 2. Expected mean squares for the source of variation

Source of variation	Degrees of freedom	Mean square	Expected mean squares
Genotype	n-1	MS_1	$\sigma^2e + r\sigma^2gy + ry\sigma^2g$
Genotype x year	(n-1) (y-1)	MS_2	$\sigma^2e + r\sigma^2gy$
Error	y (n-1) (r-1)	MS_3	σ^2e

y= year: 2
r= replication: 3
n=21

The genotypic and the phenotypic variances were estimated by using the mean squares given in table 2:

Genotypic variance (σ^2_g)= MS_1-MS_2/ ry

Genotype x year interaction (σ^2_{gxy})= MS_2-MS_3/r

Environmental variance (σ^2_e)= σ^2_e=MS_3

Phenotypic variance (σ^2_{ph}) = $\sigma^2_g+\sigma^2_{gxy}/y+\sigma^2_e/ry$

The broad sense heritability was estimated by using the genotypic and phenotypic variances as following:

H (broad-sense) = σ^2g/σ^2ph

Where σ^2g= genotypic variance

σ^2ph= phenotypic variance

Genotypic coefficient of variation (GCV) was estimated by usinggenetic standard deviation:$\frac{\sigma g}{x}100$

Genetic advance was calculated by using the formula: i.S_{ph}.H

RESULTS AND DISCUSSION

The F values pertinent to genotypes, and genotype x year interaction sources of variation given in table 3 indicated significant variations for genotypes and genotype x year interaction for all the traits studied at the p≤ 0.01 levels. Therefore, means of genotypes could be compared and the genotypic variability and the heritabilities could be estimated.

Table 3. The F values of the genotype and genotype x year interaction component of variation for quantitative traits of potato population grown at Bornova-Izmir, Turkey during the 2013 and 2014 growing seasons.

Traits	Genotype	Genotype xYear
Plant height (cm)	505.789**	118.013**
Number of stems	94.110**	84.727**
Number of leaves	1695.252**	1221.133**
Leaf width (cm)	39.595**	10.447**
Leaf length (cm)	84.212**	29.894**
Tuber number	38.849**	28.686**
Single tuber weight (g)	36.832**	9.661**
Plant yield (g)	67.400**	26.864**
Plot yield (kg)	273.092**	148.718**
Tuber width (cm)	73.941**	28.891**
Tuber length (cm)	174.581**	64.303**
Starch content (%)	25.887**	3.304**

**: denotes significance at the p≤0.01 level

Comparison of genotype means

Means of the physiological traits are shown in table 4. The genotype Agria had the highest means for plant height (54.6 cm) and number of leaves (93.9). Folva had the highest mean for number of stem (4.7). Juwell had the highest means for leaf width (2.5 cm) and length (4.1 cm).

Table 4. The means of physiological traits of the potato genotypes grown at Bornova-Izmir, Turkey during the 2013 and 2014 growing seasons.

Genotypes	Plant height (cm)	Number of stems	Number of leaves	Leaf width (cm)	Leaf length (cm)
Juwell	**43.8**	**4.0**	55.2	**2.5**	**4.1**
Shelford	34.8	2.5	55.0	**2.1**	3.3
Bettina	**46.6**	4.0	**85.6**	1.9	3.4
Soprano	36.1	2.6	46.0	2.0	3.4
Agria	**54.6**	**4.3**	**93.9**	2.1	3.3
Pentland Dell	39.4	2.8	53.8	1.6	2.6
Lady Olimpia	**43.4**	3.8	**79.7**	2.1	**3.5**
Musica	36.7	3.8	55.1	2.0	3.2
Royal	34.1	3.8	48.1	2.1	**3.5**
Lady Claire	36.2	**4.1**	**75.6**	**2.2**	**3.6**
Melody	36.3	3.5	56.1	**2.4**	**3.8**
Agata	29.1	**4.6**	53.9	2.0	3.4
Folva	**45.2**	**4.7**	**89.6**	1.9	3.4
Brooke	32.6	2.9	55.9	1.9	3.3
Atlantic	30.7	2.2	43.0	1.9	3.0
Granola	22.2	3.4	30.4	**2.2**	3.3
Hermes	25.4	3.7	44.4	1.9	3.1
Marabel	27.1	3.2	40.7	2.0	3.0
Nif	28.9	2.7	56.4	1.8	3.1
Clone 122	25.3	2.8	40.7	1.7	2.7
Resy	24.2	2.4	37.8	1.8	2.8
$LSD_{(0.05)}$	1.06	0.22	1.21	0.09	0.11

Means of the agronomical trait studied are shown in table 5. It could be seen in table 5 that Agria had the highest means for single tuber weight (140.2 g), plot yield (15.8 kg) and tuber width (5.6 cm). Musica had the highest means for tuber number (10.2), plant yield (991.6 g) and tuber length (8.3 cm). Brooke had the highest starch content (18.4 %).

Table 5. The means of the agronomical traits of the potato genotypes grown at Bornova-Izmir, Turkey during the 2013 and 2014 growing seasons.

Genotypes	Tuber number	Single tuber weight (g)	Plant yield (g)	Plot yield (kg)	Tuber width (cm)	Tuber length (cm)	Starch content (%)
Juwell	**9.4**	86.4	809.8	**13.6**	**5.1**	6.9	13.3
Shelford	7.4	82.3	622.8	8.8	4.7	6.2	**17.5**
Bettina	**9.6**	**91.4**	**867.3**	**15.8**	**5.2**	**7.0**	15.4
Soprano	5.7	73.3	404.8	5.5	4.4	**7.1**	13.2
Agria	6.5	**140.2**	**854.9**	**15.8**	**5.6**	**7.9**	13.6
Pentland Dell	5.2	71.9	379.8	5.7	4.1	6.7	14.2
Lady Olimpia	7.7	**110.8**	**847.7**	**13.2**	4.9	**7.5**	14.8
Musica	**10.2**	**100.1**	**991.6**	**15.2**	4.6	**8.3**	12.8
Royal	6.9	86.9	606.4	6.9	**5.0**	6.1	14.2
Lady Claire	8.2	81.3	675.6	11.1	4.8	5.9	**16.3**
Melody	**10.1**	84.9	**830.7**	13.0	4.8	6.5	13.9
Agata	8.9	82.9	697.1	11.3	4.7	6.4	11.5
Folva	9.0	76.3	666.6	10.5	4.6	6.6	12.8
Brooke	9.3	84.5	734.5	8.9	4.7	5.9	**18.4**
Atlantic	5.9	**105.2**	636.3	9.2	**4.9**	6.1	**17.4**
Granola	6.6	46.1	265.8	4.1	3.9	5.1	13.2
Hermes	7.5	65.7	459.6	6.3	4.5	5.6	**15.4**
Marabel	7.1	69.9	478.5	6.6	4.5	6.2	13.3
Nif	**9.5**	77.9	715.7	8.7	4.7	5.9	13.8
Clone 122	6.7	60.8	407.7	5.1	4.6	5.6	14.4
Resy	6.2	66.9	405.8	6.1	4.4	6.5	13.8
$LSD_{(0.05)}$	0.69	9.17	67.81	0.64	0.12	0.17	0.96

Genetic variability

Estimates of the variance components, heritability and genetic coefficient of variation are shown in table 6. Genotypicvariance, genotype x year interaction variance and phenotypic variance were significant at the p≤0.01 level. Genetic coefficient of variation was high for plot yield (26.2 %), plant height (21.2 %) and single tuber weight (20.4 %). The heritabilities in board sense (H) were high for starch content (0.87), plant height (0.77), single tuber weight (0.74), leaf length (0.71), leaf width (0.69);moderate for tuber length (0.61), plant yield (0.60) and plot yield (0.46), respectively. The heritability estimates for leaf number (0.28), tuber number (0.26) and stem number (0.10)had low values.Low heritability estimates could be the result of the genotype x environment interactions.

Genetic advance

Selection differential and the genetic advance obtained are shown in table 7. Estimates of the expected genetic advance are the result of the tentative selection applied in the population based on the heritability estimates. Plant yield (150.8 g), single tuber weight (18.5 g), plant height (8.2 cm), leaf number (6.3) and plot yield (2.2 kg) had comparatively high genetic advances than that of the other traits.

Heritability estimates presented in this study were in close agreement with the estimates reported previously (Yildirim and Caliskan, 1985; Eraslan and Yildirim, 1997; Regassa and Basavaraj, 2005; Mishra et al., 2006). In general estimates of heritabilities over 0.7 have been accepted as usable in plant breeding programs (Yildirim and Dere, 2005). Heritability estimates reported mostly are based on the variance component analysis. The accuracy of this kind of estimate depends on the certain assumptions about the application of the variance component analysis method. First, the validity of the expected mean square depends on the randomness. Unfortunately, in practice it is difficult to select genotypes, locations and years at random. In practice, only the replication is accepted as random. Therefore, the expected mean squares arc assumed to be in agreement with the random model.

Comstock and Moll (1963) have given the expected mean square estimates for variance components in the analysis of variance table. The variance components method applied in different studies are based on the calculation of variance components by using the expected mean squares in a random model, which is in fact a mixed model theoretically.

Table 6. Estimates of variance componentsand heritabilities for physiological and agronomical traits of the potato population grown at Bornova-Izmir, Turkey during the 2013 and 2014 growing seasons.

Traits	Mean	S^2g	S^2gy	S^2e	S^2ph	HBroad Sense	GCV (%)
Plant height (cm)	34.9	54.55**	32.92**	0.85	71.2	0.77	21.2
Number of stems	3.41	0.06**	0.01**	0.04	0.56	0.10	6.7
Number of leaves	56.9	88.0**	453.2**	1.11	314.8	0.28	16.5
Leaf width (cm)	2.0	0.03**	0.02**	0.01	0.04	0.69	8.5
Leaf length (cm)	3.3	0.08**	0.08**	0.01	0.12	0.71	8.4
Tuber number	7.8	0.62**	3.38**	0.37	2.37	0.26	10.1
Single tuber weight (g)	83.1	288**	183.7**	63.6	390.4	0.74	20.4
Plant yield (g)	636.1	23496.5**	29984.4**	3477.9	390.68	0.60	24.1
Plot yield (kg)	9.6	6.32**	15.0**	0.31	13.87	0.46	26.2
Tuber width (cm)	4.7	0.08**	0.09**	0.11	0.14	0.56	5.9
Tuber length (cm)	6.5	0.39**	0.46**	0.02	0.63	0.63	9.7
Starch content (%)	14.4	2.67**	0.55**	0.71	3.06	0.87	11.32

**: denotes significance at the $p \leq 0.01$ level
S^2g, S^2gy, S^2e, S^2ph: estimates of genotypic, genotype x year interaction, environmental and phenotypic variances respectively
H: estimate of heritability in broad sense
GCV (%): estimate of genotypic coefficient of variation

Table 7. Results of the individual selection applied for the physiological and agronomical traits of the potato population grown at Bornova, Izmir, Turkey during the 2013 and 2014 growing seasons.

Traits	Range	Mean of selected genotypes	Selection differential	Genetic advance
Plant height (cm)	22.2-54.6	46.7	11.8	8.2
Number of stems	2.2-4.7	4.4	1.0	0.1
Number of leaves	30.4-93.9	84.9	28.0	6.3
Leaf width (cm)	1.6-2.5	2.3	0.3	0.2
Leaf length (cm)	2.6-4.1	3.7	0.4	0.3
Tuber number	5.2-10.2	9.7	1.9	0.5
Single tuber weight (g)	46.1-140.2	109.5	26.4	18.5
Plant yield (g)	265.8-991.6	878.4	242.3	150.8
Plot yield (kg)	4.1-15.8	14.7	5.1	2.2
Tuber width (cm)	4.1-5.6	5.2	0.5	0.3
Tuber length (cm)	5.1-8.3	7.6	1.1	0.6
Starch content (%)	11.5-18.4	17.0	2.6	1.9

i: Selection intensity approximately 1.27 for the 25 % of the upper part of population

Besides, the estimation of heritability by the variance component methods reported by several workers had certain limitations due to the selection of genotypes, locations and years. Most of the reports have been based on the growing selected genotypes at one location for 2 or more years or at two or more location in one year. In these estimations the magnitude of heritabilities could be inflated due to second order interaction of the genotype x location x year.

Yildirim and Caliskan (1985) reported heritability estimates based on growing genotypes at 3 locations for 2 years in the target area. These estimates could be accepted as more accurate in comparison to the several heritabilities reported. The heritability estimates found in this study were also in close agreement with the reported values by these researchers.

The values regarding the selection applied for the genotypes with high mean were given in tables 6 and 7, which resulted in selection differential ranged from 0.3 (leaf width) to 242.3 g (plant yield). Based on the selection differential estimates, we should expect increase of 11.8 cm for plant height, 28 for number of leaves, 26.4 g for tuber weight, 243.3 g for plant yield, 5.1 kg for plot yield, and 26 % starch content.

Based on the estimation of genetic advances, the realized expected genetic gains for plant height, number of leaf, single tuber weight, plant yield, plot yield and starch contents would be 8.2 cm, 6.3, 18.5 g, 150.8 g, 2.2 kg and 1.9 % respectively. The actual genetic gain depends on the genetic coefficient of variation and heritability values of the traits. The genetic advances of tuber yield (0.19) was reported by Tekalign (2009), for tuber number (3.4), plant height (8.8), plant yield (117.79) and dry matter content (4.6 %) by Regassa and Basavaraj (2005) and for plant height (27.4 %), tuber number (45.3 %), tuber weight (21.1 %) and plant yield (62.1 %) by Sattar et al. (2007).

The discrepancies between the reported genetic advance values might be due to the differences in sampling the genotypes and environments. In general, it has been accepted that the Genetic Coefficient of Variation was not sufficient to test heritable variation so more reliable procedures need to estimate the heritability and the genetic advance.

Estimates of heritability, genetic coefficient of variation and genetic advance would be helpful to the breeder during the application of selection in the breeding programmes. The heritabilities estimated by the variance components method would be useful in making these decisions. Mishra et al. (2006) also concluded that high heritability and high genetic advance would be valuable in selection programmes.

To recapitulate, the estimates of heritabilities based on the growing potato genotypes at several locations for several years will help to the potato breeders in deciding about the breeding strategy. Potato clones could have the additive, dominant and epistatic type genetic variability due to their stabile genetic structure. For that reason, estimation of heritabilities by using the variance components method might be more dependable in potatoes in comparison to the generatively producing crops.

LITERATURE CITED

Anonymous, 2014a. www.varieties.potato.org.uk (accessed on 31.08.2014).

Anonymous, 2014b. www. europotato.org (accessed on 31.08.2014).

Anonymous, 2014c. www.solanum-int.com (accessed on 10.09.2014).

Comstock, R.E. and R.H. Moll. 1963. Genotype-environment interactions. In: Statistical Genetics and Plant Breeding. NAS-NRC. Publ. 982, pp.164-196.

Desai, N.C. and S.N. Jaimini. 1997. Genetic variability, heritability and genetic advance for yield in potato. Indian Journal of Potato Association. 24: 52-58.

Eraslan, M. and M.B. Yildirim. 1997. Heritability estimates and relationships of certain characteristics in potatoes. Turk. J. Field Crops. 2(1): 22-26.

Falconer, D.S. 1960. Introduction to Quantitative Genetics, Agricultural Research Council Animal Genetics, University of Edinburgh, The Ronald Press Company, Newyork.

Kavut, Y.T., R. Avcioglu, G. Demiroglu, H. Geren and B. Kir. 2014. Adaptability of newly improved creeping bentgrass (*Agrotis stolonifera* L.) genotypes in mediterranean environment. Turk. J. Field Crops. 19(1): 19-24.

Li, C.C., 1955. Population Genetics. University of Chicago Press, Chicago, USA.

Mishra, A.C., N.P. Singh, S. Kamal and V. Kumar, 2006. Studies on genetic variability, heritability and genetic advance in potato (*Solanum tuberosum* L.) Int. J. of Plant Sci. 1(1): 39-41.

Regassa, D. and N. Basavaraj, 2005. Genetic variability studies in potato (*Solanum tuberosum* L.). Karnataka J. Agric. Sci. 18(1): 87-90.

Sattar, M.A., N, Sultana, M.M. Hossain, M.H. Rashid and A.K.M.A. Islam, 2007. Genetic variability correlation and path analysis in potato (*Solanum tuberosum* L.). Bangladesh J. Pl. Breed Genet. 20(1): 33-38.

Steel, R.G.D., J.A. Torrie and D.A. Deckey. 1997.Principles and Procedures of Statistics: A Biometrical Approach. 3rd Ed., Mc Graw Hill Book INC. N.Y.

Tekalign,T. 2009. Genetic variability, correlation and heritability in tetraploid potato (*Solanum tuberosum* L.) germplasm. Tropical Agriculture. 86 (2): 52-59.

Yildirim, M. B. and C. Caliskan. 1985. Genotype x environment interactions in potato (*Solanum tuberosum* L.), American Potato J. 62: 371-375.

Yildirim, M. B. and S. Dere. 2005. Quantitative Genetics. Ege Univ. Fac. Agri. Dep. of Field Crops, Bornova, Izmir (in Turkish).

Permissions

All chapters in this book were first published in TJFC, by Society of Field Crops Science; hereby published with permission under the Creative Commons Attribution License or equivalent. Every chapter published in this book has been scrutinized by our experts. Their significance has been extensively debated. The topics covered herein carry significant findings which will fuel the growth of the discipline. They may even be implemented as practical applications or may be referred to as a beginning point for another development.

The contributors of this book come from diverse backgrounds, making this book a truly international effort. This book will bring forth new frontiers with its revolutionizing research information and detailed analysis of the nascent developments around the world.

We would like to thank all the contributing authors for lending their expertise to make the book truly unique. They have played a crucial role in the development of this book. Without their invaluable contributions this book wouldn't have been possible. They have made vital efforts to compile up to date information on the varied aspects of this subject to make this book a valuable addition to the collection of many professionals and students.

This book was conceptualized with the vision of imparting up-to-date information and advanced data in this field. To ensure the same, a matchless editorial board was set up. Every individual on the board went through rigorous rounds of assessment to prove their worth. After which they invested a large part of their time researching and compiling the most relevant data for our readers.

The editorial board has been involved in producing this book since its inception. They have spent rigorous hours researching and exploring the diverse topics which have resulted in the successful publishing of this book. They have passed on their knowledge of decades through this book. To expedite this challenging task, the publisher supported the team at every step. A small team of assistant editors was also appointed to further simplify the editing procedure and attain best results for the readers.

Apart from the editorial board, the designing team has also invested a significant amount of their time in understanding the subject and creating the most relevant covers. They scrutinized every image to scout for the most suitable representation of the subject and create an appropriate cover for the book.

The publishing team has been an ardent support to the editorial, designing and production team. Their endless efforts to recruit the best for this project, has resulted in the accomplishment of this book. They are a veteran in the field of academics and their pool of knowledge is as vast as their experience in printing. Their expertise and guidance has proved useful at every step. Their uncompromising quality standards have made this book an exceptional effort. Their encouragement from time to time has been an inspiration for everyone.

The publisher and the editorial board hope that this book will prove to be a valuable piece of knowledge for researchers, students, practitioners and scholars across the globe.

List of Contributors

Oktay ERDOGAN
Cotton Research Station, Aydin-TURKEY

Nedim OZBEK
Cotton Research Station, Aydin-TURKEY

Aydın UNAY
University of Adnan Menderes, Faculty of Agriculture, Aydin-TURKEY

M. Erhan GORE
University of Abant Izzet Baysal, Vocational Higher School, Bolu- TURKEY

Nimet KARA
Suleyman Demirel University, Agriculture Faculty, Department of Field Crops, Isparta, TURKEY

Hasan BAYDAR
Suleyman Demirel University, Agriculture Faculty, Department of Field Crops, Isparta, TURKEY

Naser Al-Suhaibani
Plant Production Department, College of Food and Agriculture Sciences, King Saud University, Riyadh, Saudi Arabia

Salah El-Hendawy
Plant Production Department, College of Food and Agriculture Sciences, King Saud University, Riyadh, Saudi Arabia
Agronomy Department, Faculty of Agriculture, Suez Canal University, Ismailia, Egypt

Urs Schmidhalter
Department of Plant Sciences, Technische Universität München, Freising-Weihenstephan, Germany

Duran KATAR
Eskisehir Osmangazi University, Ali Numan Kırac Campus, Faculty of Agriculture, Eskisehir-TURKEY

Memet Inan
Kahta Vocational School of Higher Education, Adıyaman University, Adıyaman, Turkey

Saliha Kirici
Department of Field Crops, Agriculture Faculty, Cukurova University, Adana, Turkey

E. Sultan Giray
Department of Chemistry, Letter and Science Faculty, Cukurova University, Adana, Turkey

Murat Turk
Ceyhan Vocational School of Higher Education, Cukurova University, Adana, Turkey

Hassan Taghikhani
Department of Field Crops, Institute of Science, Cukurova University, Adana, Turkey

Duran KATAR
Eskişehir Osmangazi University, Faculty of Agriculture, Ali Numan Kıraç Campus, Eskişehir, TURKEY

Yusuf ARSLAN
Central Research Institute for Field Crops, Ankara, TURKEY

Ilhan SUBAŞI
Central Research Institute for Field Crops, Ankara, TURKEY

Goran TODOROVIĆ
Maize Research Institute, Zemun Polje, Belgrade-Zemun, REPUBLIC SERBIA

Mile SEĈANSKI
Maize Research Institute, Zemun Polje, Belgrade-Zemun, REPUBLIC SERBIA

Tomislav ŽIVANOVIĆ
Faculty of Agriculture,University Belgrade, REPUBLIC SERBIA

Rade PROTIĆ
Institute for Science Aplication in Agriculture, Belgrade, REPUBLIC SERBIA

Miroslav KOSTIĆ
Institute for Medicinal Plant Research "Dr Josif Panĉić", Belgrade

Snežana JOVANOVIĆ
Maize Research Institute, Zemun Polje, Belgrade-Zemun, REPUBLIC SERBIA

Dragan BOŽOVIĆ
"Agros", Panĉevo, REPUBLIC SERBIA

Agnieszka GODLEWSKA
University of Natural Sciences and Humanities, Siedlce, POLAND

Grażyna Anna CIEPIELA
University of Natural Sciences and Humanities, Siedlce, POLAND

Ahmad SHER
College of Agriculture, Bahauddin Zakariya University, Bahadur Campus Layyah, Pakistan

Muhammad ANSAR
Department of Agronomy, PMAS-Arid Agriculture University Rawalpindi, Pakistan

Abdul MANAF
Department of Agronomy, PMAS-Arid Agriculture University Rawalpindi, Pakistan

Abdul QAYYUM
Department of Agriculture, University of Haripur, 22620, Pakistan

Muhammad Farhan SAEED
College of Agriculture, Bahauddin Zakariya University, Bahadur Campus Layyah, Pakistan

Muhammad IRFAN
Department of Agronomy, PMAS-Arid Agriculture University Rawalpindi, Pakistan

Erdem GULUMSER
Bozok University Agriculture and Natural Science Faculty, Yozgat, TURKEY

Zeki ACAR
Ondokuz Mayıs University Agricultural Faculty, Samsun, TURKEY

Robert Idziak
Poznan University of Life Sciences, Faculty of Agronomy and Bioengineering, Department of Agronomy Dojazd, Poznan

Zenon Woznica
Poznan University of Life Sciences, Faculty of Agronomy and Bioengineering, Department of Agronomy Dojazd, Poznan

Selahattin CINAR
Gaziosmanpasa University, Faculty of Agriculture, Deparment of Field Crops, Tokat, TURKEY

Rustu HATIPOGLU
Cukurova University, Faculty of Agriculture, Deparment of Field Crops, Adana, TURKEY

Feyza Dondu GUNDEL
East Mediterranean Agricultural Research Institue, Adana, TURKEY

Arif AKTAS
East Mediterranean Agricultural Research Institue, Adana, TURKEY

Mustafa AVCI
East Mediterranean Agricultural Research Institue, Adana, TURKEY

Jun LI
College of Life Science, China West Normal University, NanChong City, CHINA

Xiaohong LIU
College of Life Science, China West Normal University, NanChong City, CHINA

Nazan DAGUSTU
Uludag University Agricultural Faculty, Department of Field Crops, Bursa, TURKEY

Gamze BAYRAM
Uludag University Agricultural Faculty, Department of Field Crops, Bursa, TURKEY

Mehmet SINCIK
Uludag University Agricultural Faculty, Department of Field Crops, Bursa, TURKEY

Melek BAYRAKTAROGLU
Uludag University Agricultural Faculty, Department of Field Crops, Bursa, TURKEY

Celaleddin BARUTCULAR
Çukurova University, Faculty of Agriculture, Department of Field Crops, Adana, TURKEY

Irem TOPTAS
Çukurova University, Faculty of Agriculture, Department of Field Crops, Adana, TURKEY

Hatice TURKTEN
Ondokuz Mayıs University, Faculty of Agriculture, Department of Agricultural Economics, Samsun, TURKEY

Mehmet YILDIRIM
Dicle University, Faculty of Agriculture, Department of Field Crops, Diyarbakır, TURKEY

Mujde KOC
Çukurova University, Faculty of Agriculture, Department of Field Crops, Adana, TURKEY

Tolga YILDIRIM
Akdeniz University, Faculty of Science, Department of Biology, Antalya, TURKEY

Hüseyin CANCI
Akdeniz University, Faculty of Agriculture, Department of Field Crops, Antalya, TURKEY

Nisa ERTOY INCI
Akdeniz University, Faculty of Agriculture, Department of Field Crops, Antalya, TURKEY

Fatma Oncu CEYLAN BALOGLU
Akdeniz University, Faculty of Agriculture, Department of Field Crops, Antalya, TURKEY

Cengiz IKTEN
Akdeniz University, Faculty of Agriculture, Department of Plant Protection, Antalya, TURKEY

Cengiz TOKER
Akdeniz University, Faculty of Agriculture, Department of Field Crops, Antalya, TURKEY

Naeem Sarwar
Department of Agronomy, Bahauddin Zakariya University Multan, PAKISTAN

Hakoomat Ali
Department of Agronomy, Bahauddin Zakariya University Multan, PAKISTAN

Muhammad Maqsood
Department of Agronomy, University of Agriculture Faisalabad, PAKISTAN

Ehsan Ullah
Department of Agronomy, University of Agriculture Faisalabad, PAKISTAN

Muhammad Shahzad
Department of Agronomy, University of Agriculture Faisalabad, PAKISTAN

Khuram Mubeen
Department of Agronomy, University of Poonch Rawalakot, PAKISTAN

Ahmad Naeem Shahzad
Department of Agronomy, Bahauddin Zakariya University Multan, PAKISTAN

Muhammad Asghar Shahid
Department of Pest Warning and Quality of Pesticides Punjab, PAKISTAN

Shakeel Ahmad
Department of Agronomy, Bahauddin Zakariya University Multan, PAKISTAN

Engin YOL
Department of Field Crops, Faculty of Agriculture, Akdeniz University, Antalya, TURKEY

Seymus FURAT
Department of Field Crops, Faculty of Agriculture, Akdeniz University, Antalya, TURKEY

Bulent UZUN
Department of Field Crops, Faculty of Agriculture, Akdeniz University, Antalya, TURKEY

Ferit TURANLI
Ege Üniversitesi, Ziraat Fakültesi, Bitki Koruma Bölümü, Izmir, TURKEY

Emre ILKER
Ege Üniversitesi, Ziraat Fakültesi,Tarla Bitkileri Bölümü, Izmir, TURKEY

Firdevs ERSIN DOGAN
Ege Üniversitesi, Ziraat Fakültesi, Bitki Koruma Bölümü, Izmir, TURKEY

Lale ASKAN
Ege Üniversitesi, Ziraat Fakültesi, Bitki Koruma Bölümü, Izmir, TURKEY

Deniz ISTIPLILER
Ege Üniversitesi, Ziraat Fakültesi,Tarla Bitkileri Bölümü, Izmir, TURKEY

Gulsum OZTURK
Ege University, Faculty of Agriculture, Department of Field Crops, Bornova-Izmir, TURKEY

Zihin YILDIRIM
Ege University, Faculty of Agriculture, Department of Field Crops, Bornova-Izmir, TURKEY

Öner CANAVAR
Adnan Menderes University, Faculty of Agriculture, Department of Field Crops, Aydın, TURKEY
Humboldt University, Agriculture and Horticulture Faculty, Crop Science Department, Berlin, GERMANY

Klaus-Peter GÖTZ
Humboldt Universität, Agriculture and Horticulture Faculty, Agroclimatology, Department, Berlin, GERMANY

Yakup Onur KOCA
Adnan Menderes University, Faculty of Agriculture, Department of Field Crops, Aydın, TURKEY

Frank ELLMER
Humboldt University, Agriculture and Horticulture Faculty, Crop Science Department, Berlin, GERMANY

Maya Izar KHAIDIZAR
Strategic Resource Research Center Development Institute (MARDI), Kuala Lumpur, MALAYSIA

Kamil HALILOGLU
Department of Agronomy, Faculty of Agriculture, Ataturk University, Erzurum, TURKEY

Erdal ELKOCA
Department of Agronomy, Faculty of Agriculture, Ataturk University, Erzurum, TURKEY

Murat AYDIN
Department of Agronomy, Faculty of Agriculture, Ataturk University, Erzurum, TURKEY

Faik KANTAR
Department of Agricultural Biotechnology, Faculty of Agriculture, Akdeniz University, Antalya, TURKEY

Muhammet SAKIROGLU
Kafkas University, Department of Bioengineering, Kars, TURKEY

Muge Mavioglu KAYA
Kafkas University, Department of Biology, Kars, TURKEY

AMANULLAH
Department of Agronomy, The University of Agriculture Peshawar, PAKISTAN

Shahen SHAH
Department of Agronomy, The University of Agriculture Peshawar, PAKISTAN

Zahir SHAH
Department of Soil & Environmental Sciences, The University of Agriculture Peshawar, PAKISTAN

Shad Khan KHALAIL
Department of Agronomy, The University of Agriculture Peshawar, PAKISTAN

Amanullah JAN
Department of Agronomy, The University of Agriculture Peshawar, PAKISTAN

Muhammad Tariq JAN
Department of Agronomy, The University of Agriculture Peshawar, PAKISTAN

Muhammad AFZAL
Department of Soil & Environmental Sciences, The University of Agriculture Peshawar, PAKISTAN

Habib AKBAR
Department of Agronomy, The University of Agriculture Peshawar, PAKISTAN

Hamayoon KHAN
Department of Agronomy, The University of Agriculture Peshawar, PAKISTAN

Hidayat-ur-RAHMAN
Department of Plant Breeding & Genetics, The University of Agriculture Peshawar, PAKISTAN

Khalid NAWAB
Department of Agricultural Extension Education & Communication The University of Agriculture Peshawar, PAKISTAN

FARHATULLAH
Department of Plant Breeding & Genetics, The University of Agriculture Peshawar, PAKISTAN

Fida MUHAMMAD
Department of Plant Breeding & Genetics, The University of Agriculture Peshawar, PAKISTAN

Zahid HUSSAIN
Department of Agronomy, The University of Agriculture Peshawar, PAKISTAN

Khair Muhammad KAKAR
Department of Agronomy, The University of Agriculture Peshawar, PAKISTAN

Imran KHAN
Department of Agronomy, The University of Agriculture Peshawar, PAKISTAN

S. M. Mahdi MORTAZAVIAN
Department of Agronomy Sciences and Plant Breeding, College of Aburaihan, University of Tehran, Pakdasht, IRAN

Shiva AZIZI-NIA
Department of Agronomy Sciences and Plant Breeding, College of Aburaihan, University of Tehran, Pakdasht, IRAN

Esin SAHIN
Department of Field Crops, Faculty of Agriculture, Ataturk University. Erzurum, TURKEY

Metin TOSUN
Department of Field Crops, Faculty of Agriculture, Ataturk University. Erzurum, TURKEY

Kamil HALILOGLU
Department of Field Crops, Faculty of Agriculture, Ataturk University. Erzurum, TURKEY

Reza AMIRNIA
Urmia University, Faculty of Agriculture, Department of Agronomy, Urmia, IRAN

Mahdi BAYAT
Urmia University, Faculty of Agriculture, Department of Agronomy, Urmia, IRAN

Mahdi TAJBAKHSH
Urmia University, Faculty of Agriculture, Department of Agronomy, Urmia, IRAN

Suchittra KHAMPAS
Khon Kaen University, Faculty of Agriculture, Department of Plant Science and Agricultural Resources, Khon Kaen, THAILAND

Kamol LERTRAT
Khon Kaen University, Faculty of Agriculture, Department of Plant Science and Agricultural Resources, Khon Kaen, THAILAND
Khon Kaen University, Plant Breeding Research Center for Sustainable Agriculture, Khon Kaen, THAILAND

Khomsorn LOMTHAISONG
Khon Kaen University, Faculty of Science, Department of Biochemistry, Khon Kaen, THAILAND

Sakunkan SIMLA
Mahasarakham University, Faculty of Technology, Department of Agricultural Technology, Mahasarakham, THAILAND

Bhalang SURIHARN
Khon Kaen University, Faculty of Agriculture, Department of Plant Science and Agricultural Resources, Khon Kaen, THAILAND
Khon Kaen University, Plant Breeding Research Center for Sustainable Agriculture, Khon Kaen, THAILAND

Jinbao YAO
Provincial Key Laboratory of Agrobiology, Jiangsu Academy of Agricultural Sciences, Nanjing, CHINA

Hongxiang MA
Provincial Key Laboratory of Agrobiology, Jiangsu Academy of Agricultural Sciences, Nanjing, CHINA

Xueming YANG
Provincial Key Laboratory of Agrobiology, Jiangsu Academy of Agricultural Sciences, Nanjing, CHINA

Miaoping ZHOU
Provincial Key Laboratory of Agrobiology, Jiangsu Academy of Agricultural Sciences, Nanjing, CHINA

Dan YANG
Provincial Key Laboratory of Agrobiology, Jiangsu Academy of Agricultural Sciences, Nanjing, CHINA

Mevlut TURK
Suleyman Demirel University, Faculty of Agriculture, Department of Field Crops, Isparta, TURKEY

Sebahattin ALBAYRAK
Suleyman Demirel University, Faculty of Agriculture, Department of Field Crops, Isparta, TURKEY

Yalcin BOZKURT
Suleyman Demirel University, Faculty of Agriculture, Department of Field Crops, Isparta, TURKEY

Penko SPETSOV
Dobroudja Agricultural Institute, General Toshevo, BULGARIA

Nadia DASKALOVA
Faculty of Marine Sciences and Ecology, Technical University - Varna, BULGARIA

Dragomir PLAMENOV
Faculty of Marine Sciences and Ecology, Technical University - Varna, BULGARIA

Todor MORALIYSKI
Faculty of Marine Sciences and Ecology, Technical University - Varna, BULGARIA

Faisal ZAHOOR
Department of Agronomy, Faculty of Agriculture, The University of Poonch, Rawalakot, Azad Jammu Kashmir (AJK, Pakistan)

Mansoor AHMED
Department of Agronomy, Faculty of Agriculture, The University of Poonch, Rawalakot, Azad Jammu Kashmir (AJK, Pakistan)

M. Azim MALIK
Department of Agronomy, PMAS-Arid Agriculture Rawalpindi

Khuram MUBEEN
Department of Agronomy, PMAS-Arid Agriculture Rawalpindi

M. H. SIDDIQUI
Department of Agronomy, Faculty of Agriculture, The University of Poonch, Rawalakot, Azad Jammu Kashmir (AJK, Pakistan)

Muhammad RASHEED
Department of Agronomy, PMAS-Arid Agriculture Rawalpindi

Ramzan ANSAR
Department of Agronomy, PMAS-Arid Agriculture Rawalpindi

Khalid MEHMOOD
Department of Agronomy, Faculty of Agriculture, The University of Poonch, Rawalakot, Azad Jammu Kashmir (AJK, Pakistan)

Cengiz TOKER
Akdeniz University, Faculty of Agriculture, Department of Field Crops, Antalya, Turkey, Fatma Oncu CEYLAN Akdeniz University, Faculty of Agriculture, Department of Field Crops, Antalya, Turkey,

Adem BARDAK
Bingol University, Faculty of Agriculture, Department of Field Crops, Bingol, TURKEY

Yuksel BOLEK
Kahramanmaras Sutcu Imam University, Faculty of Agriculture, Department of Field Crops, Kahramanmaraş, TURKEY

Mustafa YILDIRIM
Kelkit Aydin Dogan Vocational High School, Gumushane University, Gumushane, TURKEY

Sertan CEVIK
Mersin University, Faculty of Art and Sciences, Mersin, TURKEY

Serpil UNYAYAR
Mersin University, Faculty of Art and Sciences, Mersin, TURKEY

Ali ERGUL
Ankara University, Institute of Biotechnology, Ankara, TURKEY

Seda NEMLI
Department of Bioengineering, Ege University, Bornova-Izmir, TURKEY

Umit SUBASI
Botanical Garden & Herbarium Research and Application Center, Ege University, Bornova-Izmir, TURKEY

Volkan EROGLU
Botanical Garden & Herbarium Research and Application Center, Ege University, Bornova-Izmir, TURKEY

Serdar Gokhan SENOL
Botanical Garden & Herbarium Research and Application Center, Ege University, Bornova-Izmir, TURKEY

Muhammed Bahattin TANYOLAC
Department of Bioengineering, Ege University, Bornova-Izmir, TURKEY

Gulsum OZTURK
Ege University, Faculty of Agriculture, Department of Field Crops, Bornova-Izmir, TURKEY

Zihin YILDIRIM
Ege University, Faculty of Agriculture, Department of Field Crops, Bornova-Izmir, TURKEY